TIME-DEPENDENT STRUCTURE AND CONTROL OF ARTERIAL BLOOD PRESSURE

ANNALS OF THE NEW YORK ACADEMY OF SCIENCES
Volume 783

TIME-DEPENDENT STRUCTURE AND CONTROL OF ARTERIAL BLOOD PRESSURE

Edited by Francesco Portaluppi and Michael H. Smolensky

The New York Academy of Sciences
New York, New York
1996

⊗ The paper used in this publication meets the minimum requirements of American National Standard for Information Sciences—Permanence of Paper for Printed Library Materials, ANSI Z39.48-1984.

Library of Congress Cataloging-in-Publication Data

Time-dependent structure and control of arterial blood pressure/
edited by Francesco Portaluppi and Michael H. Smolensky.
p. cm.—(Annals of the New York Academy of Sciences, ISSN 0077-8923; v. 783)
Includes bibliographical references and indexes.
ISBN 1-57331-008-5 (cloth: alk. paper).—ISBN 1-57331-009-3 (pbk.: alk. paper)
1. Hypertension—Pathophysiology—Congresses. 2. Chronobiology—Congresses. 3. Chronopharmacology—Congresses. 4. Clinical chronobiology—Congresses. 5. Hypotensive agents—Congresses. 6. Blood pressure—Regulation—Congresses. I. Portaluppi, Francesco, 1952–. II. Smolensky, Michael H. III. New York Academy of Sciences. IV. Series.
[DNLM: 1. Blood Pressure—physiology—congresses. 2. Circadian Rhythm—congresses. 3. Hypertension—physiopathology—congresses. W1 AN626YL v. 783 1996 / WG 106 T582 1996]
Q11.N5 vol. 783
[RC685.H8]
500 s—dc20
[616.1′32]
DNLM/DLC
for Library of Congress 96-7856
CIP

SP
Printed in the United States of America
ISBN 1-57331-008-5 (cloth)
ISBN 1-57331-009-3 (paper)
ISSN 0077-8923

ANNALS OF THE NEW YORK ACADEMY OF SCIENCES

Volume 783
August 15, 1996

TIME-DEPENDENT STRUCTURE AND CONTROL OF ARTERIAL BLOOD PRESSURE[a]

Editors and Conference Chairs
FRANCESCO PORTALUPPI AND MICHAEL H. SMOLENSKY

CONTENTS

[a] This volume is the result of a conference entitled **Time-Dependent Structure and Control of Arterial Blood Pressure** sponsored by the New York Academy of Sciences and held on September 10–12, 1995 in Ferrara, Italy.

Poster Papers

***

Financial assistance was received from:

- FERRARA RICERCHE, ITALY
- PROVINCE OF FERRARA, ITALY
- SAB-"G. MARCONI" AIRPORT OF BOLOGNA, ITALY
- UNIVERSITY OF FERRARA, ITALY
- CITY COUNCIL OF FERRARA, ITALY
- ALZA CORPORATION, USA
- SEARLE, USA
- BRISTOL-MYERS-SQUIBB PHARMACEUTICAL RESEARCH INSTITUTE, USA
- SMITHKLINE BEECHAM PHARMACEUTICALS, USA
- ASTRA FARMACEUTICI, ITALY
- LABORATORI GUIDOTTI, ITALY
- MALESCI, ITALY
- MERCK SHARP & DOHME, ITALY
- PARKE DAVIS, ITALY
- PFIZER, ITALY
- ROERIG, ITALY
- SANDOZ, ITALY
- ZENECA, ITALY

Preface

FRANCESCO PORTALUPPI[a]
AND MICHAEL H. SMOLENSKY[b]

[a] *Hypertension Unit*
First Institute of Internal Medicine
University of Ferrara
Ferrara, Italy

[b] *University of Texas-Houston*
School of Public Health
Houston, Texas 77225

Hypertension today constitutes a major health concern of industrialized nations. The findings of the Third National Health and Nutrition Examination Survey (NHANES III), which was conducted in the United States between 1988 and 1991, found the prevalence of hypertension—systolic blood pressure equal to or greater than 140 mm Hg and/or diastolic blood pressure equal to or greater than 90 mm Hg—to be considerable. The prevalence is roughly 45% in men and women 45 to 64 years of age and is much higher in those who are older. Current estimates indicate that in the United States alone, 50 million persons have high blood pressure. Unfortunately, less than 3 of 10 such persons are satisfactorily controlled by medical intervention. Untreated high blood pressure poses a risk to health; prolonged elevation of blood pressure results in injury to cardiac, renal, retinal, and other tissues of the body, and it increases one's risk of stroke, acute myocardial infarction, and sudden cardiac death, among other diseases.

Despite the great financial cost of hypertension to individuals and health care systems and its heavy toll on human productivity and life world wide, our conceptualization and understanding of blood pressure regulation, measurement, variability, and treatment are far from complete. The application of new ambulatory blood pressure assessment technology demonstrates that blood pressure is not at all constant during a 24-hour period as predicted by the concept of homeostasis. Many publications have established that blood pressure exhibits both high and low frequency variability over time. This has significant implications in clinical medicine in terms of the diagnosis of hypertension and its treatment. Moreover, in hypertensive patients the exact nature of the blood pressure profile in time is often indicative of underlying pathophysiologic mechanisms. Temporal oscillations in blood pressure represent the combined effects of several factors including behavior, activity, environment, endogenous rhythms, and disease processes.

The major goal of this New York Academy of Sciences Conference, entitled **Time-Dependent Structure and Control of Arterial Blood Pressure,** was to promote dialogue between academicians, health care specialists as well as industrial and governmental representatives of diverse points of view regarding the knowledge and understanding of temporal patterns in blood pressure and their underlying mechanisms in health and disease. Accordingly, the proceedings of the conference focus on the biological and environmental determinants of this variation as well as its impact on patient assessment and treatment, cardiovascular risk, development of new pharmaceuticals, and governmental regulatory agencies with regard to optimal methods and procedures used in new drug evaluation. The conference and its proceedings bring to the fore the current concepts of the time-dependent structure and control of arterial blood pressure in health and disease. We gratefully acknowledge the staff support of the New York Academy of Sciences, especially Ms. Renee Wilkerson and Ms. Angela Fink. We also acknowledge the industrial sponsors and all the participants who made this conference a great success.

OVERVIEW

The Rhythms of Blood Pressure in Humans

Exogenous and Endogenous Components and Implications for Diagnosis and Treatment

FRANCESCO PORTALUPPI,[a] JAMES WATERHOUSE,[b,c] AND DAVID MINORS[b]

[a]*Hypertension Unit*
First Institute of Internal Medicine
University of Ferrara
Ferrara, Italy

[b]*School of Biological Sciences*
University of Manchester
Manchester, United Kingdom

[c]*School of Human Sciences*
Liverpool John Moores University
Liverpool, United Kingdom

Researchers require ever more data, particularly if they study biological rhythms, where many time points per cycle are desirable. Accordingly, the development of portable machines that could automatically, noninvasively, and repeatedly measure blood pressure and store the data for later analysis was of considerable importance.[1–3] Through the ambulatory application of this technology, it has become increasingly evident that blood pressure variability along the time scale is a relevant physiological and clinical feature. Most scientists have focused on circadian (24-hour) and to some extent seasonal components of blood pressure (BP) variability.[4] Others are concerned with ultradian components having periodicities of less than 24 hours.[5–7] There is a growing awareness that the temporal structure of blood pressure and other cardiovascular variables may be reflected in a closely similar distribution of cardiac and cerebrovascular accidents.[8] In this regard, there has been a surge of scientific interest in the field of medical chronobiology as it pertains to cardiovascular diseases, including hypertension. Moreover, the role of circadian and other rhythms in the pharmacokinetics and pharmacodynamics of antihypertensive medications has been investigated. The fascinating new field of research, known as chronopharmacology (the study of biological rhythms and medications), has great clinical relevance in hypertension[9] because of the medical and social impact of this disease.

Unfortunately, although meaningful advances have occurred in the last few years, our understanding of the temporal structure of blood pressure, its sources and mechanisms, and its relationships to the temporal distribution of target-organ damage and treatment of hypertension is far from complete. For practical reasons, most of the data currently available pertain to the circadian component of such structure: what information about circadian rhythms do blood pressure monitoring devices give in both normotensive subjects and patients with hypertension, and how can the data be interpreted with regard to the mechanisms involved and from the viewpoint of clinical diagnosis and treatment?

CIRCADIAN RHYTHMS IN NORMOTENSION AND HYPERTENSION

What ambulatory monitoring has shown repeatedly is the lability of both diastolic and systolic blood pressure, this being closely related to the activity of the subject.[10–12] One of the most marked and persistent changes is the fall in BP at the time of retiring and the rise at the time of rising. This effect of sleep is equally clear when it occurs during a daytime nap.[13] Despite such lability, the pattern of blood pressure on successive days is generally reproducible,[14,15] but this might reflect no more than the reproducibility of lifestyle during the daily routine.

When data of several days from an individual or a group of individuals living normally are pooled, the average rhythm approximates that of a sinusoid. This enables an objective estimate of the rhythm to be made. Rhythm analysis may be performed with the method of cosinor,[16] a periodogram,[17] partial Fourier series (harmonics usually limited to the first four or five),[18–21] spline functions,[22] cumulative sums,[23] or square-wave fit.[24] The advantages and limitations of the different statistical approaches are still a matter of study and discussion.[7,25,26] Further research is needed to assess the prognostic and therapeutic implications of the different statistical parameters derived from the various analyses before the best method of analysis of the blood pressure time series can be determined.

When BP measurements are repeated in patients with hypertension, the lability of the pressures is equally clear, but some other differences from normotensive controls are present. Except for the elevated value for the 24-hour mean, the fall at night can be reduced in some patients. This nocturnal fall has given rise to the concept of "dippers," those whose blood pressure falls by at least a certain (arbitrarily defined) amount, and "non-dippers." However, a large overlap exists between normotensives and hypertensives[12] as well as in patients with autonomic neuropathy or after transplantation (see below). Also, some individuals can vary in classification when studied on several occasions.[27,28] Cosinor analysis of data from hypertensive subjects shows little difference from normotensives in acrophase (about mid-activity) but, predictably, a higher mean value (mesor) and a lower amplitude (due to the increased frequency of "non-dippers").[10]

CAUSES OF CIRCADIAN RHYTHM

Exogenous Effects of Lifestyle and Environment

The marked effects of sleep and physical activity on blood pressure are well known, as is the observation that the acrophase of the cosine curve coincides with mid-activity. Other data from both field and laboratory studies support this dominant effect of lifestyle. Thus, (1) Subjects appear to adjust fully and almost immediately to shift work, using the time of midactivity during night work/day sleep as the reference point[19,29–32]; (2) If subjects live on non-24-hour "days," 18 hours for example, blood pressure rhythms show a period equal to that of the imposed sleep/wake cycle.[33] In this respect, the rhythms differ from others such as those of core temperature and plasma melatonin, which show a period much closer to 24 hours.

Even though interpretation of the results from (1) and (2) requires further comment (see below), it is clear that lifestyle is a dominant influence on blood pressure. More detailed studies have confirmed this, but they have also stressed the complexities involved. For example, sleep is not homogeneous neurophysiologically, and different sleep stages are associated with different mean blood pressures[6,34]; stages 1 and 2 (light sleep) are associated with higher pressures than stages 3 and 4 (deep sleep) but lower and less labile values than rapid eye movement (dream) sleep.[35,36] Similarly, not only do different degrees of physical activity raise blood pressure in proportion,[37] but also mental and social stimuli can both produce large effects.[38]

Endogenous Causes due to the Body Clock

Can any of the circadian changes in blood pressure therefore be attributed to the body clock? The conventional means of establishing such an endogenous component[39] would be to show that (see also previous section): (1) a component of the rhythm did not adjust fully and immediately to the change in timing of lifestyle demanded by a time-zone transition or night work; (2) a component of the rhythm did not adjust to the imposed non-24-hour "day."

Two further methods exist for establishing the presence of an endogenous component: (1) If a subject is kept awake for 24 hours while adopting a constant posture and regular, identical snacks in an unchanging environment (with regard to light, temperature, humidity, and social and mental stimuli), then the endogenous component of a rhythm will remain, no longer "masked" by exogenous factors. (2) Mathematical models now exist that use the known effects of activity upon blood pressure to correct the measured rhythm for these effects and so "purify" the data. This process enables the endogenous component to be separated from the exogenous component.[40]

Taken together, the results of these methods indicate that a small endogenous component, driven by the body clock, exists in the measured circadian rhythm of blood pressure; its amplitude is 5–10 mm Hg. Additional evidence in favor of an endogenous component of blood pressure rhythmicity is the significant genetic effects demonstrated for the 24-hour profiles, the daytime mean, the value of the evening acrophase, and the value of the major acrophase of diastolic blood pressure[41–44] as well as the observation that lesion of the suprachiasmatic nuclei in rats abolishes circadian rhythmicity in blood pressure and heart rate without affecting the sleep-wake and motor activity cycles.[45,46]

It can be seen that the endogenous component is much weaker than the effects normally caused by lifestyle.[17,47,48] Related to this issue of the relative sizes of the two components are the results of pioneering studies of ambulatory monitoring of blood pressure using data pooled from individuals.[49–51] Such data (see above) showed a sinusoidal profile that could be interpreted as resulting from the output of the biological clock. However, this result would also be obtained if the individuals in the group had daily changes in blood pressure that were square wave in shape (high daytime and low nighttime values) coupled with different times of retiring and rising. In support of this interpretation is the result that when expressed relative to rising times, blood pressure shows an abrupt increase immediately after rising and a decrease after retiring (which confirms the importance of exogenous factors), but also showed a small rise in blood pressure before waking and a fall before retiring, which can reasonably be attributed to an endogenous component.[17,52,53] Sleep stages were not measured, and so any effects due to their changes cannot be assessed. What these results stress, therefore, are the interpretative problems associated with the circadian changes, if the exogenous component is not controlled or taken into account in some way.

The Autonomic System and Endocrines

An important role of the autonomic system in producing exogenous and endogenous effects in blood pressure is almost inevitable. Methods of study have included the use of α and β agonists and antagonists,[53–55] measurement of abnormalities in blood pressure rhythms in dysautonomic states,[56–60] congestive heart failure[61–65] or heart transplantation,[66,67] and the effects of the autonomic nervous system on baroreceptor-mediated reflexes[68] and short-term changes in blood pressure.[69,70] These approaches indicate that changes in both branches of the autonomic nervous system influence blood pressure (directly via vasomotor tone and indirectly via actions upon the heart) and its circadian rhythm and that these mechanisms can become altered in the aforedescribed patients and hypertensive subjects, often decreasing the nocturnal "dip" in blood pressure.

There are vasomotor influences other than neural ones, amongst which are the circadian rhythms of several hormones and peptides, including the endogenous opioid system, the hypothalamic-pituitary hormonal axes, the renin-angiotensin-aldosterone system, and the vasoactive peptides. Evidence in favor of this view comes from the study of many diverse pathologic conditions in which alterations in the circadian rhythm of the aforementioned neurohumoral factors are reflected by consistent modifications of the 24-hour pattern of BP. In fact, catecholamine-producing tumors,[71–75] Cushing's syndrome[76] and exogenous glucocorticoid administration,[77,78] mineralocorticoid excess syndromes,[79–81] Addison's disease,[82] hyperthyroidism,[83] and pseudohypoparathyroidism[84] are often associated with the lack of a decrease in nocturnal blood pressure values.

IMPLICATION OF THE CIRCADIAN RHYTHM IN THE DIAGNOSIS AND TREATMENT OF HYPERTENSION

For any diagnostic test there is a need to standardize the conditions under which the measurement is made. For blood pressure, which is so labile, this need is greater but more difficult to realize. Indeed, the "stress" of seeing a clinician and having blood pressure measured (white coat hypertension) is well known.[85,86] Many believe that the possibility of considerable error exists if "casual measurements" (that is, those taken over a very restricted portion of the circadian cycle, as during a visit to a clinic) are used. Accordingly, many advocate 24-hour monitoring of ambulatory blood pressure. Even if this is practicable, there is still the problem that "normal" values depend on the time of measurement.

When it comes to assessing the severity of hypertension for its prognostic value for cardiovascular morbidity, there is the problem that the most relevant measure(s) is not yet decided upon. Some possibilities are: (1) the excess above "normal," defined by casual readings, with the time of measurement having been taken into account; (2) the excess above "normal," but defined by making use of measurements throughout the 24 hours. Hypertension or "hyperbaric" indices have been calculated based on this approach[87]; (3) the fall in blood pressure on retiring and/or the values during sleep (a measurement that is directly related to the classification into "dippers" and "non-dippers," see above); (4) the rise in blood pressure between its minimum just before waking and its maximum about 2–3 hours after this event.

These methods differ in their practicality and also in the epidemiologic evidence that would support them. Thus, evidence that left ventricular size correlates more positively with nocturnal than daytime blood pressure[88,89] points to the importance of factor 3 above. By contrast, the greater incidence of cardiovascular incidents in the hours before noon[8,90] suggests that factor 4 needs consideration also, and the more general finding that hypertension per se is damaging implies that factor 2 above cannot be discounted.

The existence of various pathologic conditions in which the blood pressure patterns are altered raises the question of whether and when the goal of treatment should be resetting of the normal blood pressure pattern in essential and secondary forms of hypertension. Intriguingly enough, the time course of drug absorption and distribution need not coincide with the time course of drug effects on BP.[91] This phenomenon implies that drug effect is not simply a result of drug concentration per se; it can depend on a set of mechanisms that are also variable in time and that affect therapeutic results.

Work on the autonomic nervous system and other vasoactive substances (above) indicates possible means for reducing blood pressure,[9,92] but the time when blood pressure reduction is most important deserves consideration. Thus, treatment might be most effective if it ensures reduced blood pressure throughout the 24 hours or particularly during the night or if it lessens the morning increase. In this last case, treatment once daily after waking might be inadequate if the effects of the drug were to wear off after 20 hours at exactly the time when the greatest danger might exist. Finally, it must also be borne in mind that a risk might exist, particularly at night, if blood pressure were to be reduced too much.

CONCLUDING COMMENTS

Hypertension is associated with considerable cardiovascular morbidity and mortality. The details of these disorders remain to be elucidated, but, once established, would offer a rationale for treatment. It seems currently that a knowledge of the factors that cause the circadian rhythmicity of blood pressure will contribute to more successful diagnosis and treatment.

REFERENCES

1. KNOEBEL, S. B., M. H. CRAWFORD, M. I. DUNN *ET AL.* 1989. Guidelines for ambulatory electrocardiography. A report of the American College of Cardiology/American Heart Association Task Force on Assessment of Diagnostic and Therapeutic Cardiovascular Procedures (Subcommittee on Ambulatory Electrocardiography). Circulation **79:** 206–215.
2. GRUEN, W. 1990. Chronophysiology and ambulatory monitoring instrumentation. Chronobiol. Int. **7:** 89–91.
3. MEYER-SABELLEK, W., K. L. SCHULTE & R. GOTZEN. 1990. Non-invasive ambulatory blood pressure monitoring: Technical possibilities and problems. J. Hypertens. **8** (Suppl. 6): S3–10.
4. CORNELISSEN, G., E. HAUS & F. HALBERG. 1992. Chronobiologic blood pressure assessment from womb to tomb. *In* Biological Rhythms in Clinical and Laboratory Medicine. Y. Touitou & E. Haus, Eds.: 428–452. Springer-Verlag. Heidelberg.
5. PAGANI, M., R. FURLAN, S. DELL'ORTO *ET AL.* 1985. Simultaneous analysis of beat by beat systemic arterial pressure and heart rate variabilities in ambulatory patients. J. Hypertens. **3**(Suppl. 3): S83–85.
6. MANCIA, G. 1993. Autonomic modulation of the cardiovascular system during sleep. N. Engl. J. Med. **328:** 347–349.
7. STAESSEN, J. A., R. FAGARD, L. THIJS & A. AMERY. 1993. Fourier analysis of blood pressure profiles. Am. J. Hypertens. **6:** 184S–187S.

8. MULLER, J. E. & B. MANGEL. 1994. Circadian variation and triggers of cardiovascular disease. Cardiology **85**(Suppl. 2): 3–10.
9. LEMMER, B. & F. PORTALUPPI. 1995. Chronopharmacology of cardiovascular diseases. *In* Physiology and Pharmacology of Biological Rhythms. P. H. Redfern & B. Lemmer, Eds.: Springer-Verlag. Heidelberg.
10. DELEA, C. S. 1979. Chronobiology of blood pressure. Nephron **23:** 91–97.
11. PICKERING, T. G. 1991. Ambulatory Monitoring and Blood Pressure Variability. Science Press Ltd. London.
12. STEWART, M. J. & P. L. PADFIELD. 1992. Blood pressure measurement: An epitaph for the mercury sphygmomanometer? Clin. Sci. **83:** 1–12.
13. BURSZTYN, M., J. MEKLER, N. WACHTEL & D. BEN-ISHAY. 1994. Siesta and ambulatory blood pressure monitoring. Comparability of the afternoon nap and night sleep. Am. J. Hypertens. **7:** 217–221.
14. PORTALUPPI, F. & L. MONTANARI. 1988. Consistency of circadian blood pressure pattern assessed by non-invasive monitoring and cosinor analysis in hospitalized hypertensive patients. Acta Cardiol. **43:** 605–613.
15. WEBER, M. A., J. I. DRAYER, D. K. NAKAMURA & F. A. WYLE. 1984. The circadian blood pressure pattern in ambulatory normal subjects. Am. J. Cardiol. **54:** 115–119.
16. HALBERG, F. 1969. Chronobiology. Ann. Rev. Physiol. **31:** 675–725.
17. DEGAUTE, J. P., P. VAN DE BORNE, P. LINKOWSKI & E. VAN CAUTER. 1991. Quantitative analysis of the 24-hour blood pressure and heart rate patterns in young men. Hypertension **18:** 199–210.
18. GERMANO, G., S. DAMIANI, M. CIAVARELLA, A. APPOLLONI, A. FERRUCCI & V. CORSI. 1984. Detection of a diurnal rhythm in arterial blood pressure in the evaluation of 24-hour antihypertensive therapy. Clin. Cardiol. **7:** 525–535.
19. CHAU, N. P., J. M. MALLION, R. DE GAUDEMARIS *ET AL.* 1989. Twenty-four-hour ambulatory blood pressure in shift workers. Circulation **80:** 341–347.
20. MATTES, A., K. WITTE, W. HOHMANN & B. LEMMER. 1991. PHARMFIT—a nonlinear fitting program for pharmacology. Chronobiol. Int. **8:** 460–476.
21. THIJS, L., J. STAESSEN & R. FAGARD. 1992. Analysis of the diurnal blood pressure curve. High Blood Press. **1:** 17–28.
22. STREITBERG, B., W. MEYER-SABELLEK & P. BAUMGART. 1989. Statistical analysis of circadian blood pressure recordings in controlled clinical trials. J. Hypertens. **7**(Suppl. 3): S11–17.
23. STANTON, A., J. COX, N. ATKINS, K. O'MALLEY & E. O'BRIEN. 1992. Cumulative sums in quantifying circadian blood pressure patterns. Hypertension **19:** 93–101.
24. IDEMA, R. N., E. S. GELSEMA, G. J. WENTING *ET AL.* 1992. A new model for diurnal blood pressure profiling. Square wave fit compared with conventional methods. Hypertension **19:** 595–605.
25. PARATI, G., M. DI RIENZO, S. OMBONI, P. CASTIGLIONI, A. FRATTOLA & G. MANCIA. 1993. Spectral analysis of 24 h blood pressure recordings. Am. J. Hypertens. **6:** 188S–193S.
26. WU, M. C. 1993. Analysis of blood pressure variability. Am. J. Hypertens. **6:** 198S–200S.
27. VERDECCHIA, P., G. SCHILLACI, F. BOLDRINI, M. GUERRIERI, I. ZAMPI & C. PORCELLATI. 1991. Quantitative assessment of day-to-day spontaneous variability in non-invasive ambulatory blood pressure measurements in essential hypertension. J. Hypertens. **9**(Suppl. 6): S322–323.
28. STAESSEN, J., C. J. BULPITT, E. O'BRIEN *ET AL.* 1992. The diurnal blood pressure profile. A population study. Am. J. Hypertens. **5:** 386–392.
29. SUNDBERG, S., A. KOHVAKKA & A. GORDIN. 1988. Rapid reversal of circadian blood pressure rhythm in shift workers. J. Hypertens. **6:** 393–396.
30. BAUMGART, P., P. WALGER, G. FUCHS, K. G. DORST, H. VETTER & K. H. RAHN. 1989. Twenty-four-hour blood pressure is not dependent on endogenous circadian rhythm. J. Hypertens. **7:** 331–334.
31. PIEPER, C., K. WARREN & T. G. PICKERING. 1993. A comparison of ambulatory blood pressure and heart rate at home and work on work and non-work days. J. Hypertens. **11:** 177–183.

32. GOTO, T., K. YOKOYAMA, T. ARAKI *ET AL.* 1994. Identical blood pressure levels and slower heart rates among nurses during night work and day work. J. Hum. Hypertens. **8:** 11–14.
33. SCHAEFER, K. E., C. M. KERR, D. BUSS & E. HAUS. 1979. Effect of 18-h watch schedules on circadian cycles of physiological functions during submarine patrols. Undersea Biomed. Res. **6**(Suppl): S81–90.
34. VAN DE BORNE, P., H. NGUYEN, P. BISTON, P. LINKOWSKI & J. P. DEGAUTE. 1994. Effects of wake and sleep stages on the 24-h autonomic control of blood pressure and heart rate in recumbent men. Am. J. Physiol. **266:** H548–554.
35. COCCAGNA, G., M. MANTOVANI, F. BRIGNANI, A. MANZINI & E. LUGARESI. 1971. Laboratory note. Arterial pressure changes during spontaneous sleep in man. Electroencephalogr. Clin. Neurophysiol. **31:** 277–281.
36. SOMERS, V. K., D. PHIL, M. E. DYKEN, A. L. MARK & F. M. ABBOUD. 1993. Sympathetic-nerve activity during sleep in normal subjects. N. Engl. J. Med. **328:** 303–307.
37. KOHNO, I., H. ISHII, T. NAKAMURA & K. TAMURA. 1993. Relationship between activity levels and circadian blood pressure variations. Chronobiologia **20:** 53–61.
38. JAMES, G. D. & T. G. PICKERING. 1993. The influence of behavioral factors on the daily variation of blood pressure. Am. J. Hypertens. **6:** 170S–173S.
39. MINORS, D. S. & J. M. WATERHOUSE. 1981. Circadian Rhythms and the Human. John Wright. Bristol, UK.
40. MINORS, D. S. & J. M. WATERHOUSE. 1992. Investigating the endogenous component of human circadian rhythms: A review of some simple alternatives to constant routines. Chronobiol. Int. **9:** 55–78.
41. RAVOGLI, A., S. TRAZZI, A. VILLANI *ET AL.* 1990. Early 24-hour blood pressure elevation in normotensive subjects with parental hypertension. Hypertension **16:** 491–497.
42. VAN HOOFT, I. M., D. E. GROBBEE, H. J. WAAL-MANNING & A. HOFMAN. 1993. Hemodynamic characteristics of the early phase of primary hypertension. The Dutch Hypertension and Offspring Study. Circulation **87:** 1100–1106.
43. DEGAUTE, J. P., E. VAN CAUTER, P. VAN DE BORNE & P. LINKOWSKI. 1994. Twenty-four-hour blood pressure and heart rate profiles in humans. A twin study. Hypertension **23:** 244–253.
44. SOMES, G. W., G. A. HARSHFIELD, B. S. ALPERT, M. M. GOBLE & R. M. SCHIEKEN. 1995. Genetic influences on ambulatory blood pressure patterns. The Medical College of Virginia Twin Study. Am. J. Hypertens. **8:** 474–478.
45. JANSSEN, B. J., C. M. TYSSEN, H. DUINDAM & W. J. RIETVELD. 1994. Suprachiasmatic lesions eliminate 24-h blood pressure variability in rats. Physiol. Behav. **55:** 307–311.
46. WITTE, K., A. SCHNECKO, R. BUIJS & B. LEMMER. 1995. Circadian rhythms in blood pressure and heart rate in SCN-lesioned and unlesioned transgenic hypertensive rats (abstr.). Biol. Rhythm Res. **26:** 458–459.
47. BAUMGART, P. 1991. Circadian rhythm of blood pressure: Internal and external time triggers. Chronobiol. Int. **8:** 444–450.
48. COCA, A. 1994. Circadian rhythm and blood pressure control: Physiological and pathophysiological factors. J. Hypertens. **12**(Suppl. 5): S13–21.
49. FLORAS, J. S., J. V. JONES, J. A. JOHNSTON, D. E. BROOKS, M. O. HASSAN & P. SLEIGHT. 1978. Arousal and the circadian rhythm of blood pressure. Clin. Sci. Mol. Med. **4**(Suppl. 1): S395–397.
50. LITTLER, W. A., M. J. WEST, A. J. HONOUR & P. SLEIGHT. 1978. The variability of arterial pressure. Am. Heart J. **95:** 180–186.
51. MILLAR-CRAIG, M. W., C. N. BISHOP & E. B. RAFTERY. 1978. Circadian variation of blood-pressure. Lancet **1:** 795–797.
52. BROADHURST, P., G. BRIGDEN, P. DASGUPTA, A. LAHIRI & E. B. RAFTERY. 1990. Ambulatory intra-arterial blood pressure in normal subjects. Am. Heart J. **120:** 160–166.
53. GOULD, B. A. & E. B. RAFTERY. 1991. Twenty-four-hour blood pressure control: An intraarterial review. Chronobiol. Int. **8:** 495–505.
54. PANZA, J. A., S. E. EPSTEIN & A. A. QUYYUMI. 1991. Circadian variation in vascular tone and its relation to alpha-sympathetic vasoconstrictor activity. N. Engl. J. Med. **325:** 986–990.

55. TAYLOR, A. A. 1994. Autonomic control of cardiovascular function: Clinical evaluation in health and disease. J. Clin. Pharmacol. **34:** 363–374.
56. MARTINELLI, P., G. COCCAGNA, N. RIZZUTO & E. LUGARESI. 1981. Changes in systemic arterial pressure during sleep in Shy-Drager syndrome. Sleep **4:** 139–146.
57. MANN, S., D. G. ALTMAN, E. B. RAFTERY & R. BANNISTER. 1983. Circadian variation of blood pressure in autonomic failure. Circulation **68:** 477–483.
58. ALAM, M., D. V. PAVITT & C. J. MATHIAS. 1993. Cumulative sums analysis of twenty-four-hour blood pressure profiles in patients with sympathetic denervation. J. Hypertens. **11**(Suppl. 5): S286–287.
59. IKEDA, T., T. MATSUBARA, Y. SATO & N. SAKAMOTO. 1993. Circadian blood pressure variation in diabetic patients with autonomic neuropathy. J. Hypertens. **11:** 581–587.
60. TORFFVIT, O. & C. D. AGARDH. 1993. Day and night variation in ambulatory blood pressure in type 1 diabetes mellitus with nephropathy and autonomic neuropathy. J. Intern. Med. **233:** 131–137.
61. CARUANA, M. P., A. LAHIRI, P. M. CASHMAN, D. G. ALTMAN & E. B. RAFTERY. 1988. Effects of chronic congestive heart failure secondary to coronary artery disease on the circadian rhythm of blood pressure and heart rate. Am. J. Cardiol. **62:** 755–759.
62. PORTALUPPI, F., L. MONTANARI, M. FERLINI, L. VERGNANI, B. BAGNI & E. C. DEGLI UBERTI. 1992. Differences in blood pressure regulation of congestive heart failure, before and after treatment, correlate with changes in the circulating pattern of atrial natriuretic peptide. Eur. Heart J. **13:** 990–996.
63. VAN DE BORNE, P., M. ABRAMOWICZ, S. DEGRE & J. P. DEGAUTE. 1992. Effects of chronic congestive heart failure on 24-hour blood pressure and heart rate patterns: A hemodynamic approach. Am. Heart J. **123:** 998–1004.
64. SUZUKI, Y., I. KUWAJIMA, A. KANEMARU *ET AL.* 1992. The cardiac functional reserve in elderly hypertensive patients with abnormal diurnal change in blood pressure. J. Hypertens. **10:** 173–179.
65. KASTRUP, J., H. WROBLEWSKI, J. SINDRUP, H. ROLIGHED CHRISTENSEN & N. WIINBERG. 1993. Diurnal blood pressure profile in patients with severe congestive heart failure: Dippers and non-dippers. Scand. J. Clin. Lab. Invest. **53:** 577–583.
66. LANUZA, D. M., K. GRADY, M. HETFLEISCH & M. R. JOHNSON. 1994. Circadian rhythm changes in blood pressure and heart rate during the first year after heart transplantation. J. Heart Lung Transplant. **13:** 614–623.
67. VON POLNITZ, A., C. BRACHT, B. KEMKES & B. HOFLING. 1990. Circadian pattern of blood pressure and heart rate in the longer term after heart transplantation. J. Cardiovasc. Pharmacol. **16**(Suppl. 5): S86–89.
68. HARTIKAINEN, J., I. TARKIAINEN, K. TAHVANAINEN, M. MANTYSAARI, E. LANSIMIES & K. PYORALA. 1993. Circadian variation of cardiac autonomic regulation during 24-h bed rest. Clin. Physiol. **13:** 185–196.
69. TURJANMAA, V., S. KALLI, S. MAJAHALME, N. SARANUMMI & A. UUSITALO. 1987. Diurnal blood pressure profiles and variability in normotensive ambulant subjects. Clin. Physiol. **7:** 389–401.
70. FURLAN, R., S. GUZZETTI, W. CRIVELLARO *ET AL.* 1990. Continuous 24-hour assessment of the neural regulation of systemic arterial pressure and RR variabilities in ambulant subjects. Circulation **81:** 537–547.
71. ISSHIKI, T., N. AKATSUKA, H. TSUNEYOSHI & H. OKA. 1986. Periodic fluctuation of blood pressure in a case of norepinephrine secreting extra-adrenal pheochromocytoma. Jpn Heart J. **27:** 437–442.
72. OISHI, S., M. SASAKI, M. OHNO, T. UMEDA & T. SATO. 1988. Periodic fluctuation of blood pressure and its management in a patient with pheochromocytoma. Case report and review of the literature. Jpn Heart J. **29:** 389–399.
73. IMAI, Y., K. ABE, Y. MIURA *ET AL.* 1988. Hypertensive episodes and circadian fluctuations of blood pressure in patients with phaeochromocytoma: Studies by long-term blood pressure monitoring based on a volume-oscillometric method. J. Hypertens. **6:** 9–15.
74. DABROWSKA, B., T. FELTYNOWSKI, B. WOCIAL, W. SZPAK & W. JANUSZEWICZ. 1990. Effect of removal of phaeochromocytoma on diurnal variability of blood pressure, heart rhythm and excretion of catecholamines. J. Hum. Hypertens. **4:** 397–399.

75. STATIUS VAN EPS, R. G., A. H. VAN DEN MEIRACKER, F. BOOMSMA, A. J. MAN IN'T VELD & M. A. D. H. SCHALEKAMP. 1994. Diurnal variation of blood pressure in patients with catecholamine-producing tumors. Am. J. Hypertens. **7:** 492–497.
76. IMAI, Y., K. ABE, S. SASAKI *ET AL.* 1988. Altered circadian blood pressure rhythm in patients with Cushing's syndrome. Hypertension **12:** 11–19.
77. IMAI, Y., K. ABE, S. SASAKI *ET AL.* 1989. Exogenous glucocorticoid eliminates or reverses circadian blood pressure variations. J. Hypertens. **7:** 113–120.
78. VAN DE BORNE, P., M. GELIN, J. VAN DE STADT & J. P. DEGAUTE. 1993. Circadian rhythms of blood pressure after liver transplantation. Hypertension **21:** 398–405.
79. IMAI, Y., K. ABE, S. SASAKI *ET AL.* 1992. Circadian blood pressure variation in patients with renovascular hypertension or primary aldosteronism. Clin. Exp. Hypertens [A] **14:** 1141–1167.
80. WHITE, W. B. & C. MALCHOFF. 1992. Diurnal blood pressure variability in mineralocorticoid excess syndrome. Am. J. Hypertens. **5:** 414–418.
81. VEGLIO, F., G. PINNA, R. MELCHIO *ET AL.* 1993. Twenty-four-hour power spectral analysis by maximum entropy method of blood pressure in primary hyperaldosteronism. Blood Press. **2:** 189–196.
82. FALLO, F., G. FANELLI, A. CIPOLLA, C. BETTERLE, M. BOSCARO & N. SONINO. 1994. 24-Hour blood pressure profile in Addison's disease. Am. J. Hypertens. **7:** 1105–1109.
83. MIDDEKE, M. & J. SCHRADER. 1994. Nocturnal blood pressure in normotensive subjects and those with white coat, primary, and secondary hypertension. Br. Med. J. **308:** 630–632.
84. BRICKMAN, A. S., N. STERN & J. R. SOWERS. 1990. Circadian variations of catecholamines and blood pressure in patients with pseudohypoparathyroidism and hypertension. Chronobiologia **17:** 37–44.
85. RICHARDSON, D. W., A. J. HONOUR, G. W. FENTON, F. H. STOTT & G. W. PICKERING. 1964. Variation in arterial pressure throughout the day and night. Clin. Sci. **26:** 445–460.
86. PICKERING, T. G. & G. D. JAMES. 1989. Some implications of the differences between home, clinic and ambulatory blood pressure in normotensive and hypertensive patients. J. Hypertens. **7** (Suppl. 3): S65–72.
87. WEGMANN, R., A. WEGMANN, M. A. WEGMANN-GODDIJN, W. MARZ & F. HALBERG. 1987. Hyperbaric indices (HBI) assess the extent and timing of deviant blood pressure in patients under treatment. Chronobiologia **14:** 27–30.
88. VERDECCHIA, P., G. SCHILLACI, M. GUERRIERI *ET AL.* 1990. Circadian blood pressure changes and left ventricular hypertrophy in essential hypertension. Circulation **81:** 528–536.
89. PALATINI, P., M. PENZO, A. RACIOPPA *ET AL.* 1992. Clinical relevance of nighttime blood pressure and of daytime blood pressure variability. Arch. Intern. Med. **152:** 1855–1860.
90. WROE, S. J., P. SANDERCOCK, J. BAMFORD, M. DENNIS, J. SLATTERY & C. WARLOW. 1992. Diurnal variation in incidence of stroke: Oxfordshire community stroke project. Br. Med. J. **304:** 155–157.
91. LEMMER, B., B. SCHEIDEL & S. BEHNE. 1991. Chronopharmacokinetics and chronopharmacodynamics of cardiovascular active drugs. Propranolol, organic nitrates, nifedipine. Ann. N.Y. Acad. Sci. **618:** 166–181.
92. RAFTERY, E. B. 1991. Circadian variation in blood pressure: Considerations for therapy. Postgrad. Med. J. **67**(Suppl. 5): S9–14.

Low and High Frequency Components of Blood Pressure Variability

M. PAGANI,[a] D. LUCINI, O. RIMOLDI, R. FURLAN,
S. PIAZZA, A. PORTA,[b] AND A. MALLIANI

Centro Ricerche Cardiovascolari
CNR
Medicina Interna II
Centro Ricerca Neurovegetativa
Ospedale "L. Sacco"
Università di Milano
Milan, Italy

[b]*Dipartimento di Automazione Industriale*
Università di Brescia
Brescia, Italy

The dynamic nature of arterial blood pressure was clearly recognized after the initial observations by Hales who, as early as 1733, performed direct arterial cannulation in animals and recognized the presence of oscillations of arterial pressure related to both heart beat and respiration.[1] The subsequent development of instruments that could produce permanent recordings of arterial pressure as a function of time permitted better definition of the temporal properties of cardiogenic and respiratory oscillations (indicated as first- and second-order pressure waves) and recognition of the additional presence of other oscillations, with a period slower than respiration (i.e., third order) (FIG. 1).[2–4] In recent decades the availability of smaller, more accurate, battery-operated, portable recorders, subsequently combined with powerful computer techniques, permitted assessment of the temporal properties of arterial pressure in a variety of dynamic, real-life conditions, even in space. As evident in FIGURE 1, oscillations in arterial pressure are usually classified according to their average period by either strip chart recordings or computer techniques. Spontaneous blood pressure variability may depend on the simultaneous presence of several oscillations of different frequencies associated, to a variable extent, with nonoscillatory (i.e., nonsinusoidal) components as well.

Changes in the level of physical and mental activity modulate arterial pressure and its variability by way of complex regulatory mechanisms, mostly reflecting various levels of activity of the autonomic nervous system, combined with other direct mechanical or humoral influences.[5]

In the last 12 years we have pursued the hypothesis that computer analysis of beat-by-beat cardiovascular variability could provide a means to infer the dynamic changes in the balance between the two autonomic outflows (vagal and sympathetic) controlling the circulation in various pathophysiologic conditions.[5,6]

In this paper we analyze in detail the second- and third-order components of

[a]Address for correspondence: Massimo Pagani, MD, Medicina Interna II, Ospedale "L. Sacco," Università di Milano, Via G. B. Grassi, 74, 20157 Milan, Italy (Tel: 39 2 38210532; Fax: 39 2 3564630).

arterial pressure variability (henceforth indicated as high frequency [HF] and low frequency [LF] in both the experimental and the clinical laboratory. The main emphasis is on the LF component as a marker of sympathetic modulation of peripheral vasculature.

METHODOLOGY

Because of the tight link between cardiovascular and respiratory systems, it is important to consider heart period and arterial pressure variability simultaneously with respiratory activity.

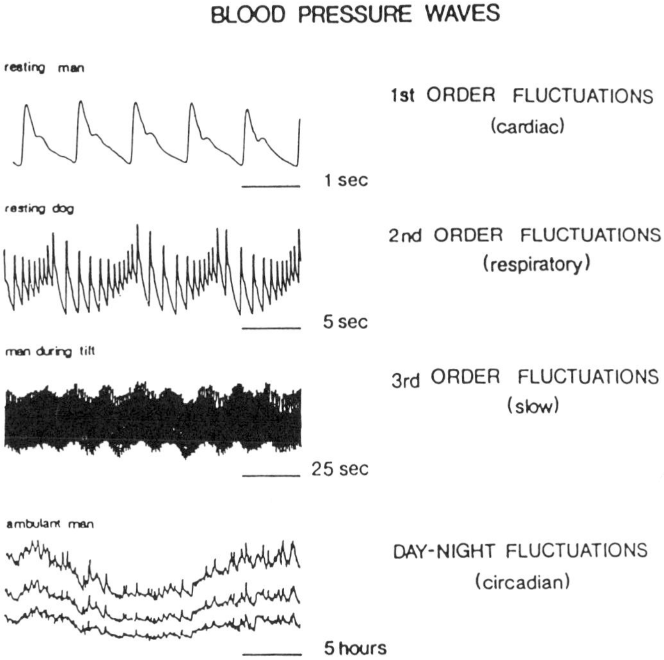

FIGURE 1. Example of spontaneous oscillations that can be observed in the blood pressure signal.

FIGURE 2 is an example of spectral analysis of the three variability signals as usually performed in our laboratory. Details of this procedure were described previously.[5–7]

Briefly, at rest, two major oscillatory components are consistently observed in both heart period and systolic blood pressure variabilities. A low-frequency (LF, 0.1 Hz) oscillatory component, if extracted from RR interval variability and expressed in normalized units, provides an index of sympathetic efferent activity to the sinoatrial node, whereas if obtained from systolic blood pressure variability and expressed in mm Hg^2, it furnishes an index of sympathetic vasomotor control. A second high-frequency (HF, 0.25 Hz) oscillatory component, synchronous with respiration, is a marker of the vagal efferent activity to the heart if extracted from RR variability, whereas if obtained from systolic blood pressure variability, it represents mostly the mechanical effects of respiration on blood pressure.[5,7]

Finally, cross-spectral analysis of the variability signals provides a means to formally quantify the statistical link between any two of these periodical phenomena by way of phase and coherence diagrams (FIG. 2).[8] More complex models considering heart period, systolic arterial pressure, and respiration simultaneously have also been described.[9] Note from FIGURE 2 that high coherence between tachogram and

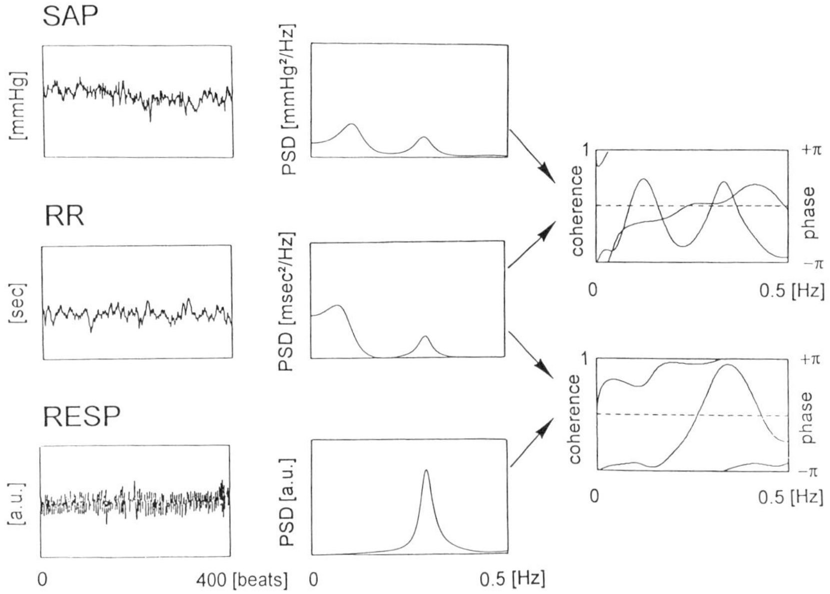

FIGURE 2. Schematic representation of spectral and cross-spectral analysis of systolic arterial pressure (SAP) and RR interval variabilities and of respiration. From the time series (**left panels**) of consecutive RR intervals (tachogram) and systolic pressure values (systogram), power spectral density is computed (**central panels**). In a normal subject two main oscillatory components at low (LF, about 0.10 Hz) and high (HF, synchronous with respiration) frequency are usually recognized. Cross-spectral analysis (**right panels**) of RR interval and SAP variabilities and of respiration provides the relationship between these periodic phenomena by means of phase and coherence diagrams.

systogram is present in correspondence with both LF and HF, whereas between tachogram and respiration high coherence is present only in correspondence with respiratory frequency. The effects of changes in respiratory frequency are complex and have been addressed mostly as regards RR variability. In brief, an inverse relationship exists, so that increases in respiratory rate lead to a reduction in HF power (particularly in the frequency range of 7.5–15 breaths/min), whereas the reverse happens with reductions in respiratory rate.[10] If respiration is sufficiently slow (~ 0.1 Hz, i.e., 6 breaths per minute) the respiratory component will fuse with the LF component, leading to a single very large oscillation (i.e., entrainment occurs).[11] Obviously in this case the model of sympathovagal balance is not applicable, and spectral analysis cannot be employed to infer the state of autonomic interaction.

At this point it is appropriate to focus on some technical details of the requirements of arterial blood pressure measurement for subsequent spectral analysis.

The frequency content of peripheral arterial pressure is such that about 7–10 harmonics are sufficient to accurately reproduce arterial waveforms even in dynamic conditions.[12,13] Therefore, in the resting state (with a heart period of about 1,000 ms, corresponding to 60 beats/min) a frequency response of the recording reproducing system of about 7–10 Hz is sufficient. Accordingly, both fluid-filled catheter manometer systems[14] and the noninvasive Finapres apparatus[15–17] may provide adequate signals. In dynamic conditions, such as exercising, with an increasing heart rate (up to 180 beats/min) the frequency content of the pressure signal increases to approximately 30–40 Hz. Considering additionally the possible direct effects of body motion on the pressure recording apparatus, the use of a micromanometer becomes ideal, as this guarantees an artifact-free signal throughout.[12,13]

Fluid-filled catheter manometer and noninvasive systems may also be used in dynamic conditions with caution and only after careful editing of the entire recording. The arterial pressure waveform also contains information on pulse interval length; hence, the tachogram can be obtained from appropriate handling of the pressure signal.[18] However, interdiastolic measurements are preferred to intersystolic measurements because the latter estimate heart period with less accuracy, particularly in dynamic conditions.[19]

ANIMAL STUDIES

In conscious dogs the autospectra of both RR interval and systolic arterial pressure contain a single major component synchronous with respiration.[7,20] FIGURE 3 shows the effects of an increase in sympathetic efferent activity as induced reflexly by bilateral common carotid artery occlusion[18] in a conscious dog that previously had hydraulic occluders placed around both common carotid arteries.

During the transient hypertension accompanying carotid baroreceptor unloading (produced by inflating the implanted occluders, leading to carotid artery hypotension) there is a clear rise in LF_{SAP} (as well as in systolic arterial pressure [SAP] variability).

Marked increases in the LF component of SAP variability were observed also in another experimental condition characterized by enhanced sympathetic activity, that is, in conscious dogs during treadmill exercise.[18,21] Importantly, as depicted in FIGURE 4, chronic α_1-adrenergic blockade drastically abated such an exercise-induced rise in LF_{SAP} power.[21] This finding further supports the concept that these oscillations are mediated by an α_1-adrenergic mechanism. By contrast, the HF component of SAP variability was only slightly modified by α_1-adrenergic blockade, in keeping with the hypothesis of its mostly mechanical nature. Such a hypothesis had previously been suggested in experiments with ganglionic transmission blockade, which abolished the LF component of SAP variability (and only reduced the power of the HF_{SAP}).[20] Furthermore, the LF component of SAP variability increased in all other conditions associated with enhanced vascular sympathetic activity tested so far. Data are available for arterial hypotension induced by nitroglycerine,[20] transient coronary artery occlusion,[18,20] and situational stress.[17] These disparate conditions, in which baroreceptor afferent activity could be either reduced or increased, indicate that the LF component of arterial pressure oscillations is not simply dependent on the baroreflex loop which, nevertheless, has an important modulatory effect. The complexity of the neural mechanisms underlying LF oscillation of arterial pressure is further indicated by the observation that selective sympathetic cardiac denervation

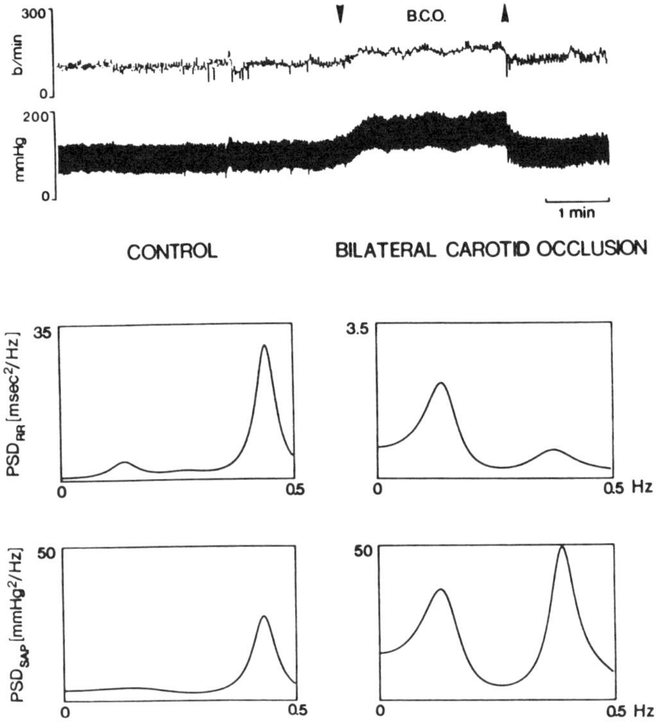

FIGURE 3. Example of computer analysis of RR interval and systolic arterial pressure (SAP) variabilities at rest and during bilateral carotid artery occlusion (BCO, indicated by *arrows*) in a conscious dog. **Upper panel:** Effects of BCO (lasting 5 minutes) on heart rate and systemic arterial pressure. **Bottom panels:** Autospectra of both RR interval and SAP variabilities. During BCO there is a marked increase in the low frequency component of SAP variability (LF_{SAP}). In this example, BCO also produces a clear increase in LF_{RR}.

did not modify the LF_{SAP} component or its increase in response to sympathetic activation, while selectively interfering with LF_{RR}.[20]

It should also be added that analysis of beat-by-beat arterial pressure and heart period variability has been employed in modeling studies of the lumped mechanical properties of the arterial load,[22] and will not be addressed in detail in this report. Suffice it to recall that changes in RR interval variability, such as the respiratory related HF_{RR}, lead to variations in stroke volume, inducing corresponding changes in pulsatile arterial pressure. Hence, respiratory oscillations of arterial pressure apparently depend on the simultaneous action of direct (i.e., mechanical) and reflex mechanisms that are complexly intertwined.[9]

HUMAN INVESTIGATIONS

Spectral profiles from normal human volunteers at rest, as depicted in FIGURES 2 and 5, contain two major oscillatory LF and HF components of similar amplitude, in both RR and SAP variability signals.[5,7]

A variety of physiologic conditions leading to different levels of sympathovagal interaction have been explored in human subjects to furnish observational support to the hypothesis that spectral analysis can be used to examine changing levels of autonomic modulation. Also, data are more detailed in the case of RR interval

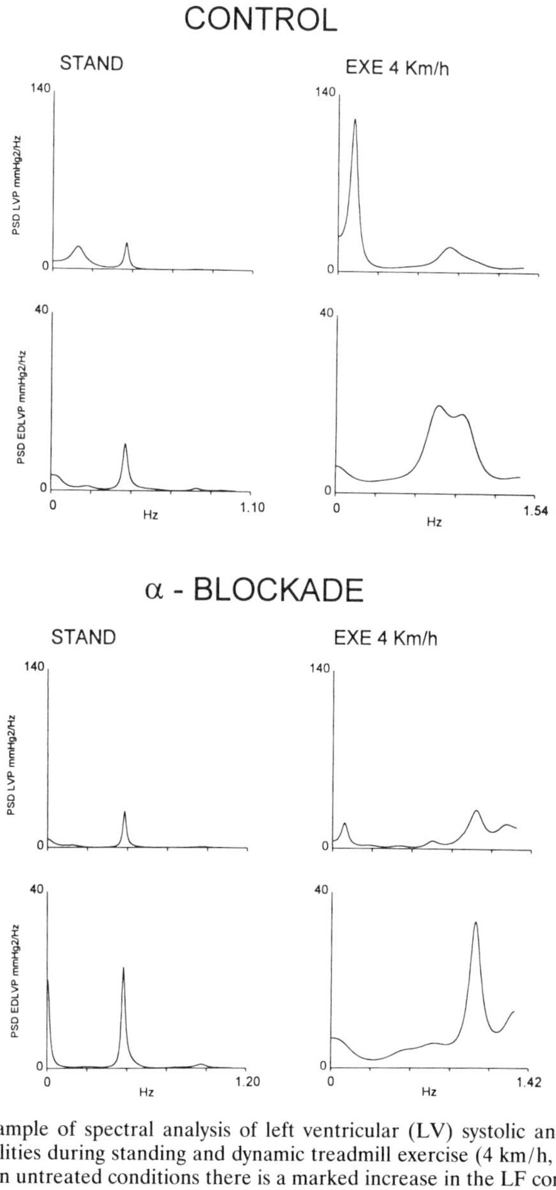

FIGURE 4. Example of spectral analysis of left ventricular (LV) systolic and end-diastolic pressure variabilities during standing and dynamic treadmill exercise (4 km/h, 0° incline) in a conscious dog. In untreated conditions there is a marked increase in the LF component of LV systolic pressure variability. Conversely, the increase in LF during exercise is markedly blunted after pharmacologic α_1-adrenergic blockade with prazosin.

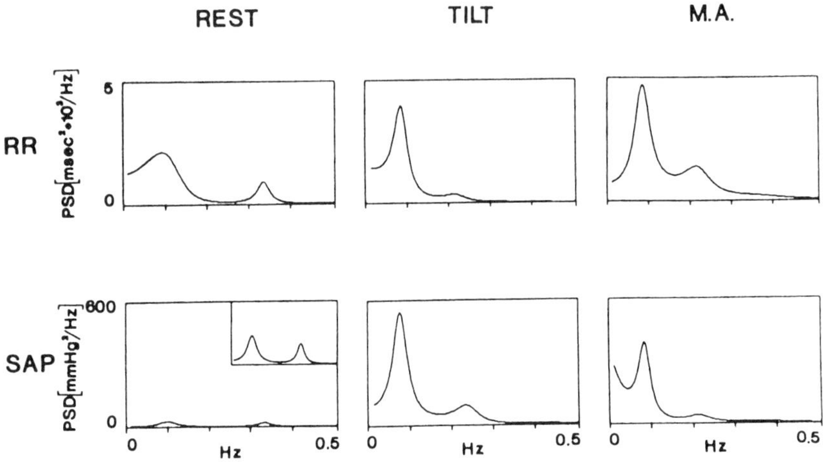

FIGURE 5. Example of spectral analysis of RR interval and systolic arterial pressure (SAP) variabilities in a human subject at rest, during passive tilt, and during a mental arithmetic test (MA). At rest two main spectral components can be observed in both RR interval and SAP variabilities (the autospectrum obtained from SAP variability is magnified by 10 in the *small inset*). Passive tilt and MA induce a marked increase in the low frequency component which becomes predominant in the autospectra of both RR interval and SAP variabilities. (Reprinted with permission from Malliani *et al.*[5])

variability; however, as shown in FIGURES 5 and 6 various physiologic conditions of enhanced sympathetic drive, such as tilt[7] and mental[17] and physical exercise,[21] are all accompanied by elevations in the LF component of systolic arterial pressure variability. Recently, Mukai and Hayano[23] reported that progressive tilt induced a graded increase in LF_{SAP} in healthy human volunteers. The effects of physical activity require more details,[19] such as the type and intensity of exercise as well as the extent of muscular masses that are involved. An additional confounder, particularly in the case of heavy dynamic exercise, may relate to the presence of oscillatory components reflecting the mechanical effects of movement on the blood pressure column.[24,25] These effects, which are not of major importance, can be appreciated with certainty only with high fidelity recordings of arterial pressure, which avoids the risk of motion artifacts in the recorded waveforms.[25] Motion artifacts might be particularly marked when fluid-filled catheter manometer (or non invasive) systems are used in dynamic conditions.[26] Microtip technology also provides more consistent quality when arterial pressure data are obtained from 24-hour ambulant subjects, thus rendering subsequent computer analysis easier to perform.[13,27]

Spectral analysis of 24-hour arterial pressure recordings disclosed an important circadian variation in the LF component of systolic arterial pressure variability with a maximum during working hours and a minimum at night[27] (FIG. 7). A rapid rise in LF_{SAP} at wake-up time may represent additional, but indirect, evidence that a rise in sympathetic vasomotor control in the morning may be part of the complex mechanisms underlying the preferential distribution of acute cardiovascular events observed at that time of day. Baroreceptor mechanisms can easily be assessed by computer analysis of a continuous series of RR interval and systolic arterial pressure variability in both time[28] and frequency domain.[14,29]

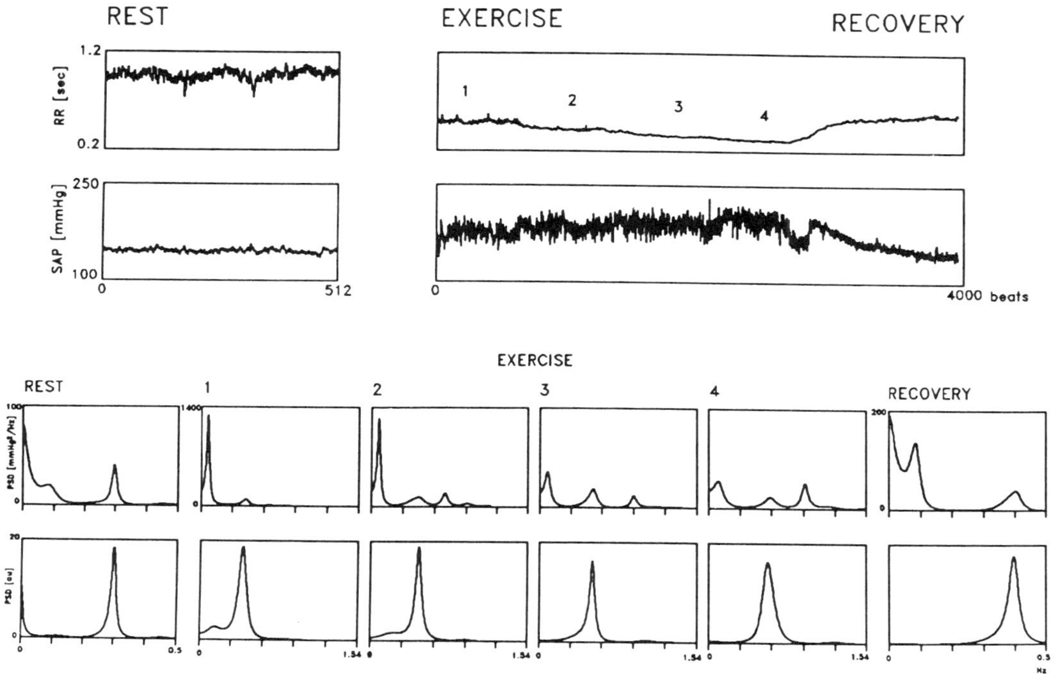

FIGURE 6. Example of the effects of treadmill exercise (Bruce protocol up to the 4th stage) on the time series of RR interval and systolic arterial pressure (SAP) in a young subject (*upper panels*). Arterial pressure was measured invasively by a high-fidelity microtip pressure transducer (Millar Ø3F) inserted into the radial artery. *Lower panels* show the autospectra obtained from SAP variability and respiration. A clear LF predominance is apparent in SAP autospectra during all exercise levels. However, two HF components become progressively more evident during exercise: one is synchronous with respiration, while the one at the highest frequency corresponds to the step rate. (Reprinted with permission from Rimoldi *et al.*[21])

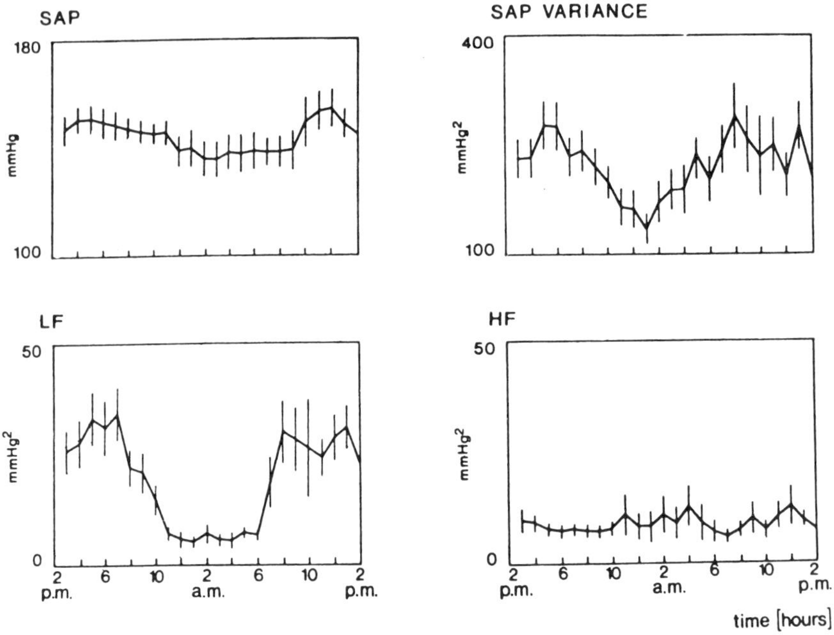

FIGURE 7. Plots of average hourly values of systolic arterial pressure (SAP), of its variance, and of its low and high frequency components during 24 hours in a group of hospitalized patients free to move. As in FIGURE 6, arterial pressure was measured invasively by a high-fidelity microtip pressure transducer (Millar Ø3F). Spectral analysis was performed by a recursive algorithm working on consecutive data segments of a 512-beat length. (Reprinted with permission from Furlan *et al.*[27])

In our laboratory we currently employ spectral and cross-spectral analysis of both LF and HF components to obtain an α index,[14] which provides results similar to those of the more common phenylephrine slope approach, without the need for artificial disturbances in arterial pressure or for preselection of data, also using noninvasive arterial pressure measurements.[30]

With this technique we reported clear day-night changes in the arterial pressure-heart period baroreflex, with maximal values during the night.[14] Furthermore, this gain is under an important sympathetic restraint and is significantly correlated with total heart rate variability (HRV).[30] This latter point may provide an experimental link to the observation that both HRV[31] and baroreflex gain[32] are important prognostic indicators in various cardiovascular conditions such as heart failure and postmyocardial infarction. An elegant new approach to the study of the influence of baroreceptor afferents on cardiovascular oscillations was reported recently by Bernardi *et al.*,[33] who employ a computer-controlled neck/collar to apply oscillatory pressure stimuli to the carotid baroreceptor area.

Clinical studies specifically addressing the role of oscillations in arterial pressure are relatively scarse, whereas emphasis is more on RR interval variability. We have shown that the LF component of SAP variability tends to be greater in hypertensive subjects than in normotensive volunteers.[34] In preliminary observations in patients with an implanted pacemaker, we also reported that the LF component of SAP does

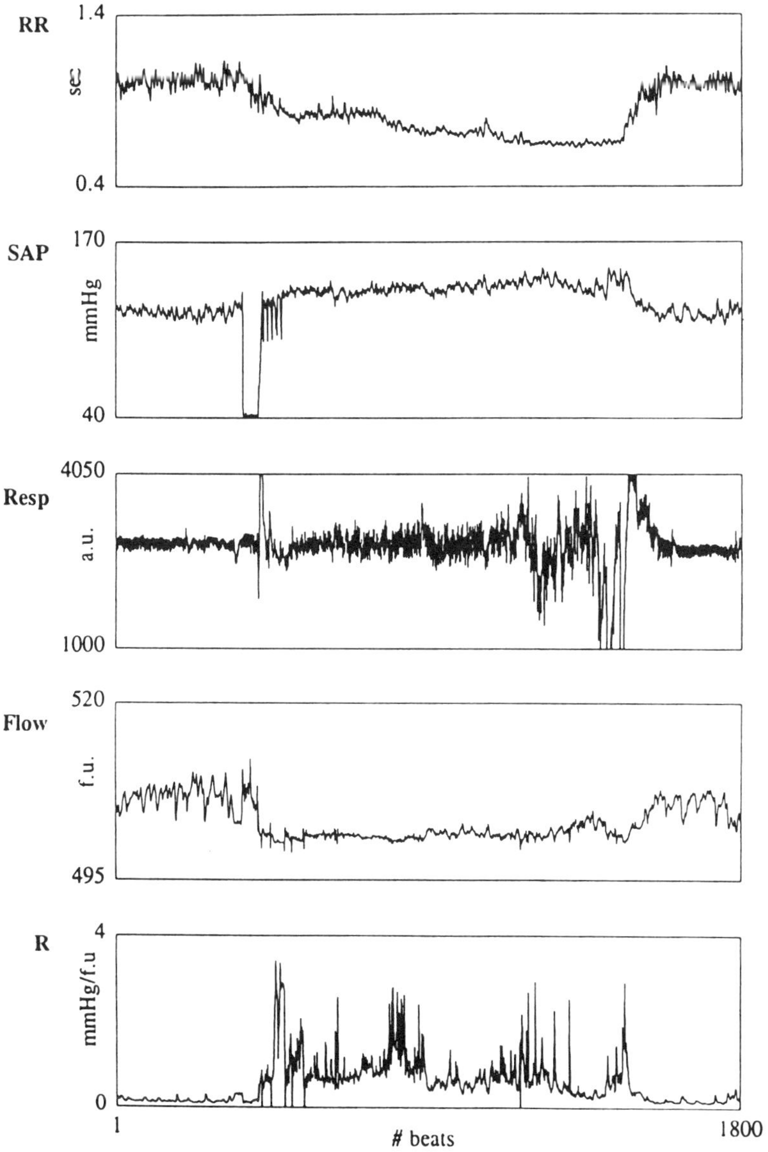

FIGURE 8. Example of the stepwise effects of moderate dynamic exercise (up to 30% of nominal maximum) using the horizontal bicycle in a control subject. Note the progressive reduction in RR interval (**top panel**) and the increase in systolic arterial pressure (SAP) and respiration (Resp). During exercise brachial artery flow velocity (Flow) is reduced and nominal computed resistance (R) increased.

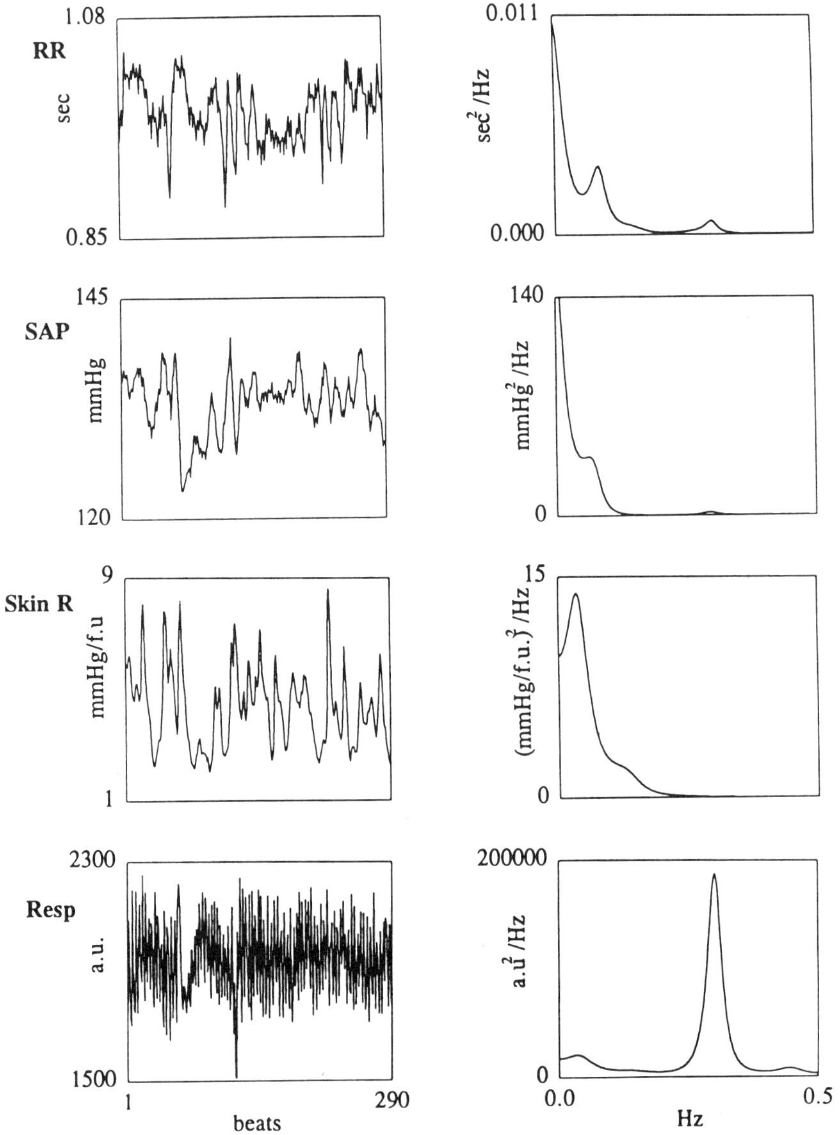

FIGURE 9. Example of the time series of RR interval, systolic arterial pressure (SAP), skin resistance (R), and respiration (Resp) in a control resting subject (**left panels**). The corresponding autospectra are shown in the **right panels.**

not require the presence of a simultaneous similar oscillation of the RR interval. Also in a collaborative study with the Ochsner Institution in patients with an orthotopic cardiac transplant who almost totally lack RR interval variability owing to cardiac denervation, the LF component of SAP appears to be present at rest and to respond normally to orthostatic challenges.

CONCLUSIONS

Beat-by-beat changes in arterial pressure can be analyzed with appropriate computer techniques in both animals and man during a variety of physiologic techniques in circumstances characterized by various levels of autonomic modulation. By analogy with spectral analysis of respiratory rate interval variability, the power of spectral components of arterial pressure can also be used to infer changes in the state of the sympathetic outflows modulating peripheral vessels. Major differences, however, should be recognized. Neural control of the sinoatrial node depends on the continuous interaction of an inhibitory and an excitatory outflow, that is, the combined effects of vagal and sympathetic efferent activity. The state of their interaction can best be appreciated by using normalized units to quantify the power of either component or by using their ratio (i.e., LF/HF).[5,7,35] This approach takes into account the presence of large interindividual variations in total power, which could cloud key information, provided by the relative distribution of power across the frequency axis. The correct application of this approach, using monovariate analysis, requires that the direct influence of respiration on RR interval variability be minimal, as happens in resting human subjects breathing at a frequency of approximately 15–20 breath per minute. By contrast, neural control of peripheral vasomotion is based essentially on single vasoconstrictor regulation furnished by sympathetic efferent activity. Local factors (either metabolic, hormonal, or endothelial) play additional, very important roles in regulating the instantaneous activity of peripheral vessels (FIG. 8). The pattern of activity of the peripheral vasculature is complex and contains, however, a clear LF component (although frequently with a slower center frequency than that observed for LF_{RR} and LF_{SAP}) (FIG. 9).[36] In the framework of ongoing research we[7,18] as well as others[23] feel that the LF component of SAP provides a convenient marker of the changes in sympathetic vasomotor control in a variety of pathophysiologic conditions.

REFERENCES

1. HALES, S. 1733. Statistical Essays: Containing Haemastaticks. Vol. 2. Innys, Manby and Woodward. London.
2. MAYER, S. 1876. Studien zur Physiologie des Herzens und der Blutgefässe. 5. Abhandlung: Über spontane Blutdruckschwankungen. Sber. Akad. Wiss. Wien. **74:** 281–307.
3. PEÑÁZ, J. 1978. Mayer waves: History and methodology. Automedica **2:** 135–141.
4. KOEPCHEN, H. P. 1984. History of studies and concepts of blood pressure waves. *In* Mechanisms of Blood Pressure Waves. K. Miyakawa, H. P. Koepchen & C. Polosa, Eds.: 3–23. Japan Science Society Press/Springer-Verlag. Tokyo/Berlin.
5. MALLIANI, A., M. PAGANI, F. LOMBARDI & S. CERUTTI. 1991. Cardiovascular neural regulation explored in the frequency domain. Circulation **84:** 482–492.
6. BROVELLI, B., G. BASELLI, S. CERUTTI, S. GUZZETTI, D. LIBERATI, F. LOMBARDI, A. MALLIANI, M. PAGANI & P. PIZZINELLI. 1983. Computerized analysis for an experimen-

tal validation of neurophysiological models of heart rate control. Comput. Cardiol. 205–208.
7. PAGANI, M., F. LOMBARDI, S. GUZZETTI, O. RIMOLDI, R. FURLAN, P. PIZZINELLI, G. SANDRONE, G. MALFATTO, S. DELL'ORTO, E. PICCALUGA, M. TURIEL, G. BASELLI, S. CERUTTI & A. MALLIANI. 1986. Power spectral analysis of heart rate and arterial pressure variabilities as a marker of sympathovagal interaction in man and conscious dog. Circ. Res. **59:** 178–193.
8. BASELLI, G., S. CERUTTI, S. CIVARDI, D. LIBERATI, F. LOMBARDI, A. MALLIANI & M. PAGANI. 1986. Spectral and cross-spectral analysis of heart rate and arterial blood pressure variability signals. Comp. Biomed. Res. **19:** 520–534.
9. BASELLI, G., S. CERUTTI, S. CIVARDI, A. MALLIANI & M. PAGANI. 1988. Cardiovascular variability signals: Towards the identification of a closed-loop model of the neural control mechanisms. IEEE Trans. Biomed. Eng. **35:** 1033–1046.
10. BROWN, T. E., L. A. BEIGHTOL, J. KOH & D. L. ECKBERG. 1993. Important influence of respiration on human R-R interval power spectra is largely ignored. J. Appl. Physiol. **75:** 2310–2317.
11. KITNEY, R. I. & O. ROMPELMAN. 1980. The study of heart rate variability. Clarendon Press. Oxford.
12. PAGANI, M., R. FURLAN, S. DELL'ORTO, P. PIZZINELLI, G. LANZI, G. BASELLI, C. SANTOLI, S. CERUTTI, F. LOMBARDI & A. MALLIANI. 1986. Continuous recording of direct high-fidelity arterial pressure and electrocardiogram in ambulant patients. Cardiovasc. Res. **20:** 384–388.
13. FURLAN, R., M. PAGANI, S. DELL'ORTO, E. GENTILE, S. PIAZZA, A. CASTELLI, J. P. MURGO, F. LOMBARDI & A. MALLIANI. 1993. Continuous 24 h ambulatory recording of systemic and right ventricular pressures. J. Amb. Monit. **6:** 47–54.
14. PAGANI, M., V. SOMERS, R. FURLAN, S. DELL'ORTO, J. CONWAY, G. BASELLI, S. CERUTTI, P. SLEIGHT & A. MALLIANI. 1988. Changes in autonomic regulation induced by physical training in mild hypertension. Hypertension **12:** 600–610.
15. DE BOER, R. W. 1985. Beat-to-beat blood pressure fluctuations and heart rate variability in man: Physiological relationships, analysis techniques and a simple model (Thesis). University of Amsterdam. Amsterdam.
16. PARATI, G., P. CASTIGLIONI, M. DI RIENZO, S. OMBONI, A. PEDOTTI & G. MANCIA. 1990. Sequential spectral analysis of 24-hours blood pressure and pulse interval in humans. Hypertension **19:** 414–421.
17. PAGANI, M., O. RIMOLDI, P. PIZZINELLI, R. FURLAN, W. CRIVELLARO, D. LIBERATI, S. CERUTTI & A. MALLIANI. 1991. Assessment of the neural control of the circulation during psychological stress. J. Auton. Nerv. Syst. **35:** 33–42.
18. RIMOLDI, O., M. R. PAGANI, S. PIAZZA, M. PAGANI & A. MALLIANI. 1994. Restraining effects of captopril on sympathetic excitatory responses in dogs: A spectral analysis approach. Am. J. Physiol. **267:** H1608–H1618.
19. PAGANI, M., D. LUCINI, O. RIMOLDI, R. FURLAN, S. PIAZZA & L. BIANCARDI. 1995. Effects of physical and mental exercise on heart rate variability. *In* Heart Rate Variability. M. Malik & A. J. Camm, Eds. Futura Publishing Company, Inc. Armonk, NY. 245–266.
20. RIMOLDI, O., S. PIERINI, A. FERRARI, S. CERUTTI, M. PAGAN & A. MALLIANI. 1990. Analysis of short-term oscillations of RR and arterial pressure in conscious dogs. Am. J. Physiol. **258:** H967–H976.
21. RIMOLDI, O., R. FURLAN, M. R. PAGANI, S. PIAZZA, M. GUAZZI, M. PAGANI & A. MALLIANI. 1992. Analysis of neural mechanisms accompanying different intensities of dynamic exercise. Chest **101:** 226S–230S.
22. BASELLI, G., S. CERUTTI, S. CIVARDI, A. MALLIANI, G. ORSI, M. PAGANI & G. RIZZO. 1988. Parameter extraction from heart rate and arterial blood pressure variability signal in dogs for the validation of a physiological model. Comput. Biol. Med. **18:** 1–16.
23. MUKAI, S. & J. HAYANO. 1995. Heart rate and blood pressure variabilities during graded head-up tilt. J. Appl. Physiol. **78:** 212–216.
24. FURLAN, R., S. PIAZZA, S. DELL'ORTO, E. GENTILE, M. SALATI, F. LOMBARDI, M. PAGANI, C. SANTOLI & A. MALLIANI. 1992. Effects of respiratory activity on heart rate and blood pressure variabilities during physical exercise. J. Amb. Monit. **5**(2&3): 87–94.

25. PIAZZA, S., R. FURLAN, S. DELL'ORTO, A. PORTA, F. LOMBARDI, M. PAGANI & A. MALLIANI. 1995. Mechanical influences of respiration and stepping on systolic arterial pressure variability during treadmill exercise. J. Hypertens. **13:** 1643–1647.
26. PALATINI, P., L. MOS, P. MORMINO, A. DI MARCO, L. MUNARI, G. FAZIO, G. GIULIANO, A. PESSINA & C. DAL PALÙ. 1989. Blood pressure changes during running in humans: The "beat" phenomenon. J. Appl. Physiol. **67:** 52–59.
27. FURLAN, R., S. GUZZETTI, W. CRIVELLARO, S. DASSI, M. TINELLI, G. BASELLI, S. CERUTTI, F. LOMBARDI, M. PAGANI & A. MALLIANI. 1990. Continuous 24h assessment of the neural regulation of systemic arterial pressure and RR variabilities in ambulant subjects. Circulation **81:** 537–547.
28. BERTINERI, G., M. DI RIENZO, A. CAVALLAZZI, A. U. FERRARI, A. PEDOTTI & G. MANCIA. 1985. A new approach to analysis of the arterial baroreflex. J. Hypertens. **3:** S79–S81.
29. ROBBE, H. W. J., L. J. M. MULDER, H. RUDDEL, W. A. LANGEWITZ, J. B. P. VELDMAN & G. MULDER. 1987. Assessment of baroreceptor reflex sensitivity by means of spectral analysis. Hypertension **10:** 538–543.
30. LUCINI, D., M. PAGANI, G. S. MELA & A. MALLIANI. 1994. Sympathetic restraint of baroreflex control of heart period in normotensive and hypertensive subjects. Clin. Sci. **86:** 547–556.
31. KLEIGER, R. E., J. P. MILLER, J. T. BIGGER, A. R. MOSS & MULTICENTER POST-INFARCTION RESEARCH GROUP. 1987. Decreased heart rate variability and its association with increased mortality after acute myocardial infarction. Am. J. Cardiol. **59:** 256–262.
32. LA ROVERE, M. T., G. SPECCHIA, A. MORTARA & P. J. SCHWARTZ. 1988. Baroreflex sensitivity, clinical correlates, and cardiovascular mortality among patients with a first myocardial infarction. Circulation **78:** 816–824.
33. BERNARDI, L., B. BIANCHINI, G. SPADACINI, S. LEUZZI, F. VALLE, E. MARCHESI, C. PASSINO, A. CALCIATI, M. VIGANÒ, M. RINALDI, L. MARTINELLI, G. FINARDI & P. SLEIGHT. 1995. Demonstrable cardiac reinnervation after human heart transplantation by carotid baroreflex modulation of RR interval. Circulation. In press.
34. FURLAN, R., E. GENTILE, S. PIAZZA, S. DELL'ORTO, F. BARBIC, M. R. PAGANI, F. LOMBARDI, M. PAGANI & A. MALLIANI. 1991. Increased vascular sympathetic activity at rest and reduced responsiveness to excitatory stimuli in essential hypertension. J. Hypertens. **9:** S60–S61.
35. MONTANO, N., T. GNECCHI RUSCONE, A. PORTA, F. LOMBARDI, M. PAGANI & A. MALLIANI. 1994. Power spectrum analysis of heart rate variability to assess the changes in sympathovagal balance during graded orthostatic tilt. Circulation **90:** 1826–1831.
36. PORTA, A., G. BASELLI, L. MAINARDI, D. LUCINI, M. PAGANI, A. MALLIANI & S. CERUTTI. 1995. Beat-to-beat variability of microvascular peripheral resistances assessed with a non-invasive approach. Computers in Cardiology, Vienna, September 10–13. In press.

Impeachment of Casual Blood Pressure Measurements and the Fixed Limits for Their Interpretation and Chronobiologic Recommendations[a]

GERMAINE CORNÉLISSEN[b] AND FRANZ HALBERG

Chronobiology Laboratories
University of Minnesota
5-187 Lyon Laboratories
420 Washington Ave. SE
Minneapolis, Minnesota 55455

The fifth report of the Joint National Committee on Detection, Evaluation and Treatment of High Blood Pressure[1] recommends a recheck in 2 years for adults aged 18 and older with a casual systolic (S)/diastolic (D) blood pressure (BP) measurement below 130/85 mm Hg and a recheck in 1 year for those with a SBP/DBP measurement of 130–139/85–89 mm Hg. A more immediate follow-up is only recommended for patients with SBP/DBP values above 140/90 mm Hg. Such measurements, as a rule, are taken in a physician's office. The validity of this current approach is here questioned in view of two of its major limitations: reliance on casual BP measurements during office hours that ignore the large variability of BP and reliance on fixed thresholds of acceptability (such as 140/90 mm Hg) that neglect the (statistically deterministically) periodic behavior of BP, notably along the circadian scale, now amply documented from womb to tomb.[2]

LIMITATIONS FROM RELIANCE ON CASUAL BLOOD PRESSURE MEASUREMENTS

Reliance on single readings is not desirable.[3,4] Not only can BP abnormality occur at odd times when it is unlikely to be checked (FIG. 1), but also even a 24-hour profile obtained by ambulatory monitoring is not necessarily sufficient to correctly diagnose BP disorders which may be detectable on some days but not every day (FIG. 2).[5,6] Whereas some attention has been given to sampling requirements for a 24-hour profile by ambulatory monitoring,[7] the single 24-hour record has remained the "gold standard." Because of infradian variability, circadian characteristics may change from day to day. It is therefore important to determine an observation span of appropriate length for monitoring BP. Longitudinal records are valuable to answer this question. Data from a clinically healthy woman, monitored for 26 days, showed the desirability of monitoring for at least 48 hours (and preferably for 1 week or longer) rather than 24 hours at a time.[8] Reliability of the estimates of the circadian parameters increased greatly (35.4% reduction of the error term), whereas the gain

[a] Dedicated to Hélène Minne Cornélissen with deepest grateful affection that words cannot convey.

[b] Tel: 612/624-6976; fax: 612/624-9989; e-mail: corne001@maroon.tc.umn.edu.

was much less from the further lengthening of the monitoring span from 2 to 5 days (10.5% reduction of the error term).

It is not often realized by how much the circadian characteristics of BP can vary from day to day. A comparison of the MESOR and circadian amplitude of SBP and DBP from two 24-hour profiles obtained about 1 month apart on 40 presumably healthy men 20–60 years of age[9] reveals average differences in MESOR of 7 and 5 mm Hg, respectively (maximal differences of 31 and 19 mm Hg, respectively), and differences in circadian double amplitude averaging 5.3 and 3.6 mm Hg, respectively (maximal differences of 22.3 and 13.6 mm Hg, respectively) (TABLE 1). These large

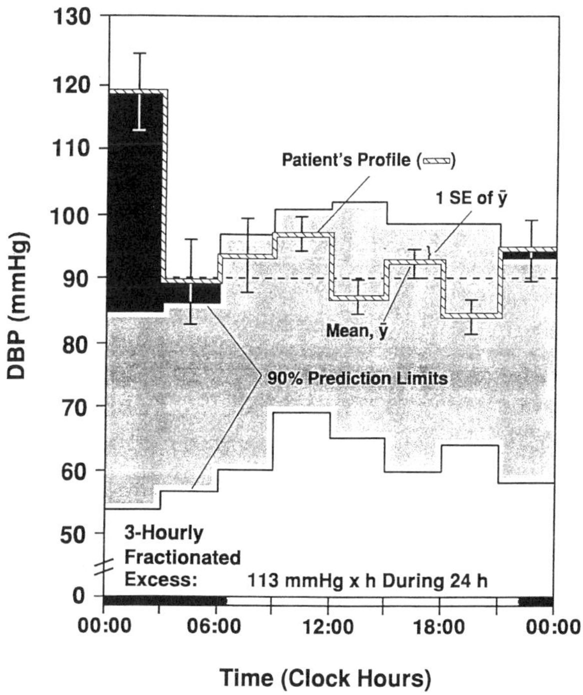

FIGURE 1. Odd-time diastolic blood pressure (DBP) excess (■) with treatment in the morning. Patient was taking Vasotec (an ACE inhibitor) 10 mg daily. From Halberg, with permission.

differences are readily observed, notwithstanding a good correlation between the two profiles ($0.58 \leq r \leq 0.74; p < 0.001$).

LIMITATIONS FROM RELIANCE ON FIXED THRESHOLDS

Despite much evidence for large-amplitude circadian (and other) rhythms in BP, so far the medical community has mostly remained unconvinced of their clinical importance, considering that BP measured during office hours exhibits only minimal circadian variation. Evidence has been forthcoming, however, for the clinical signifi-

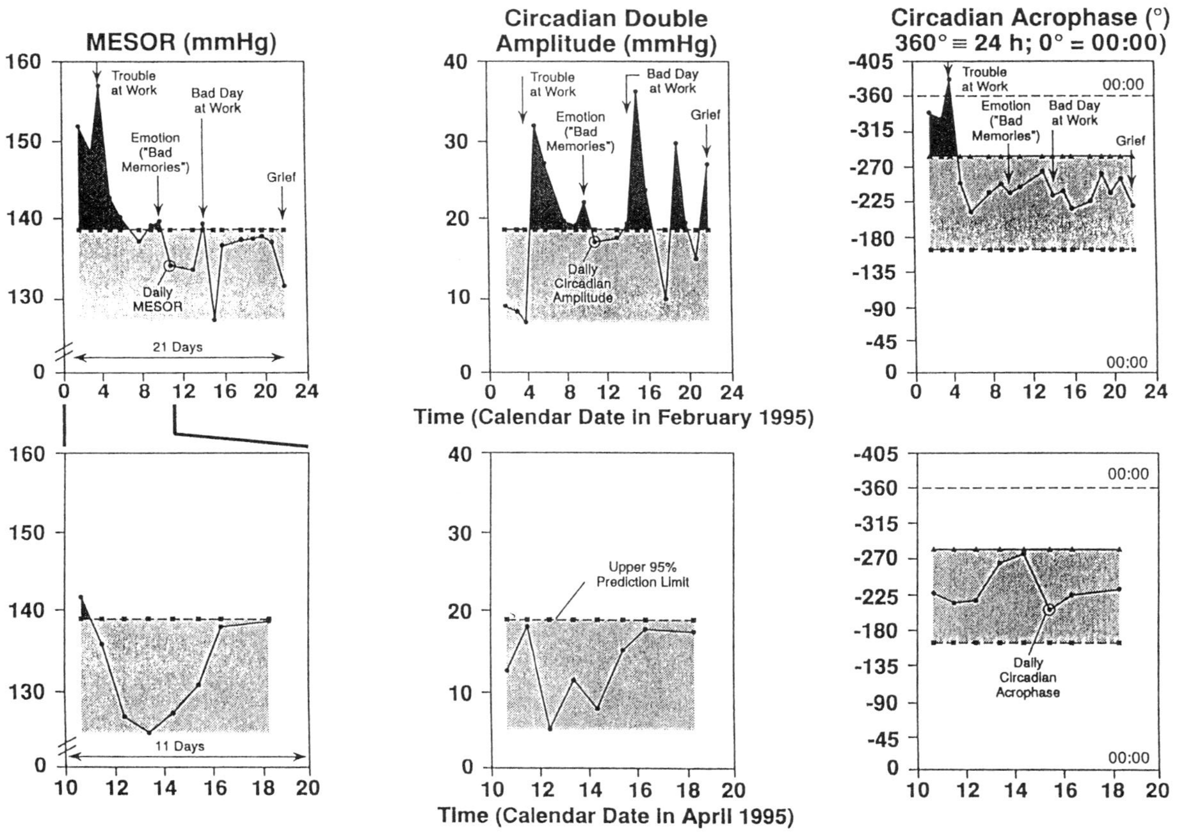

FIGURE 2. Infradian systolic blood pressure (SBP) variability associated with psychophysiologic responses. MESOR hypertension was evident only for 5 consecutive days at the beginning of the record but not for the ensuing 16 days (*top left*), or for the 11-day span in April (*bottom left*), or the 11-day span in July (not shown). The circadian amplitude is acceptable for the first 3 days, but intermittently excessive thereafter in February (but not in April or July). Acrophase is deviant only initially. From Halberg, with permission.

cance of BP disorders affecting characteristics other than the 24-hour mean value,[10–12] particularly those involving the circadian BP amplitude.

On the basis of self-measurements carried out at about 4-hour intervals during waking, Scarpelli *et al.*[13] found that among cases with accelerated hypertension those who died had a statistically significantly larger circadian BP amplitude than did those who survived. In the absence of any elevation of the rhythm-adjusted 24-hour BP mean (MESOR), Kumagai *et al.*[14] found that a large circadian amplitude of BP is associated with left ventricular mass index values higher than those in amplitude- and MESOR-normotension, but lower than those found in patients with an elevated BP MESOR. The fact that too large a circadian BP amplitude is associated with changes involving the heart and also with excess morbidity was shown by Otsuka *et al.*[15–17] on 297 patients, some with a SBP/DBP MESOR lower than 130/80 mm Hg.

These results suggest the urgency to properly diagnose an excessive circadian BP amplitude, a task that cannot be done by single measurements, but one that requires both around-the-clock monitoring and a chronobiologic analysis and interpretation.

TABLE 1. Comparison of Cardiovascular Rhythm Characteristics Derived from Two 24-Hour Profiles, about 1 Month Apart, on 40 Presumably Healthy Men 20–60 Years of Age

Endpoint	r	p	Δ	SD	Min.	Max.
SBP						
M	0.586	<0.001	7.13	7.09	0.18	31.09
A	0.739	<0.001	3.16	2.66	0.08	11.16
DBP						
M	0.652	<0.001	5.11	4.50	0.17	19.42
A	0.742	<0.001	2.56	1.78	0.18	6.82
HR						
M	0.927	<0.001	3.16	2.54	<0.01	9.69
A	0.768	<0.001	2.61	2.24	0.34	8.66

Abbreviations: SBP = systolic blood pressure; DBP = diastolic blood pressure; HR = heart rate; SBP and DBP expressed in mm Hg, HR expressed in beats/min; M = MESOR; A = amplitude of 24-hour component; r = correlation coefficient; p = associated p value; Δ = average absolute difference between the two profiles; SD = standard deviation of Δ.

From the viewpoint of diagnosis, the large excursion of BP values even during the active span is a potential source of error. Single values may indicate either that the patient is normotensive (in which case no treatment is provided and the risk of vascular disease is great) or that the patient is hypertensive (in which case it is likely that treatment will include long-acting drugs intended for 24-hour coverage, even if it is not invariably achieved).[3,4,18] From the viewpoint of treatment, it is critical to administer antihypertensive medication at the opportune time, because in circadian hyper-amplitude-tension (CHAT) high BP values are only encountered during part of the day and BP should not be lowered further when it is already naturally low.

THEORETIC EVIDENCE

The desirability of replacing fixed thresholds by time-specified reference limits and by reference values for rhythm characteristics (chronodesms) is illustrated theoretically in TABLE 2 and FIGURE 3. These results recognize the abundantly established fact that BP follows a circadian rhythm with a timing of overall high values occurring in mid-afternoon around 4 PM, as it does for most individuals in good health as well as in patients with high BP as long as they are awake by day and asleep by night.[3,4,19–21] A circadian rhythm of BP with an amplitude near the upper 95% prediction limit of clinically healthy peers (namely, a double amplitude of 40 mm Hg) is assumed. The MESOR is then allowed to vary in a relatively wide range of values, namely 120–140/70–90 mm Hg, to investigate, with a 24-hour cosine model, the time intervals during which SBP/DBP can be anticipated to assume values above 140/90 mm Hg. A single cosine curve is here considered as a first approximation only. It is anticipated that similar results would be obtained with models including higher-order harmonics that approximate the circadian waveform of BP more closely.

For measurements taken at equal intervals during a 24-hour cycle, the percentage of values that can be anticipated to be above 140/90 mm Hg can be derived considering either the whole 24-hour span or a span restricted to office hours (e.g., from 08 AM to 5 PM). For SBP/DBP MESORs in the range of 125–140/75–90 mm Hg, during office hours, it can be anticipated that 42–78% of the measurements will be too high (while, of course, 58 to 22% are acceptable). Thus, on the basis of a fixed threshold, even in the absence of any measurement error, there is always a large uncertainty and inaccuracy in the diagnosis when fixed limits are used that disregard the large predictable circadian variation in BP, demonstrated overwhelmingly in health as well as in disease.

TABLE 2. Fixed Thresholds[a] Can Render the Office Blood Pressure (BP) Equivalent to Flipping a Coin

SBP/DBP MESOR (mm Hg)	Daily Span (hr) with SBP/DBP > 140/90 mm Hg	Measurements > 140/90 mm Hg (%)[b]	
		During 24 Hours	During Office Hours
120/70	0	0	0
125/75	5.52	23.00	41.78
130/80	8.00	33.33	44.44
135/85	10.07	42.00	67.06
140/90	12.00	50.00	77.78

[a]Best replaced by chronodesms, i.e., 90% peer prediction limits for single samples and chronome characteristics.

[b]Percentage of BP values above or below NIH-WHO-WHL limits of 140/90 mm Hg (systolic/diastolic BP) expected to occur within a 24-hour span or during office hours (8 AM–5 PM), assuming a circadian amplitude of 20 mm Hg, near the upper 95% prediction limit of healthy peers and an average acrophase in mid-afternoon (around 4 PM), with SBP MESORs ranging from 120–140 mm Hg and DBP MESORs from 70–90 mm Hg.

CLINICAL EVIDENCE

The uncertainty associated with the conventional diagnosis and treatment of a patient's BP status is indirectly apparent from nonchronobiologic clinical trials involving relatively few casual measurements on large numbers of individuals; reports concerned with the merits of ambulatory monitoring or home self-measurements; chronobiologic pilots; and longitudinal chronobiologic (N-of-1) case

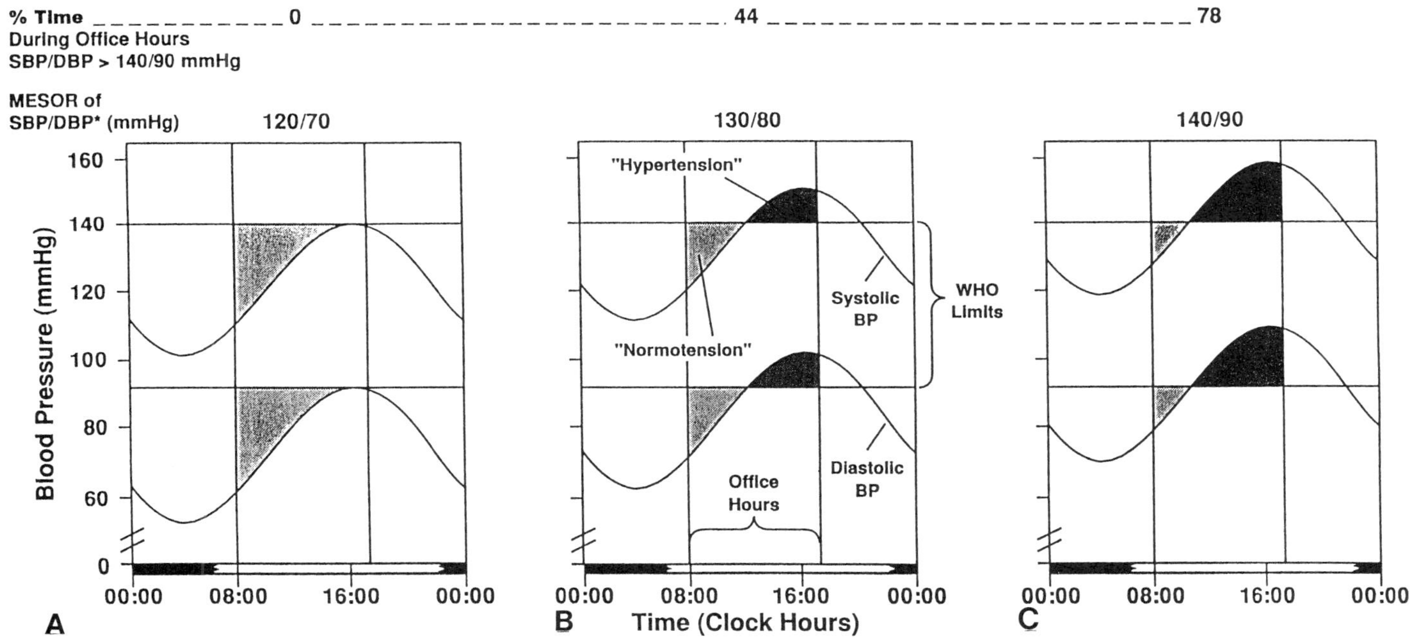

FIGURE 3. Limitations in dealing with blood pressure (BP) interpreted by fixed limits (and casual measurements or automatic office hour profiles). Theoretic evidence points to the need to replace fixed thresholds by time-varying reference standards (chronodesms) and to rely on more than one or a few casual blood pressure measurements. In the absence of a measurement error, the circadian rhythm in BP results in contradictory diagnoses. (**A**) Only in cases with SBP/DBP MESORs <120/70 (or >150/100) mm Hg is the diagnosis consistent with current criteria, but in any range of BPs failure to assess day-to-day variability and to use time-varying limits flaws current directives of BP assessment. (**B**) The diagnosis of "borderline hypertension" based on casual office measurements assessed by WHO limits is no better than flipping a coin. (**C**) Even for patients with a 24-hour mean of SBP/DBP around 140/90 mm Hg, "hypertension" is not detected 22% of the time because of the lack of consideration of circadian variation. Computations based on a circadian amplitude of 20 mm Hg (near, but still below the upper 95% prediction limit in clinically healthy adults) and acrophase of −240° (4 PM).

reports that explore the statistical significance of a change for the given subject examined.

Large Clinical Trials and Metaanalyses

A 60-million-dollar Australian study of mild hypertension,[22] wherein 48% of the 1,943 patients assigned to the placebo group were "cured" 3 years later, constitutes a hint of false positives at entry, false negatives after completion of the study, and placebo responders.[19,23] Metaanalysis[24] shows only marginal benefit (with wide confidence limits) from treating patients with an elevated BP (FIG. 4).

Ambulatory Monitoring and Home Self-Measurement Studies

The above 40% error in diagnosis based on casual BP measurements found in the Australian study,[22] as in several chronobiologic pilots[5,25] (see below), is supported by a report that after 24-hour ambulatory monitoring, the diagnosis was changed in 41% of the patients and antihypertensive therapy was changed in 46%.[26] White coat hypertension, that is, high readings in the clinic but acceptable readings at home, was reported in 21–58% of "hypertensive" subjects without end-organ injury, as determined by means of home BP monitoring.[27] In Tecumseh, white coat hypertension is reportedly present in 7.1% of the whole population and in 58% of all subjects with an elevated BP in the clinic.[28] The use of nonpharmacologic methods in the treatment of

#	Study	# and Gender of Patients	Blood Pressure Systolic/Diastolic (mmHg)	Reference
1	U.S. Veterans Administration	523 M	/ ≥90	JAMA 202: 1028-1034, 1967 JAMA 213: 1143-1152, 1970
2	U.S. Hypertension Detection and Follow-Up Program	10,940 M & F	/ > 90	JAMA 242: 2562-2571, 1979 JAMA 247: 633-638, 1982
3	Oslo	785 M	150-179 / ≥100	Am. J. Med. 69: 725-732, 1980
4	Australian	3,427 M & F	/ > 95-109	Lancet I: 1261-1267, 1980 Circulation 69: 668-676, 1984
5	U.S. Multiple Risk Factor Interventions Trial	12,866 M	/ > 90-115	JAMA 248: 1465-1477, 1982
6	European Elderly	940 M & F	/ > 90-119	Lancet I: 1349-1354, 1985
7	International Prospective Primary Prevention Study	6,537 M & F	/ > 100-125	J. Hypertension 3: 379-392, 1985
8	British MRC	17,354 M & F	/ > 90-109	Br. Med. J. 291: 97-104, 1985
9	Gothenburg	635 M	> 175 / or / > 115	Lancet I: 1-5, 1978

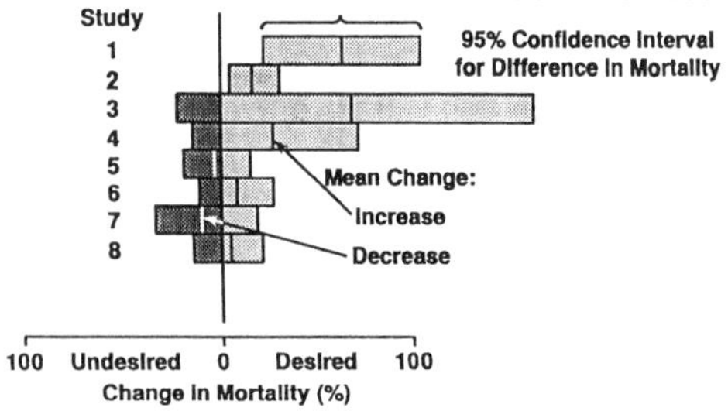

FIGURE 4. Quo usque tandem: Treatment trials of "blood pressure elevation." The traditional approach is unsatisfactory in view of the metaanalysis of Wilcox *et al.*[24] who suggest that clinical practice should not be based on results of such clinical trials. This should be more than a hint to include minimal chronobiologic provisions in future studies before justifying a total cost of hundreds of millions of US dollars. From Halberg, with permission.

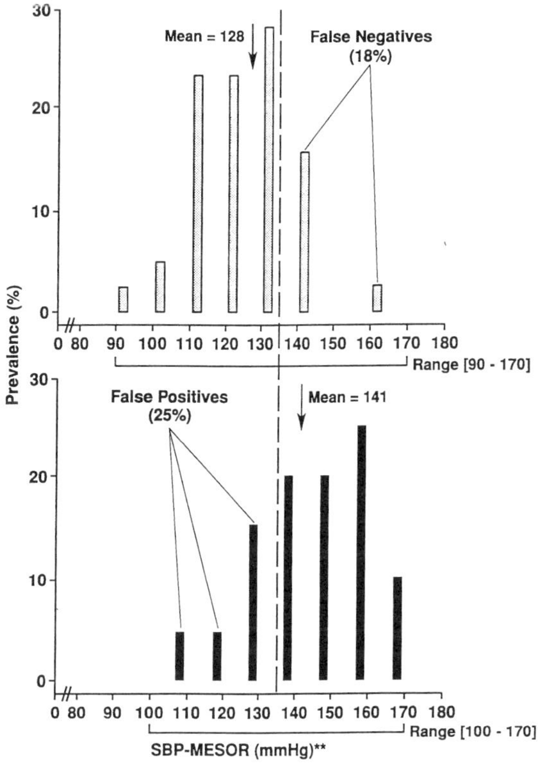

FIGURE 5. There is a large overlap between the distributions of MESORs (rhythm-adjusted means) of systolic blood pressure of elderly patients conventionally diagnosed as normotensive (□; $n = 39$) or hypertensive (■; $n = 20$).[25] False-positive and false-negative results account for about 43% of all conventional diagnoses in this particular study; the percentage uncertainty is higher in another larger study.[22,23] *Misdiagnoses by loose WHO criteria applied to casual measurements of 62–90-year-old men and women. Systematic chronobiologic self-monitoring (manually or preferably automatically) by everyone is needed. **MESOR = rhythm-adjusted 24-hour mean. From Halberg, with permission. Data from I. Kuwajima.

hypertension in general practice enabled some investigators to reduce the dose of antihypertensive drugs in 55% of the participants, with 46% of them entirely free from antihypertensive drugs at the end of study.[29]

Chronobiologic Pilots

A study by Kuwajima *et al.*[25] on 84 patients shows diagnostic disagreement between single measurements taken in the physician's office and the SBP MESOR from ambulatory monitoring in 42% of the cases investigated (FIG. 5). A false-negative diagnosis on casual measurements has a subsequent SBP MESOR of 160 mm Hg, even when the low readings during rest and sleep enter into the computation of the MESOR, a consideration requiring appropriate time-specified reference standards.[19] A false-positive diagnosis on casual measurements has a subsequent SBP MESOR of 110 mm Hg.

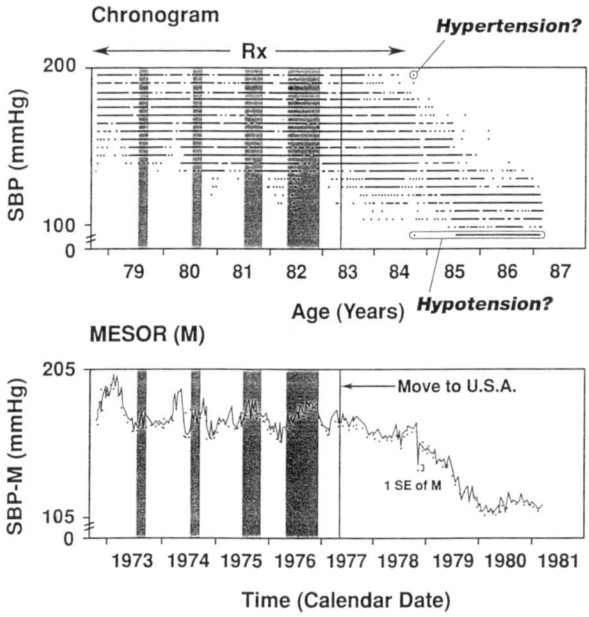

FIGURE 6. Self-measurements in elderly woman reveal drastic changes in systolic blood pressure (SBP) with age, warranting the cessation of antihypertensive medication. There were 14,046 SBP measurements taken. Serial section at trial period of 24 hours over nonoverlapping 168-hour intervals. Transmeridian flights over 7 time zones are *shaded;* antihypertensive medication was used up to 1978. From Halberg, with permission. Data from E. Haus.

A study by Raab *et al.*[5] on 23 patients yields diagnostic discrepancies among three consecutive records from ambulatory monitoring, separated by at most a few weeks, each record based on around-the-clock measurements every 5–15 minutes for spans of 4, 2, and 3 days, in this order. Considering all 9 days of around-the-clock BP monitoring, only 12 (52%) of the 23 patients, conventionally diagnosed to be normotensive, were correctly diagnosed (insofar as 9 days of monitoring can tell). Of the remaining 11 patients (48%) who showed occasional BP excess warranting follow-up, 3 (13%) were consistently found to be MESOR-hypertensive.

Treatment with some beta-blockers may result in an increase in the circadian BP amplitude,[19,30] even when accompanied by a decrease in MESOR. Such CHAT may be treated by increasing the dose of the drug, as demonstrated for atendol.[19]

Chronobiologic Case Reports

A longitudinal placebo-controlled N-of-1 study by Little *et al.,*[31] with around-the-clock monitoring at 15-minute intervals for over 6 months with few interruptions, documents that there was no undue BP elevation when, after 10 years of antidiuretic treatment, a patient was taken off antihypertensive drugs and followed beyond the monitored 6-month span on placebo by added ambulatory monitoring profiles.

Self-measurements by a physician document the spontaneous transition from hypertension (with waking SBP measurements averaging about 200 mm Hg under treatment) to normotension (with waking SBP readings averaging about 110 mm Hg in the absence of treatment) (FIG. 6).[32] Treatment was stopped in a properly timed fashion only because of the surveillance offered by the self-monitoring of BP.

Self-measurements by a physician, 46 years old at the start of study, document that during 26 years, by conventional criteria of the World Health Organization (WHO), World Hypertension League (WHL), and National Institutes of Health (NIH),[1] he is "hypertensive" in August and "normotensive" in March, with average SBP/DBP readings of about 120/70 mm Hg (FIG. 7).[6] Chronobiologic analysis reveals a steady increase in the extent of the about-yearly change (circannual amplitude) in BP. Accordingly, whereas he may still occasionally be hypertensive in the light of a fixed limit drawn at 140 mm Hg, all his 1994 readings are below the upper limit of the time-specified 90% prediction interval computed from data of peers of his ethnicity, gender, and age. On the other hand, he is occasionally found to be hypotensive by chronobiologic criteria with readings below the lower 5% prediction limit of peers.

A neurosurgeon who monitored his BP around the clock for 23 days at 15–60-minute intervals illustrates the shortcomings of the current diagnostic procedure of BP status. Conventional interpretation of his data collected during office hours (9 AM–5 PM) leaves one in a quandary, because 56% of his SBP/DBP measurements are

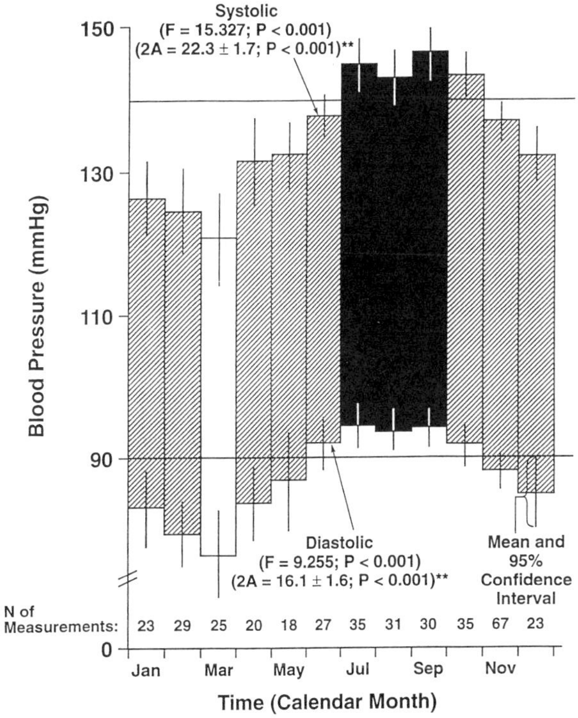

FIGURE 7. Case of unusually large circannual BP amplitude (summer hypertension) in a 70-year-old man. Circannual rhythm was revealed to the naked eye by stacking self-measurements (n = 363) for 26 years into a single idealized calendar year. **2A = double circannual amplitude from cosinor analysis. From Halberg, with permission. Data from G. Northrup analyzed by A. Portela.

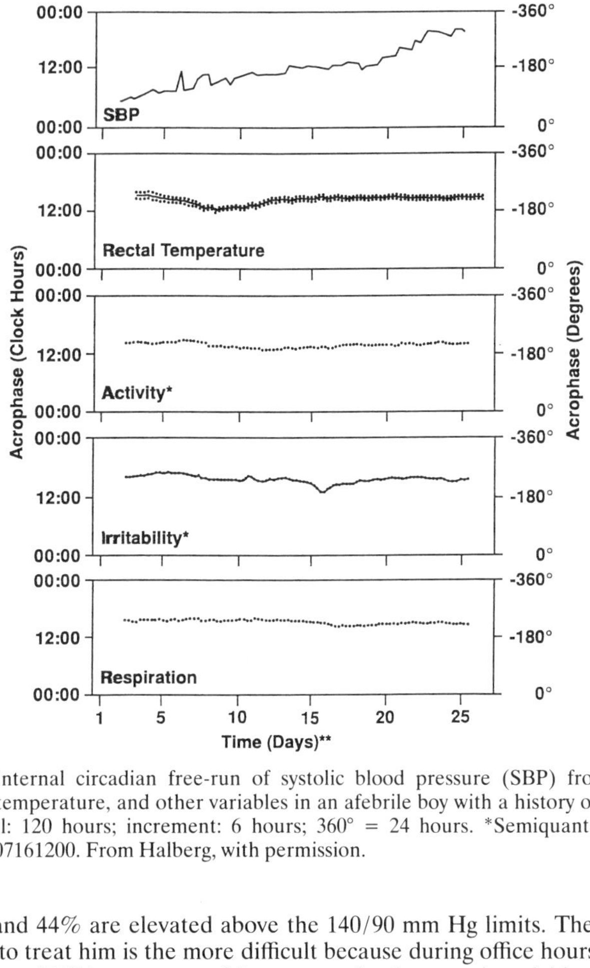

FIGURE 8. Internal circadian free-run of systolic blood pressure (SBP) from locomotor activity, core temperature, and other variables in an afebrile boy with a history of intermittent fever. Interval: 120 hours; increment: 6 hours; 360° = 24 hours. *Semiquantitative rating. **Start: 196407161200. From Halberg, with permission.

acceptable and 44% are elevated above the 140/90 mm Hg limits. The decision to treat or not to treat him is the more difficult because during office hours most of his measurements (77%) are acceptable on one day but are invariably too high on another day.[6]

Around-the-clock BP monitoring at 30-minute intervals for 21 days in February and 11 days in April of a 60-year-old woman reveals very large day-to-day variability in MESOR and circadian amplitude and acrophase (FIG. 1). Her daily MESORs of SBP and DBP vary from above 150/85 to below 130/70 mm Hg, whereas her circadian double amplitudes vary from 35/25 to less than 10/5 mm Hg, and her circadian acrophases usually occurring around 3 PM can deviate towards 00:00. Deviations outside chronobiologic reference standards are associated with conflicts, grief, and perhaps anxiety for the first few days of monitoring. Accordingly, her syndrome is likely a psychophysiologic circadian amplitude excess accompanied by monitor-novelty-MESOR-hypertension and monitor-novelty-ecphasia. Her condition may be an "automatic cuff-inflation hypertension"[33]

Even after a diagnosis of MESOR-hypertension when a patient is treated with antihypertensive drugs, casual BP measurements in the physician's office may be insufficient. This is the case in a 78-year-old man on 10 mg Vasotec (an ACE inhibitor) in the morning who was found to have drastically elevated DBP readings by night, usually between midnight and 3 AM on 11 of 12 days of ambulatory monitoring at 15-minute intervals (FIG. 2).[3,4,34] Such a DBP excess for several hours when it is unlikely to be checked should prompt concern and added treatment targeted in time.

ENDOGENICITY AND REACTIVITY OF THE CIRCADIAN BLOOD PRESSURE RHYTHM

Changes in BP have long been considered to be mere responses to movement, rest, sleep, food intake, and emotions. A genetic basis of the circadian BP rhythm,

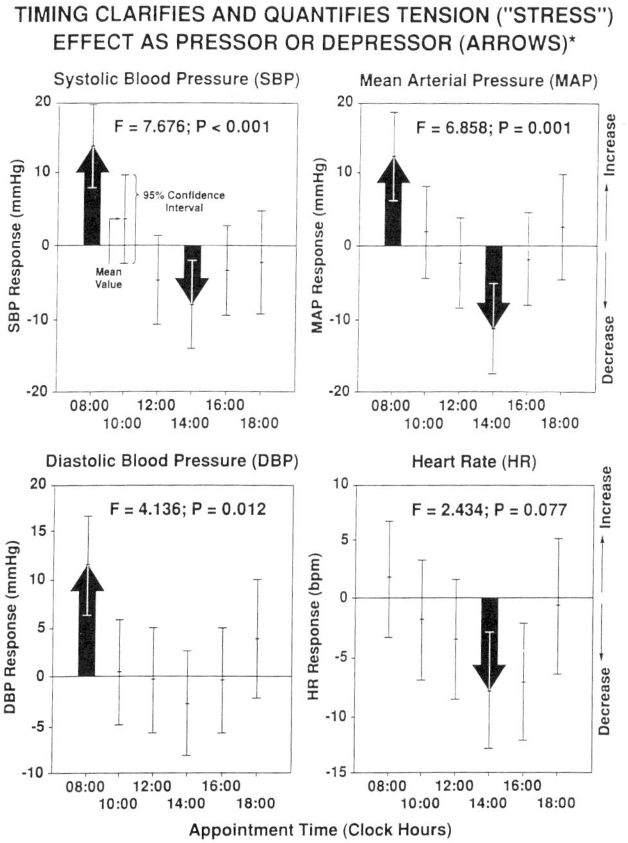

FIGURE 9. Timing clarifies and quantifies tension (stress) effect as pressor or depressor (*arrows*). The same kind of surgery in the morning on the average raises and in the afternoon lowers blood pressure and heart rate. Circadian stage determines predictable increase or decrease in human blood pressure and heart rate during surgery, assessed as the difference, corrected for posture, between the mean value during the appointment and the value predicted from the circadian profile provided by each patient. Results are from ambulatory monitoring at 5- to 15-minute intervals for 3 days bracketing surgery in 23 patients (n = 254 sets of measurements per patient; total n = 5,849; including monitoring on two previous occasions: total n = 17,758). From Halberg, with permission. Data of F. Raab and E. Schaffer.

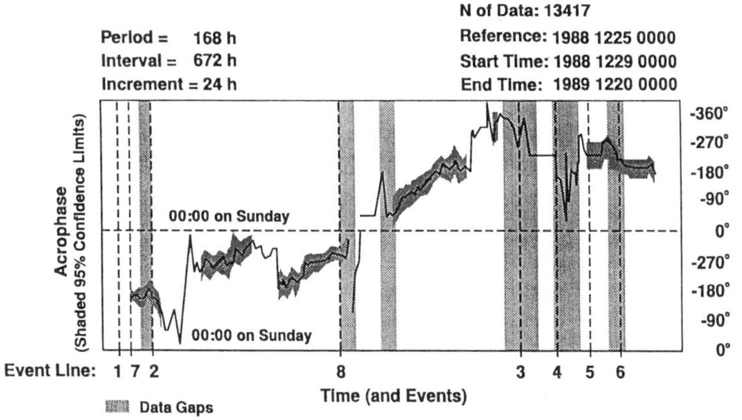

FIGURE 10. Time course of the circaseptan acrophase of systolic blood pressure in a 27-year-old woman (SF) before, during (from event line 7 to event line 8), and after isolation from society in a cave. Woman became amenorrheic during isolation. A drift in acrophase to later and later times during the week indicates that during isolation, the circaseptan rhythm of systolic blood pressure free-runs. It continued to free-run after SF emerged from the cave, while SF remained amenorrheic until menstruation was medically reinduced (at event line 3; medroxyprogesterone acetate, 20 mg/day for 6 days; after reinduction of menstruation, cloniphene citrate, 50 mg/day, was given from the 5th to the 9th day after onset of menses for 2 months). Neurohormonal or other events at menstruation may be a condition for the synchronization of a built-in circaseptan component with societal schedules. From Halberg, with permission.

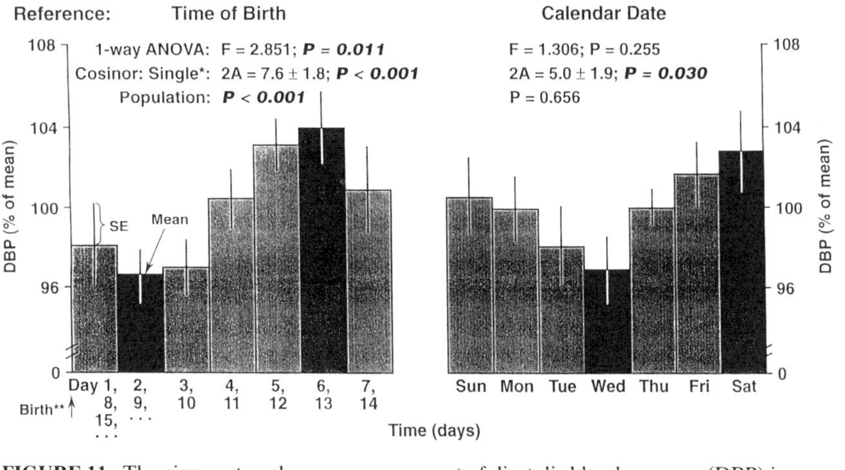

FIGURE 11. The circaseptan chronome component of diastolic blood pressure (DBP) is more endogenous (*left*) than exogenous (*right*); its phase may be set by the stimulus of birth. The environment may also exert an influence, because the results show weak synchronization with the social schedule (once inter-individual variability in MESOR is reduced by expressing data relative to each baby's mean value). *2A = double circadian amplitude; **does not carry 7-day information. From Halberg, with permission. Data from J. Siegelová *et al.*

however, was shown in studies on twins and by their ability to free-run[35–37]; in the absence of known environmental synchronizers, circadian rhythmicity persists while assuming a period differing slightly but with statistical significance from exactly 24 hours. While synchronized by the socioecologic environment, the circadian BP rhythm is genetically anchored and is not dictated solely by changes in activity. This is shown, for instance, by the free-run of SBP from other physiologic variables (including overall body activity measured concomitantly) which remain synchronized

TABLE 3. Within-Hour Variability in Blood Pressure (BP) and Heart Rate (HR) Estimated by Hourly Standard Deviations Computed from Samples at 7.5-Minute Intervals over 24 Hours on 40 Healthy Men, 20–60 Years of Age, Monitored on Two Occasions

Time Interval	Systolic BP (mm Hg)			Diastolic BP (mm Hg)			HR (beats/min)		
	Mean	SD	Median	Mean	SD	Median	Mean	SD	Median
0–1	7.87	3.84	6.83	6.50	3.65	5.87	5.08	3.18	4.54
1–2	7.01	3.09	7.10	5.62	2.98	5.27	4.45	2.92	3.74
2–3	7.54	3.68	6.78	7.01	4.34	5.49	4.63	3.08	3.58
3–4	7.74	3.43	7.18	6.78	3.78	5.99	4.85	3.55	3.83
4–5	8.19	3.75	7.11	6.49	3.48	5.55	6.21	4.19	5.43
5–6	8.56	3.99	8.02	8.12	4.00	7.54	8.09	4.35	7.29
6–7	9.54	4.27	8.28	7.58	4.33	7.08	9.62	4.35	9.52
7–8	7.96	3.71	7.24	6.27	2.96	5.87	7.75	3.01	7.16
8–9	6.88	2.68	6.19	5.59	2.61	5.17	6.58	3.72	5.76
9–10	6.97	2.24	6.65	5.57	2.41	4.95	6.99	3.28	6.52
10–11	6.48	2.05	6.43	5.61	1.81	5.46	6.20	3.14	5.31
11–12	7.16	2.41	6.89	5.10	1.99	4.96	6.86	3.18	6.10
12–13	7.40	2.99	7.04	6.89	2.88	6.07	7.48	2.82	7.50
13–14	6.66	2.28	6.70	5.43	2.10	4.92	6.11	3.36	5.14
14–15	6.95	2.48	6.75	5.38	2.13	4.76	5.87	1.85	5.93
15–16	6.68	2.30	6.53	5.66	2.24	5.27	6.55	3.66	5.72
16–17	7.34	2.95	6.72	5.89	2.90	5.46	6.53	2.69	6.25
17–18	7.04	2.70	6.47	6.18	2.96	5.77	7.00	3.32	6.39
18–19	7.72	3.81	7.13	6.58	2.33	6.49	6.78	3.66	6.07
19–20	7.54	2.82	6.99	6.35	2.87	5.62	6.75	2.87	6.45
20–21	6.91	2.67	6.65	5.63	2.52	5.23	5.82	3.16	5.08
21–22	6.58	2.81	5.87	5.65	3.05	5.15	5.91	2.98	5.63
22–23	7.99	3.68	7.25	7.56	3.30	6.84	5.87	3.18	5.55
23–24	7.85	3.93	6.99	7.02	3.01	6.45	5.60	3.10	4.84
Overall	7.44	3.23	6.88	6.27	3.11	5.63	6.40	3.49	5.83
Kruskal-Wallis[a] χ^2	58.5; $p < 0.001$			105.7; $p < 0.001$			221.7; $p < 0.001$		

[a]Test for differences of location in ranked data grouped by single classification (R estimator for one-way ANOVA); χ^2 test with 23 degrees of freedom. This test is preferred to the one-way ANOVA in view of the skewness of the distributions (i.e., departure from normality; $p < 0.05$).

to 24 hours in the case of an afebrile boy with a history of intermittent fever (FIG. 8).[38] It is also supported by the finding that BP rises before awakening and cannot therefore result only from a change in posture or activity[39] and by the demonstration of a statistically significant circadian rhythm in BP assessed concomitantly with (and separated in a model from) the effect of BP changes in response to changes in activity, monitored at the same time.[40] Moreover, the same external stimulus can have drastically different effects on BP depending on the circadian stage at which it is applied. A case in point is provided by periodontal surgery which, on the average, raises BP when it is performed in the morning but lowers it when it is done in the afternoon (FIG. 9).[5]

INFRADIAN VARIATION

By free-running, endogenicity is documented for components other than circadian rhythms. The biologic week and half-week are cases in point, demonstrated in

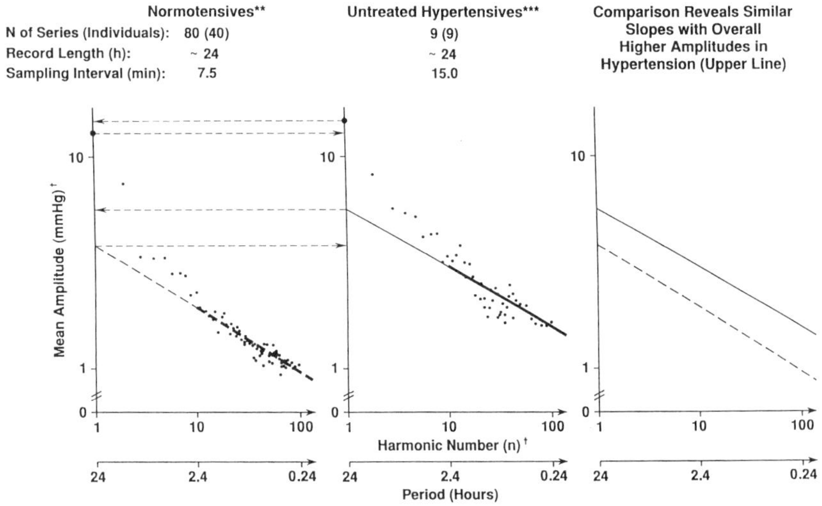

FIGURE 12. Comparison of spectral characteristics of systolic blood pressure in normotensive versus untreated hypertensive patients shows a similar spectral structure but an overall larger variability at all frequencies in hypertension. †Log scale; ● = circadian component (and next 9 harmonics) excluded from computation of slope. *From Cornéllssen *et al.*[30] Hypertension was diagnosed conventionally; Results were confirmed in 7-day profiles at 30-minute intervals from 78 MESOR-normotensives and 45 untreated MESOR-hypertensives studied by Y. Wantanabe ($p < 0.05$). **Data from J.I.M. Drayer and M.A. Weber. ***Data from V.R. McCall and W.C. McCall. From Halberg, with permission.

the social isolation of a cave for 4 months (FIG. 10).[36] Other evidence of endogenicity is single-stimulus induction, perhaps best illustrated in early extrauterine life.[41] The biologic week, prominently expressed at birth, shows a much higher degree of synchronization by reference to the time of birth than by calendar date (FIG. 11).[42]

Circatrigintan and circannual rhythms also prominently characterize BP variation. A circatrigintan component peaking around the time of ovulation was reported in a clinically healthy woman 28 years of age.[43] This component was also consistently found in another woman (who had had a hysterectomy) to modulate the MESOR, as well as the circadian amplitude and acrophase in three separate records spanning several months to 2 years (with data collected at 15- to 60-minute intervals). About-yearly changes in BP are already seen in neonates; the circannual amplitude

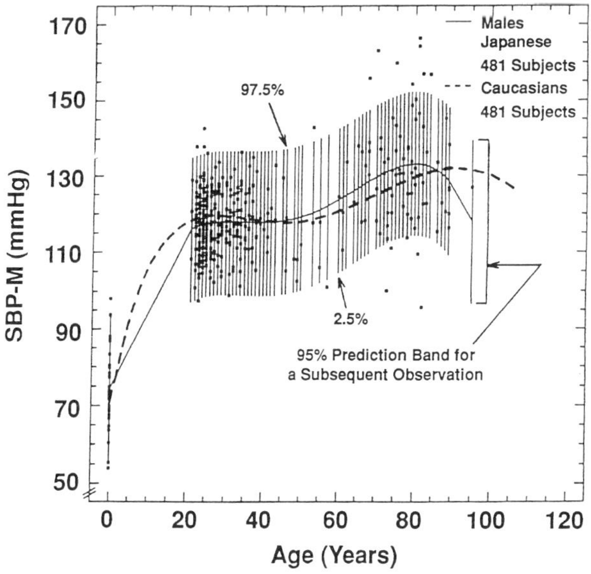

FIGURE 13. Changes in circadian MESOR (M) of systolic blood pressure (SBP) of caucasian and Japanese men as a function of age. *Each dot represents an estimate from about 100 measurements. R^2 from 4th degree polynomial fit: 0.683; $p < 0.001$. From Halberg, with permission. Data from International Womb-to-Tomb Chronome Initiative.

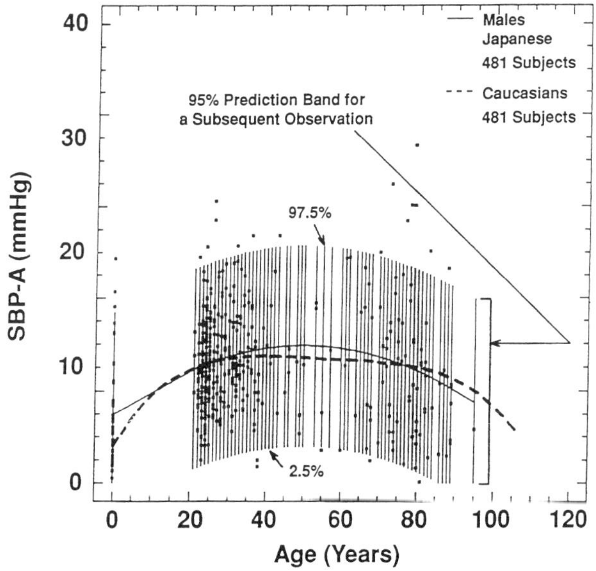

FIGURE 14. Changes in circadian amplitude (A) of systolic blood pressure (SBP) of caucasian and Japanese men as a function of age. *Each dot represents an estimate from about 100 measurements. R^2 from 2nd degree polynomial fit: 0.097; $p < 0.001$. From Halberg, with permission. Data from International Womb-to-Tomb Chronome Initiative.

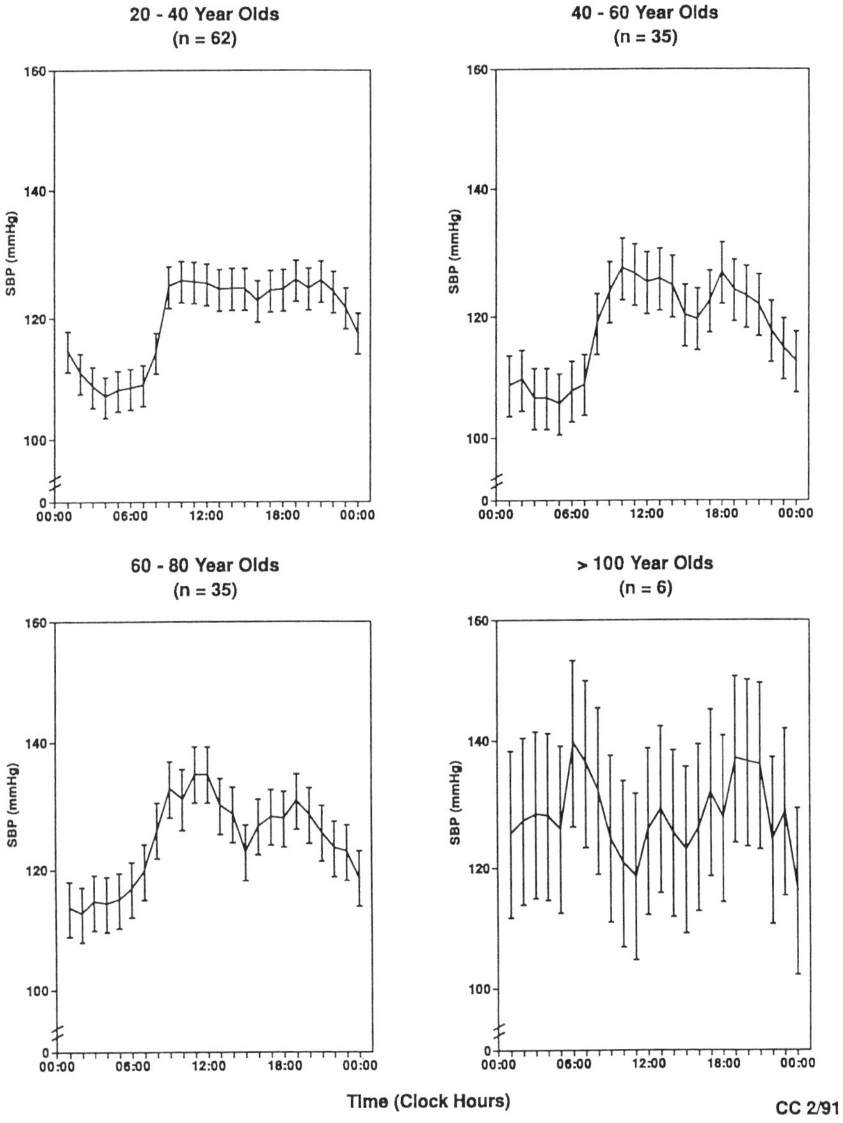

FIGURE 15. Changes in circadian waveform of systolic blood pressure (SBP) of caucasian men as a function of age. Means and 95% confidence limits. From Halberg, with permission. Data from International Womb-to-Tomb Chronome Initiative.

of babies with a patroclinous positive history of high BP and/or other vascular diseases is larger than that of babies with a negative history.[44] Circannual changes characterize not only BP but also a variety of hormones and their receptors, electrophysiologic and other cardiovascular variables, as well as morbidity and mortality statistics.[45]

ULTRADIAN VARIATION

On the basis of data collected at 7.5-minute intervals for 24 hours on two occasions on 40 clinically healthy men 20–60 years of age,[23] the characteristics of random variability of BP and heart rate (HR) were estimated; the short-term variability in the data was approximated by computing for each 24-hour profile the mean and standard deviation of BP and HR within each hour (TABLE 3). The error on SBP/DBP and HR measurements averages about 7/6 mm Hg and 6 beats/min. The within-hour standard deviations are not evenly distributed over 24 hours, being larger around the time of awakening and smaller during the rest span.[9]

FIGURE 12 (left) illustrates the least-squares spectrum of SBP, averaged over the 80 profiles. In this representation of log amplitudes versus log frequency, the slope in the high-frequency region of the spectrum (harmonic $n > 10$) provides a measure of the serial correlation characterizing consecutive measurements while the intercept represents a measure of overall variability. A comparison of these characteristics between normotensive and untreated hypertensive subjects (FIG. 12) reveals a similar spectral structure of SBP but an increased variability of SBP at all frequencies in hypertensive versus normotensive subjects.[9,30] The contribution of 24-hour BP variability has already been documented for the cardiovascular complications of hypertension.[11] Our own studies suggest that the circadian BP amplitude may constitute an even more sensitive index of risk elevation, at least for ischemic stroke and nephropathy,[17] while also being more sensitive in detecting treatment effects.[16]

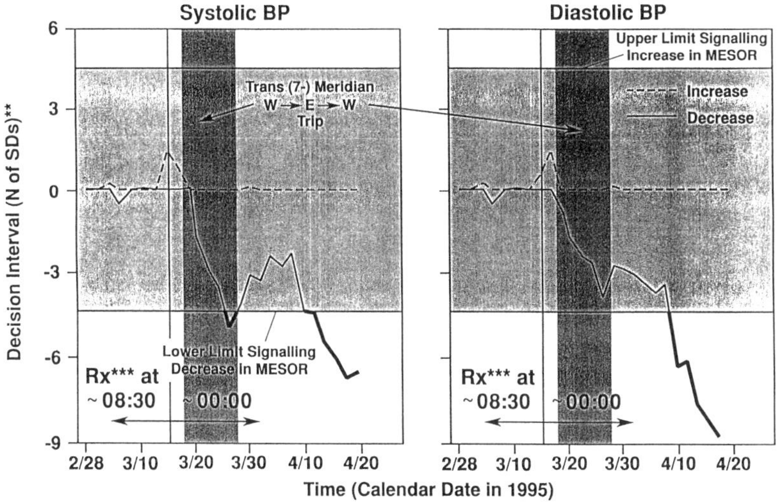

FIGURE 16. Control charts are useful to assess effects of individualized anti-MESOR hypertensive chronotherapy. The method can be used to optimize the timing of medication. The same dose of the same drug can have different effects on the MESOR of systolic and diastolic blood pressure. In the case shown herein, diltiazem HCl is more effective when taken at bedtime than upon awakening for this 76-year-old patient. *Blood pressure (BP) MESOR lowering by change in timing diltiazem HCl (240 mg/day) assessed by self-starting CUSUM. **Standard deviation from CUSUM; if there is significant displacement of 1 SD, it would be diagnosed by a slope of (1 – 0.5 =) 0.5 SD. ***After awakening (~08:30) or at bedtime (~00:00). From Halberg, with permission.

CHRONOME ASSESSMENT FROM WOMB TO TOMB

Maps of our time structure of multifrequency rhythms, trends, and noise (chronomes; from *chro*nos = time, *no*mos = rule and chromos*ome*) are being mapped for BP and HR at different ages to obtain refined reference standards and for an eventual assessment of outcomes. FIGURES 13 and 14 illustrate chronome maps in clinical health for the MESOR and circadian amplitude of SBP in two populations, caucasian and Japanese. FIGURE 15 shows changes in the circadian waveform of SBP as a function of age.

Chronome maps obtained so far stem from linked cross-sectional (hybrid) studies in the sense that each individual is sampled around the clock. An additional longitudinal element along the age scale derived from low-risk, long-lived individuals, validated by outcomes, will be indispensable for a definite chronome map to distinguish between (1) rhythm alterations indicative of a heightened vascular disease risk versus (2) naturally occurring changes as a function of age.[2]

CONCLUDING REMARKS

The status quo is not likely to improve until the two culprits, reliance on casual time-unspecified measurements and fixed limits, are reformed. Reliance on systematically collected time series and on chronodesms for rhythm parameters and on reference standards for nonparametric indices is indispensable "up front." Thus, conditions such as an excessive circadian BP amplitude can be screened for an important task, because among a set of risk factors it represents the factor with the largest relative risk for ischemic stroke or nephropathy.[6,17] The recommendation to account for rhythms should not be viewed as equivalent to "controlling" for them by sampling at a fixed time of day. Whereas it is always important to make sure that BP is properly taken[46] and to acknowledge the merit of home measurements, as it was advocated, for a long time, among others, from the Mayo Clinic,[47,48] it is mandatory that the attitude also change. Circadian rhythms need not be looked upon as a confounder[49,50] when they should be assessed and exploited. What indeed is confounding is the replacement of around-the-clock monitoring by a 6-hour profile. The critical endpoints that can provide new, invaluable information on the largest risk factor for ischemic stroke are then irretrievably lost.

The usefulness of a true chronobiologic approach is further supported by the demonstration that certain antihypertensive drugs, when given at the appropriate time, can lower the circadian BP amplitude.[51,52] Moreover, for MESOR-normotensive subjects with an excessive circadian BP amplitude, who may not need a pharmacologic intervention, nonpharmacologic methods such as autogenic training are also shown to reduce the circadian BP amplitude.[53]

The practicality of a chronobiologic approach lies in the availability of analytic inferential statistical methods for the rigorous assessment of intervention effects applicable to the individual patient (FIG. 16). Even based on self-measurements taken a few times each day with affordable and accurate digital home monitors, control charts, notably those based upon self-starting cumulative sums, are a particularly powerful tool,[54–56] to check not only on changes in mean but also on conditions such as CHAT[16,17] or a reduced circadian BP amplitude seen for instance in sleep apnea[57] or in Cushing's syndrome.[58]

Available technology involves continued surveillance in many aspects of modern life. Surveillance is also indicated when current endeavors to prolong the window after a stroke[59] can be readily complemented by a chronobiologic approach to stroke prevention. This was the unanimously endorsed recommendation of several recent international consensus meetings.[6,60]

REFERENCES

1. THE FIFTH REPORT OF THE JOINT NATIONAL COMMITTEE ON DETECTION, EVALUATION, AND TREATMENT OF HIGH BLOOD PRESSURE (JNC V). 1993. Arch. Intern. Med. **153:** 154–183. NIH Publ. No. 93-1088, March 1994. Reprinted by Citizens for Public Action on Blood Pressure and Cholesterol.
2. CORNÉLISSEN, G., E. HAUS & F. HALBERG. 1992. Chronobiologic blood pressure assessment from womb to tomb. *In* Biological Rhythms in Clinical and Laboratory Medicine. Y. Touitou & E. Haus, Eds.: 428–452. Springer-Verlag. Berlin.
3. HALBERG, F., G. CORNÉLISSEN, A. CARANDENTE, E. BAKKEN & E. YOUNG. 1993. Chronobiologic perspectives of international health care reform for the future of children. Chronobiologia **20:** 269–275.
4. CORNÉLISSEN, G., P. DELMORE, C. BINGHAM, G. RUTLEDGE, Y. KUMAGAI, I. KUWAJIMA, Y, SUZUKI, K, KURAMOTO, K, OTSUKA, P. T. SCARPELLI, B. TARQUINI, M. CAGNONI, L. GARCIA, R. M. ZASLAVSKAYA, E. SYUTKINA, F. CARANDENTE, S. I. RAPOPORT, Y. A. ROMANOV, K. TAMURA, E. BAKKEN & F. HALBERG. 1993. A response to the health care crisis: A 'health start' from 'womb to tomb'. Chronobiologia **20:** 277–291.
5. RAAB, F., E. SCHAFFER, G. CORNÉLISSEN & F. HALBERG. 1993. More on the dentist's role in the prevention of high blood pressure. Chronobiologia **20:** 245–250.
6. HALBERG, F., G. CORNÉLISSEN & INTERNATIONAL WOMB-TO-TOMB CHRONOME INITIATIVE GROUP. 1995. Resolution from a meeting of the International Society for Research on Civilization Diseases and the Environment (New SIRMCE Confederation), Brussels, Belgium, March 17–18, 1995: Fairy tale or reality? Medtronic Chronobiology Seminar 8, April 1995: 1–30. (Available free of charge while supplies last from Patrick Delmore, Director of Communications, Medtronic Inc., 7000 Central Ave. N.E., Minneapolis, MN 55432.)
7. THIJS, L., J. STAESSEN, R. FAGARD, P. ZACHARIAH & A. AMERY. 1994. Number of measurements required for the analysis of diurnal blood pressure profile. J. Hum. Hypertens. **8:** 239–244.
8. HALBERG, F., L. E. SCHEVING, E. LUCAS, G. CORNÉLISSEN, R. B. SOTHERN, E. HALBERG, J. HALBERG, F. HALBERG, J. CARTER, K. D. STRAUB & D. P. REDMOND. 1984. Chronobiology of human blood pressure in the light of static (room-restricted) automatic monitoring. Chronobiologia **11:** 217–247.
9. CORNÉLISSEN, G. 1987. Instrumentation and data analysis methods needed for blood pressure monitoring in chronobiology. *In* Chronobiotechnology and Chronobiological Engineering. L. E. Scheving, F. Halberg & C. F. Ehret, Eds.: 241–261. Martinus Nijhoff. Dordrecht, The Netherlands.
10. FLORAS, J. S. 1991. Will knowing the variability of ambulatory blood pressure improve clinical outcome? An additional consideration in the critical evaluation of this technology. Clin. Invest. Med. **14:** 231–240.
11. FRATTOLA, A., G. PARATI, C. CUSPIDI, F. ALBINI & G. MANCIA. 1993. Prognostic value of 24-hour blood pressure variability. J. Hypertens. **11:** 1133–1137.
12. PARATI, G., A. RAVOGLI, A. FRATTOLA, A. GROPPELLI, L. ULIAN, C. SANTUCCIU & G. MANCIA. 1994. Blood pressure variability; clinical applications and effects of antihypertensive treatment. J. Hypertens. (Suppl.) **12:** S35–S40.
13. SCARPELLI, P. T., S. ROMANO, R. LIVI, L. SCARPELLI, G. CORNÉLISSEN, M. CAGNONI & F. HALBERG. 1987. Instrumentation for human blood pressure rhythm assessment by self-measurement. *In* Chronobiotechnology and Chronobiological Engineering. L. E. Scheving, F. Halberg & C. F. Ehret, Eds.: 304–309. Martinus Nijhoff. Dordrecht, The Netherlands.
14. KUMAGAI, Y., T. SHIGA, K. SUNAGA, G. CORNÉLISSEN, A. EBIHARA & F. HALBERG. 1992.

Usefulness of circadian amplitude of blood pressure in predicting hypertensive cardiac involvement. Chronobiologia **19:** 43–58.

15. OTSUKA, K., M. SHINAGAWA & Y. KUBO. 1994. Circadian BP profiles and target organ damage. Cardiac Pract. **5:** 197–204.
16. OTSUKA, K., G. CORNÉLISSEN, F. HALBERG, J. HALBERG, F. HALBERG, E. M. SCHAFFER, F. J. RAAB, II, L. GARCIA ALONSO, A. PORTELA, J. C. MENENDEZ, P. DELMORE, E. V. SYUTKINA, A. E. GRIGORIEV, A. S. ABRAMIAN, M. D. MITISH, G. V. YATSYK, I. LOGVINENKO & R. M. ZASLAVSKAYA. 1996. Circadian amplitude-hypertension, an independent clinical entity. Bull. Exp. Biol. Med. In press.
17. OTSUKA, K., G. CORNÉLISSEN, F. HALBERG & G. OEHLERT. 1996. Excessive circadian amplitude of blood pressure increases risk of ischemic stroke and nephropathy. J. Ambulatory Monitoring. *9:* in press.
18. MILLAR-CRAIG, M. W., D. KENNY, S. MANN, V. BALASUBRAMANIAN & E. B. RAFTERY. 1979. Effect of once-daily atenolol on ambulatory blood pressure. Br. Med. J. **27:** 237–238.
19. HALBERG, F., G. CORNÉLISSEN, E. HALBERG, J. HALBERG, P. DELMORE, M. SHINODA & E. BAKKEN. 1988. Chronobiology of human blood pressure. Medtronic Continuing Medical Education Seminars, 4th ed.: 1–242. (Available free of charge while supplies last from Patrick Delmore, Director of Communications, Medtronic Inc., 7000 Central Ave. N.E., Minneapolis, MN 55432.)
20. HALBERG, F., E. BAKKEN, G. CORNÉLISSEN, J. HALBERG, E. HALBERG, J. Y. WU, S. SANCHEZ DE LA PEÑA, P. DELMORE & B. TARQUINI. 1990. Chronobiologic blood pressure assessment with a cardiovascular summary, the sphygmochron. *In* Blood Pressure Measurements. W. Meyer-Sabellek, M. Anlauf, R. Gotzen & L. Steinfeld, Eds.: 297–326. Steinkopff Verlag. Darmstadt, FRG.
21. PICKERING, T. G. & G. D. JAMES. 1989. Some implications of the differences between home, clinic and ambulatory blood pressure in normotensive and hypertensive patients. J. Hypertens. (Suppl.) **7:** S65–72.
22. AUSTRALIAN NATIONAL BLOOD PRESSURE STUDY MANAGEMENT COMMITTEE. 1980. The Australian Therapeutic Trial in Mild Hypertension. Lancet **i:** 1261–1267.
23. WEBER, M. A., J. I. M. DRAYER & D. D. BREWER. 1987. Repetitive blood pressure measurements: Clinical issues, techniques, and data analysis. *In* Chronobiotechnology and Chronobiological Engineering. L. E. Scheving, F. Halberg & C. F. Ehret, Eds.: 270–277. Martinus Nijhoff. Dordrecht, The Netherlands.
24. WILCOX, R. G., J. R. A. MITCHELL & J. R. HAMPTON. 1986. Treatment of high blood pressure: Should clinical practice be based on results of clinical trials? Br. Med. J. **293:** 433–437.
25. KUWAJIMA, I., Y. SUZUKI, K. KURAMOTO, K. OTSUKA, Y. KUMAGAI, G. CORNÉLISSEN & F. HALBERG. 1993. A chronobiologic vs. a conventional approach to blood pressure measurements. *In* Toward phase zero preclinical and clinical trials: Chronobiologic designs and illustrative applications. University of Minnesota Medtronic Chronobiology Seminar Series #6, 2nd extended edition. Part I. February 1993. G. Cornélissen, E. Halberg, E. Bakken, P. Delmore & F. Halberg, Eds.: 33–45.
26. GRIN, J. M., E. J. MCCABE & W. B. WHITE. 1993. Management of hypertension after ambulatory blood pressure monitoring. Ann. Intern. Med. **118:** 833–837.
27. KJELDSEN, S. E., A. MOAN, M. ROSTRUP, G. NORDBY, I. OS & I. EIDE. 1993. White-coat hypertension and blood pressure measurement at home. Tidsschr. Norske Laegeforening **113:** 1839–1843.
28. JULIUS, S. 1991. Home blood pressure monitoring: Advantages and limitations. J. Hypertens. (Suppl.) **9:** S41–S46.
29. GRAN, B. 1991. Non-pharmacological methods reduce drug use in the treatment of hypertension. A two-year trial in general practice. Scand. J. Prim. Health Care **9:** 121–128.
30. CORNÉLISSEN, G., P. T. SCARPELLI, F. HALBERG, J. HALBERG, F. HALBERG & E. HALBERG. 1988. Cardiovascular rhythms: Their implications and applications in medical research and practice. *In* Trends in Chronobiology, W. Th. J. M. Hekkens, G. A. Kerkhof & W. J. Rietveld, Eds.: 335–355. Pergamon Press. Oxford.

31. LITTLE, J., S. SÁNCHEZ DE LA PEÑA, G. CORNÉLISSEN, F. ABRAMOWITZ, N. TUNA & F. HALBERG. 1990. Longitudinal chronobiologic blood pressure monitoring for assessing the need and timing of antihypertensive treatment. Prog. Clin. Biol. Res. **341B:** 601–611.
32. HAUS, E., M. HAUS, SR., G. CORNÉLISSEN, J. Y. WU & F. HALBERG. 1993. A longitudinal view with 16,944 sets of self-measurements of the aging human circulation. *In* Chronocardiology and Chronomedicine: Humans in Time and Cosmos. K. Otsuka, G. Cornélissen & F. Halberg, Eds.: 99–102. Life Science Publishing. Tokyo.
33. MEJIA, A. D., B. M. EGAN, N. J. SCHORK & A. J. ZWEIFLER. 1990. Artefacts in measurement of blood pressure and lack of target organ involvement in the assessment of patients with treatment-resistant hypertension. Ann. Intern. Med. **112:** 270–277.
34. HALBERG, F., G. CORNÉLISSEN, C. BINGHAM, F. DEL POZO, M. T. ARREDONDO, K. OTSUKA, O. QUADENS, T. BREUS, S. RAPOPORT & F. KOMAROV. 1994. Telehygiene system for preventive chronopharmacology in space. J. Clin. Pharmacol. **34:** 552–557.
35. ZASLAVSKAYA, R. M. 1993. Chronodiagnosis and Chronotherapy of Cardiovascular Diseases. 2nd Ed. Translation into English from Russian. Medicina, Moscow.
36. HALBERG, F., K. TAMURA & G. CORNÉLISSEN. 1994. Chronobioengineering toward a cost-effective quality health care. Front. Med. Biol. Eng. **6:** 83–102.
37. DEGAUTE, J. P., E. VAN CAUTER, P. VAN DE BORNE & P. LINKOWSKI. 1994. Twenty-four-hour blood pressure and heart rate profiles in humans. A twin study. Hypertension **23:** 244–253.
38. HALBERG, F., R. A. GOOD & H. LEVINE. 1966. Some aspects of the cardiovascular and renal circadian system. Circulation **34:** 715–717.
39. HALBERG, E., F. HALBERG & K. SHANKARAIAH. 1981. Plexo-serial linear-nonlinear rhythmometry of blood pressure, pulse and motor activity by a couple in their sixties. Chronobiologia **8:** 351–366.
40. MARLER, M. R., R. G. JACOB, J. P. LEHOCZKY & A. P. SHAPIRO. 1988. The statistical analysis of treatment effects in 24-hour ambulatory blood pressure recordings. Stat. Med. **7:** 697–716.
41. CORNÉLISSEN, G., & F. HALBERG. 1994. Introduction to chronobiology. Medtronic Chronobiology Seminar #7, April 1994: 1–52. (Available free of charge while supplies last from Patrick Delmore, Director of Communications, Medtronic Inc., 7000 Central Ave. N.E., Minneapolis, MN 55432.)
42. SIEGELOVA, J., G. CORNÉLISSEN & F. HALBERG. 1995. Circaseptan component of blood pressure and heart rate in neonates. 1995. Arch. Physiol. Biochem. **103:** D83.
43. PORTELA, A., J. C. ILLERA, G. CORNÉLISSEN & F. HALBERG. 1993. Partial infradian but not circadian allophase of female human blood pressure (BP) and heart rate (HR) chronomes. Chronobiologia **20:** 130–131.
44. SAFIN, S. R., E. SYUTKINA, A. E. GRIGORIEV, Y. A. POLYAKOV, A. S. ABRAMIAN, E. HALBERG & F. HALBERG. 1991. Human neonatal blood pressure (BP) and heart rate (HR) chronome and patroclinous BP history. Biochim. Clin. **15:** 155–156.
45. HALBERG, F., G. CORNÉLISSEN, E. HAUS, G. NORTHRUP, A. PORTELA, H. WENDT, K. OTSUKA, Y. KUMAGAI, Y. WATANABE & R. ZASLAVSKAYA. 1996. Clinical relevance of about-yearly changes in blood pressure and the environment. Int. J. Biometeorol. In press.
46. OTSUKA, K., G. CORNÉLISSEN & F. HALBERG, EDS. 1993. Chronocardiology and Chronomedicine: Humans in Time and Cosmos. Life Science Publishing. Tokyo.
47. BROWN, G. E. 1930. Daily and monthly rhythm in the blood pressure of a man with hypertension: A three-year study. Ann. Intern. Med. **3:** 177–189.
48. MUELLER, S. C. & G. E. BROWN. 1930. Hourly rhythms in blood pressure in persons with normal and elevated pressure. Ann. Intern. Med. **3:** 1190–1200.
49. SCHULTZ, J. F. & S. G. SHEPS. 1994. Management of patients with hypertension: A hypertension clinic model. Mayo Clin. Proc. **69:** 997–999.
50. SHEPS, S. G. & V. J. CANZANELLO. 1994. Current role of automated ambulatory blood pressure and self-measured blood pressure determinations in clinical practice. Mayo Clin. Proc. **69:** 1000–1005.
51. TAMURA, K., I. KOHNO, Y. SAITO, K. WAKASUGI, S. ACHIWA, Y. IMANISHI, P. CUGINI & F.

HALBERG. 1991. Antihypertensive individualized therapeutic strategy. Dif. Soc. **6:** 109–124.

52. MUNAKATA, M., Y. IMAI, J. HASHIMOTO, H. SAKUMA, H. SEKINO, K. ABE & K. YOSHINAGA. 1992. The influence of antihypertensive agents on circadian rhythms of blood pressure and heart rate in patients with essential hypertension. Tohoku J. Exp. Med. **166:** 217–227.
53. WATANABE, Y., G. CORNÉLISSEN, F. HALBERG, Y. SAITO, K. FUKUDA, M. REVILLA, C. RODRIGUEZ, D. HAWKINS, K. OTSUKA & T. KIKUCHI. 1996. Method and need for continued assessment of autogenic training effect upon blood pressure: case report. New Trends Exp. Clin. Psychiatry. In press.
54. HAWKINS, D. M. 1987. Self-starting CUSUM charts for location and scale. The Statistician **36:** 299–315.
55. HAWKINS, D. M. 1993. Cumulative SUM control charting: An underutilized SPC tool. Qual. Eng. **5:** 463–477.
56. CORNÉLISSEN, G., F. HALBERG, D. HAWKINS, K. OTSUKA & W. HENKE. 1996. Objective individualized assessment of antihypertensive response. J. Ambulatory Monitoring. In press.
57. SIEGELOVA, J., M. MORAN, B. FISER, Z. KADANKA, J. DUSEK, M. AL-KUBATI, F. HALBERG & G. CORNÉLISSEN. 1996. Circadian variations in blood pressure in patients with sleep apnea and essential hypertension. Proc. 23rd Congress of the International Society of Internal Medicine, Manila, The Philippines, Feb. 1–6, 1996. :273–276.
58. KREZE, A., E. SPIROVA, G. CORNÉLISSEN, M. MIKULECKY & F. HALBERG. 1996. Circadian blood pressure and cortisol variation in untreated patients with Cushing's syndrome. 2nd Int. Cong. Chronobiology and Chronomedicine, Nanchang, China, Sept. 7–12, 1996. In press.
59. CAMARATA, P.J., R. C. HEROS & R. E. LATCHAW. 1994. "Brain attack": The rationale for treating stroke as a medical emergency. Neurosurgery **34:** 144–158.
60. HALBERG, F., G. CORNÉLISSEN & F. CARANDENTE. 1991. Chronobiology leads toward preventive health care for all: Cost reduction with quality improvement. A challenge to education and technology via chronobiology. Chronobiologia **18:** 187–193.

Neural Cardiovascular Regulation and 24-Hour Blood Pressure and Heart Rate Variability

G. PARATI,[a,d,e] M. DI RIENZO,[b] AND G. MANCIA[c,d]

[a]*Istituto Scientifico Ospedale S. Luca*
Fondazione Centro Auxologico italiano
Milan, Italy

[b]*LaRC*
Centro di Bioingengeria
Fondazione pro Juventute
Milan, Italy

[c] *Cattedra di Medicina Interna I*
Ospedale S. Gerardo
Monza and University of Milan
Milan, Italy

[d]*Centro di Fisiologia Clinica e Ipertensione*
University of Milan
Milan, Italy

A major step forward in studying blood pressure variability in humans was made in the 1960s when Bevan and coworkers[1,2] at Oxford described a method for continuous intraarterial blood pressure monitoring in ambulatory subjects. For the first time blood pressure variations over 24 hours could be described together with 24-hour heart rate variations. This approach also allowed us to precisely address for the first time the blood pressure changes that take place under the different behavioral conditions of daily life.[3] However, the invasive nature of the Oxford method represented a major limitation, because its inherent risk prevented its use in normal subjects and limited its application in severely ill patients.

Recently, a technique for noninvasive beat-to-beat blood pressure monitoring in ambulatory subjects was described. It is based on a principle originally reported by Peñaz[4] in the 1960s and implemented a few years later by Wesseling and coworkers.[5] The device for laboratory use is called Finapres (Ohmeda, Colorado). A portable version of this device, called Portapres, is now available[6] and consists of a recorder for 24-hour ambulatory blood pressure monitoring. This device makes use of (1) the Peñaz's volume-clamp approach combined with an infrared photoplethysmograph to keep the volume of the finger artery at a preset value during the recording time, (2) a fast servosystem and autocalibration procedure to provide a quantitatively reliable blood pressure signal, and (3) a hydrostatic/height correction system to account for the continuous changes in hydrostatic difference between the heart and the instrumented hand, with a time constant of 2 seconds. Both the Finapres and the Portapres devices closely reflect intraarterial blood pressure fluctuations not only in resting

[e]Address for correspondence: Gianfranco Parati, MD, Istituto Scientifico Ospedale S. Luca, Centro Auxologico Italiano, via Spagnoletto 3, Milan 20149, Italy. tel: +39-2-58216 704 fax: +39-2-58216 712.

conditions, but also under fast and pronounced blood pressure changes induced by either laboratory stimuli[7] or behavioral alterations in daily life conditions.[8]

When these devices are used to quantify blood pressure variability, however, caution is needed in interpreting the results. The reason is that analysis of finger blood pressure tracings may lead to overestimation of the amplitude of intraarterial systolic (but not diastolic and mean) blood pressure variations recorded at the brachial artery level, that is, at a more proximal site within the arterial tree. This is the case either when blood pressure variability is quantified in the time domain, as the overall variance of the recorded signal, or when the powers of the different frequency components that contribute to the overall variance are quantified by spectral analysis in the frequency domain.[9] The overestimation is not a major one, however, and both variance and spectral powers of diastolic and mean blood pressure are precisely reflected. Thus, both Finapres- and Portapres-derived data on blood pressure variability are acceptable, having the invaluable advantage of noninvasiveness over the estimates of blood pressure variability based on intraarterial blood pressure monitoring.

The use of both invasive and noninvasive beat-by-beat 24-hour ambulatory blood pressure recorders offers a deep insight into blood pressure and heart rate variations in daily life. In particular, they show that these variations are characterized by fast and slow components, whose fluctuation period ranges from a few seconds to several hours.[10] These fluctuations originate from factors as different as physical activity, influences of vasoactive substances, and mechanical effects of ventilation, but evidence is available that neural factors of either a central or peripheral reflex nature exert a prominent role.[11]

METHODOLOGIC ASPECTS

The method most commonly employed to assess blood pressure and heart rate variability over 24 hours is calculation of the standard deviation of the 24-hour average values. This is a simple and straightforward approach, because it summarizes the overall variance of a given signal in a single number. The standard deviation of a mean value, however, does not provide any detailed information on the patterns of the variability phenomena, and indeed similar standard deviation values are compatible with widely different types of blood pressure or heart rate variations.[12] In particular, expressing blood pressure or heart rate variability by the standard deviation does not allow the separation of fast and slow fluctuations, which prevents investigation of their different mechanisms, clinical importance, and alterations in cardiovascular diseases.[13,14]

The first attempt to separately quantify fast and sustained fluctuations in blood pressure or heart rate in man over 24 hours was made in the 1980s by a time domain approach to computer analysis of 24-hour intraarterial blood pressure recordings.[3] The approach was based on calculation of the average blood pressure and heart rate values for each half hour of the 24-hour intraarterial blood pressure recording. The standard deviation of a half-hour mean value was thought to reflect the relatively fast fluctuations occurring throughout this relatively short period, and the average of the 48 half-hour standard deviations so obtained (within half-hour standard deviations) was taken as a measure of relatively short-term blood pressure variability. This was complemented by calculation of the standard deviation of the average of the 48 half-hour blood pressure and heart rate mean values which was thought to reflect more long-lasting fluctuations, that is, fluctuations occurring between half hours. Thus, the between half-hour standard deviation so obtained was taken as a measure of longer term blood pressure or heart rate variability (FIG. 1).

A more effective method of splitting 24-hour blood pressure or heart rate variability into its different frequency components is spectral analysis of blood pressure and heart rate signals.[15–17] The blood pressure and heart rate spectral components that have been considered most frequently are those with frequencies ranging from 0.025 to 0.35 Hz (that is, those with periods from 30 seconds to 3 seconds). These components are usually integrated over three frequency regions, defined as low frequency (LF, 0.025–0.07 Hz), mid frequency (MF, 0.07–0.14 Hz), and high frequency (HF, 0.14–0.35 Hz), or more simply over two frequency regions defined as LF (0.04–0.14 Hz) and HF (0.14–0.35 Hz), respectively.[17] This is based on evidence that blood pressure and heart rate powers at these frequencies may reflect autonomic cardiovascular influences and thus demonstrate their derangement in cardiovascular or noncardiovascular diseases.[17–19] Through algorithms designed to

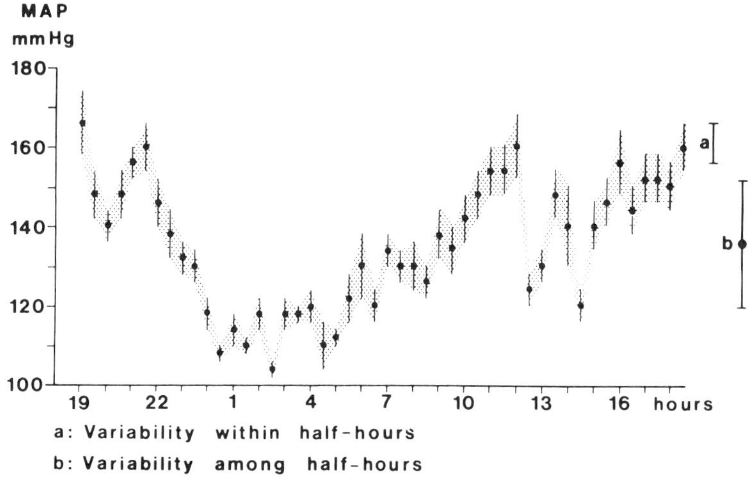

FIGURE 1. Example of the separate quantification of short- and long-term blood pressure variability from analysis of a 24-hour intraarterial blood pressure recording. *Full circles* refer to average mean arterial pressure values for each half hour of the recording (± standard deviation). The average of these half-hour standard deviations (**a**) represents an index of a relatively short-term blood pressure variability, reflecting fluctuations that occur within half hours. The average of half-hour mean values results in 24-hour average mean pressure ± a standard deviation (**b**) that represents an index of relatively long-term blood pressure variability, reflecting fluctuations that occur among half hours. From Mancia *et al.*[3] with permission.

assess the time-varying features of blood pressure and heart rate spectral powers, these spectral components now can also be measured throughout the 24 hours, which allows assessment of their normal and abnormal circadian variations and of their relationship with behavior and environmental stimuli.[20–23] (FIG. 2).

Recently, broad-band spectral analysis techniques have been developed to quantify not only fast but also slow blood pressure and heart rate fluctuations. These are obtained by calculating a single spectrum over the whole recording period,[23,24] thus simultaneously identifying all frequency components of the recorded signal from the slowest to the fastest ones. Preliminary broad-band spectral analysis data have confirmed the frequent occurrence, in the blood pressure and heart rate spectrum, of peaks at or near 0.1 Hz and at the respiratory frequency (around 0.3–0.5

Hz). These data have also shown, however, that at lower frequencies, blood pressure and heart rate powers are characterized by a 1/f distribution, which means that the lower the frequency of a given blood pressure or heart rate variability component, the greater its power and contribution to total variance (FIG. 3). This 1/f distribution may be quantitatively described by the parameters (κ, the intercept, and α, the slope) of the 1/f regression line modeling the raw spectra, facilitating between-condition and between-subject comparisons. So far, evidence from long-lasting blood pressure and heart rate recordings does indeed indicate significant changes in the intercept and slope of the 1/f model lines of blood pressure and heart rate spectra with aging, primary dysautonomias, hypertension, and other diseases.[23–27]

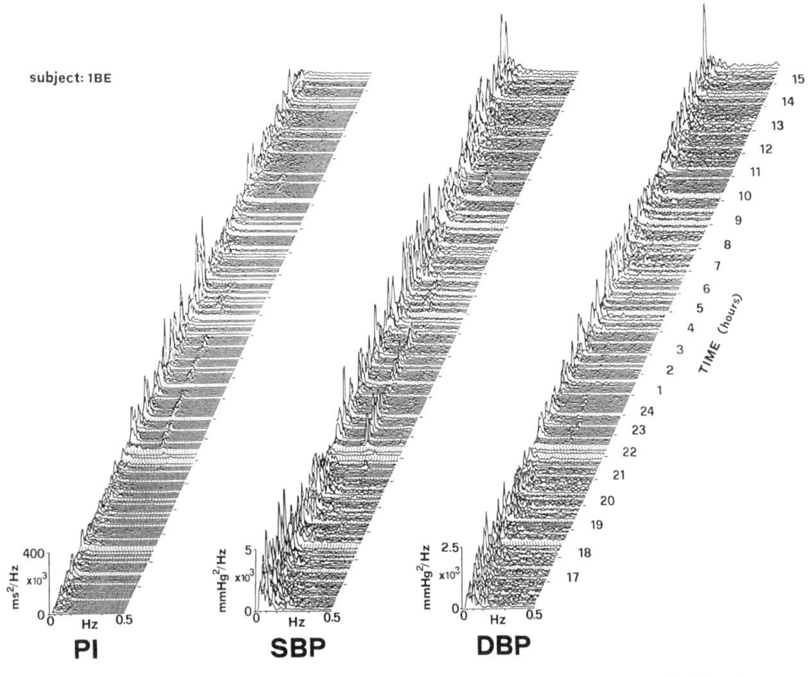

FIGURE 2. Sequential power spectral densities of pulse interval (PI), systolic blood pressure (SBP), and diastolic blood pressure (DPB) throughout the 24 hours. *Dotted lines* refer to segments in which power spectral densities could not be estimated because of the occurrence of non-stationarities. From Di Rienzo *et al.*[20] with permission.

NEURAL MECHANISMS OF BLOOD PRESSURE AND HEART RATE VARIABILITY

Several mechanisms are likely to play a role in the determination of blood pressure and heart rate variability over 24 hours.[11,13,28] Behavioral factors mediated through influences stemming from the central nervous system are unquestionably a major mechanism, as shown by the pronounced increases in blood pressure and heart rate associated with dynamic or isometric exercise or triggered by stressful events.

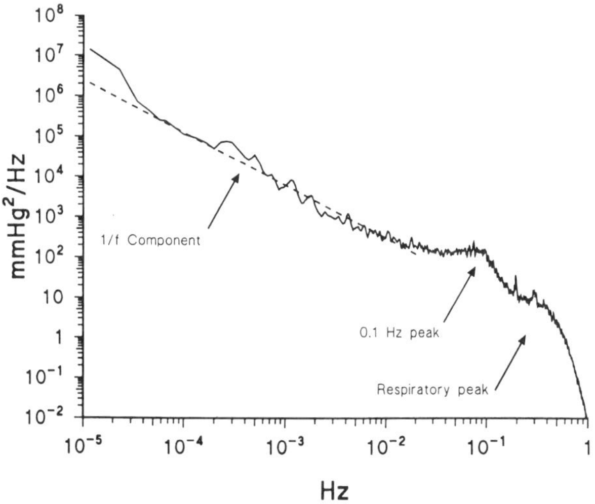

FIGURE 3. Plot shows broad band spectrum of systolic blood pressure obtained from analysis of a 24-hour ambulatory intraarterial blood pressure recording in a normotensive volunteer. Spectral components with frequencies ranging from 1 to approximately 0.000023 Hz (i.e., with periods ranging from 1 second to 12 hours) are considered. *Continuous line* refers to the actual spectra; *discontinuous line* is the 1/f line modeling the spectrum in the frequency region where the 1/f model is suitably applicable. From Di Rienzo *et al.*[26] with permission.

This is exemplified by the pressor and tachycardic responses induced in the patient by the visit of an unfamiliar physician (FIG. 4).[29,30] It is also attested to by the pronounced pressor and tachycardic reactions triggered by daily life situations such as a university examination (FIG. 5), a card game (FIG. 6), or a joke telling (FIG. 7).

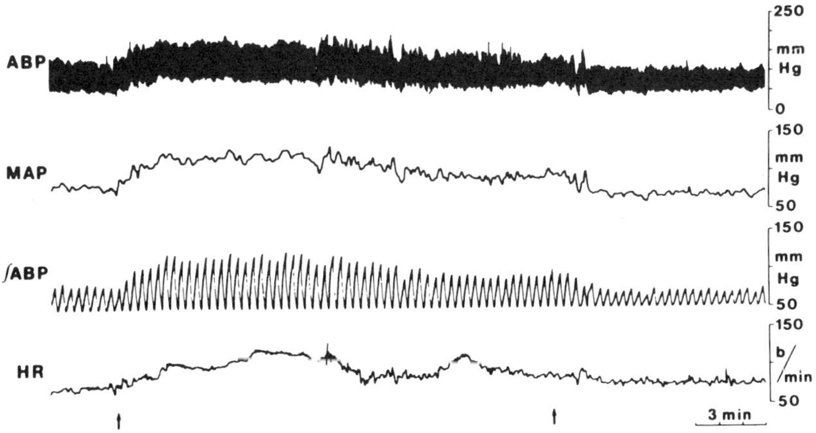

FIGURE 4. Original tracing of intraarterial blood pressure and heart rate in one patient over a 15-minute period when a doctor was at the bedside and measured blood pressure by the cuff method four times. From Mancia *et al.*[29] with permission.

Conversely, behavioral influences may also induce marked blood pressure and heart rate reductions, as shown on a regular basis by night sleep and afternoon siesta.[3,11] Interestingly, night sleep influences also affect fast components of blood pressure and heart rate variability, as shown by the nocturnal reduction of the half-hour blood pressure and heart rate standard deviations (FIG. 8),[3] the nocturnal reduction of blood pressure spectral powers near 0.1 Hz, and the nocturnal increase in the spectral powers of the heart rate at the respiratory frequency[20–22] (FIG. 9). The latter two changes may reflect the decrease in sympathetic and the increase in parasympathetic cardiovascular modulation which take place during night sleep; however, the incomplete specificity of these powers as indices of autonomic cardiovascular influences open the way to additional explanations.[11,17,22]

Blood pressure and heart rate variabilities are also importantly influenced by arterial baroreflexes. This can clearly be seen in experimental animals in which surgical sinoaortic denervation (SAD) is followed by a pronounced increase in overall blood pressure variability and a pronounced decrease in overall heart rate variability.[31,32] SAD, however, also deeply affects the different blood pressure and pulse interval (reciprocal of heart rate) spectral components.

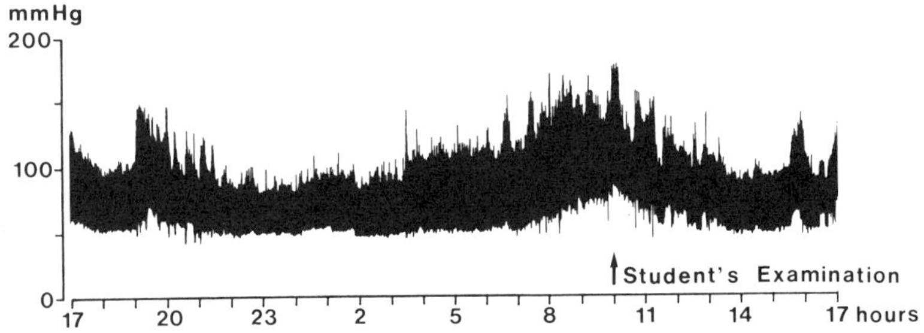

FIGURE 5. A 24-hour intraarterial blood pressure tracing performed in a medical student who at the time indicated by the *arrow* took an examination. The recording was started at 5 PM and ended at 5 PM the following day. From Parati *et al.*[13] with permission.

By recording intraarterial blood pressure in cats both before and after SAD, we showed that this intervention produced (1) a homogeneous reduction in all the spectral powers of pulse interval, (2) a reduction of blood pressure spectral components between 0.05 and 0.2 Hz and below 0.0003 Hz, (3) an increase in blood pressure spectral components in an intermediate frequency region between 0.0003 and 0.05 Hz, (4) almost no change in blood pressure powers at the respiratory frequency, and (5) a change in the slope of the 1/f distribution of blood pressure spectral powers[33,34] (FIG. 10). Similar results were obtained in rats by a comparable experimental protocol,[35] leading to the following conclusions. First, the arterial baroreflex exerts a prooscillatory influence on all pulse interval spectral components, but it has a variable influence on blood pressure spectral components. Second, these influences are antioscillatory for low frequency blood pressure spectra but prooscillatory for blood pressure spectra around 0.1 Hz. This supports the hypothesis that blood pressure spectral peaks in this frequency region are due to resonance of the baroreflex loop mediated through sympathetic efferents.[36]

Evidence on the importance of the arterial baroreflex in modulating blood pressure and heart rate variability has also been obtained in humans,[37] in whom

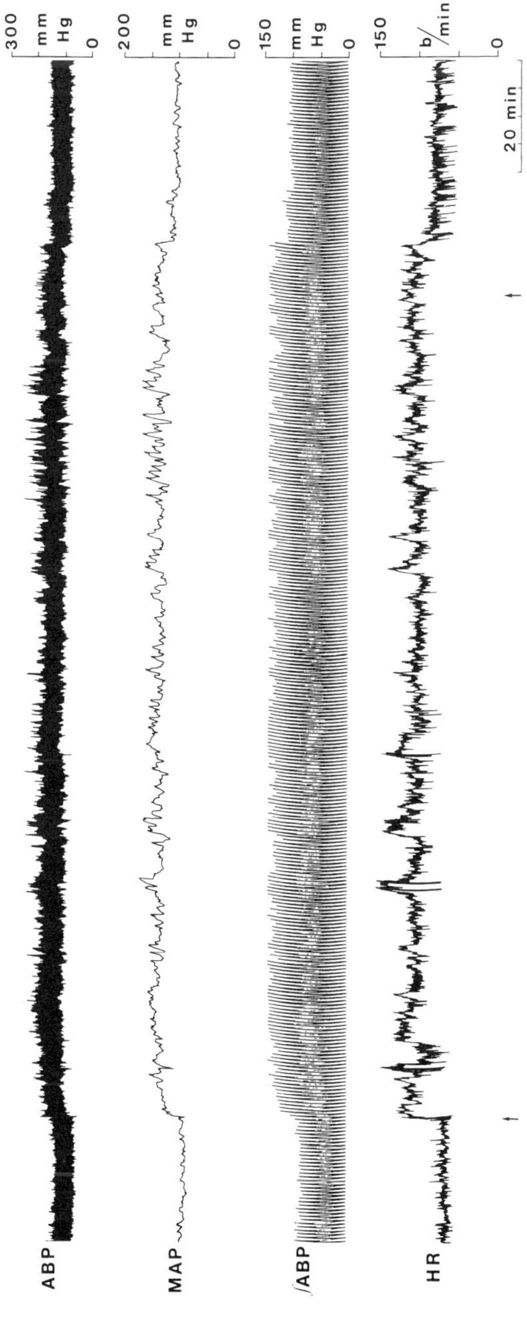

FIGURE 6. Blood pressure and heart rate increases during a poker game of several hours' duration. Original tracing from one subject during a 24-hour intraarterial recording. ABP = pulsatile arterial blood pressure; MAP = mean arterial pressure; ∫ABP = arterial blood pressure integrated at regular intervals of several seconds; HR = heart rate. From Mancia & Zanchetti[11] with permission. *Arrows* at bottom indicate beginning and end of poker game

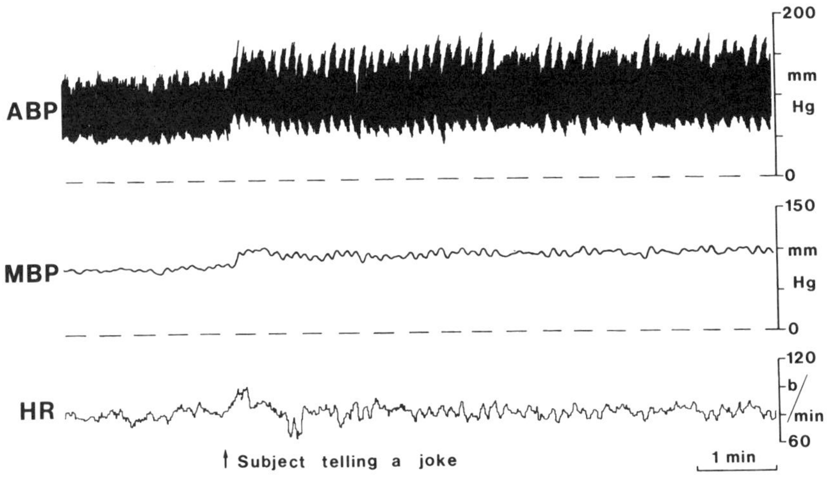

FIGURE 7. Increases in blood pressure and heart rate recorded in a subject telling a joke. Original tracing from one subject during a 24-hour intraarterial recording. ABP = pulsatile arterial blood pressure; MBP = mean arterial pressure; HR = heart rate.

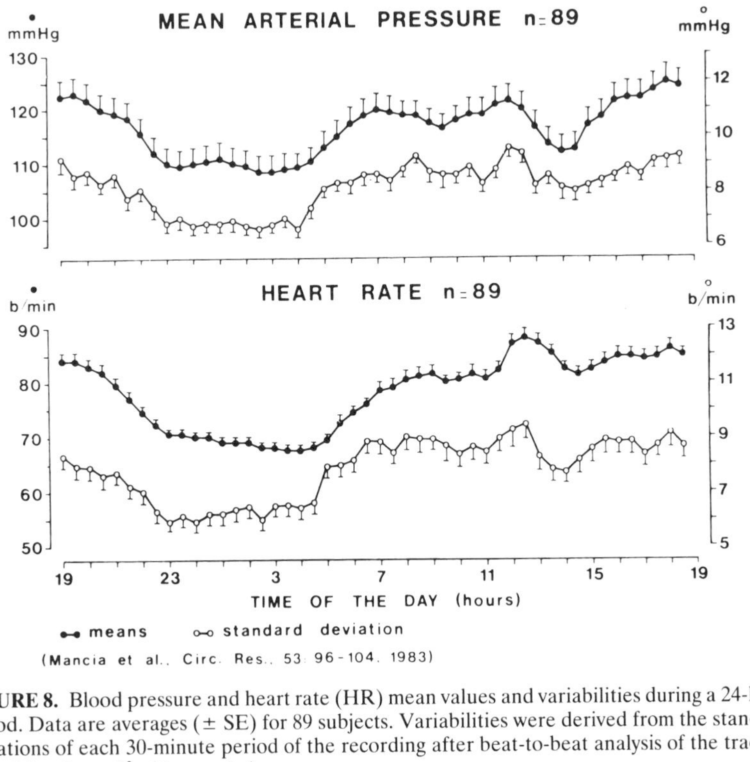

FIGURE 8. Blood pressure and heart rate (HR) mean values and variabilities during a 24-hour period. Data are averages (± SE) for 89 subjects. Variabilities were derived from the standard deviations of each 30-minute period of the recording after beat-to-beat analysis of the tracing. From Mancia *et al.*[3] with permission.

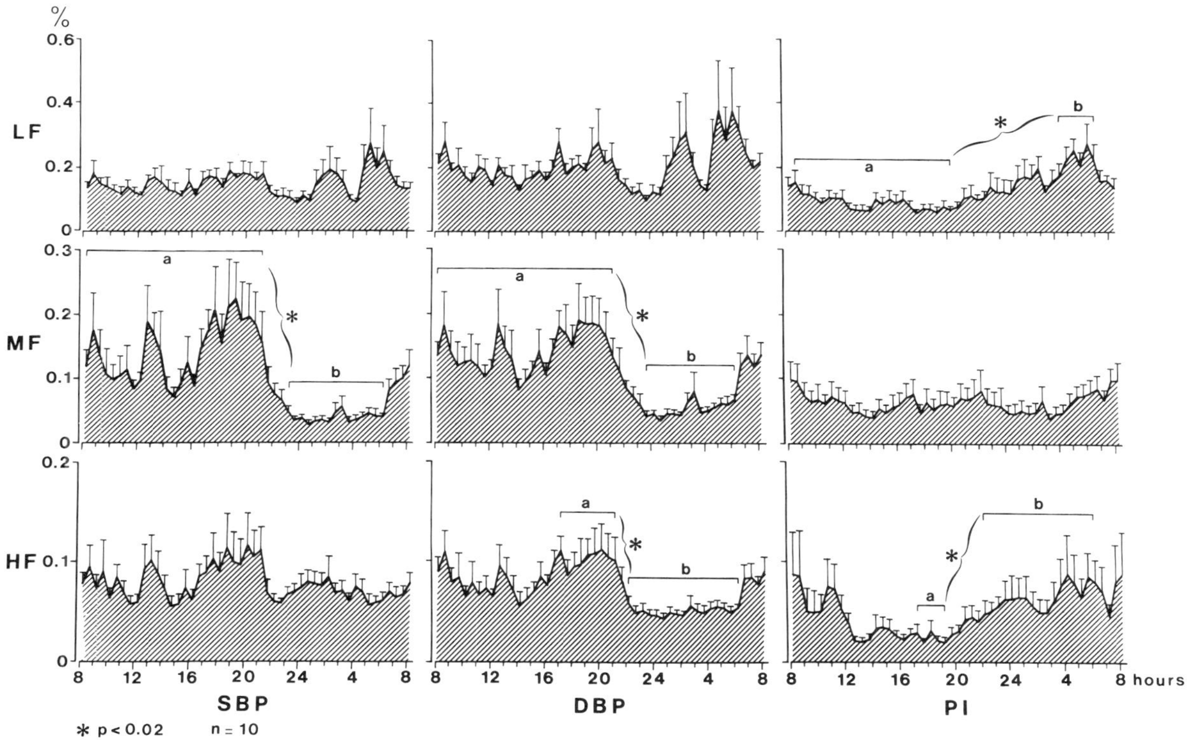

FIGURE 9. Graphs showing low-(LF), mid-(MF), and high-(HF) frequency powers of systolic blood pressure (SBP), diastolic blood pressure (DBP), and pulse interval (PI) for 10 normotensive subjects undergoing 24-hour ambulatory intraarterial blood pressure monitoring. Data are averaged for each half hour and shown as means ± SE. Powers are represented as a percentage of the 24-hour variance. *Asterisks* show statistical significance of differences between subperiods of day (**a**) and night (**b**). Subperiod duration is indicated by *horizontal bars.* From Parati *et al.*[21] with permission.

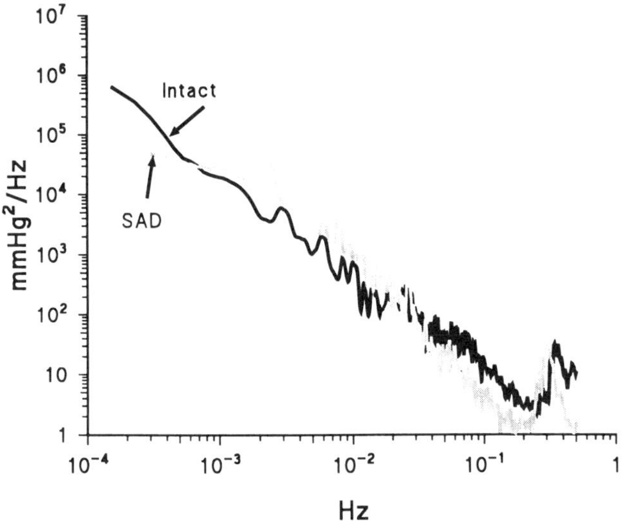

FIGURE 10. Plot shows broad band systolic blood pressure spectra obtained from a free moving cat in the intact condition (*dark line*) and 3 weeks after sinoaortic denervation (SAD, *light line*). Blood pressure was continuously recorded for 3 hours by means of an intraarterial catheter inserted into the abdominal aorta through a femoral artery. Systolic pressure spectral components ranging from approximately 0.5 to 0.0001 Hz are considered. From Parati *et al.*[17] with permission.

baroreflex sensitivity, as assessed by laboratory tests, was inversely related to 24-hour blood pressure variability and directly related to 24-hour heart rate variability (FIG. 11). Thus, the arterial baroreflex exerts a buffering influence on blood pressure variations; however, this does not extend to all specific frequency components in which these variations can be subdivided. The buffering effect is obtained in part by enhancing heart rate (and thus cardiac output) variations.

COMPUTER ANALYSIS OF BLOOD PRESSURE AND HEART RATE VARIABILITY AND DYNAMIC ASSESSMENT OF BAROREFLEX SENSITIVITY

Computer analysis of spontaneous blood pressure and heart rate variations has also been proposed as a method of dynamically assessing the sensitivity of the arterial baroreflex control of heart rate over 24 hours.[38] This method was aimed at overcoming the limitations of traditional laboratory methods for assessing this sensitivity (injection of vasoactive drugs, neck chamber devices, Valsalva maneuver, etc.), that is, the artificial nature of their stimuli, the need for the patient's cooperation, and the limited amount of data collected.[39] It was also aimed at providing a means to assess baroreflex function in daily life conditions.[38]

These new methods are based on time domain or frequency domain (spectral analysis) assessment of the interaction between blood pressure and pulse interval (reciprocal of heart rate) variations occurring during the recording period. The time domain method, known as the sequence method, makes use of computer analysis of continuous blood pressure and heart rate recordings to identify sequences of consecutive beats characterized by progressive increases in systolic blood pressure and pulse interval (hypertension/bradycardia sequences, +PI/+SBP) or by progressive decreases in systolic blood pressure and pulse interval (hypotension/tachycardia sequences, −PI/−SBP) (FIG. 12). As with the laboratory technique based on injection of vasoactive drugs, the slope of the regression line between changes in systolic blood pressure and pulse interval is taken as an index of the sensitivity of baroreflex control of the heart.[40–42] This is further justified by the observation that in cats either sequence type almost completely disappears following SAD,[38,40,41] which demonstrates their specific dependence on the baroreflex.

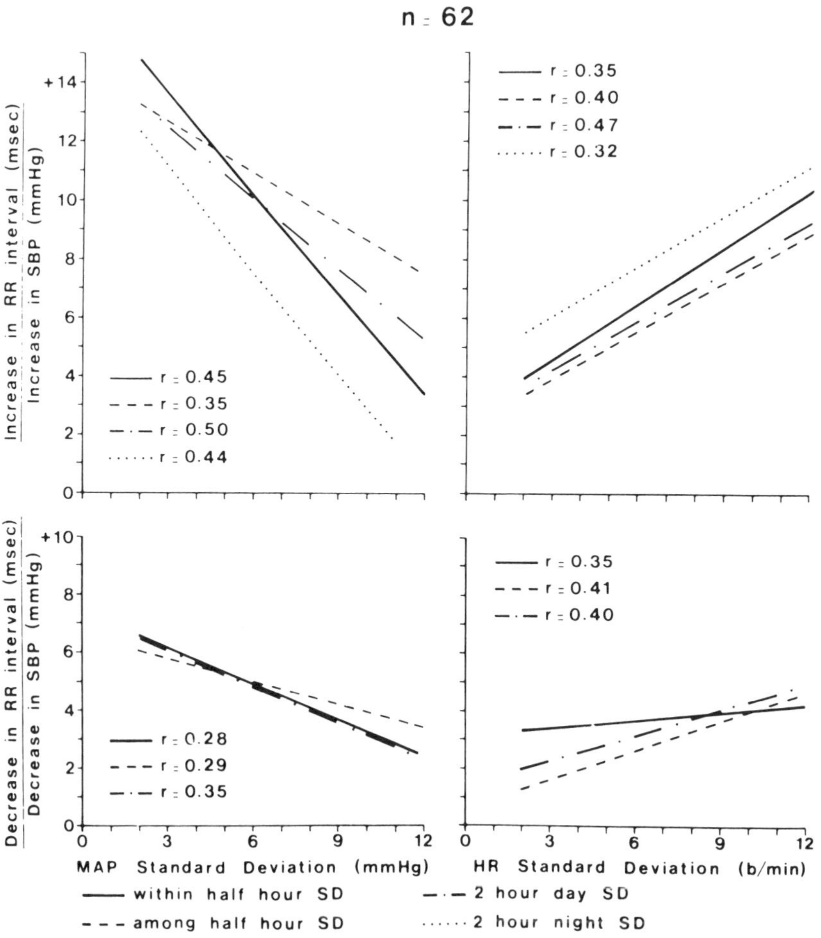

FIGURE 11. Regression lines of mean arterial pressure (MAP) or heart rate (HR) standard deviations and baroreflex sensitivities measured by changes in the R-R interval induced by increasing and decreasing systolic blood pressure through phenylephrine and nitroglycerine injections. Only regression lines achieving statistical significance are shown. From Mancia *et al.*[37] with permission.

The sequence approach has been used to analyze the changes in baroreflex sensitivity that occur with various daily behaviors[38,42–44] in ambulatory subjects under intraarterial blood pressure recording. In normotensive individuals, several hundreds of both +PI/+SBP and −PI/−SBP sequences were found over 24 hours. The slope of these sequences showed marked changes in relation to different behaviors and to the time of day or night. In particular, the sequence slope increased markedly during sleep, whether at night or during the afternoon siesta. In essential hypertensive patients, the sequence number was much lower as was the sequence slope compared to that of normotensive individuals. Furthermore, the wake-sleep modulation of the baroreflex was reduced, that is, sensitivity was similarly low during sleep and wakefulness (FIG. 13).[42] Similar alterations in baroreflex sensitivity throughout the 24 hours characterize elderly normotensive subjects compared with young control subjects.[44]

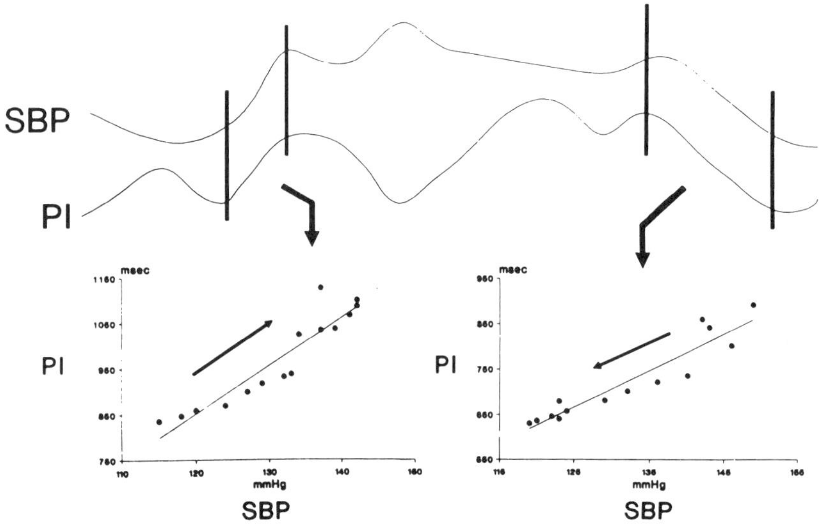

FIGURE 12. Schematic drawing of the hypertension/bradycardia and hypotensio/tachycardia sequences automatically identified by the computer to quantify baroreflex sensitivity. The corresponding regression lines between changes in systolic blood pressure (SBP) and changes in pulse interval (PI) are also shown. From Parati *et al.*[38] with permission.

The frequency-domain method is based on calculation of the modulus of the cross-spectrum[45] or the squared ratio[38,44,46,47] between blood pressure and heart rate powers in the frequency regions where these powers are coherent. This is usually the case near 0.1 and 0.35 Hz. Blood pressure–heart rate coherence near 0.35 Hz, however, does not seem to be specifically due to baroreflex-mediated blood pressure–heart rate coupling, because coherence at this frequency in cats partially survives SAD, limiting a more specific index of baroreflex sensitivity to joint analysis of blood pressure and heart rate spectra near 0.1 Hz. As in the case of sequence slope, in normotensive individuals also, this spectral index (termed alpha coefficient) displays a clearcut day-night modulation that is largely lost in hypertensive patients and in the elderly.[38,44,47] The information provided by the time domain and frequency domain methods on 24-hour baroreflex function is

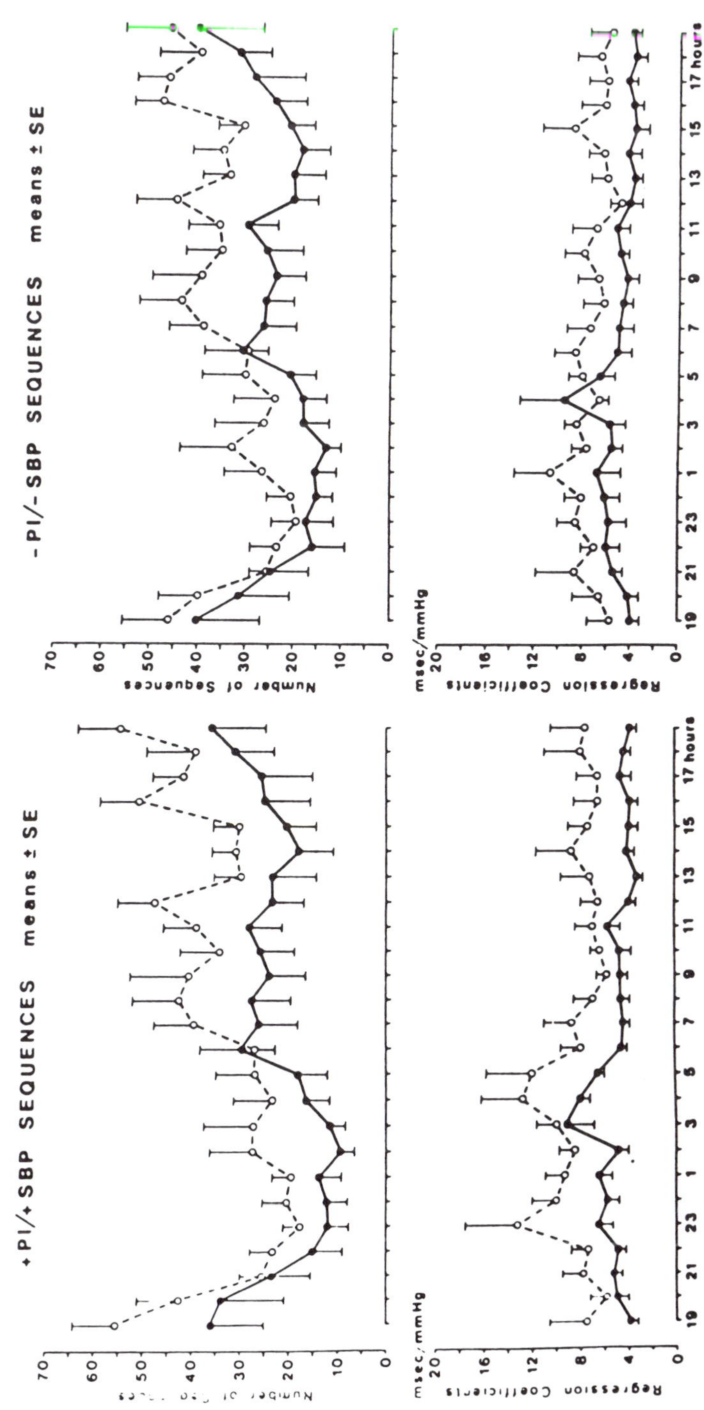

FIGURE 13. Frequency and mean regression coefficients (or slopes) of +PI/+SBP and −PI/−SBP sequences during each hour of 24-hour ambulatory recordings. Data are shown separately as means ± SE for normotensive and hypertensive subjects. Sequences of different duration are pooled. From Parati *et al.*[42] with permission.

comparable in several conditions, making the use of these methods interchangeable according to their practical advantage in specific instances.[38,44,47]

CONCLUSIONS

Analysis of 24-hour ambulatory intraarterial blood pressure recordings has considerably widened our knowledge of blood pressure and heart rate variability in daily life. It has also clarified the importance of central and reflex mechanisms in the modulation of blood pressure and heart rate variations. Reflex mechanisms, in particular, have been thoroughly investigated by techniques for the dynamic evaluation of baroreflex sensitivity based on computer analysis of the interaction between blood pressure and heart rate fluctuations in either the time or the frequency domain.

The introduction of a device (Portapres) that can continuously monitor 24-hour blood pressure noninvasively in ambulatory subjects now allows blood pressure and heart rate variability to be analyzed in the clinical setting, where invasive procedures cannot be employed routinely. This device provides the opportunity to noninvasively derive dynamic information on neural cardiovascular control in many clinical conditions. This may offer deeper insights into the pathophysiologic role of neural factors in daily cardiovascular control. It may also offer data of clinical importance in diseases (diabetes, myocardial infarction, heart failure, etc.) in which assessment of autonomic cardiovascular regulation may have diagnostic and prognostic implications.[48–54]

SUMMARY

The availability of ambulatory intraarterial blood pressure monitoring techniques offers a deeper insight into the features of blood pressure variability over 24 hours and allows better understanding of the mechanisms responsible for the continuous and marked blood pressure changes that occur throughout the day and at night. Among these mechanisms, central and reflex neural influences play a major role. This has led to the development of techniques for the assessment of 24-hour "spontaneous" baroreflex sensitivity through combined computer analysis of blood pressure and heart rate variations in the time or frequency domain. The recent introduction of continuous noninvasive ambulatory blood pressure monitors offers the unique possibility of obtaining dynamic information on neural cardiovascular control in clinical conditions in which the assessment of autonomic cardiovascular regulation may have diagnostic and prognostic implications.

REFERENCES

1. Bevan, A. T., A. J. Honour & F. H. Stott. 1969. Direct arterial pressure recording in unrestricted man. Clin. Sci. **36:** 329–344.
2. Stott, F. D., V. G. Teny & A. J. Honour. 1976. Factors determining the design and construction of a portable pressure transducer system. Postgrad. Med. J. **52**(suppl. 7): 97–99.
3. Mancia, G., A. Ferrari, L. Gregorini, G. Parati, G. Pomidossi, G. Bertinieri. *et al.* 1983. Blood pressure and heart rate variabilities in normotensive and hypertensive human beings. Circ. Res. **53:** 96–104.

4. PENAZ, J. 1973. Photoelectric measurement of blood pressure, volume and flow in the finger. Digest of the International Conference on Medicine and Biological Engineering. Dresden, Conference Committee of the 10th International Conference of Medicine and Biological Engineering. p. 104.
5. WESSELING, K. H., B. DE WITT, J. J. SETTELS & W. H. KLAVER. 1982. On the indirect registration of the finger blood pressure after Penaz. Funkt. Biol. Med. **1:** 245–250.
6. LANGEWOUTERS, G. J., B. DE WITT, G. M. A. VAN DER HOEVEN, G. A. VAN MONTFRANS, B. P. IMHOLZ, W. W. WIELING & K. H. WESSELING. 1990. Feasibility of continuous noninvasive 24 h ambulatory measurements of finger arterial blood pressure with Portapres. J. Hypertens. **8**(suppl. 3): S88.
7. PARATI, G., R. CASADEI, A. GROPPELLI, M. DI RIENZO & G. MANCIA. 1989. Comparison of finger and intra-arterial blood pressure monitoring at rest and during laboratory testing. Hypertension **13:** 647–655.
8. IMHOLZ, B. P., G. J. LANGEWOUTERS, G. A. VAN MONTFRANS, G. PARATI, J. VAN GOUDOEVER, K. H. WESSELING *et al.* 1993. Feasibility of ambulatory continuous 24h finger arterial pressure recording. Hypertension **21:** 65–73.
9. OMBONI, S., G. PARATI, A. FRATTOLA, E. MUTTI, M. DI RIENZO, P. CASTIGLIONI & G. MANCIA. 1993. Spectral and sequence analysis of finger blood pressure variability. Comparison with analysis of intra-arterial recordings. Hypertension **22:** 26–33.
10. MANCIA, G., M. DI RIENZO & G. PARATI. 1993. Ambulatory blood pressure monitoring in hypertension research and clinical practice. Hypertension **21:** 510–524.
11. MANCIA, G. & A. ZANCHETTI. 1986. Blood pressure variability. *In* Handbook of Hypertension, vol. 7: 125–152. A. Zanchetti & R. C. Tarazi, Eds. Elsevier Amsterdam.
12. DE BOER, R. W., J. M. KAREMAKER & J. STRACKEE. 1987. Haemodynamic fluctuations and baroreflex sensitivity in humans. A beat-to-beat model. Am. J. Physiol. **253** (Heart Circ. Physiol. 22): H680–H689.
13. PARATI, G., E. MUTTI, S. OMBONI & G. MANCIA. 1992. How to deal with blood pressure variability. *In* Ambulatory Blood Pressure Recording. H. Brunner & B. Waeber, Eds. 71–99. Raven Press, Ltd. New York.
14. PARATI, G., M. DI RIENZO, S. OMBONI, L. ULIAN & G. MANCIA. 1995. Blood pressure variability over 24 hours: Its different components and its relationship to the arterial baroreflex. J. Sleep Res. **4**(Suppl. 1): 21–29.
15. KITNEY, R. I. & O. ROMPELMAN. 1980. The Study of Heart Rate Variability. Claredon Press. Oxford, UK.
16. AKSELROD, S., D. GORDON, J. B. MADWED, N. C. SNIDMAN, D. C. SHANNON & R. J. COHEN. 1985. Haemodynamic regulation: Investigation by spectral analysis. Am. J. Physiol. **249** (Heart Circ. Physiol. 18): H867–875.
17. PARATI, G., J. P. SAUL, M. DI RIENZO & G. MANCIA. 1995. Spectral analysis of blood pressure and heart rate variability in evaluating cardiovascular regulation. A critical appraisal. Hypertension **25:** 1276–1286.
18. PAGANI, M., F. LOMBARDI, S. GUZZETTI, O. RIMOLDI, R. FURLAN, P. PIZZINELLI, G. SANDRONE, G. MALFATTO, S. DELL'ORTO, E. PICCALUGA, M. TURIEL, G. BASELLI, S. CERUTTI & A. MALLIANI. 1986. Power spectral analysis of heart rate and arterial pressure variabilities as a marker of sympatho-vagal interaction in man and conscious dog. Circ. Res. **59:** 178–193.
19. APPEL, M. I., R. D. BERGER, J. P. SAUL, J. M. SMITH & R. J. COHEN. 1989. Beat-to-beat variability in cardiovascular variables. Noise or music? J. Am. Coll. Cardiol. **14:** 1139–1148.
20. DI RIENZO, M., P. CASTIGLIONI, G. MANCIA, G. PARATI & A. PEDOTTI. 1989. 24-hour sequential spectral analysis of arterial blood pressure and pulse interval in free-moving subjects. IEEE Trans. Biomed. Eng. **36:** 1066–1074.
21. PARATI, G., P. CASTIGLIONI, M. DI RIENZO, S. OMBONI, G. MANCIA & A. PEDOTTI. 1990. Sequential spectral analysis of 24-h blood pressure and pulse interval in humans. Hypertension **16:** 414–421.
22. FURLAN, R., S. GUZZETTI, W. CRIVELLARO, S. DAZZI, M. TINELLI, G. BASELLI, S. CERUTTI, F. LOMBARDI, M. PAGANI & A. MALLIANI. 1990. Continuous 24-hour assessment of the

neural regulation of systemic arterial pressure and RR variabilities in ambulant subjects. Circulation **81:** 537–547.
23. Di Rienzo, M., P. Castiglioni, A. J. Ramirez, G. Mancia & A. Pedotti. 1992. Sequential spectral analysis of blood pressure and heart rate in humans and animals. *In* Blood Pressure and Heart Rate Variability. M. Di Rienzo, G. Mancia, G. Parati, A. Pedotti & A. Zanchetti, Eds. 24–38. IOS Press. Amsterdam.
24. Kobayashi, M. & T. Musha. 1982. 1/f fluctuations of heart beat period. IEEE Trans. Biomed. Eng. BME-29 **6:** 465–467.
25. Saul, P. J., P. Albrecht, R. D. Berger & R. Cohen. 1992. Analysis of long term heart rate variability: Methods, 1/f scaling and implications: 465–466. Proceedings of the 14th Annual Conference of the IEEE Engineering in the Medicine and Biology Society, Paris.
26. Di Rienzo, M., G. Parati, P. Castiglioni, G. Mancia & A. Pedotti. 1995. The wide band spectral analysis: A new insight into modulation of blood pressure, heart rate and baroreflex sensitivity. *In* Computer Analysis of Cardiovascular Signals. M. Di Rienzo, G. Mancia, G. Parati, A. Pedotti & A. Zanchetti, Eds.: 67–74. Amsterdam. IOS Press.
27. Frattola, A., G. Parati, M. Di Rienzo, P. Castiglion, A. Pedotti & G. Mancia. 1994. Blood pressure and heart rate variability in the elderly: Evaluation by computer analysis of ambulatory 24 hour intra-arterial recordings. High Blood Pressure **3:** 1–7.
28. Conway, J., N. Bonn, C. Davies, J. V. Jones & P. Sleight. 1984. Neural and humoral mechanisms involved in blood pressure variability. J. Hypertens. **2:** 203–208.
29. Mancia, G., G. Bertinieri, G. Grassi, G. Parati, G. Pomidossi, A. Ferrari *et al.* 1983. Effects of blood pressure measurements by the doctor on patient's blood pressure and heart rate. Lancet **2:** 695–698.
30. Mancia, G., G. Parati, G. Pomidossi, G. Grassi, R. Casadei & A. Zanchetti. 1987. Alerting reaction and rise in blood pressure during measurement by physician and a nurse. Hypertension **9:** 209–215.
31. Cowley, A. W., L. F. Liard & A. C. Guyton. 1973. Role of the baroreceptor reflex in daily control of arterial blood pressure and other variables in dogs. Circ. Res. **32:** 564–570.
32. Ramirez, A. J., G. Bertinieri, L. Belli *et al.* 1985. Reflex control of blood pressure and heart rate by arterial baroreceptors and by cardiopulmonary receptors in the unanesthetized cat. J. Hypertens. **3:** 327–335.
33. Di Rienzo, M., G. Parati, P. Castiglion, S. Omboni, A. U. Ferrari, A. J. Ramirez, A. Pedotti & G. Mancia. 1991. Role of sinoaortic afferents in modulating BP and pulse interval spectral characteristics in unanesthetized cats. Am. J. Physiol. **261:** 1811–1818.
34. Di Rienzo, M., P. Castiglioni, G. Parati, G. Mancia & A. Pedotti. 1996. Effects of sino-aortic denervation on spectral characteristics of blood pressure and pulse interval variability: Wide-band approach. Med. Biol. Eng. Comput. **34:** 133–141.
35. Cerrutti, C., M. P. Gustin, C. Z. Paultre, M. Lo, C. Julien, M. Vincent & J. Sassard. 1991. Autonomic nervous system and cardiovascular variability in rats: A spectral analysis approach. Am. J. Physiol. **261:** H1292–H1299.
36. Wesseling, K. H., J. J. Settels, H. G. Walstra, H. J. van Esch & J. J. H. Donders. 1983. Baromodulation as the cause of short-term blood pressure variability. *In* Applications of Physics to Medicine and Biology. G. Alberi *et al.,* Eds.: 247–276. Word Scientific, Singapore.
37. Mancia, G., G. Parati, G. Pomidossi, R. Casadei, M. Di Rienzo & A. Zanchetti. 1986. Arterial baroreflex and blood pressure and heart rate variabilities in humans. Hypertension **8:** 147–153.
38. Parati, G., S. Omboni, A. Frattola, M. Di Rienzo, A. Zanchetti & G. Mancia. 1992. Dynamic evaluation of the baroreflex in ambulant subjects. *In* Blood Pressure and Heart Rate Variability. M. Di Rienzo, G. Mancia, G. Parati, A. Pedotti & A. Zanchetti, Eds.: 123–137. IOS Press. Amsterdam.
39. Mancia, G. & A. L. Mark. 1983. Arterial baroreflex in humans. *In* J. T. Sheperd & F. M. Abboud, Eds. The Cardiovascular System IV; sect. 2, part 2, 755–793. American Physiological Society. Bethesda. (Handbook of Physiology, vol 3).
40. Di Rienzo, M., G. Bertinieri, G. Mancia & A. Pedotti. 1985. A new method for

evaluating the baroreflex role by a joint pattern analysis of pulse interval and systolic blood pressure series. Proceedings of the XIV International Conference on Medical and Biological Engineering. Med. Biol. Eng. Comput. **23** (Suppl. 1): 313–314.

41. BERTINIERI, G., M. DI RIENZO, A. CAVALLAZZI, A. U. FERRARI, A. PEDOTTI & G. MANCIA. 1988. Evaluation of baroreceptor reflex by blood pressure monitoring in unanesthetized cats. Am. J. Physiol. **254:** H377–H383.
42. PARATI, G., M. DI RIENZO, G. BERTINIER, G. POMIDOSSI, R. CASADEI, A. GROPPELLI, A. PEDOTTI, A. ZANCHETTI & G. MANCIA. 1988. Evaluation of the baroreceptor-heart rate reflex by 24-hour intra-arterial blood pressure monitoring in humans. Hypertension **12:** 214–222.
43. PARATI, G., E. MUTTI, A. FRATTOLA, P. CASTIGLIONI, M. DI RIENZO & G. MANCIA. 1994. Beta-adrenergic blocking treatment and 24-hour baroreflex sensitivity in essential hypertensive patients. Hypertension **23:** 992–996.
44. PARATI, G., A. FRATTOLA, M. DI RIENZO, P. CASTIGLIONI, A. PEDOTTI & G. MANCIA. 1995. Effects of aging on 24 hour dynamic baroreceptor control of heart rate in ambulant subjects. Am. J. Physiol. **268** (Heart Circ. Physiol. 37): H1606–H1612.
45. ROBBE, H. W. J., L. J. M. MULDER, H. RUDDEL, W. A. LANGEWITZ, J. B. P. VELDMAN & J. MULDER. 1987. Assessment of baroreceptor reflex sensitivity by means of spectral analysis. Hypertension **10:** 538–543.
46. PAGANI, M., V. SOMERS, R. FURLAN, S. DELL'ORTO, J. CONWAY, G. BASELLI, S. CERUTTI, P. SLEIGHT & A. MALLIANI. 1988. Changes in autonomic regulation induced by physical training in mild hypertension. Hypertension **12:** 600–610.
47. PARATI, G., M. DI RIENZO, P. CASTIGLIONI, A. FRATTOLA, S. OMBONI, A. PEDOTTI & G. MANCIA. 1995. Daily-life baroreflex modulation: New perspectives from computer analysis of blood pressure and heart rate variability. *In* Computer Analysis of Cardiovascular Signals. M. Di Rienzo, G. Mancia, G. Parati, A. Pedotti & A. Zanchetti, Eds.: 209–218. IOS Press. Amsterdam.
48. KLEIGER, R. E., J. P. MILLER, J. T. BIGGER & A. R. MOSS. 1987. Multicenter Post-Infarction Research Group: Decreased heart rate variability and its association with increased mortality after acute myocardial infarction. Am. J. Cardiol. **59:** 256–262.
49. LOMBARDI, F., G. SANDRONE, S. PERNPRUNER, R. SALA, R. GARIMOLDI, S. CERUTTI, G. BASELLI, M. PAGANI & A. MALLIANI. 1987. Heart rate variability as an index of sympathovagal interaction after acute myocardial infarction. Am. J. Cardiol. **60:** 1239–1245.
50. SCHWARTZ, P. J., A. ZAZA, M. PALA, E. LOCATI, G. BERIA & A. ZANCHETTI. 1988. Baroreflex sensitivity and its evolution during the first year after myocardial infarction. J. Am. Coll. Cardiol. **12:** 629–636.
51. LA ROVERE, M. T., G. SPECCHIA, A. MORTARA & P. J. SCHWARTZ. 1988. Baroreflex sensitivity, clinical correlates, and cardiovascular mortality among patients with a first myocardial infarction. A prospective study. Circulation **78:** 816–824.
52. OSCULATI, G., G. GRASSI, C. GIANNATTASIO, G. SERAVALLE, F. VALAGUSSA, A. ZANCHETTI & G. MANCIA. 1990. Early alterations of the baroreceptor control of heart rate in patients with acute myocardial infarction. Circulation **81:** 939–948.
53. SCHWARTZ, P. J., E. VANOLI, M. STRAMBA BADIALE, G. M. DE FERRARI, G. E. BILLMAN & R. D. FOREMAN. 1988. Autonomic mechanisms and sudden death. New insights from analysis of baroreceptor reflexes in conscious dogs with and without myocardial infarction. Circulation **78:** 669–679.
54. DE FERRARI, G. M., M. MANTICA, E. VANOLI, S. S. HULL, JR. & P. J. SCHWARTZ. 1993. Scopolamine increases vagal tone and vagal reflexes in patients after myocardial infarction. J. Am. Coll. Cardiol. **22:** 1327–1334.

Respiratory Inputs to Central Cardiovascular Neurons[a]

PAUL PILOWSKY,[b] LEONARD ARNOLDA,
JOHN CHALMERS, IDA LLEWELLYN-SMITH,
JANE MINSON, TAKASHI MIYAWAKI, AND QI-JIAN SUN

Department of Medicine and Centre for Neuroscience
Flinders University
Bedford Park
South Australia 5042, Australia

Precise coordination of respiration and circulation is essential to obtain optimal delivery of oxygen and removal of carbon dioxide and to ensure normal homeostasis. One way that this coordination occurs is through connections between respiratory and sympathetic neurons in the brain stem and spinal cord. We know that these connections exist because sympathetic nerve activity often has a pronounced respiratory modulation. Respiratory modulation of sympathetic nerve activity was first noted by Adrian *et al.*[1] in 1932. In the subsequent 60 years, much was discovered about the relationship between central inspiratory activity and sympathetic nerve activity. It is now clear that respiratory modulation of sympathetic nerve activity is a result of connections between central respiratory neurons and central cardiovascular neurons. It is not due simply to phasic changes in baroreceptor activation that might occur because of ventilation-induced changes in hemodynamics. This information has frequently been obtained by comparing the phasic discharges recorded from the phrenic nerve in paralyzed, ventilated, and often vagotomized animals in order to remove sources of external inputs to central autonomic neurons. Experiments of this type have shown, first, that respiratory rhythmicity, as measured by synchronization with phrenic nerve discharge, is present in animals that are ventilated at a rate different from phrenic nerve discharge.[2,3] This means that changes in the activity of sensory afferent nerves are not the sole cause of the coupling between respiration and sympathetic nerve activity. Second, not all sympathetic nerves have the same type of respiratory rhythmicity. This means that different central sympathetic pathways must receive inputs from different components of the central respiratory rhythm generating network or at least that the proportion of such inputs received is variable.

Despite some progress it is still not clear which neurons provide direct or indirect inputs to presympathetic neurons. Several possible mechanisms of cardiorespiratory coupling have been canvassed in recent reviews.[4–7] Initially, it was proposed that central cardiorespiratory coupling occurred as a result of "irradiation" from central respiratory rhythm-generating neurons whose primary role was to control the activity

[a]Work in the authors' laboratories is supported by Grants-in-Aid from the National Sudden Infant Death Syndrome Council of Australia, the National Heart Foundation of Australia and the National Health and Medical Research Council of Australia, and the Flinders Medical Centre Foundation.

[b]Address for correspondence: Dr. Paul Pilowsky, Department of Medicine, Flinders Medical Centre, Bedford Park, South Australia 5042, Australia (tel: 61-8-2044293; fax: 61-8-2045268; e-mail: paul.pilowsky@flinders.edu.au).

of phrenic, and other somatic, motoneurons to presympathetic neurons that regulate the heart, adrenal medulla, and blood vessels.[4] Later, more complex models were developed that suggest the existence of a common "cardiorespiratory oscillator." This common oscillator receives inputs from the full range of peripheral inputs, processes them, and then provides appropriate outputs to both presympathetic and respiratory premotor neurons.

Work in our laboratory is aimed at investigating the types of respiratory neurons that form connections with neurons involved in the control of central sympathetic pathways. We also aim to determine which neurotransmitters are present in the respiratory and presympathetic nerves.

RESPIRATORY RHYTHMICITY IN PERIPHERAL SYMPATHETIC NERVES

Several laboratories have described different types of respiratory rhythmicity in sympathetic pre- and postganglionic neurons. In the rat, Numao *et al.*[8] found a postinspiratory activation pattern in cervical and lumbar sympathetic nerves: by contrast, cardiac, splanchnic, adrenal, and renal nerves showed an early inspiratory peak in their activity. These findings indicate an underlying diversity in the inputs to different types of sympathetic preganglionic neurons, a possibility that has been extensively explored. In a series of studies on the activity patterns of single fibers in sympathetic nerves, Jänig and co-workers[9] found that certain respiratory-related activity patterns are characteristic of particular types of sympathetic nerves. Thus, nerves in muscle vasoconstrictor pathways display a strong inspiratory-related increase in their firing rate. In combination with passive properties such as axon conduction velocity and other features such as reflex responses to a variety of stimuli it is possible to classify sympathetic preganglionic neurons. The source of this respiratory rhythmicity in sympathetic preganglionic neurons could be either a direct input from medullary respiratory-related premotoneurons or modulation of the activity of barosensitive presympathetic neurons. In a recent intracellular study we examined the respiratory modulation of sympathetic preganglionic neurons in the cat lumbar spinal cord. Seven of 85 neurons that had a respiratory modulation were identified. All seven had axonal conduction velocities of less than 2.0 m/s. None of 54 neurons with conduction velocities ranging from 2.1–9.0 m/s had a respiratory modulation. Six of the seven respiratory modulated cells were activated during inspiration, suggesting that they were of the vasoconstrictor type described by Jänig and colleagues.[4,9] Respiratory modulation, when present, consisted of a short burst of excitatory postsynaptic potentials or a small depolarization at the peak of phrenic nerve discharge. We found no evidence for synaptic inhibition during expiration in the inspiratory modulated neurons. This type of respiratory modulation is most consistent with modulation of an antecedent input rather than a direct input from a respiratory-related neuron in the brain stem. A direct input from a respiratory-related neuron would instead be expected to produce large respiratory-related swings in membrane potential.

Respiratory modulation of the activity of spinally projecting barosensitive neurons in the rostral ventrolateral medulla that are believed to be sympathoexcitatory has also been the subject of recent studies.[10–13] Recently, we investigated the nature of the interaction between baroreceptor and respiratory-related inputs in barosensitive neurons in the rostral ventrolateral medulla of rats. Barosensitive neurons exhibited either no respiratory modulation or one of four distinct types, largely confirming the findings of previous workers.[10] These four types consist of early inspiratory activation, postinspiratory activation, inspiratory depression, or expiratory depression.[13] To some extent, the particular types of respiratory modulation seen in bulbospinal barosensitive neurons correspond to the respiratory modulations seen in peripheral sympathetic nerves. In the rat, type II neurons have an activity pattern that is similar to that of cardiac and adrenal nerves, while the activity seen in type III neurons is similar to the pattern that can be recorded from the cervical sympathetic nerve.[8,10,13] It is tempting to speculate that these activity patterns provide a functional fingerprint for the different sympathetic pathways up to the level of the premotor neurons. However, this idea must be treated with some caution, because it has also been reported that the characteristic activity patterns found in the cat are not as obvious those in the rat.[14]

In the rat, respiratory inputs appear to summate with baroreceptor inputs in bulbospinal neurons in the rostral ventrolateral medulla so that in most cases aortic nerve stimulation is most effective in inspiration when respiratory inputs are minimal. This effect is also seen as a reduction in the amplitude of the cardiac related modulation of unit firing rate.[13] Thus, our data, and those of others, indicate that respiratory neurons may form direct connections with spinally projecting barosensitive neurons. Furthermore, these connections are likely to arise from more than one type of respiratory neuron in order to account for the diversity of respiratory modulations seen.

A major difficulty in establishing which types of respiratory neurons form connections with barosensitive cardiovascular neurons is the marked heterogeneity of both respiratory and cardiovascular neurons. In addition, the two cell types are both located in the ventrolateral medulla and are often intermingled. These features make standard neuroanatomic tracing techniques difficult to apply because extracellularly applied tracers could be taken up and transported by a range of neurons in addition to those being investigated.

PROJECTIONS FROM RESPIRATORY NEURONS TO PRESYMPATHETIC NEURONS IN THE ROSTRAL VENTROLATERAL MEDULLA

Recently, we have started to determine which respiratory neurons project towards presympathetic cardiovascular nuclei in the ventrolateral medulla. To do this we use a combination of techniques. First, we use intracellular recording and dye-filling to identify neurons with a prominent respiratory modulation of their membrane potential. Second, we use tyrosine hydroxylase immunoreactivity to identify spinally projecting catecholamine-synthesizing neurons in the rostral ventrolateral medulla. Intracellular injection of Lucifer yellow into inspiratory and expiratory neurons in the ventral medulla of the adult rat revealed that they are not catecholamine-synthesizing. However, when their filled axons were reconstructed, some of these respiratory neurons were found to have axonal projections to that part of the ventrolateral medulla where the cell bodies of bulbospinal presympathetic neurons are known to be located.[15] This finding suggested that respiratory neurons in the ventral respiratory group may form connections with cardiovascular neurons in the rostral ventrolateral medulla. More recently, in experiments in cats, we found that a specific subclass of neurons in the ventral respiratory group, the inspiratory neurons in the rostral ventral respiratory group, have axonal varicosities within the subretrofacial nucleus.[16] In the cat, the subretrofacial nucleus is a well-defined compact cluster of spinally projecting barosensitive neurons, many of which are tyrosine hydroxylase immunoreactive. The subretrofacial nucleus is homologous to the rostral ventrolateral medulla in the rat.[17]

Taken together, these results suggest that the commonly observed respiratory modulation of sympathetic nerve discharge occurs as a result of a direct projection from neurons in the ventral respiratory group towards presympathetic neurons in the rostral ventrolateral medulla. We are presently exploring which of the specific types of respiratory neuron are involved in this projection.

NEUROTRANSMITTERS IN RESPIRATORY INPUTS TO CARDIOVASCULAR NEURONS

The neurotransmitters used by excitatory and inhibitory respiratory neurons are assumed to be glutamate and GABA, respectively. Physiologic and pharmacologic studies have repeatedly demonstrated that interfering with glutamatergic and GABAergic neurotransmission can affect inputs to medullary cardiovascular neurons.[5,17–20] However, direct anatomic evidence that these neurotransmitters are present in particular types of respiratory neurons that project to cardiovascular neurons is lacking.

NEUROTRANSMITTERS IN INPUTS TO RESPIRATORY NEURONS

Synaptic inputs to respiratory neurons may indirectly affect cardiovascular neurons by changing the activity of central respiratory rhythm-generating neurons. Candidate respiratory neurons that might play a role in an indirect mechanism of this type would include those that are known to have intramedullary connections. Bötzinger neurons in the most rostral part of the ventral respiratory group often have spinal axons and also form widespread synaptic connections within the brain stem. Bötzinger neurons are generally active in expiration and, from studies in the cat, are believed to be inhibitory.[21] Intracellular labeling in combination with immunohistochemistry has been used to identify neurotransmitters that may modulate the activity of respiratory neurons. By dye-filling Bötzinger neurons and then using immunocytochemistry for tyrosine hydroxylase we found that Bötzinger neurons receive a catecholaminergic input.[22] This finding indicates one site where catecholamines may act to affect phrenic nerve discharge. If Bötzinger neurons form connections with cardiovascular neurons, then this may also form an indirect route through which centrally acting drugs such as clonidine might affect sympathetic nerve activity.[23–25]

DO CARDIOVASCULAR NEURONS AFFECT RESPIRATORY NEURONS?

More contentious than the question of respiratory modulation of sympathetic nerve activity is the question of whether or not cardiovascular neurons can affect respiratory neurons. Cardiovascular neurons are generally defined as central neurons whose activity is increased or decreased by activation of baroreceptors. Respiratory neurons are defined as neurons that have a large respiratory modulation of their activity. This may seem confusing because some cardiovascular neurons have respiratory modulation, but in practice the respiratory modulation in respiratory neurons is much larger than that seen in cardiovascular neurons. There are several possible ways in which cardiovascular neurons could affect respiratory neurons. First, respiratory neurons could themselves be barosensitive. If so, then activation of baroreceptors could indirectly affect cardiovascular neurons by changing the level of respiratory modulation received by cardiovascular neurons. Second, there could be common antecedent neurons that affect the activity of both respiratory and cardiovascular neurons. Third, cardiovascular neurons might affect the activity of respiratory neurons.

Regarding the first possibility, some experimenters reported that changes in baroreceptor inputs can affect the rate or amplitude of ventilation or phrenic nerve discharge.[26,27] Unfortunately, stimulation of arterial baroreceptors in a manner that does not concomitantly activate other types of receptors is a difficult task. Thus, for example, electrical stimulation of the carotid sinus nerve will activate chemoreceptor as well as baroreceptor nerves. In some species, the aortic depressor nerve also contains nociceptive or chemoreceptive fibers.[28] However, in the rat the aortic depressor nerve is thought to be an essentially pure baroreceptor nerve.[29,30] Electrical stimulation of the aortic depressor nerve, at intensities that suppress sympathetic nerve discharge and lower blood pressure, often has no effect on phrenic nerve discharge.[13,31] This result suggests that baroreceptor activation, in isolation from other peripheral receptors, does not affect central respiratory rhythm generation. If this is the case, then baroreceptor-mediated inhibition of sympathetic activity is occurring at a site, such as the barosensitive bulbospinal neurons in the ventrolateral medulla, that is not involved in the control of breathing. In other words, baroreceptor inputs do not affect blood pressure through effects on respiratory rhythm-generating neurons.

The second possibility is that baroreceptor inputs affect common antecedent neurons. If this was so, then baroreceptor activation would affect respiration. Again, although this issue is still not resolved, evidence indicates that this is unlikely.[13,31] Instead, the sympathoinhibitory action of baroreceptors appears to be due solely to inhibition of the activity of presympathetic sympathoexcitatory neurons. Finally, cardiovascular neurons may possibly project towards respiratory neurons. There is currently no physiologic or morphologic evidence to support this idea. Thus, respiratory modulation of sympathetic activity most likely occurs by activation of pathways from central respiratory neurons towards presympathetic neurons.

Another occasional feature of sympathetic activity proposed as evidence for two independent but entrained oscillators is the finding of sympathetic slow waves in the absence of phrenic nerve discharge.[2,32] Although such findings may indeed indicate the existence of two tonically active oscillators, a simpler explanation may be that stimuli which inhibit phrenic nerve discharge[33] also activate other pattern-generating systems, such as swallowing. These other systems may also affect presympathetic pathways.[34–36]

PHYSIOLOGIC SIGNIFICANCE

Clearly, much remains to be discovered about the manner in which respiratory and cardiovascular neurons interact. Furthermore, the reasons for the interaction are also not clear. Häbler *et al.*[4] suggest several possibilities. First, by grouping sympathetic action potentials into bursts, transmission at sympathetic ganglia may be more effective. Second, during exercise, local mechanisms in muscle may cause vasodilatation. In this situation, respiratory-sympathetic coupling would be useful in preventing falls in arterial blood pressure. Finally, respiratory patterning of the activity to specific populations of sympathetic neurons may be important in the regulation of temperature and extracellular fluid volume.[4] A further possibility is that since respiratory and central autonomic pathways have as common goals the delivery of oxygen and the removal of carbon dioxide, respiratory activation of sympathetic nerves would be essential in producing an appropriate circulatory response during mild chemoreceptor stimuli and critical during asphyxia. This idea is similar to that proposed by Adrian *et al.*[1] who suggested that the sensitivity of the respiratory center would enable the vasomotor center to react promptly to changes "in the blood" that were too small for vasomotor cells to detect. Sixty years later this explanation still has appeal.

REFERENCES

1. ADRIAN, E. D., D. W. BRONK & G. PHILLIPS. 1932. Discharges in mammalian sympathetic nerves. J. Physiol. (Lond.) **74:** 115–133.
2. GOOTMAN, P. M., M. R. GHANDI, A. M. STEELE, B. W. HUNDLEY, H. L. COHEN, L. P. EBERLE & A. L. SICA. 1991. Respiratory modulation of sympathetic activity in neonatal swine. Am. J. Physiol. Regul. Integr. Comp. Physiol. **261:** R1147–1154.
3. PILOWSKY, P., I. J. LLEWELLYN-SMITH, L. ARNOLDA, J. MINSON & J. CHALMERS. 1994. Intracellular recording from sympathetic preganglionic neurons in cat lumbar spinal cord. Brain Res. **656:** 319–328.
4. HÄBLER, H.-J., W. JÄNIG & M. MICHAELIS. 1994. Respiratory modulation in the activity of sympathetic neurones. Prog. Neurobiol. **43:** 567–606.
5. PILOWSKY, P. 1995. Good vibrations? Respiratory rhythms in the central control of blood pressure. Clin. Exp. Pharmacol. Physiol. **22:** 594–604.
6. KOEPCHEN, H.-P., D. KLUSSENDORF & D. SOMMER. 1981. Neurophysiological background of central neural cardiovascular-respiratory coordination: Basic remarks and experimental approach. J. Auton. Nerv. Syst. **3:** 335–368.
7. FELDMAN, J. L. & H. H. ELLENBERGER. 1988. Central coordination of respiratory and cardiovascular control in mammals. Ann. Rev. Physiol. **50:** 593–606.
8. NUMAO, Y., N. KOSHIYA, M. P. GILBEY & K. M. SPYER. 1987. Central respiratory drive-related activity in sympathetic nerves of the rat: The regional differences. Neurosci. Lett. **81:** 279–284.
9. BOCZEK-FUNCKE, A., H.-J. HÄBLER, W. JÄNIG & M. MICHAELIS. 1992. Respiratory modulation of the activity in sympathetic neurones supplying muscle, skin and pelvic organs in the cat. J. Physiol. (Lond.) **449:** 333–361.
10. HASELTON, J. R. & P. G. GUYENET. 1989. Central respiratory modulation of medullary sympathoexcitatory neurons in rat. Am. J. Physiol. **256:** R739–750.
11. MCALLEN, R. M. 1987. Central respiratory modulation of subretrofacial bulbospinal neurones in the cat. J. Physiol. (Lond.) **388:** 533–545.
12. TERUI, N., Y. SAEKI & M. KUMADA. 1986. Barosensory neurons in the ventrolateral medulla in rabbits and their responses to various afferent inputs from peripheral and central sources. Jpn. J. Physiol. **36:** 1141–1164.
13. MIYAWAKI, T., P. PILOWSKY, Q.-J. SUN, J. MINSON, S. SUZUKI, L. ARNOLDA, I. LLEWELLYN-SMITH & J. CHALMERS. 1995. Central inspiration increases barosensitivity of neurons in rat rostral ventrolateral medulla. Am. J. Physiol. **268:** R909–918.
14. HÄBLER, H.-J., W. JÄNIG, M. KRUMMEL & O. A. PETERS. 1993. Respiratory modulation of the activity in postganglionic neurons supplying skeletal muscle and skin of the rat hindlimb. J. Neurophysiol. **70:** 920–921.
15. PILOWSKY, P. M., C. JIANG & J. LIPSKI. 1990. An intracellular study of respiratory neurons in the rostral ventrolateral medulla of the rat and their relationship to catecholamine-containing neurons. J. Comp. Neurol. **301:** 604–617.
16. PILOWSKY, P., I. J. LLEWELLYN-SMITH, J. LIPSKI, J. MINSON, L. ARNOLDA & J. CHALMERS. 1994. Projections from inspiratory neurons of the ventral respiratory group to the subretrofacial nucleus of the cat. Brain Res. **633:** 63–71.
17. DAMPNEY, R. A. L. 1994. The subretrofacial vasomotor nucleus: Anatomical, chemical and pharmacological properties and role in cardiovascular regulation. Prog. Neurobiol. **42:** 197–227.
18. MINSON, J. B., I. J. LLEWELLYN-SMITH, L. F. ARNOLDA, P. M. PILOWSKY, J. R. OLIVER & J. P. CHALMERS. 1994. Disinhibition of the rostral ventral medulla increases blood pressure and Fos expression in bulbospinal neurons. Brain Res. **646:** 44–52.

19. Guyenet, P. G., J. R. Halselton & M.-K. Sun. 1989. Sympathoexcitatory neurons of the rostroventrolateral medulla and the origin of the sympathetic vasomotor tone. Prog. Brain Res. **81:** 105–116.
20. Chalmers, J. P. & P. M. Pilowsky. 1991. Brainstem and bulbospinal neurotransmitter systems in the control of blood pressure. J. Hypertens. **9:** 675–694.
21. Ezure, K. 1990. Synaptic connections between medullary respiratory neurons and considerations on the genesis of respiratory rhythm. Prog. Neurobiol. **35:** 429–450.
22. Sun, Q., P. Pilowsky, J. Minson, L. Arnolda, J. Chalmers & I. J. Llewellyn-Smith. 1994. Close appositions between tyrosine hydroxylase immunoreactive boutons and respiratory neurons in the rat ventrolateral medulla. J. Comp. Neurol. **340:** 1–10.
23. Drolet, G., V. Aslanian, J. B. Minson, M. Morris & J. P. Chalmers. 1990. Differences in the central hypotensive actions of alpha-methyldopa and clonidine in the spontaneously hypertensive rat: Contribution of neurons arising from the B3 and the C1 areas of the rostral ventrolateral medulla. J. Cardiovasc. Pharmacol. **15:** 118–123.
24. Ockert, D. M., J. Czachurski, K. Dembowsky & H. Seller. 1992. Central action of α-adrenoceptor agents on the baroreceptor reflex. J. Auton. Nerv. Syst. **41:** 93–102.
25. Koshiya, N. & P. G. Guyenet. 1995. Sympatholytic effect of clonidine depends on the respiratory phase in rat splanchnic nerve. J. Auton. Nerv. Syst. **53:** 82–86.
26. Richter, D. W. & H. Sellar. 1975. Baroreceptor effects on medullary respiratory neurones of the cat. Brain Res. **86:** 168–171.
27. Wasicko, M. J., R. W. Giering, S. L. Knuth & J. C. Leiter. 1993. Hypoglossal and phrenic nerve response to carotid baroreceptor stimulation. J. Appl. Physiol. **75:** 1395–1403.
28. Numao, Y., M. Saito, N. Terui & M. Kumada. 1983. Physiological and pharmacological properties of the three subsystems constituting the aortic nerve-renal sympathetic reflex in rabbits. J. Auton. Nerv. Syst. **9:** 361–380.
29. Numao, Y., M. Siato, N. Terui & M. Kumada. 1985. The aortic nerve-sympathetic reflex in the rat. J. Auton. Nerv. Syst. **13:** 65–79.
30. Sapru, H. N. & A. J. Krieger. 1977. Carotid and aortic chemoreceptor function in the rat. J. Appl. Physiol. **42:** 344–348.
31. Hayashi, F., S. K. Coles, K. B. Bach, G. S. Mitchell & D. R. McCrimmon. 1993. Time-dependent phrenic nerve responses to carotid afferent activation: Intact vs. decerebellate rats. Am. J. Physiol. **265:** R811–R8119.
32. Barman, S. M. & G. L. Gebber. 1976. Basis for synchronization of sympathetic and phrenic nerve discharges. Am. J. Physiol. **231:** 1601–1607.
33. Jiang, C. & J. Lipski. 1992. Synaptic inputs to medullary respiratory neurons from superior laryngeal afferents in the cat. Brain Res. **584:** 197–206.
34. König, S., J. Czachurski & K. Dembowsky. 1990. Inhibition of cardiac sympathetic nerve activity during swallowing evoked by laryngeal afferent stimulation in the cat. Neurosci. Lett. **118:** 265–268.
35. Huangfu, D. & P. G. Guyenet. 1991. Sympatholytic response to stimulation of superior laryngeal nerve in rats. Am. J. Physiol. **260:** R290–R297.
36. Faber, J. E. & M. J. Brody. 1983. Reflex hemodynamic response to superior laryngeal nerve stimulation in the rat. J. Auton. Nerv. Syst. **9:** 607–622.

Signal Transduction in Animal Models of Normotension and Hypertension

KLAUS WITTE[a] AND BJÖRN LEMMER

Institute of Pharmacology and Toxicology
Faculty of Clinical Medicine Mannheim
Ruprecht-Karls-University Heidelberg
Maybachstraße 14-16
D-68169 Mannheim, Germany

It is now well established that human blood pressure and heart rate exhibit significant time-dependent variation. Peaks in both parameters are usually observed in the daily activity period; troughs normally occur at night. Time-dependent changes in the signal transduction processes of several neurohumoral factors are proposed to be involved in the circadian regulation of blood pressure and heart rate. As there is no doubt about the pertinent role of the sympathetic nervous system in modulating and controlling cardiovascular function, circadian variation in sympathetic signaling is of great interest to both basic research and clinical pharmacology. The main neurotransmitter within the sympathetic nervous system is the catecholamine noradrenaline. Noradrenaline is synthesized and stored in vesicles in the presynaptic nerve ending, and upon nerve impulse, it is released into the synaptic cleft. Once released, noradrenaline binds to a postsynaptic receptor, which then activates the receptor-coupled signal transduction cascade. Thus, binding of noradrenaline to β-adrenoceptors induces the dissociation of a stimulatory G-protein by promoting the release of guanosine-5′-diphosphate (GDP) and subsequent binding of guanosine-5′-triphosphate (GTP) at the guanine nucleotide binding domain of the stimulatory $G_s\alpha$-subunit. GTP-bound $G_s\alpha$ activates adenylyl cyclase, resulting in enhanced formation of cAMP. The intracellular effects of the second messenger are terminated by hydrolysis of cAMP to 5′-AMP by a group of cAMP-degrading enzymes, the phosphodiesterases.

In laboratory animals almost all steps of sympathetic signaling exhibit circadian rhythmicity. Thus, rhythmic changes in the concentration of noradrenaline have been reported in rat pineal, salivary glands,[1] and several brain regions.[2,3] By contrast, noradrenaline content in rat hearts did not show significant daily variation.[4] However, because most of the noradrenaline measured in tissue homogenates is accumulated in storage granules and only a small functional pool of the transmitter is required for neuronal function, tissue concentrations of noradrenaline may not fully reflect its functional role. Moreover, the substrate for biosynthesis of noradrenaline, tyrosine, and the enzymes involved, tyrosine hydroxylase, DOPA decarboxylase, and dopamine β-hydroxylase, are also present in the terminal nerve endings. Endogenous levels of a biogenic amine are the resultants of biosynthesis, storage, and release. Therefore, a lack of daily variation in the actual concentration of a transmitter does not exclude rhythmicity at different steps in the transmitter system and vice versa. In rat heart tissue, where no rhythmicity in noradrenaline content was observed, the turnover of noradrenaline, which is a more appropriate measure of sympathetic activity, was significantly greater in the rat's nightly activity period than

[a]Tel: 621/331067; fax: 621/377601.

during the day.[4,5] In addition, postsynaptic events such as the formation,[6–8] content,[9] and degradation of cAMP[6] exhibited significant daily variation in heart ventricles from Wistar rats, whereas β-adrenoceptor density did not vary with the circadian time.[6] In humans, circadian rhythms have been observed in the incidence of sudden cardiac death,[10] myocardial infarction,[11,12] and myocardial ischemia,[13] which are thought to be due, at least partly, to the morning increase in sympathetic tone. Consequently, the morning increase in cardiac risk can be prevented by treatment with β-adrenoceptor antagonists.[10,12] Human vascular resistance as a major determinant of blood pressure also exhibits diurnal variation, which was abolished by injection of the α-adrenoceptor antagonist phentolamine.[14] Moreover, infusion of noradrenaline led to a more pronounced blood pressure increase in the early morning hours and during the day than at night,[15] indicating that besides circadian rhythmicity in sympathetic tone, time-dependent variation must exist in the sensitivity of target tissues towards the agonist noradrenaline. This hypothesis is supported by the observation that rat aortic rings *in vitro* showed pronounced ultradian rhythmicity in the sensitivity towards vasoconstriction by noradrenaline.[16] As just outlined, the circadian variation in tissue sensitivity towards noradrenaline could be due to daily rhythmicity at different steps of the signal transduction cascade. This article, therefore, addresses the question of whether circadian rhythmicity in different second messenger pathways could contribute to circadian regulation of blood pressure and heart rate.

ANIMAL MODELS OF PRIMARY AND SECONDARY HYPERTENSION

Several authors have observed that in some forms of human secondary hypertension circadian blood pressure profiles are frequently flattened or even inverse, whereas in primary hypertension the nocturnal blood pressure fall is typically preserved.[17] As biochemical experiments on cellular signal transduction require animal tissue, appropriate animal models of primary and secondary hypertension had first to be defined. Using radiotelemetry for continuous blood pressure monitoring, we observed that spontaneously hypertensive rats (SHR), which are considered an animal model of human primary hypertension, indeed showed normal circadian rhythms in blood pressure, heart rate, and motor activity with peaks in the rat's nocturnal activity period[18] (FIG. 1). By contrast, transgenic hypertensive TGR(mRen-2)27 rats (TGR) exhibited circadian rhythms in blood pressure inverse to those in heart rate and motor activity and were therefore considered to represent an animal model of circadian blood pressure patterns in human secondary hypertension[18] (FIG. 2). Studies on circadian rhythms in cardiovascular signal transduction were then performed in four different inbred strains: primary hypertensive SHR, secondary hypertensive TGR, and in their normotensive controls, that is, Wistar-Kyoto (WKY) and Sprague-Dawley rats (SPRD), respectively.

The methods used for determination of β-adrenoceptor density, adenylyl cyclase activity, and soluble guanylyl cyclase activity have been described in detail elsewhere.[19,20] Results of our studies in WKY and SHR, which are part of this review, are taken from the respective original publications. Recent studies in SPRD and TGR have not been published yet, but methods were the same as those in the earlier studies in WKY and SHR.[19,20] Rhythmicity in cardiovascular as well as biochemical parameters was analyzed using PHARMFIT[21] as described.

RHYTHMS IN SIGNAL TRANSDUCTION

Cardiac β-Adrenoceptor Adenylyl Cyclase System

In ventricles from WKY and SHR the density of total β-adrenoceptors as well as of the β_1-subtype exhibited significant 12-hour rhythms, with peaks occurring at 1 AM and 1 PM, that is, the middle of the light and dark periods. For the β_2-subtype no significant rhythmicity could be detected (FIG. 3). In SHR, cardiac β-adrenoceptor density, calculated as the 24-hour mean, was approximately 20% lower than that in WKY because of a 25% loss in the β_1-subtype, whereas the density of β_2-adrenoceptors was similar in both strains (FIG. 3). By contrast, formation of cAMP

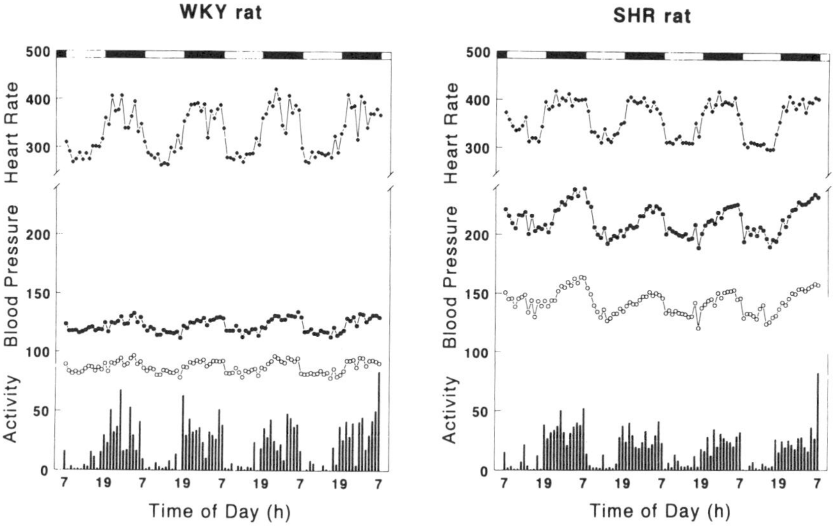

FIGURE 1. Circadian rhythms in motor activity (*filled bars*), systolic blood pressure (*closed circles*), diastolic blood pressure (*open circles*), and heart rate (*closed diamonds*) in a single WKY (**left panel**) and SHR rat (**right panel**). In both animals, all parameters exhibited significant circadian rhythmicity with dominant period lengths of 24 hours and peaks occurring in the animals' nightly activity period.

was 20–30% higher in SHR than in WKY under all experimental conditions except after uncoupling of G-proteins by Mn^{2+} (FIG. 4). In cardiac tissue from both strains, significant 12-hour rhythms were observed in basal adenylyl cyclase activity and after stimulation by GTP, isoprenaline (IPN) and GTP, and a water-soluble forskolin derivative (FOR), respectively. Peaks in basal and stimulated cAMP formation occurred around 7 AM and 7 PM, being 6 hours apart from those in β-adrenoceptor density. The addition of Mn^{2+} instead of Mg^{2+} abolished the circadian rhythmicity in adenylyl cyclase activity (FIG. 4), indicating that G-proteins must be involved in the circadian regulation of cardiac cAMP formation. The discrepancy between the effects of FOR and those of Mn^{2+}, both of which directly activate the catalytic subunit of the adenylyl cyclase, can be explained by the fact that coupling of $G_s\alpha$ to

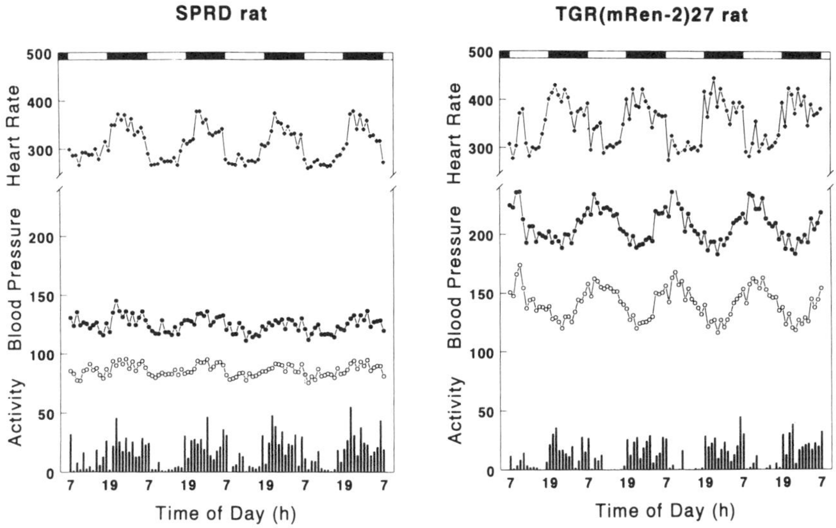

FIGURE 2. Circadian rhythms in motor activity (*filled bars*), systolic blood pressure (*closed circles*), diastolic blood pressure (*open circles*), and heart rate (*closed diamonds*) in a single SPRD (**left panel**) and TGR rat (**right panel**). In both animals, all parameters exhibited significant circadian rhythmicity with dominant period lengths of 24 hours. In the SPRD rat all parameters peaked in the animal's nightly activity period, whereas in the TGR rat rhythms in blood pressure were inverse to those in heart rate and motor activity.

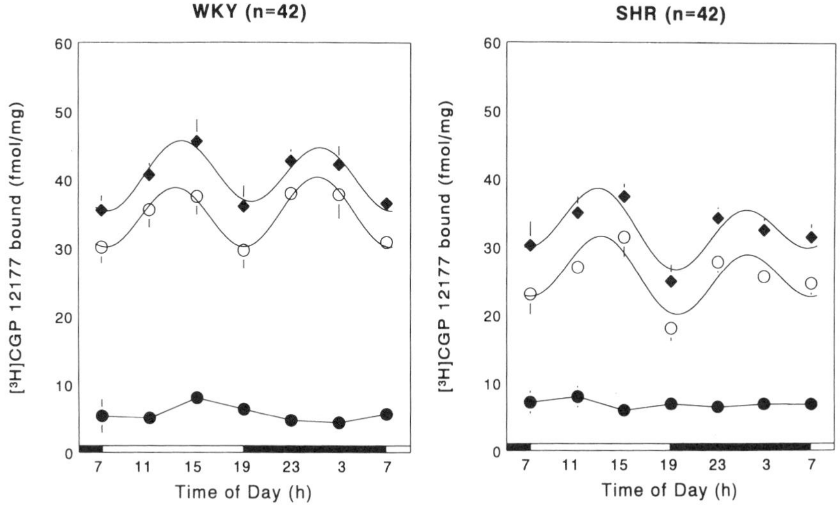

FIGURE 3. Ultradian rhythmicity in the densities of total β-adrenoceptors (*closed diamonds*) and of the β_1-subtype (*open circles*) in heart ventricles from WKY (**left panel**) and SHR rats (**right panel**) sacrificed at different times of day. The β_2-subtype (*closed circles*) did not show significant time-dependent variation. The 24-hour means in cardiac β- and β_1-adrenoceptor densities were significantly lower in SHR than WKY rats, whereas those in β_2-adrenoceptor density did not differ between strains. Redrawn from ref. 19 with permission.

the catalytic subunit potentiates the stimulatory effect of forskolin, probably by increasing the number of high affinity forskolin binding sites.[22] Because coupling of G-proteins is disrupted by the addition of Mn^{2+} ions,[23] the resulting stimulation of cAMP formation solely reflects the direct activation of the catalytic subunit.

The observation that peaks in cAMP coincided with troughs in β-adrenoceptor density could indicate that cAMP-dependent desensitization and downregulation of β-adrenoceptors are involved in the circadian regulation of receptor density. In TGR the densities of cardiac β-adrenoceptors and the β_1- and β_2-subtypes did not differ from those in normotensive SPRD. Neither β-adrenoceptor density (FIG. 5) nor adenylyl cyclase activity (FIG. 6) displayed significant time-dependent variation in

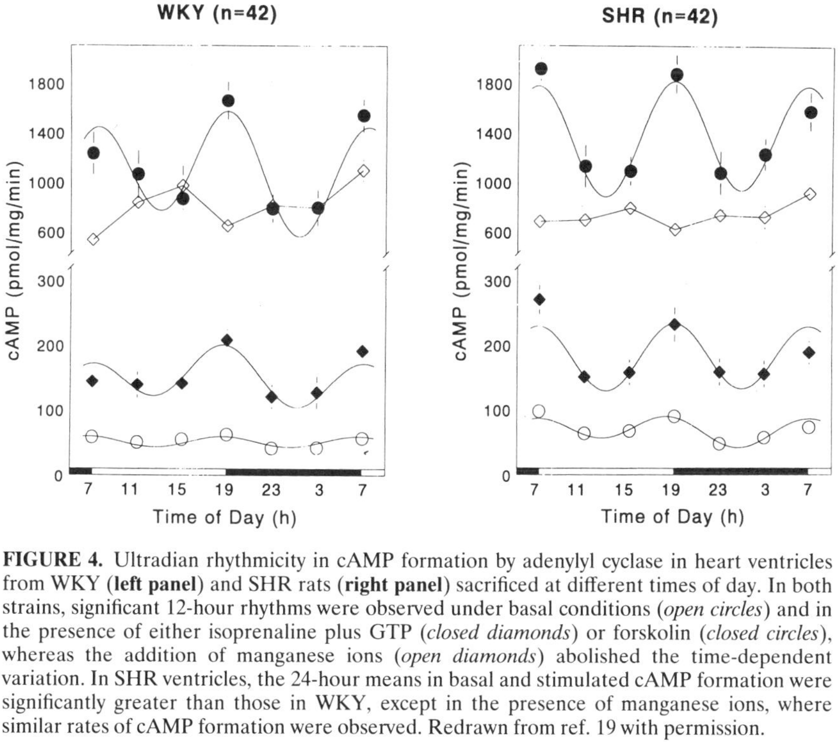

FIGURE 4. Ultradian rhythmicity in cAMP formation by adenylyl cyclase in heart ventricles from WKY (**left panel**) and SHR rats (**right panel**) sacrificed at different times of day. In both strains, significant 12-hour rhythms were observed under basal conditions (*open circles*) and in the presence of either isoprenaline plus GTP (*closed diamonds*) or forskolin (*closed circles*), whereas the addition of manganese ions (*open diamonds*) abolished the time-dependent variation. In SHR ventricles, the 24-hour means in basal and stimulated cAMP formation were significantly greater than those in WKY, except in the presence of manganese ions, where similar rates of cAMP formation were observed. Redrawn from ref. 19 with permission.

ventricles from TGR and SPRD. Interestingly, despite similar β-adrenoceptor densities basal and stimulated cAMP formation was significantly reduced by 20–30% in cardiac tissue from TGR than from SPRD.

In summary, 12-hour rhythms in myocardial β-adrenoceptor density and adenylyl cyclase activity were observed in WKY and SHR, but no rhythmic variation was found in SPRD and TGR. There is no obvious explanation for these conflicting data in different strains of rats, but it is of interest that of all strains studied, SPRD rats, showing no significant rhythmicity in cardiac β-adrenergic signaling, exhibited the lowest degree of circadian rhythmicity in systolic and diastolic blood pressure.[24] However, in all four strains studied blood pressure and heart rate display dominant

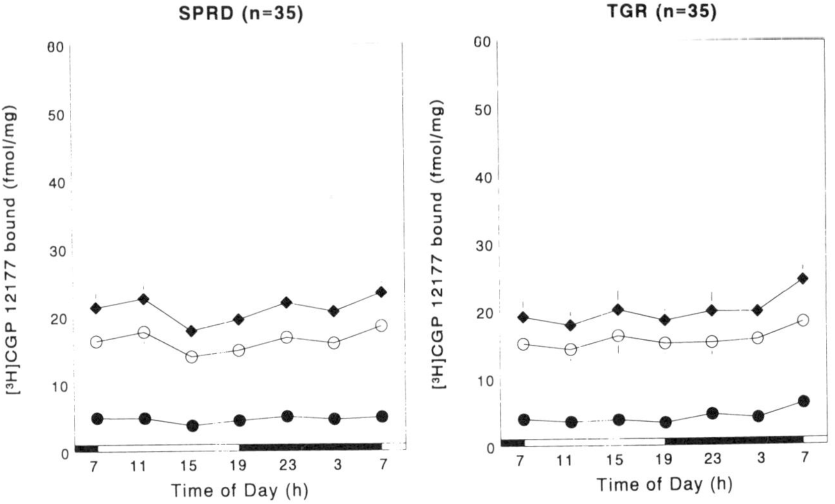

FIGURE 5. Lack of time-dependent variation in the densities of total β-adrenoceptors (*closed diamonds*) and of the β_1- (*open circles*) and β_2-subtypes (*closed circles*) in heart ventricles from SPRD (**left panel**) and TGR rats (**right panel**) sacrificed at different times of the day. The 24-hour means in cardiac β-, β_1-, and β_2-adrenoceptor densities did not differ between strains.

24-hour rather than 12-hour rhythms, making it unlikely that 12-hour rhythms in the cardiac β-adrenoceptor adenylyl cyclase system of WKY and SHR could be correlated directly to circadian variation in blood pressure. The observation that peaks in cAMP formation in ventricles from WKY and SHR were observed at the times of lights on and off points to the importance of these critical time points, when the most rapid and pronounced changes in blood pressure and heart rate occur. In a chronopharmacologic study of antihypertensive agents in SHR, Janssen *et al.*[25] observed that blood pressure reductions by propranolol and metoprolol were significantly greater after acute intravenous injection at 4 PM than at 10 AM. With regard to the results of our *in vitro* experiments in heart ventricles from SHR, injection of a β-adrenoceptor antagonist at 4 PM would be equivalent to 3 hours before a peak in cardiac cAMP formation, and injection at 10 AM would be 3 hours before a trough. Although it is difficult to combine the results obtained by different researchers in different sets of SHR, the correlation between our *in vitro* data and the findings reported by Janssen *et al.*[25] could indicate a possible linkage between rhythms in cardiac cAMP formation and *in vivo* sensitivities of the cardiovascular system towards β-adrenoceptor antagonists.

Down-regulation of cardiac β-adrenoceptors was observed in primary hypertensive SHR, which is in agreement with results obtained by other groups,[26–29] whereas in secondary hypertensive TGR, cardiac β-adrenoceptor density was unaffected. Interestingly, secondary hypertensive TGR exhibited a decrease in basal and stimulated myocardial cAMP formation despite unchanged β-adrenoceptor densities, whereas SHR showed decreased β-adrenoceptor densities but increased rates of cAMP formation. These findings clearly demonstrate that changes in one step of the complex signal transduction cascade do not necessarily reflect parallel changes in other parts of the system.

VASCULAR ADENYLYL CYCLASE

The vascular cAMP system could be another candidate for the circadian regulation of blood pressure. Unfortunately, no data on vascular cAMP content and/or formation with regard to the circadian time have been published.

In our studies in vascular tissue from WKY and SHR as well as SPRD and TGR, adenylyl cyclase activity showed neither circadian variation nor disease-related differences (data not shown). Due to the need for a certain amount of tissue for the *in vitro* experiments, these studies were performed in the descending part of the thoracic aorta. However, because other parts of the vascular bed, especially the small arteries and arterioles, are of greater importance in the regulation of peripheral resistance, the lack of circadian variation in aortic cAMP formation does not definitely rule out involvement of vascular adenylyl cyclase in circadian blood pressure regulation. Additional data obtained in more peripheral vascular regions are needed to elucidate the final contribution of vascular cAMP to the circadian regulation of the cardiovascular system.

VASCULAR SOLUBLE GUANYLYL CYCLASE

Vascular tissue has been found to contain both a soluble guanylyl cyclase, which can be stimulated by nitric oxide, and a particular guanylyl cyclase activated by atrial natriuretic peptide.[30,31] In our studies in primary and secondary hypertensive rats, we

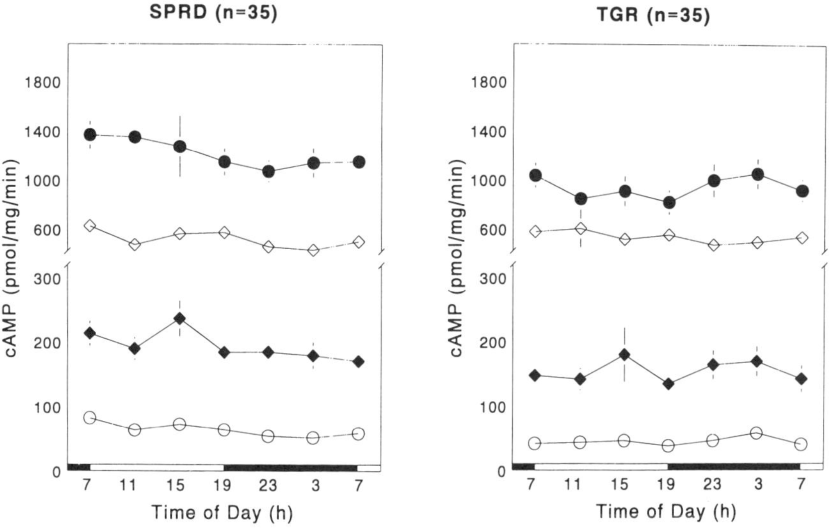

FIGURE 6. Lack of time-dependent variation in cAMP formation by adenylyl cyclase in heart ventricles from SPRD (**left panel**) and TGR rats (**right panel**) sacrificed at different times of the day. In TGR ventricles, 24-hour means in adenylyl cyclase activity were significantly lower than in SPRD ventricles under basal conditions (*open circles*) and in the presence of either isoprenaline plus GTP (*closed diamonds*) or forskolin (*closed circles*), whereas the addition of manganese ions (*open diamonds*) abolished the strain-dependent differences.

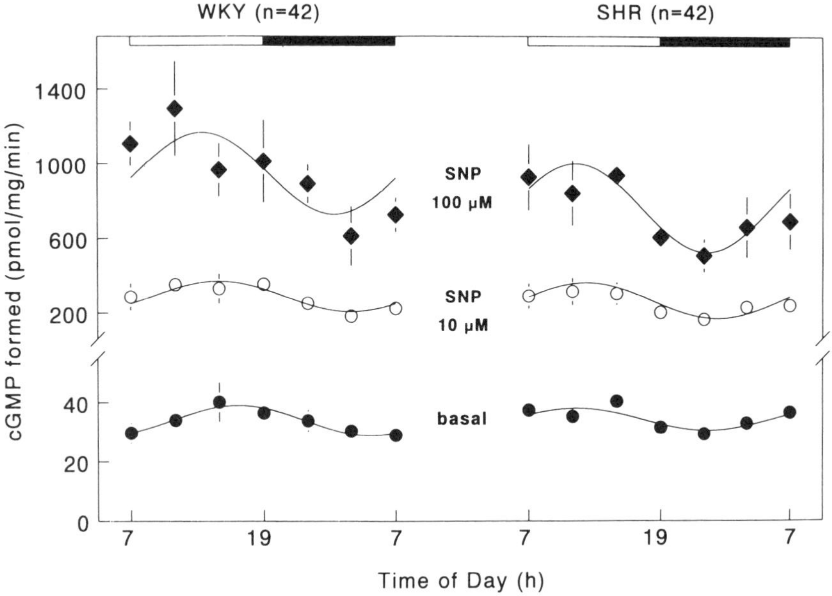

FIGURE 7. Circadian rhythms in basal (*closed circles*) and stimulated cGMP formation by soluble guanylyl cyclase in aortic tissue from WKY (**left**) and SHR rats (**right**). Soluble guanylyl cyclase was stimulated by the addition of the nitric oxide donor sodium nitroprusside (SNP) in concentrations of 10 μM (*open circles*) and 100 μM (*closed diamonds*). In both strains, basal and stimulated cGMP formation peaked in the animals' daily resting period. In the presence of SNP 100 μM, the 24-hour mean in guanylyl cyclase activity was significantly lower in SHR than in WKY.

focused on vascular soluble guanylyl cyclase, which was studied under basal conditions and after stimulation by the nitric oxide donor sodium nitroprusside (SNP).

In aortic tissue from SHR, basal formation of cGMP did not differ from that in normotensive WKY controls. After the addition of 100 μM SNP the 24-hour mean in guanylyl cyclase activity was significantly lower in vascular tissue from the hypertensive strain, indicating a lower sensitivity of SHR's aortic guanylyl cyclase towards nitric oxide (FIG. 7). In both strains basal and stimulated cGMP formation exhibited significant circadian rhythmicity with a period length of 24 hours and acrophases in the middle of the animals' daily resting period (FIG. 7). Because cGMP is a vasodilating second messenger, increased formation of cGMP during the day could result in lower blood pressure and thus contribute to its circadian regulation. To test the hypothesis that nitric oxide-stimulated guanylyl cyclase could be involved in blood pressure rhythmicity, we studied the effects of the nitric oxide synthase inhibitor Nω-nitro-L-arginine methyl ester (L-NAME) after acute and chronic *in vivo* treatment in WKY rats.[20] Acute inhibition of endogenous nitric oxide synthesis in WKY resulted in pronounced and dose-dependent increases in systolic and diastolic blood pressure (FIG. 8). After the highest dose of L-NAME (30 mg/kg, ip) a markedly greater increase in blood pressure was observed when the drug was given at 7 AM than at 7 PM. Chronic application of L-NAME via drinking water led to a dose-dependent upward shift of the circadian blood pressure profile in WKY rats.

Only at the highest dose of L-NAME applied chronically (100 mg/kg/d) could a reduction in the morning blood pressure fall be observed.[20] Thus, in WKY rats circadian time dependency in the effects of acute inhibition of nitric oxide synthesis was correlated to rhythmicity in the *in vitro* sensitivity of soluble guanylyl cyclase towards nitric oxide. These results support the hypothesis that the nitric oxide—guanylyl cyclase pathway could indeed be involved in circadian blood pressure regulation. Our finding that during chronic application of L-NAME circadian blood pressure patterns were only slightly flattened instead of being reversed could, at least partly, be due to circadian drinking behavior of the rats, which are known to drink more than 90% of their daily water intake at night.[32] Thus, fluctuating plasma concentrations of L-NAME, which in our study was applied in drinking water, will probably have resulted. Nevertheless, increases in blood pressure by chronic L-NAME were similar during the day, the time of low drug intake, and at night, when 90% of the drug was presumably ingested. This observation could indicate that during chronic inhibition of nitric oxide synthesis a time-dependent variation also exists in the sensitivity of the target system. Further studies in which different, clearly defined circadian times of chronic L-NAME application are compared will help to definitely answer that question.

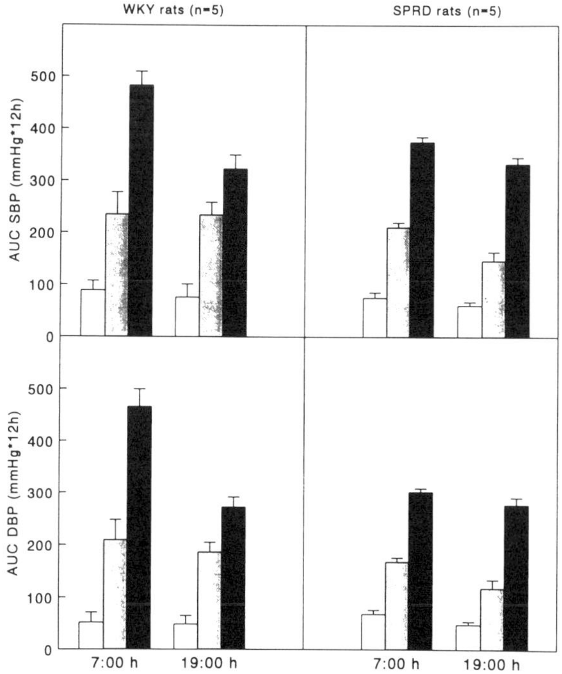

FIGURE 8. Effects of the nitric oxide synthase inhibitor Nω-nitro-L-arginine methyl ester (L-NAME) on systolic (**top**) and diastolic (**bottom**) blood pressure in normotensive WKY (**left**) and SPRD rats (**right**). L-NAME was injected ip in doses of 3 (*open bars*), 10 (*gray bars*), and 30 mg/kg (*black bars*) either at 7 AM or at 7 PM. Bivariate analysis of variance revealed that in WKY the effects of L-NAME depended on both the dose and the circadian time of injection, whereas in SPRD injection of L-NAME resulted in a dose-dependent increase in blood pressure, which did not depend on the dosing time.

In SPRD and TGR neither basal nor nitric oxide-stimulated soluble guanylyl cyclase exhibited significant circadian variation (FIG. 9). Interestingly, the 24-hour mean in cGMP formation was lower in TGR than in SPRD under basal conditions as well as after stimulation by SNP (FIG. 9). Thus, besides differences between SPRD and TGR in the activity of the renin-angiotensin system,[33] additional regulatory systems such as the vascular nitric oxide—guanylyl cyclase pathway seem to be affected by the transgene and/or by the secondary hypertension itself. With regard to circadian blood pressure regulation, however, the vascular soluble guanylyl cyclase is less likely to play an important role in SPRD and TGR. In the light of these *in vitro* findings one could reasonably expect that in SPRD rats injection of L-NAME at

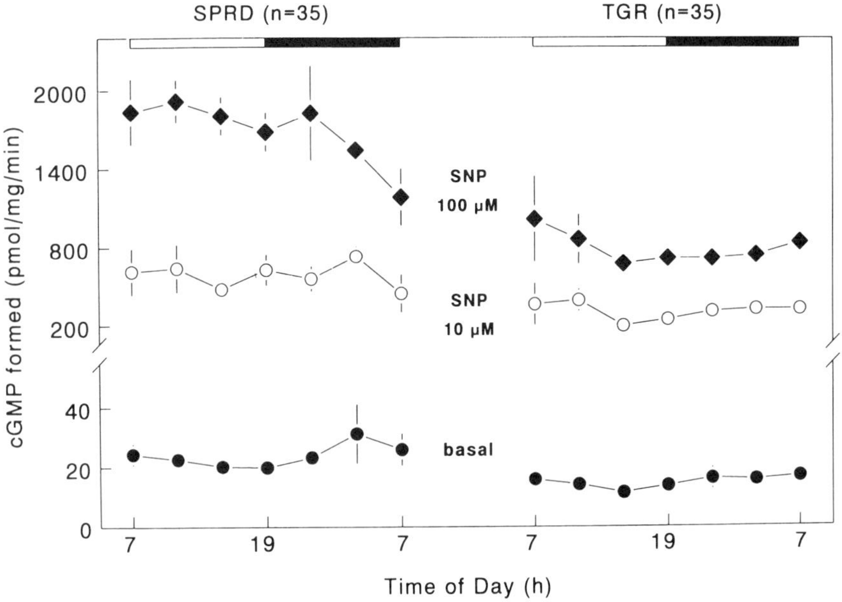

FIGURE 9. Lack of time-dependent variation in basal (*closed circles*) and stimulated cGMP formation by soluble guanylyl cyclase in aortic tissue from SPRD (**left**) and TGR rats (**right**). Soluble guanylyl cyclase was stimulated by the addition of the nitric oxide donor sodium nitroprusside (SNP) in concentrations of 10 μM (*open circles*) and 100 μM (*closed diamonds*). In aortic tissue from TGR, 24-hour means in basal and stimulated cGMP formation were significantly lower than in that from SPRD rats.

different circadian times should result in similar blood pressure increases. Indeed, application of the same doses of L-NAME as those given to WKY rats resulted in a dose-dependent increase in blood pressure that was not influenced by the time of injection (FIG. 8).

VASCULAR PHOSPHOLIPASE C

Because several transmitters and hormones elicit their intracellular effects via inositol-trisphosphate mediated increases in intracellular calcium, such as noradrenaline via α_1-adrenoceptors and angiotensin II via AT_1-receptors, changes in the formation of inositol-trisphosphate by phospholipase C could be involved in cardiovascular circadian regulation. Unfortunately, no data on circadian variation in this signal transduction system are available in the literature. Indirect evidence, however, is given by the aforementioned study on circadian time-dependency in the vasoconstrictor effects of noradrenaline in rat aortic tissue. In their study, Gohar *et al.*[16] determined the sensitivity towards noradrenaline in rat aortic rings, which were prepared at different times of day. They found significant ultradian rhythmicity in the vasoconstrictor potency of noradrenaline with a period length of approximately 12 hours, peaks occurring around 4 PM and 2 AM. Since in their *in vitro* experiments the agonist was exogenously added, only changes in α_1-adrenoceptors and/or their signal transduction pathway can be responsible for the rhythms in drug sensitivity observed.[16]

In our animal model of secondary hypertension, TGR, the converting enzyme inhibitor enalapril[34] and the AT_1-receptor antagonist losartan[35] reduced blood pressure in hypertensive transgenic TGR(mRen-2)27 rats more effectively during the day than at night, which could be due to circadian rhythmicity in the activity of the renin-angiotensin system and/or to time-dependent variation in vascular AT_1-receptor density and functional efficacy.

Thus, studies on circadian regulation of inositol-trisphosphate linked signal transduction systems should be performed to get more insight into the complex temporal regulation of vascular resistance.

SUMMARY AND CONCLUSIONS

Experiments in inbred strains of normotensive and hypertensive rats have clearly demonstrated circadian rhythms in blood pressure and heart rate. Pre- and postsynaptic signal transduction processes *in vitro* can, but need not, vary with circadian time, greatly depending on the strain of rats investigated. These data highlight the notion of a strain-dependent, and thus genetic, regulation of the cardiovascular system. Obviously, circadian rhythms in blood pressure cannot be explained by single biochemical parameters, but results from both *in vitro* and *in vivo* studies give first evidence that the vascular nitric oxide—cGMP system may be involved in the circadian regulation of blood pressure in WKY and SHR rats. In secondary hypertensive TGR and in their normotensive controls, SPRD, the guanylyl cyclase system does not seem to play a role in circadian blood pressure regulation. In neither of the four strains studied did aortic adenylyl cyclase show any time-dependent variation. Because vascular tissue was taken from the thoracic aorta of the rats, a contribution of adenylyl cyclase to circadian blood pressure regulation in small resistance arteries cannot be ruled out. Further studies in different parts of the vascular tree are needed to definitely answer that question. No data are available on time-dependent variation in the activity of phospholipase C, the second messenger pathway of vascular α-adrenoceptors and angiotensin II AT_1-receptors, both of which mediate vasoconstriction. Future research into this system will be helpful in identifying mechanisms involved in blood pressure regulation in SPRD and TGR.

ACKNOWLEDGMENT

The authors are indebted to Anja Müller, Stephan Höfer, and Karsten Perbandt, who performed the biochemical studies in Sprague-Dawley and transgenic hypertensive rats as part of their doctoral thesis.

REFERENCES

1. WURTMANN, R. J. & J. AXELROD. 1966. A 24 hr rhythm in the content of norepinephrine in the pineal and salivary glands of the rat. Life Sci. **5:** 665–669.
2. FRIEDMAN, A. H. & C. A. WALKER. 1968. Circadian rhythms in rat midbrain and caudate nucleus biogenic amine levels. J. Physiol. (London) **197:** 77–85.
3. SCHEVING, L. E., W. H. HARRISON, P. GORDON & J. E. PAULY. 1968. Daily fluctuation (circadian and ultradian) in biogenic amines of the rat brain. Am. J. Physiol. **214:** 166–173.
4. LEMMER, B. & K. WENDA. 1979. Diurnal rhythms in noradrenaline turnover and motility after reserpine and 6-hydroxydopamine. Pharmacol. Biochem. Behav. **10:** 361–368.
5. LEMMER, B. & R. SALLER. 1974. Influence of light and darkness on the turnover of noradrenaline in the rat heart. Naunyn-Schmiedeberg's Arch. Pharmacol. **282:** 75–84.
6. LEMMER, B., P. H. LANG, S. SCHMIDT & H. BÄRMEIER. 1987. Evidence for circadian rhythmicity of the β-adrenoceptor-adenylate cyclase-cAMP-phosphodiesterase system in the rat. J. Cardiovasc. Pharmacol. **10** (Suppl. 4): S138–S140.
7. LEMMER, B. & K. WITTE. 1989. Circadian variation of the in vitro stimulation of adenylate cyclase in rat heart tissue. Eur. J. Pharmacol. **159:** 311–315.
8. LEMMER, B. & C. WALD. 1990. Influence of age on the rhythm in basal and forskolin-stimulated adenylate cyclase activity of the rat heart. Chronobiol. Int. **7:** 107–111.
9. LANG, P. H., H. BISSINGER & B. LEMMER. 1985. Circadian rhythm and seasonal variations in basal cAMP content of rat heart ventricles. Chronobiol. Int. **2:** 41–45.
10. PETERS, R. W., J. E. MULLER, S. GOLDSTEIN, R. BYINGTON & L. M. FRIEDMAN FOR THE BHAT STUDY GROUP. 1989. Propranolol and the morning increase in the frequency of sudden cardiac death (BHAT study). Am J Cardiol **63:** 1518–1520.
11. MULLER, J. E., P. H. STONE, Z. G. TURI, J. D. RUTHERFORD, C. A. CZEISLER, C. PARKER, W. K. POOLE, E. PASSAMANI, R. ROBERTS, E. BRAUNWALD & THE MILIS STUDY GROUP. 1985. Circadian variation in the frequency of onset of acute myocardial infarction. N. Engl. J. Med. **313:** 1315–1322.
12. WILLICH, S. N., T. LINDERER, K. WEGSCHEIDER, A. LEIZOROWICZ, I. ALAMERCERY, R. SCHRODER & THE ISAM STUDY GROUP. 1989. Increased morning incidence of myocardial infarction in the ISAM study: Absence with prior β adrenergic blockade. Circulation **80:** 853–858.
13. MULCAHY, D., D. CUNNINGHAM, P. CREAN, C. WRIGHT, J. KEEGAN, A. QUYYUMI, A. PARK & K. FOX. 1988. Circadian variation of total ischaemic burden and its alteration with anti-anginal agents. Lancet **II:** 755–759.
14. PANZA, J. A., S. E. EPSTEIN & A. A. QUYYUMI. 1991. Circadian variation in vascular tone and its relation to α-sympathetic vasoconstrictor activity. N. Engl. J. Med. **325:** 986–990.
15. HOSSMANN, V., G. A. FITZGERALD & C. T. DOLLERY. 1980. Circadian rhythm of baroreflex reactivity and adrenergic vascular response. Cardiovasc. Res. **14:** 125–129.
16. GOHAR, M., P. DALEAU, J. ATKINSON & Y. M. GARGOUIL. 1992. Ultradian variations in sensitivity of rat aorta rings to noradrenaline. Eur. J. Pharmacol. **229:** 69–73.
17. MIDDEKE, M. & J. SCHRADER. 1994. Nocturnal blood pressure in normotensive subjects and those with white coat, primary, and secondary hypertension. Br. Med. J. **308:** 630–632.
18. LEMMER, B., A. MATTES, M. BÖHM & D. GANTEN. 1993. Circadian blood pressure variation in transgenic hypertensive rats. Hypertension **22:** 97–101.
19. WITTE, K., R. PARSA-PARSI, M. VOBIG & B. LEMMER. 1995. Mechanisms of the circadian regulation of β-adrenoceptor density and adenylyl cyclase activity in cardiac tissue from normotensive and spontaneously hypertensive rats. J. Mol. Cell. Cardiol. **27:** 1195–1202.
20. WITTE, K., A. SCHNECKO, P. ZUTHER & B. LEMMER. 1995. Contribution of the nitric oxide-guanylyl cyclase system to circadian regulation of blood pressure in normotensive Wistar-Kyoto rats. Cardiovasc. Res. **30:** 682–688.
21. MATTES, A., K. WITTE, W. HOHMANN & B. LEMMER. 1991. Pharmfit—A nonlinear fitting program for pharmacology. Chronobiol. Int. **8:** 460–476.
22. SUTKOWSKI, E. M., W. J. TANG, C. W. BROOME, J. D. ROBBINS & K. B. SEAMON. 1994.

Regulation of forskolin interactions with type I, II, V, and VI adenylyl cyclases. Biochemistry **33:** 12852–12859.
23. LIMBIRD, L. E., A. R. HICKEY & R. J. LEFKOWITZ. 1979. Unique uncoupling of the frog erythrocyte adenylate cyclase system by manganese ions. J. Biol. Chem. **254:** 2677–2683.
24. LEMMER, B., K. WITTE, D. MINORS & J. WATERHOUSE. 1995. Circadian rhythms of heart rate and blood pressure in four strains of rats: Differences due to, and separate from, locomotor activity. Biol. Rhythm Res. **26:** 493–504.
25. JANSSEN, B. J. A., J. OOSTING, C. M. TYSSEN & H. A. J. STRUYKER-BOUDIER. 1993. Time-dependent efficacy of antihypertensive agents in spontaneously hypertensive rats. Chronobiol. Int. **10:** 420–434.
26. ROBBERECHT, P., J. WINAND, P. CHATELAIN, P. POLOCZEK, J. C. CAMUS, P. DE NEEF & J. CHRISTOPHE. 1981. Comparison of β-adrenergic receptors and the adenylate cyclase system with muscarinic receptors and guanylate cyclase activities in the heart of spontaneously hypertensive rats. Biochem. Pharmacol. **30:** 385–387.
27. YAMADA, S., T. ISHIMA, T. TOMITA, M. HAYASHI, T. OKADA & E. HAYASHI. 1984. Alterations in cardiac alpha and beta adrenoceptors during the development of spontaneous hypertension. J. Pharmacol. Exp. Ther. **228:** 454–460.
28. CASTELLANO, M., M. BESCHI, D. RIZZONI, M. PAUL, M. BÖHM, G. MANTERO, G. BETTONI, E. PORTERI, A. ALBERTINI & E. AGABITI-ROSEI. 1993. Gene expression of cardiac $beta_1$-adrenergic receptors during the development of hypertension in spontaneously hypertensive rats. J. Hypertens. **11:** 787–791.
29. BÖHM, M., M. CASTELLANO, M. PAUL & E. ERDMANN. 1994. Cardiac norepinephrine, β-adrenoceptors, and Giα-proteins in prehypertensive and hypertensive spontaneously hypertensive rats. J. Cardiovasc. Pharmacol. **23:** 980–987.
30. ANAND-SRIVASTAVA, M. B. & G. J. TRACHTE. 1993. Atrial natriuretic factor receptors and signal transduction mechanisms. Pharmacol. Rev. **45:** 455–497.
31. SCHMIDT, H. H. H. W., S. M. LOHMANN & U. WALTER. 1993. Minireview. The nitric oxide and cGMP signal transduction system: Regulation and mechanism of action. Biochem. Biophys. Acta **1178:** 153–175.
32. LEMMER, B. & L. HOLLE. 1991. Chronopharmacokinetics of imipramine and desipramine in rat forebrain and plasma after single and chronic treatment with imipramine. Chronobiol. Int. **8:** 176–185.
33. TOKITA, Y., R. FRANCO-SAENZ, E. M. REIMANN & P. J. MULROW. 1994. Hypertension in the transgenic rat TGR(mRen-2)27 may be due to enhanced kinetics of the reaction between mouse renin and rat angiotensinogen. Hypertension **23:** 422–427.
34. LEMMER, B., K. WITTE, T. MAKABE, D. GANTEN & A. MATTES. 1994. Effects of enalaprilat on circadian profiles in blood pressure and heart rate of spontaneously and transgenic hypertensive rats. J. Cardiovasc. Pharmacol. **23:** 311–314.
35. SCHNECKO, A., K. WITTE & B. LEMMER. 1995. Effects of the angiotensin II receptor antagonist losartan on 24-hour blood pressure profiles of primary and secondary hypertensive rats. J. Cardiovasc. Pharmacol. **26:** 214–221.

Vascular Mechanisms of Blood Pressure Rhythms

EDOARDO CASIGLIA

Department of Clinical and Experimental Medicine
University of Padua
Padua, Italy

The circadian rhythm of arterial blood pressure (BP) with higher values during the day than at night is mainly, but not exclusively, related to physical and mental activity. Although it is more marked in subjects who are active during the day[2] and inverted in shift workers,[3,4] it can be detected even after 24 hours of bedrest.[5,6]

The circadian rhythm of peripheral resistance (PR) is practically unknown, as PR calculation requires the recording of peripheral flow, which is difficult to obtain. Capillary diameter and erythrocyte speed may be an indirect measure of arterial flow in animals[7] and radioactive isotope clearances in man,[8–12] but these techniques can only detect cutaneous and subcutaneous flow and are unsuitable for muscular flow, which is predominant at rest.

FIRST PILOT STUDY WITH PLETHYSMOGRAPHY

Our group measures arterial flow with an automatic venous-occlusion strain-gauge plethysmographic fluximeter (Angiomed, Microlab, Padua, Italy),[13] whose main advantage is that once occlusion pressure is programmed, it is automatically repeated at each determination without any further intervention by the operator. Occlusion and deflation times and intervals between occlusions are also set up at the beginning of the procedure. At each 15-minute interval, four measurements are performed and averaged automatically.

To standardize the results of our experiments, all activities, such as reading, talking, watching television, and listening to music, always take place at the same time of day for all subjects, changes of position in bed are minimized (and not allowed at all during flow measurements), and subjects are not allowed to get out of bed to void. We also use an orthopedic device to keep still the arm and leg employed for flow measurement.

Blood pressure is usually measured by a TM-2420 model 7 ambulatory blood pressure monitoring system (A&D, Tokyo, Japan)[14,15] on the left arm, providing at 15-minute fixed intervals the automatic inflation of the cuff synchronous with flow measurement. Forearm resistance is then calculated at each 15-minute step as mean BP/flow ratio and expressed as units of resistance ($UR = mm\ Hg \cdot min \cdot dl \cdot min^{-1}$).

The study is begun at 2 PM, but the period between 2 and 4 PM is usually not taken into consideration in the analysis of results because it is disturbed by the alert reaction during application and setting up of equipment. All subjects are closely observed during the 22 hours of the experiment, and the sleeping and waking times are recorded.

Using this method, we previously showed that a circadian rhythm of PR does exist in normal subjects confined to bed.[13] FIGURE 1 plots data of our first experience in eight normotensive volunteers. As can be seen, BP was on average lower at night (systolic -9%, diastolic -4.5%, both $p < 0.0001$ vs daytime), and leg PR tracing paralleled that of BP (nighttime −36% in comparison with daytime BP, $p < 0.0001$). By contrast, leg arterial flow averaged 28% higher during sleep than during waking time. These findings confirm that a wake/sleep rhythm independent of physical activity also exists for peripheral arterial flow, at least in normotensive subjects.

Findings similar to those we described were obtained by Sindrup *et al.*[12] in healthy subjects and by Jelnes *et al.*[11] in patients with obliterant arteriopathy, although they could only measure cutaneous and subcutaneous blood flow and their subjects were unrestricted. By contrast, in our study, leg flow was detected by a method that measures total arterial flow and subjects were constantly in a clinostatic posture; the hemodynamic differences that we found could therefore only be attributed to sleep. The higher daytime PR was due purely to neural and humoral factors and appeared to be a counterregulatory adjustment fit to maintain the BP levels characteristic of waking hours. As cardiac output decreased during sleep, the increase in leg blood flow observed at night implied blood redistribution to the legs

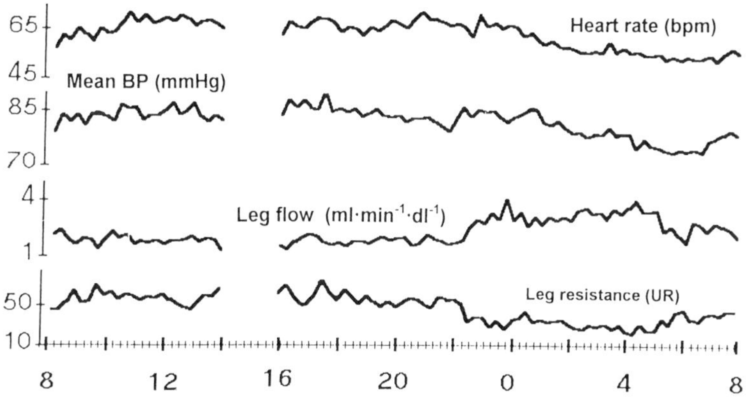

FIGURE 1. Mean blood pressure (BP), heart rate, measured leg arterial flow (calf), and calculated leg resistance in eight normal volunteers confined to bed for 24 hours. Time to 2–4 PM was not taken into consideration because it was disturbed by alert reaction due to application and setting up of equipment. Sleep period started on average at 11 PM and finished at 4:30 AM.

due to vasodilation. Hoobler *et al.*[16] in 1955 demonstrated that a sharp, pharmacologically induced decline in BP and cardiac output may actually be accompanied by an increase, rather than a decrease, in limb arterial flow,[16] and continuous flow redistribution was experimentally demonstrated in both men (at least in the kidney[17,18] and liver[19]) and animals.[20]

VALIDATION OF METHOD AND PROCEDURES AND REPEAT PERIPHERAL RESISTANCE MONITORING

The venous occlusion plethysmographic method has generally been widely validated.[13,21–23] In the Angiomed device that we employ, values of forearm flow are cleared from abnormal ones by inner software.[13]

To test the reproducibility of the method throughout the 24 hours, we also

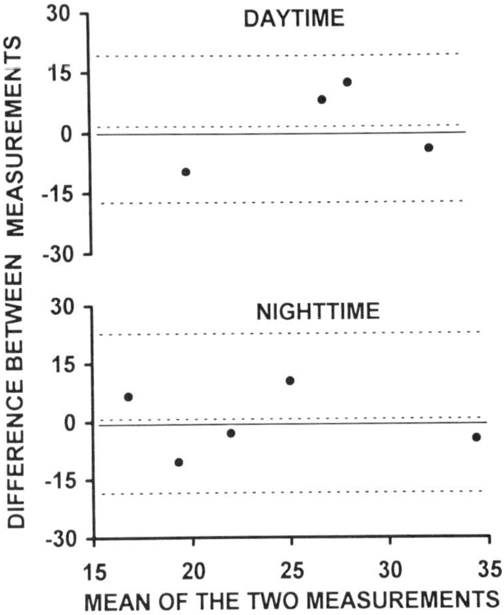

FIGURE 2. Reproducibility of daytime and nighttime forearm resistance in two measurements at a 1-week interval (Bland-Altman method)[24,25] in four normotensive volunteers.

repeated bed-resting forearm flow monitoring at 1-week intervals in four normotensive subjects and assessed the agreement of the two recordings. In FIGURE 2, for each patient the difference between the values obtained in the two days are plotted against the mean of the two measurements according to the Bland-Altman method.[24,25] Good agreement between baseline and repeat recordings is shown. In FIGURE 3, the

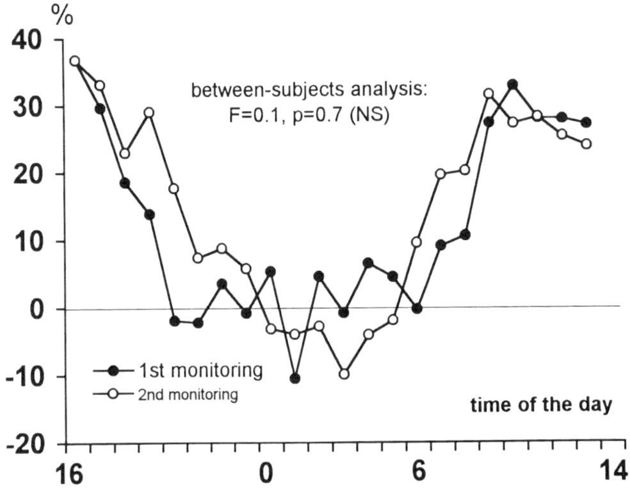

FIGURE 3. Twenty-two-hour forearm resistance variations from baseline (2 PM) in the same subjects as in FIGURE 2. Curves are similar, and between-subject analysis showed no difference between the first and the repeat monitoring.

two 24-hour PR recordings at 1-week intervals are also plotted as percentage variations from baseline, demonstrating almost perfect coincidence of the two tracings.

Some doubt remained over whether the BP and PR reduction that we observed was due to an aspecific effect of rest rather than to sleep. To clarify this question, we repeat the 24-hour BP and leg flow monitoring in seven subjects starting at 8 AM rather than 2 PM. As shown in FIGURE 4, the wake/sleep flow trend was the same in the two groups, independently of the time of day the study was begun, indicating that sleep is really the cause of the PR nocturnal decrease.

The TM-2420 system provides two microphones, one over the brachial artery on the antecubital region for detecting the Korotkow sounds and the other over the brachial muscular groups for rejecting the muscular noises. Its accuracy has been widely demonstrated.[14,15]

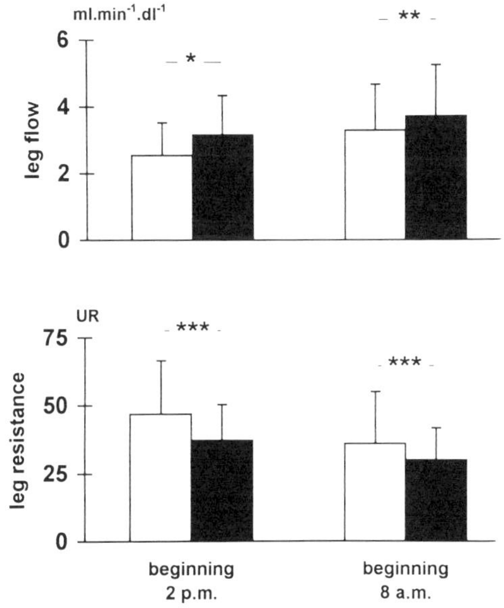

FIGURE 4. Mean ± SD of leg arterial flow and leg resistance during waking (□) and sleeping (■) in seven subjects who lay down at 8 AM and in 37 who lay down at 2 PM. The wake/sleep trend was the same in both.

NORMOTENSIVE VERSUS HYPERTENSIVE SUBJECTS: THE DIPPER/NON-DIPPER PROBLEM

The possibility that circadian rhythm of PR also exists in hypertensive patients and is different from that of normotensive subjects was examined in 43 persons, 13 normotensive volunteers (7 males and 6 females) and 30 consecutive mild-to-moderate essential hypertensive patients (21 males and 9 females) never treated for hypertension, whose general characteristics are summarized in TABLE 1. Causes of secondary hypertension were previously excluded as was diabetes. Each one gave individual informed consent to the study, which began at 2 PM.

In recent years a category of hypertensive patient not showing the "physiologic" nighttime BP reduction was identified; they are usually called "non-dippers."[26,27] The general opinion is that non-dipper hypertensive patients have more severe hypertension or more pronounced target organ damage (TOD) than do dippers.[28]

Because of the small daytime/nighttime BP differences observed in subjects confined to bed,[13] we could not apply the usual criteria for definition of dipper status.[25,26,28] Preliminary ambulatory BP monitoring, during which subjects were free to move around and go home, was therefore performed 1 week before the in-bed study. On the basis of this BP recording, subjects were considered dippers if at night systolic BP decreased by 10 mm Hg or more and diastolic BP by 5 mm Hg or more in comparison with daytime BP levels.[25] When the results of this recording were compared with those of in-bed monitoring, all ambulatory dippers had lower in-bed BP values at night than during the day (i.e., in-bed dippers), whereas ambulatory non-dippers showed no difference or even a higher BP at night than during the day (in-bed non-dippers).

Electrocardiography, chest roentgenography, and fundoscopic examination were performed to attribute a score of increasing severity of TOD by means of the objective criteria described elsewhere.[29]

In normotensive subjects the average waking period was 16 ± 1 hours and the sleeping period 6 ± 1 hours ($p < 0.001$); in hypertensive subjects it was 15 ± 1 and

TABLE 1. General Characteristics of 13 Normotensive Controls and 30 Hypertensive Patients

		Hypertensive Subjects ($n = 30$)		
	Normotensive Subjects ($n = 13$)	All ($n = 30$)	Dippers ($n = 21$)	Non-dippers ($n = 9$)
Age (yr)	30.1 ± 3.7	40.9 ± 13.0*	39.8 ± 12.6	43.3 ± 14.2
BMI (kg/m^2)	23.1 ± 3.1	25.0 ± 3.6	24.9 ± 3.9	25.2 ± 2.8
Casual SBP (mm Hg)	124.1 ± 11.4	151.7 ± 18.2*	152.5 ± 19.6	149.9 ± 15.4
Casual DBP (mm Hg)	80.6 ± 7.1	97.4 ± 0.3*	98.6 ± 10.2	94.4 ± 6.6
TOD score (see text)	0	1.05 ± 0.4	0.9 ± 0.3	1.4 ± 0.7†

Abbreviations: SBP and DBP = systolic and diastolic BP (mm Hg); BMI = body mass index calculated from weight/height^2; TOD = target organ damage. Post-hoc Bonferroni's analysis: *$p < 0.001$ vs normotensive subjects. † $p = 0.01$ dipper hypertensive subjects.

7 ± 1 hours, respectively ($p < 0.001$), without any difference between normotensive and hypertensive subjects. As the time in which subjects fell asleep and woke up was variable, two truncated periods of certain wakefulness (4–8 PM of day 1; 8–11 PM of day 2) and one of certain sleep (1–4 AM) were also considered; this is a usual procedure.[30]

All normotensive volunteers were dippers and had a TOD score of 0 (TABLE 1). Average 22-hour peripheral flow was 3.7 ± 1.6 ml · min^{-1} · dl^{-1} (21% higher at night than during the day) and PR was 27 ± 12.3 UR (24% higher awake than during sleep).

In hypertensive patients, average 22-hour systolic BP was 20% higher ($p < 0.001$) and diastolic 23% higher ($p < 0.001$) than in normotensive subjects. Their daytime/nighttime BP trend was not homogeneous (TABLE 2); 21 of them (dippers) showed higher BP values during the two waking periods with a trend similar to that of the normotensive subjects, whereas the other nine turned out to be non-dippers, showing no BP difference between sleeping and waking.

Twenty-two–hour peripheral flow measured at the right forearm in hypertensive

TABLE 2. Systolic (SBP) and Diastolic Blood Pressure (DBP) in 13 Normotensive Subjects and 21 Dipper and 9 Non-dipper Hypertensive Patients[a]

	Normotensive Subjects (n = 13)			Dipper Hypertensive Subjects (n = 21)			Non-dipper Hypertensive Subjects (n = 9)			
	5–8 PM	1–4 AM	9–11 AM	5–8 AM	1–4 AM	9–11 AM	5–8 AM	1–4 AM	9–11 AM	ANOVA
SBP (mm Hg)	118.7 ± 9.9	111.7 ± 10.8*	115.2 ± 11.8†	147.6 ± 21.6	135.4 ± 18.2*	145.7 ± 10.6†	153.2 ± 16.6	147.6 ± 21.3	135.8 ± 21.3	F = 5.5 $p < 0.01$
DBP (mm Hg)	70.9 ± 7.0	64.9 ± 6.6*	70.0 ± 9.2†	90.3 ± 12.6	83.6 ± 13.0*	91.3 ± 12.7†	95.2 ± 7.9	91.7 ± 8.3	91.2 ± 7.4	F = 6.5 $p < 0.01$

[a]One truncated period of certain sleep (1–4 AM) and two of certain wakefulness (5–8 PM and 9–11 AM) are considered. *$p < 0.001$ vs 5–8 PM; †$p < 0.001$ vs 1–4 AM.

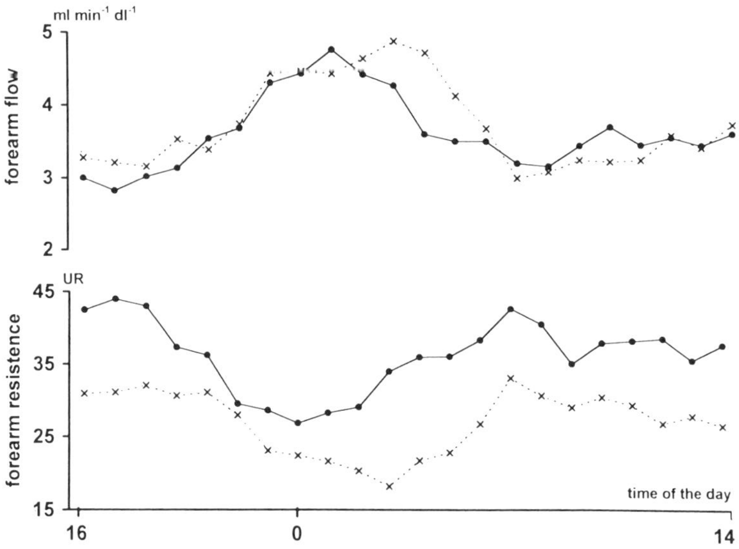

FIGURE 5. Twenty-two–hour forearm flow and resistance in 21 dipper hypertensive patients (—●—) and in 13 normotensive controls (—X—). A comparable diurnal rhythm is evident, with a sharp increase in arterial flow and a decrease in resistance during sleep in both groups.

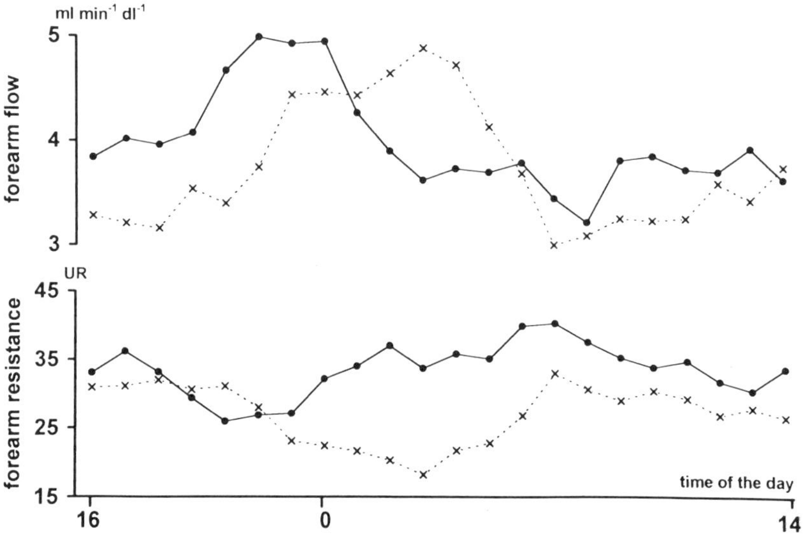

FIGURE 6. Twenty-two–hour forearm flow and resistance in 9 non-dipper hypertensive patients (—●—) and in 13 normotensive controls (—X—; same as in FIG. 5). The highest values of flow were seen in the normotensive subjects during sleep, in the hypertensive subjects during waking, and the resistance trend was opposite.

subjects was on average 3.8 ± 1.9 ml · min^{-1} · dl^{-1}, and PR was 35 ± 17.8 UR (23% higher than in normotensive subjects, $p < 0.001$). Nevertheless, the trend of flow and PR was different in dippers and non-dippers. As shown in FIGURE 5, the former (upper panel) had a flow trend similar to that of the normotensive subjects with higher values at night than during the day, whereas the behavior of non-dippers (FIG. 6, upper panel) was different from that of normotensive subjects, showing higher flow values during the day than at night. PR trend was opposite in the two groups (FIGS. 5 and 6, lower panels). The different trend of the two groups was confirmed by the significant interaction between flow and time (F = 15.9, $p < 0.0001$) and between resistance and time (F = 8.01, $p < 0.0001$).

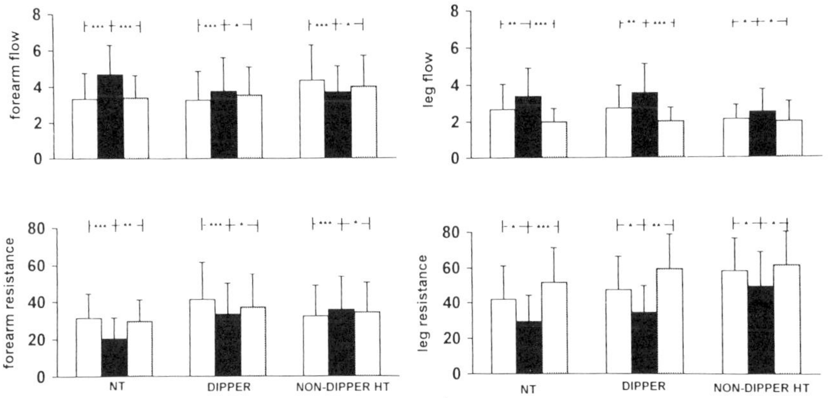

FIGURE 7. Mean ± SD of forearm and leg flow (*upper panels*), forearm and leg resistance (*lower panels*) in 13 normotensive, 21 dipper, and 9 non-dipper hypertensive subjects during two periods of certain wakfulness (□ from 5–8 PM; □ from 9–11 AM) and one of certain sleep (■ from 1–4 AM). $^*p < 0.05$; $^{**}p < 0.01$; $^{***}p < 0.001$. In the forearm (*left panels*), arterial flow significantly increased and resistance decreased during sleep, but not in the non-dipper hypertensive patients who showed an opposite trend. In the leg (*right panels*), flow increased and resistance decreased in the same way in all subjects.

Analysis of the three truncated periods provides similar results (FIG. 7, left panels). In normotensive subjects as well as in dipper hypertensive subjects, the lowest forearm flow values were detected during the two waking periods and the highest during sleep, whereas in non-dipper hypertensive subjects the highest flow was recorded during the day. In each truncated period, forearm flow was higher in both dipper and non-dipper hypertensive subjects than in normotensive subjects.

Significantly different TOD was present in the two groups of hypertensive patients (TABLE 1). Although we cannot determine if the increase in hypertensive lesions is the cause or the result of the altered hemodynamic circadian rhythm,[31–33] the more severe vascular involvement in non-dipper hypertensive subjects could well justify their inability to decrease peripheral resistance during sleep.

LEG VERSUS FOREARM

In the same subjects just described, we also measured 24-hour leg blood flow and calculated leg PR. Leg and forearm hemodynamics are compared in TABLE 3.

The aim of this study was to determine if the two limbs had a different circadian pattern.

As shown in FIGURE 7 (upper panels), leg flow increased during sleep in dipper as well as non-dipper hypertensive patients. This situation differs from that in the forearm where arterial flow, as already described, decreased during sleep in non-dipper hypertensive patients while it increased in the dippers. The PR pattern was specular to that of flow (FIG. 7, lower panels). This confirms that, as shown by Ferrari *et al.*[20] in animals, arterial blood flow and vascular resistance may vary moment to moment and in the opposite way in different regional circulations, so as to contribute to the stabilization of systemic hemodynamics. This is the first demonstration of this phenomenon in humans.

TABLE 3. Forearm and Leg Flow and Resistance in Males and Females during Two Periods of Certain Wakefulness (5–8 PM of day 1 and 8–11 AM of day 2) and One of Certain Sleep (1–4 AM)[a]

Sex	5–8 PM (Wakefulness)	1–4 AM (Sleep)	8–11 AM (Wakefulness)	22 Hours
		Forearm Flow ($ml \cdot min^{-1} \cdot dl^{-1}$)		
Males	4.1 ± 1.7	4.4 ± 1.8	3.9 ± 1.5	4.3 ± 1.8
Females	2.6 ± 1.2*	3.4 ± 1.5*,†	2.8 ± 1.2*,†	2.9 ± 1.3*
		Forearm Resistance (UR)		
Males	31.1 ± 14.1	27.0 ± 12.9†	30.9 ± 13.9†	29.1 ± 14.1
Females	44.2 ± 21.1*	35.4 ± 21.7*,†	39.7 ± 18.1*,†	39.0 ± 19.2*
		Leg Flow ($ml \cdot min^{-1} \cdot dl^{-1}$)		
Males	2.5 ± 1.1	3.3 ± 1.6†	2.0 ± 0.9†	2.7 ± 1.1
Females	2.5 ± 1.3	3.0 ± 1.3†	1.9 ± 0.6†	2.5 ± 1.1
		Leg Resistance (UR)		
Males	48.6 ± 19.9	37.9 ± 19.2†	56.8 ± 19.2†	46.3 ± 20.5
Females	48.0 ± 19.9	33.8 ± 15.0†	58.1 ± 20.4†	46.3 ± 20.9

[a] *$p < 0.001$ vs males; †$p < 0.001$ vs the immediately preceding period.

MALES VERSUS FEMALES

Finally, we compared blood flow and peripheral resistance in 23 males and 12 females. Because of the low number of non-dipper females, only normotensive and dipper hypertensive subjects were included in this last analysis.

Average 22-hour forearm flow was 32% higher ($p < 0.001$) and 22-hour forearm PR 25% lower ($p < 0.001$) in males than in females, whereas leg flow and leg PR were not different. The cause of these differences is unknown.

When the three aforementioned truncated periods were taken into consideration (TABLE 3), the usual flow increase and PR decrease during sleep were compared with those in the two waking periods in both males and females; however, in the forearm they were more pronounced in the latter than in the former.

CONCLUSIONS

Strain gauge plethysmography is a good tool to study the circadian rhythms of flow and resistance. A sleep/wake rhythm of PR as well as BP, with higher values during the day and lower at night, does exist in normotensive subjects resting continuously in bed, being therefore largely independent of physical activity. The nocturnal fall in BP and PR is attributable to an increase in peripheral flow due to redistribution. Some hypertensive subjects (dippers) have a similar rhythm to that of normotensive subjects, whereas others (non-dippers) manifest flattened or inverted rhythm due to their inability to modulate PR, probably because of structural arterial changes. The leg and forearm can have different circadian behavior; in particular, no separation between dipper and non-dipper hypertensive subjects can be made by leg hemodynamics. Males generally have a higher 22-hour forearm flow and a lower forearm PR, whereas leg flow and PR are similar in males and females.

REFERENCES

1. Pickering, T. G. & G. D. James. 1993. Determinants and consequences of the diurnal rhythm of blood pressure. Am. J. Hypertens. **6:** 1665–1695.
2. Van Den Meiracker, A. H., A. J. Man In'T Veld, H. J. Van Eck, G. J. Wenting & M. A. Schalekamp. 1988. Determinants of short term blood pressure variability. Effects of bed rest and sensory deprivation in essential hypertension. Am. J. Hypertens. **1:** 22–26.
3. Sundeberg, S., A. Kohvakka, & A. Gordin. 1988. Rapid reversal of circadian blood pressure rhythm in shift workers. J. Hypertens. **6:** 393–396.
4. Reinberg, A., Y. Motohashi, P. Bourdeleau, Y. Touitou, J. Nouguier, F. Levi & N. Annonciade. 1989. Internal desynchronization of circadian rhythms and tolerance of shift work. Chronobiologia **16:** 21–34.
5. Pessina, A. C., P. Palatini, P. Trevi, P. Benussi, P. Veronese, M. Hlede & C. Dal Palù. 1977. Circadian variations of blood pressure in patients with different degrees of hypertension. Biotel. Pat. Monitg. **4:** 141–150.
6. Pickering, T. G. 1992. The ninth Sir George Pickering memorial lecture: Ambulatory monitoring and the definition of hypertension. J. Hypertens. **10:** 401–409.
7. Hertel, R. F. 1987. Daily pattern of skin microcirculation in conscious laboratory rats. Naturwissenschaften **74:** 597–599.
8. Kety, S. S. 1949. Measurement of regional circulation by the local clearance of radioactive sodium. Am. Heart J. **38:** 321–328.
9. Sejersen, P. 1969. Blood flow in cutaneous tissue in man studied by washout of radioactive xenon. Circ. Res. **25:** 215–229.
10. Bojsen, J., B. Staberg & K. Kylendorf. 1984. Subcutaneous measurements of ^{133}Xe disappearance with portable CdTe(CI) detectors: Elimination of interference from combined convection and diffusion. Clin. Physiol. **4:** 309–320.
11. Jelnes, R. & K. Mtonnesen. 1984. Nocturnal foot blood flow in patients with arterial insufficiency. Clin. Sci. **67:** 89–95.
12. Sindrup, J. H., J. Kasstrup, B. Jørgensen, J. Bulow & N. A. Lassen. 1991. Nocturnal variations in subcutaneous blood flow rate in lower leg of normal human subjects. Am. J. Physiol. **260:** H480–485.
13. Casiglia, E., P. Palatini, M. S. Baccillieri, G. Colangeli, S. Petucco & A. C. Pessina. 1993. Circadian rhythm of peripheral resistance: A non-invasive 24-hour study in young normal volunteers confined to bed. High Blood. Press. **1:** 249–255.
14. Gerin, W., M. Rosofsky, C. Pieper & T. G. Pickering. 1993. A test of reproducibility of blood pressure and heart rate variability using a controlled ambulatory procedure. J. Hypertens. **11:** 1127–1131.
15. Clark, S., S. Fowlie, A. Coats, A. Radaelli, M. Van der Putt, R. Bird & J. Conway. 1991. Ambulatory blood pressure monitoring: Validation of the accuracy and reliability of the TM-2420 according to the AAMI recommendations. J. Hum. Hypertens. **5:** 77–82.
16. Hoobler, S. W., T. G. Kabza & R. W. Corley. 1955. The effect of protoveratrine on the cardiac output and on some regional circulation in man. J. Clin. Invest. **4:** 559–564.
17. Smith, H. W. 1951. The Kidney: Structure and Function in Health and Disease. Oxford University Press, New York.

18. SMITH, H. W. 1939. Physiology of renal circulation. Harvey Lectures **35:** 166.
19. ZEEH, J., H. LANGE, J. BOSCH, S. POHL, H. LOESGEN, R. EGGERS, M. NAVASA, J. CHESTA & J. BIRCHER. 1988. Steady-state extrarenal sorbitol clearance as a measure of hepatic plasma flow. Gastroenterology **95:** 749–759.
20. FERRARI, A. U., A. DAFFONCHIO, S. GEROSA, C. FRANZELLI, P. PALEARI, C. VENTURA, M. DI RIENZO & G. MANCIA. 1993. Spontaneous variability of regional haemodynamics in unanaesthetized rats. J. Hypertens. **11:** 535–541.
21. GRAF, K. 1964. Auswertung un Messfehler okklusion—plethysmographischer Durchblutungsregistrierungen. Acta Physiol. Scand. **60:** 120–134.
22. STRANDNESS, D. E. & J. W. BELL. 1965. Peripheral vascular disease. Diagnosis and objective evaluation using a mercury strain-gauge. Ann. Surg. **161**(Suppl.):1–4.
23. WITHNEY, R. J. 1953. The measurements of volume changes in human limbs. J. Physiol. **121:** 1–5.
24. BLAND, J. M. & D. G. ALTMAN. 1986. Statistical methods for assessing agreement between two methods of clinical measurements. Lancet **1:** 307–310.
25. O'BRIEN, E., J. SHERIDAN & K. O'MALLEY. 1988. Dippers and non-dippers. Lancet **2:** 397.
26. VERDECCHIA, P., G. SCHILLACI & C. PORCELLATI. 1991. Dippers versus non-dippers. J. Hypertens. **9**(Suppl. 8):42–44.
27. VERDECCHIA, P., G. SCHILLACI, C. GATTESCHI, I. ZAMPI, M. BATTISTELLI, C. BARTOCCINI & C. PORCELLATI. 1993. Blunted nocturnal fall in blood pressure in hypertensive women with future cardiovascular morbid events. Circulation **88:** 986–992.
28. SHIMADA, K., A. KAWAMOTO, K. MATSUBAYASHI, M. NISHINAGA, S. KINURA & T. OZAWA. 1992. Diurnal blood pressure variations and silent cerebrovascular damage in elderly with hypertension. J. Hypertens. **10:** 875–878.
29. PALATINI, P., M. PENZO, A. RACIOPPA, E. ZUGNO, G. GUZZARDI, M. ANACLERIO & A. C. PESSINA. 1992. Clinical relevance of nighttime blood pressure and of daytime blood pressure variability. Arch. Intern. Med. **152:** 1855–1860.
30. KASTRUP, J., H. WROBLEWSJI, J. SINDRUP, H. ROLIGHED CHRISTENSEN & N. WINBERG. 1993. Diurnal blood pressure profile in patients with severe congestive heart failure: Dippers and non-dippers. Scand. J. Clin. Lab. Invest. **53:** 577–583.
31. VERDECCHIA, P., G. SCHILLACI, M. GUERRIERI, C. GATTESCHI, G. BENEMIO, F. BOLDRINI & C. PORCELLATI. 1990. Circadian blood pressure changes and left ventricular hypertrophy in essential hypertension. Circulation **81:** 528–536.
32. BIANCHI, S., R. BIGAZZI, G. BALDARI & V. M. CAMPESE. 1994. Diurnal variations of blood pressure and microalbuminuria in essential hypertension. Am. J. Hypertens. **7:** 23–29.
33. VERDECCHIA, P., G. SCHILLACI, F. BOLDRINI, M. GUERRIERI & C. PORCELLATI. 1992. Sex, cardiac hypertrophy and diurnal blood pressure variations in essential hypertension. J. Hypertens. **10:** 683–692.

Renal Hemodynamic Mechanisms of Blood Pressure Rhythms

M. PONS,[a] J. CAMBAR,[a] AND J. M. WATERHOUSE[b]

[a]*Groupe d'Etude de Physiologie et Physiopathologie Rénales*
UFR des Sciences Pharmaceutiques
146 rue Léo Saignat
33076 Bordeaux Cedex, France

[b]*School of Biological Sciences*
University of Manchester
Oxford Road
M139PT Manchester, UK

For decades the most widely described temporal variations were those of blood pressure and urinary excretion, first reported respectively by Zadek in 1881 and Vogel in 1854. Exhaustive reviews of these topics were recently reported by Lemmer[1] and Cambar *et al.*[2] Although variability in blood pressure in physiologic and pathologic conditions is now obvious, the different mechanisms involved in these temporal changes need to be discussed.

Neural, respiratory, cardiac, vascular, renal, endocrine, or behavioral mechanisms all have a regulatory role in blood pressure rhythmicity. Because of the anatomic arrangement of capillaries, the kidneys receive a large proportion of the cardiac output. As a consequence, filtration at the glomeruli is about 25% of the plasma. Therefore, the contribution of the kidneys in the regulation of plasma volume as well as in fluid and electrolyte balance is of great importance.

The present review deals with recent knowledge of the renovascular chronostructures, with specific emphasis on the glomerular ones. Circadian changes in renal function, such as the glomerular filtration rate (GFR) and renal plasma flow (RPF), have largely been documented in man as well as in rodents.[2] After a brief survey of glomerular vessel structure, the relative role of the different parameters contributing to the regulation of GFR and RPF are discussed. We then review, from a theoretical and practical point of view, the different techniques used in patients and rats to precisely assess renal function.

Finally, we describe the reality of the variability of renal hemodynamics in humans and rodents. We also discuss the main factors involved in renal hemodynamics variability, such as systemic hemodynamics, glomerular ultrafiltration including the major role of the glomerular basement membrane, water and electrolyte urinary excretion, renin-angiotensin-aldosterone system, and autonomic nervous system.

INSIGHTS INTO RENAL STRUCTURE AND FUNCTION

To understand the temporal aspects of renal function, the chronobiologist must be familiar with the structure and function of the kidney. Although a detailed analysis of renal anatomy and physiology cannot be given here, some minimal information, which is required for an understanding of the chronophysiology and chronopathology of renal hemodynamics and excretion, is presented. We concentrate on renal vascularization and hemodynamics, with particular interest in the juxtaglomerular apparatus, glomerular structure and cytophysiology, and briefly urine formation by tubular reabsorption-secretion processes. (For further details, consult refs. 3, 4, and 5.)

The kidney is surrounded by a fibrous capsule and can be either unilobular (rat, mouse) or multilobular (dog, humans). A transverse section through the kidney shows clearly distinct zones. The outer layer (or cortex) which appears dark with glomeruli is present in this zone only. Internally adjacent to the cortex, the medulla is subdivided into an outer medulla, a richly vascular area, and an inner medulla. The part of the medulla that projects into the pelvis of the ureter is called the papilla.

One of the most important considerations in studies of renal hemodynamics is the renal blood supply to the nephrons. Kidneys receive nearly 25% of the cardiac output. Blood enters the kidney through the renal artery and invades the interlobar arteries. The latter give rise to numerous arcuate arteries, which pass between the medulla and cortex. From these, the interlobular arteries have a long straight course ending near the surface of the kidney.

Each interlobular artery gives rise to many branches, the afferent arterioles, which pass directly to glomeruli. At this point, the renal circulation develops into two capillary networks. On the one hand, the efferent arterioles which leave the midcortical or superficial glomeruli form a cortical capillary plexus surrounding the proximal and distal tubules. On the other hand, the efferent arterioles from the juxtamedullary glomeruli break up into the vasa recta or directly into capillary networks around the loops of Henle in the medulla, forming a medullary capillary plexus. The cortical capillary plexus is drained into interlobular veins and ends in the arcuate veins. Most of the medullary capillaries, in contrast, open into the ascending vasa recta, are collected together into bundles, and then drain in the arcuate veins.

The zones of the kidney (cortex and medulla) result not only from its vascular organization, but also from the arrangement of nephrons, the basic unit of the kidney. Their number varies depending on the species (about 2 million in humans, 500,000 in the dog, and 40,000 in the rat), but also according to the age (number decreases in aged patients).

Each nephron consists of a glomerulus and a long tubule with a very heterogeneous structure, which ends by joining a collecting duct common to numerous nephrons. These collecting ducts unite to form the large ducts of Bellini, which run through the papilla.

To understand the complexity of renal physiology and morphology, we need to stress the heterogeneity of the nephron population and the structural and functional heterogeneity of each nephron. Indeed, three groups of nephrons can be described: (1) superficial cortical nephrons, with the glomeruli located just below the renal surface; (2) deep or juxtamedullary nephrons, with the glomeruli located immediately above the corticomedullary junction; and (3) the midcortical nephrons, with the glomeruli located between the former two.

The cortical location of the glomeruli of each group of nephrons correlates well with the length of the loop of Henle, this being longer in juxtamedullary nephrons. All nephrons conform to a basic plan. The glomerulus produces selective filtration of certain blood components, and a tubule modifies the composition of the glomerular ultrafiltrate by reabsorption and secretion. The glomerulus (or renal corpuscule) consists of a capillary tuft surrounded by Bowman's capsule. Blood enters the glomerulus through the afferent arteriole and leaves it through the efferent arteriole, the difference between the entering and exiting blood volume being the glomerular ultrafiltrate.

As explained in FIGURE 1, the glomerular filtration rate (GFR), essentially the transglomerular water flow, is proportional to the glomerular filtration pressure

(GFP) and the glomerular filter permeability coefficient (Kf). GFP is the resultant of osmotic and hydrostatic forces; effective GFP is the difference between glomerular intracapillary hydrostatic pressure (P_{GC}) and the colloid osmotic pressure of the capillary (πc) and glomerular ultrafiltrate in the urinary space (πus). The ultrafiltration coefficient of the glomerular capillaries (Kf) is a term that describes quantitatively the water and solute pathway through the three different components of the glomerular filter: the endothelial and epithelial walls and the glomerular basement membrane (GBM). Kf depends on the caliber of glomerular capillaries which is controlled by mesangial cell size. Glomerular mesangium is an important effector for renal hormones and cytokines due to its content of smooth muscle cell structures.

The glomerular basement membrane (GBM), a key component in the process of glomerular filtration, can be considered an extracellular matrix and a mucopolysac-

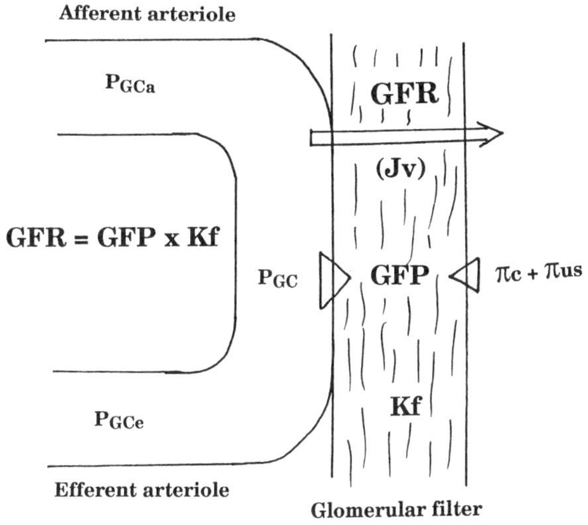

FIGURE 1. Schematic representation of the glomerular filtration rate (GFR) through the glomerular filter. P_{GCA} = afferent glomerular capillary pressure; P_{GCE} = efferent glomerular capillary pressure; P_{GC} = glomerular capillary pressure; GFP = glomerular filtration pressure; GFR = glomerular filtration rate; Jv = fluid movement; Kf = ultrafiltration coefficient; πc = capillary colloid osmotic pressure; πus = urinary space colloid osmotic pressure.

charide film located along the luminal surface, consisting of collagen, glycoproteins, proteoglycans, and mostly glycosaminoglycans.[6] The GBM acts as both a size and a charge barrier in the ultrafiltration of plasma. Under physiologic conditions, most watersoluble solutes can be filtered with little or hardly any restriction. This polyanionic electrostatic barrier provides a selective role, restricting the transglomerular passage of components with high molecular weights, especially blood proteins.[7] GAGs are mucopolysaccharide macromolecules, one of the constituents of the group of macromolecules called proteoglycans, and are normally found in urine. These major components of the GBM may have important roles in maintaining the integrity of the glomerular filtration apparatus and the selective permeability of the glomerular capillary wall.[8]

In the first step, the enzyme renin reacts with angiotensinogen to produce angiotensin I. Angiotensin I is transformed into angiotensin II by angiotensin-converting enzyme found in various organs, especially lung tissue (ACE). Angiotensin II is an active octapeptide and is the main effector hormone of the RAS. This peptide is the most potent vasoconstrictor of the system, but it has a short half-life because of its rapid metabolism into inactive fragments. At the kidney level, angiotensin II interferes with the regulation of renal blood flow, glomerular filtration, and sodium excretion.

Aldosterone is synthesized in the zona glomerulosa of the adrenal cortex. Aldosterone biosynthesis is controlled by many other factors. Thus, ACTH stimulating effects act mainly on glucocorticoid synthesis and increase aldosterone production. ACTH secretion is secondarily stimulated by angiotensin II, catecholamines, and potassium. In the kidneys, aldosterone stimulates the Na^+ transepithelial pathway at the distal tubule and also increases the tubular secretion of K^+ and H^+. Antidiuretic hormone (ADH), closely associated with aldosterone, controls blood osmotic pressure and so maintains blood volume and pressure.

Atrial natriuretic factor is a vasodilatory hormone located in granules of arterial cardiocytes. ANF and related peptides possess natriuretic and vasorelaxant effects, counteracting pressor effects of angiotensin II. These peptides affect renal hemodynamics by increasing the glomerular filtration rate. They exert powerful vasorelaxant agents on many vascular beds and can induce a fall in cardiac output. ANF effects are characterized by a decrease in renin release, a reduction of basal aldosterone secretion, and an inhibition of aldosterone secretion induced by angiotensin II, ACTH, and potassium. This system can be described as "the plasma or circulating RAS" which involves many hormones and enzymes and includes multiple target organs, mainly vessels, heart, and kidney.

Recent data suggest that the RAS is also present locally in tissues.[11] For instance, evidence of high specific angiotensin II receptors and even mRNAs has recently been provided. Furthermore, coding for renin or angiotensin has been found in 12 different extrahepatic (for angiotensinogen) and extrarenal (for renin) tissues. Renin and angiotensin have been demonstrated in small and large vessels such as the aorta, the carotid, coronary, or mesenteric arteries, and even veins.

Nevertheless, even under physiologic conditions, many milligrams of plasma proteins can be filtered, but more than 99% are reabsorbed at the tubular level. Other macromolecules, such as molecules secreted by cells along the entire nephron and urinary tract (enzymes, specific proteins), can be excreted in urine.

Renal hemodynamics is regulated by many factors, the renal circulation being controlled by hormonal and nervous processes. Most nervous renal fibers are adrenergic, inducing vasoconstriction by activation of α intrarenal receptors (only few β receptors are found in renal vessels).

Many hormones, especially those of the renin-angiotensin and kallikrein-kinin systems, and prostaglandins and cytokines play a major role in renal vasoreactivity by the presence of their receptors at the endothelial level of the vasculature and at the endothelial, epithelial, and mesangial levels in glomeruli and juxtaglomerular apparatus.

The juxtaglomerular apparatus has an important role in the regulation of the renin-angiotensin system (RAS). This complex consists of those portions of the afferent and efferent arterioles that lie near the glomerulus together with the macula densa of the distal tubule belonging to the same nephron. This unique specialized structure, juxtaposing the early distal tubule and glomerular arterioles, suggests complex mechanisms which link tubular function and renal hemodynamics.[9] The RAS plays a major role in the maintenance of water and electrolyte homeostasis and is associated with the regulation of blood pressure. The circulating RAS is an endocrine system which mediates, through the systemic circulation, the effect of renal renin release on angiotensin production. The system has two effector hormones, angiotensin II and aldosterone.

Renin biosynthesis occurs in granulated or epithelioid cells found in the media of glomerular arterioles.[10] Renin secretion from juxtaglomerular cells of the kidney is controlled by several effector systems such as renal perfusion pressure, renal nerve activity, salt load at distal tubule, and various humoral agents, especially angiotensin II, antidiuretic hormone (ADH), and the atrial natriuretic factor (ANF).

Locally produced angiotensin may influence the vascular tone and contractile state of the blood vessels. These data suggest that "cardiac RAS" may play an important role in local regulation of the coronary circulation. The role of the kidneys in the plasma RAS is well established, but several studies have suggested the presence of a tissue RAS in this organ. Indeed, angiotensinogen and renin mRNA were demonstrated in the renal cortex, whereas colocalization of renin-converting enzyme and angiotensin has been shown in juxtaglomerular cells as well as in afferent and efferent renal arterioles. Isolated renin granules have been reported to contain angiotensin I and angiotensin II. Thus, the local intrarenal production of angiotensin may play a role in regulating the glomerular contractile state, tubuloglomerular feedback, vasa recta blood flow, and reabsorption of sodium. The existence of these numerous local RAS could explain the local regulation of vascular resistance or blood flow without participation of the circulating components of RAS. This review does not justify large developments about the tubular structure and function, even if tubules contribute largely to the regulation of renal hemodynamics by the glomerulo-tubular feedback.

Whatever the population of nephrons may be, most recent studies report the existence of numerous different segments, grossly subdivided into proximal tubules (further subdivided into three parts), limb of Henle (subdivided into four parts), distal tubule (recently subdivided into three parts), collecting tubule (subdivided into three), and finally collecting tubules forming the papillary collecting duct. All of these different segments play a particular role in forming urine. The differential role of each nephron part is due to a large extent to great differences in hormonal control and in passive and active transmembrane flux.

In brief, as the fluid moves out of the proximal tubule into the loop of Henle, the classic process of urinary concentration begins, the loop of Henle acting as a countercurrent multiplier. We must consider at each nephron level the solute and water flux between tubular lumen, tubular cell, and peritubular blood.

The kidneys are the main route by which excess watersoluble substances are removed from the body. Under normal circumstances, a great amount of filtered materials are absorbed by the proximal convoluted tubule. This essentially clears the luminal fluid of glucose, fatty acids, amino acids, and proteins. In addition, approximately 80% of electrolytes are reabsorbed, followed by osmotic water absorption. Subsequent changes in water and electrolytes in the luminal fluid are under hormonal control, the result being that most of the filtered load is reabsorbed.

OVERVIEW OF RENAL HEMODYNAMICS ASSESSMENT TECHNIQUES

For physiologic and toxicologic studies, we can distinguish two major types of investigation to assess renal function: (1) techniques to assess renal plasma flow (RPF) and glomerular filtration rate (GFR) to evaluate renal hemodynamics; and (2) techniques to assess tubular function based on analysis of the urinary excretion of several substances. Histologic changes in renal parenchyma, especially at the tubular level, complete this procedure.

Analysis of urine, considered a "liquid biopsy," gives a good image of the functional and structural integrity of the kidney. Urinalysis thus offers a valuable diagnostic tool for kidney disorders and also provide a clinical advantage over other organ pathologies. With progress in analytic techniques, the number of solutes that can be determined in urine is becoming larger and larger; for example, water, electrolytes, proteins, specific tubular enzyme components of the basement membranes, blood, glucose, and renal cells are the more common ones. An exhaustive review presents the findings on experimental and clinical urinalysis.[12]

Determination of renal plasma flow (RPF) and glomerular filtration rate (GFR) are the most useful measurements of renal function, such as renal hemodynamics. The choice of technique can be made according to the aim of the experiment and the experimental model.

In physiologic investigations, the most accurate techniques involve constant infusions of reference substances, such as inulin and paraamino hippuric acid (PAH) and the withdrawal of multiple blood and urine samples. Indeed, inulin and PAH clearance techniques can be considered as the standard for GFR and RPF estimation.[13]

The clearance formula was first suggested by Addis in 1917, and the term "clearance" was first used by van Slyke in 1929, who wrote that "The blood urea clearance can be defined as the volume of blood which one minute's excretion suffices to clear of urea when the urine output is large enough to permit a maximum urea output."

If U is the concentration of the substance in the urine and V the rate of urine formation, then UV is the amount of the substance excreted in the urine per unit time. If P is the concentration of the substance in plasma, then UV/P indicates the volume of plasma required to supply the amount of substance in the urine in unit time. This formula has since been widely used. The accuracy of clearance methods depends on careful attention to numerous details of technique in such matters as collection of blood and urine samples and infusion of reference materials.

A good marker proposed as an index of GFR should meet the following criteria: (a) physiologically inert and nontoxic; (b) not subject to destruction, synthesis, or storage into the kidney; (c) not protein bound in the plasma; (d) completely filterable at the glomerulus; (e) not reabsorbed or secreted at the tubular level; (f) easily assayed by standard precise methods; and (g) have a constant clearance over a wide range of plasma concentrations. Inulin, a fructose polymer of molecular weight 5200 and molecular radius 15A, was introduced into renal physiology simultaneously by Richards and Homer Smith in 1935. Inulin meets all of the foregoing criteria and is today considered the reference index of GFR by its superiority over other markers.

For RPF measurement, the clearance of *p*-amino-hippurate (PAH) is used. PAH is partly filtered in the glomeruli and the remainder is actively secreted from the peritubular blood vessels into the urine by the proximal tubular cells. In healthy subjects, PAH is almost totally (around 90%) extracted from the blood. Theoretic considerations and experimental evidence indicate that the constant infusion technique is more accurate than the other methods, the concentration of test substance in the plasma needing to be constant. However, this method is difficult to use in man as well as in the rat.

Because of their complexity, other easier techniques have been reported (see a recent review in ref. 14). Thus, a single injection clearance technique has been proposed that needs only a single injection of reference testing substance, multiple blood samples, but no urine collection. More recently, renal GFR has been calculated from analysis of the manner in which the kidney takes up radioactive tracer using direct radionuclide imaging studies.

In prolonged experiments in patients and experimental animals extending over many days, only noninvasive blood and urine sampling can be used. The GFR can be evaluated by this approach with two commonly used procedures: when only blood samples are collected, evaluation of blood creatinine is the most commonly used procedure for monitoring GFR in humans and animals. When blood and urine are collected, creatinine clearance can be determined, giving an approximation of GFR.

Chronic Studies of Renal Hemodynamics in Animals and Humans

As the clearance method requires constant intravenous infusion of inulin and PAH plus concomitant determination of their plasma and urine levels, intravenous and intraarterial cannulations must therefore be performed. In the past, most experiments involved anesthetized animals, especially dogs and rats, during short-term studies (some hours only). However, renal hemodynamics and vasoactive systems may strongly be perturbed by anesthesia and/or the stress of acute surgical preparation, suggesting the need for a better approach.

Animals

A chronic arterial and venous catheterization method has been used with rats to measure arterial blood pressure[15] and to study the renin-angiotensin system.[16] Smith *et al.*[17] described systemic hemodynamic monitoring for 24 hours in unanesthetized rats. Published data on GFR and RPF estimation in conscious animals, especially rodents, are scarce. The rat, because of its small size and relatively low cost, is one of the most commonly used animals for scientific study; however, vascular access is difficult and presents many surgical drawbacks.

Nevertheless, in physiologic terms the chronically venous and arterial cannulated rat is a better experimental model for the study of renal hemodynamics. It is possible to chronically administer intravenous infusions and sample arterial blood repeatedly over time with minimal stress. This model is particularly well adapted to chronophysiologic investigations, because it permits, continuous determination of circadian changes in inulin and PAH clearance without external surgical disturbance.[18] Overall, it makes possible better evaluation of 24-hour variations in renal hemodynamics than does the method involving endogenous creatinine clearance as previously used in normal and obese rats.[19,20]

We present here the experimental model of chronically catheterized unanesthetized and freely moving rats developed in our laboratory to assess GFR and RPF throughout the day. Reproducible circadian changes exist between animals and in the same animal assessed for many consecutive days.

Male Sprague-Dawley rats weighing 200–220 g were housed for 1 week before experimentation in polycarbonate cages with water and food freely available. In anesthetized rats, calibrated polyethylene tubing (PE1 and PE2, respectively, 0.58 and 0.30 mm in interior diameter) was placed into the right femoral artery and left external jugular vein on day 0. Animals were connected to a device employing a protective stainless steel coiled spring with an anchoring flange and a watertight swivel with different arterial and venous ports to allow both continuous venous infusion of solutions and simultaneous intermittent sampling of arterial blood. This device prevents the rat from destroying the catheter by chewing and permits the animal to turn without producing kinks in the lines.

After surgery, each animal was housed in an individual clear plastic metabolism cage outfitted to separately collect urine and feces. Rats were allowed 48 hours for recovery from surgery and adaptation to the infusion and sampling setup. Water and food were freely supplied. Animals were housed in a sound-dampened chronobiologic room with controlled conditions of temperature (24 ± 1°C), atmospheric pressure (750–770 mm Hg), and humidity (55 + 5%). The animal chamber had a 12-hour light/12-hour dark schedule with light from 8 AM to 8 PM.

After surgery a constant rate (0.67 ml/h through the jugular vein catheter) of 0.1% heparinized saline solution was administered by a programmable syringe pump (Harvard apparatus pump 22). On days 3–10 of the experiment, PAH (1 mg/ml; Sigma) and inulin (50 mg/ml; Sigma) were added to the heparinized saline solution.

Urine samples were collected daily at 4-hour intervals, centrifuged for purification, and frozen until analysis. Blood sampling (0.5 ml) was performed on the swivel in heparinized tubes, also at 4-hour intervals, each time corresponding to the midpoint of urine collections. Blood withdrawal volume was replaced with an equivalent volume of normal saline solution. Plasma was separated by centrifugation at 4°C and then frozen until analysis.

Urine volume for each 4-hour span was directly measured and urine flow rate determined (ml/4 hours). Urinary electrolytes, inulin, and PAH were assessed by classic spectrophotometric methods. Moreover, urinary sodium and potassium electrolytes were assessed by flame photometry.

Humans

We present the protocol used for clinical estimation of GFR and RPF described by Buijsen *et al.*[21] The study was carried out over a period of 28 hours after admission to the metabolic ward of the hospital. Throughout the study all subjects remained in bed and had identical intake of fluid every 3 hours to achieve ample urine flow. They also consumed identical meals every 3 hours, determined by a dietician. Subjects received meals with a precisely defined daily sodium content. Daily potassium intake was 60 mmol and protein intake was 65 g in both study groups.

After bedrest of 5 hours starting at noon, a cannula for intravenous infusion was placed in one arm and a sampling device in the other. A priming intravenous injection of inulin was then given followed by continuous infusion by a microvolumetric infusion pump. The dose of inulin was adjusted for body weight and creatinine clearance. At 9 PM, after an equilibration period of 4 hours, a peripheral venous blood sample was taken every 3 hours till 9 PM the next day. At the same time, urine was collected by spontaneous voiding at the end of each 3-hour period after which the precise volume was noted. Subjects were allowed to sleep from 11 PM to 8 AM between sample and meal times.

We are awaiting data from these two approaches.

REALITY OF THE VARIABILITY OF RENAL HEMODYNAMICS

In man circadian variations in GFR and RPF have been reported.[22–27] Indeed, GFR is generally lowest during sleep at night and shows a low amplitude of about 20% of the daily mean.[28] This rhythmicity can be due to day-night differences in lifestyle (posture, food intake, and physical activity) or to endogenous oscillators. Both components play a part, because the rhythm does not disappear during constant bedrest and identical meals every 3 hours.[28,29]

The implication of these observations (inulin and PAH clearance circadian changes) is that removal of a watersoluble substance (urea, amino acid, or electrolyte, for example) from the plasma by glomerular ultrafiltration will be higher diurnally than nocturnally. We will discuss the possible factors involved in the temporal changes of glomerular filter structure and function.

Likewise, circadian rhythm in PAH clearance was reported by many (but not all) investigators.[29] This rhythm also presents a daily maximum, but it is not exactly identical to renal blood flow rhythm. The tendency of this rhythm to be in phase with the well-documented rhythm of blood pressure will be discussed.

The search for a correlation between these temporal changes in renal hemodynamics and those in electrolyte excretion has been done in normotensive healthy[22–24,27] as well as in hypertensive subjects[24] or in diseased patients such as those with heart failure,[25] end-stage renal disease[30,31] or kidney transplantation.[21] However, a recent article reported the absence of low-amplitude circadian rhythmicity in creatinine clearance in humans.[32]

Rhythms in renal hemodynamics, on the other hand, are poorly documented in rats. Circadian changes in the glomerular filtration rate (GFR) were initially documented by creatinine clearance.[19] We noted an increase during the daily dark span. Such circadian changes in local tissue hemodynamics have been found in the intestine, liver, and kidney.[33]

A recent study in our laboratory[18] clearly showed circadian rhythmicity in renal hemodynamics, GFR and renal plasma flow in chronically cannulated, unanesthetized, and unrestrained rats.

GFR circadian rhythmicity, assessed by inulin clearance, was detected with a nightly acrophase at midnight. In our study, nighttime values were significantly increased over daytime ones ($p < 0.001$). The GFR between midnight and 4 AM was 1.6 ± 0.3 ml/min vs 0.9 ± 0.2 ml/min between noon and 4 PM.

Likewise, circadian rhythmicity in PAH clearance was evidenced by increased renal plasma flow during the dark span with an acrophase at midnight. The difference between the maximum between midnight and 4 AM (4.3 ± 0.5 ml/min) and the minimum between 4 PM and 8 PM (2.5 ± 0.3 min) was highly significant ($p < 0.001$).

We believe, that from a biologic point of view this experimental approach represents the best means of assessing renal hemodynamics and urinary electrolyte excretion rhythms over an extended period of time, well adjusted for chronobiologic investigations.

In conclusion, in all these studies, renal hemodynamics, both renal plasma flow and GFR, in man as well as in the rat have demonstrated circadian rhythms of low amplitude with a maximum during the activity span. Similar temporal variations have been described in factors influencing renal hemodynamics such as systemic blood pressure,[34] plasma renin activity,[35,36] catecholamines,[37] or prostaglandins.[38]

FACTORS IN RENAL HEMODYNAMICS VARIABILITY

We have just shown that renal hemodynamics are controlled by neural, hormonal, and intrarenal factors. We discussed from a temporal point of view, why and how we can explain such temporal changes (FIG. 2).

The word *correlation* must be used carefully. Indeed, in many cases, when mechanisms causing circadian rhythmicity are discussed, some reviews evoke *coincidence* that exists between two different rhythms. If this *coincidence* does not show that a causal link exists, we cannot evoke a correlation.

Correlation between Circadian Changes in Renal Hemodynamics and Systemic Hemodynamics

Blood pressure fluctuates continuously throughout the day and night and from one subject to another.[39] These fluctuations may be of short duration, lasting a few seconds or minutes, or of larger duration, including circadian and even circannual changes. The most important consistent sources of blood pressure variation are the sleep-wake cycle and physical diurnal activity (posture, exercise, and the like).[40]

Biologic rhythms in heart rate and blood pressure with a prominent fall at night were described in the 19th century. Following these early observations, numerous studies evidenced circadian changes in cardiovascular function.[1] Recent technical

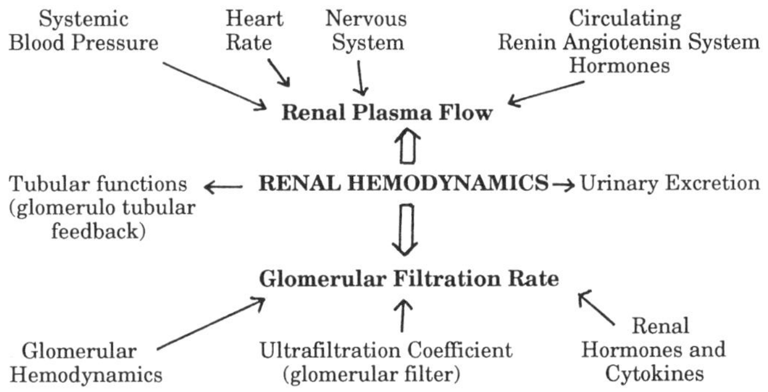

FIGURE 2. Different parameters involved in the regulation of renal hemodynamics. As all these parameters present temporal changes, it is obvious that renal hemodynamics is time dependent.

progress in measuring 24-hour ambulatory blood pressure, namely, noninvasive recording methods, has permitted measurement of blood pressure and heart rate in man and experimental animals all day and all night long for many days. In patients with essential hypertension, the day-night pattern of blood pressure change is generally similar to that of normotensive subjects with a significant fall in nocturnal blood pressure, except that the entire profile is shifted upwards.

Nevertheless, in some cases of essential hypertension or in several forms of secondary hypertension, the nocturnal decrease in blood pressure can be absent or reversed despite a nocturnal decrease in heart rate.[41] A relation seems to exist even between the absence of a nocturnal fall in arterial blood pressure and the severity of cardiovascular target organ damage.[42]

As we just noted, as kidneys receive a large proportion of cardiac output, circadian changes in systemic hemodynamics have important repercussions on the variability of renal hemodynamics.

In the normal rat, a fair correlation exists between systemic hemodynamics (blood pressure and heart rate) and renal hemodynamics with a maximum during the activity span of the animal. Similar data can be noted in healthy patients.

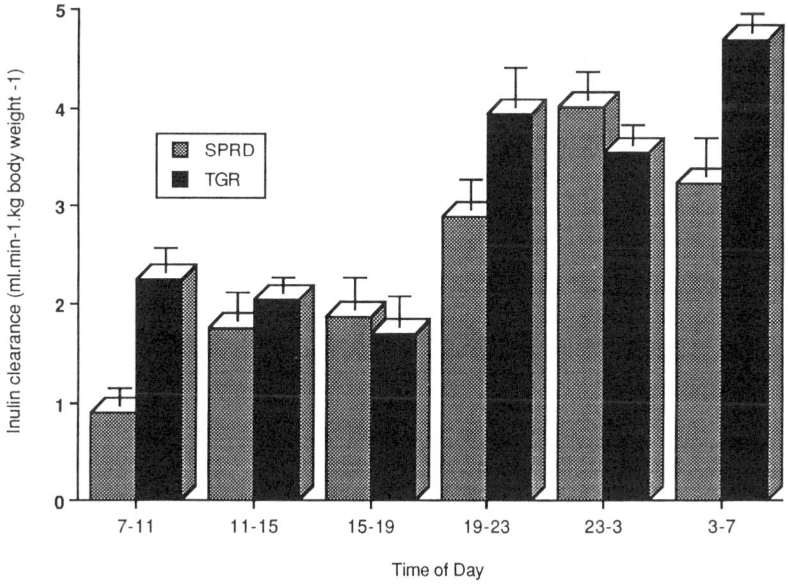

FIGURE 3. Circadian rhythm in glomerular filtration rate (inulin clearance) in transgenic hypertensive rats (TGR) and Sprague-Dawley normotensive rats (SPRD).

However, from our experience (FIGS. 3 and 4), surprising data could be reported in transgenic TGR(mRen2)27 hypertensive rats derived from the Sprague-Dawley line. The introduction of the mouse salivary gland renin gene into the genome of the Sprague-Dawley rat not only resulted in severe hypertension,[43] but also greatly

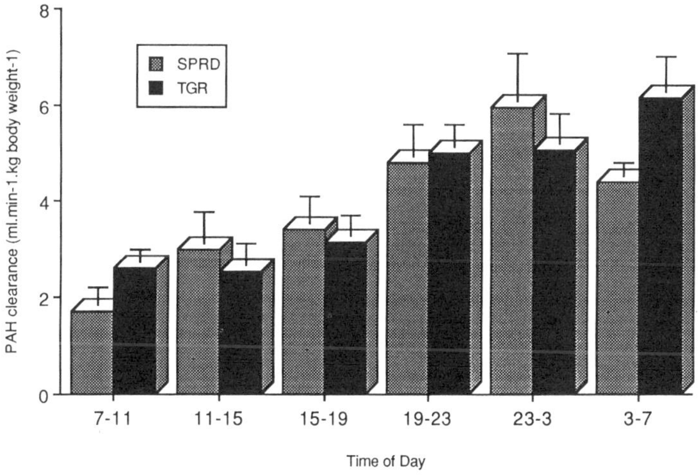

FIGURE 4. Circadian rhythm in renal plasma flow (PAH clearance) in transgenic hypertensive rats (TGR) and Sprague-Dawley normotensive rats (SPRD).

altered the usual circadian blood pressure pattern, producing changes in the phase of the rhythm of blood pressure, with peak values occurring during the resting light period of the animals.[44,45] On the other hand, circadian shapes of heart rate and motor activity remained unchanged.[44] An important finding in our study is that the inverted circadian blood pressure profile affected circadian rhythmicity in renal function in transgenic rats surprisingly little, if at all. According to our results, circadian rhythms in renal hemodynamics as well as in renal excretion remarkably similar to those in Sprague-Dawley controls were found. As previously discussed, changes in systemic blood pressure have long been suspected as a key factor in the circadian pattern of renal hemodynamics. However, our results support the notion that temporal variations in systemic blood pressure are not the major factor contributing to circadian rhythmicity in renal function. Arterial pressure regulation and renal function are not as tightly linked as they were once believed to be.[26] In transgenic rats, the timing of circadian rhythms of pressure changes in the renal and systemic circulation are very different. Autoregulatory mechanisms are suspected to account for the protection of the kidney from the transmission of systemic hypertension to the renal microcirculation in transgenic rats.[43] This autoregulation presumably includes afferent arteriolar vasoconstriction.

Our observations, opening the way to future investigations, suggest unimpaired renal autoregulation in transgenic animals, indicating that time-dependent changes in systemic blood flow (cardiac output, heart rate) may be of greater importance in the regulation of renal circadian rhythmicity than systemic blood pressure.

Correlation between Circadian Changes in Glomerular Filtration Rate and Glomerular Basement Membrane Components

We have shown that the glomerular filtration rate is proportional to the glomerular filtration pressure (GFP) and to the glomerular filter permeability coefficient (Kf). No circadian rhythmicity in GFP has been investigated. The Kf can be changed if the diameter of endothelial pores or epithelial podocyte slit-pores is increased or if the glomerular electronegative barrier is damaged by the loss of components such as glycoproteins or glycosaminoglycans largely described in the first chapter "insight into renal structure and function."

Circadian changes in GAGs, a major component of the GBM, were described for the first time in our laboratory with a fair correlation of changes with GFR and renal plasma flow. This correlation can be explained by the observation that the (water) ultrafiltrate takes with it watersoluble components of the GBM modifying then the electrostatic negative glomerular filter.

As we proposed in FIGURE 5, during the night in the rat (the activity period of all mammals), an increase in water and solute excretion occurs; quantitatively there is an increase in numerous solutes such as urea, electrolytes, proteins, or GBM components and qualitatively a change in index selectivity of macromolecules. It seems to be the consequence of a nightly GFR increase in water and solute flow and in GBM permselectivity.

Further studies are needed to show a correlation between circadian changes in urinary GBM components and serum protein renal clearance (albumin, transferrin, and immunoglobulins).

Nevertheless, circadian changes in macromolecule excretion, namely, proteinuria in normal subjects,[46,47] proteinuria[26] or β_2-microglobulin[48] in patients with nephrotic syndrome, or microalbuminuria in essential hypertension[49,50] have already been reported.

Correlation between Circadian Changes in Glomerular Filtration Rate and Diuresis and Correlation between Circadian Changes in Glomerular Filtration Rate and Electrolyte Excretion

We have already reported a fair correlation between circadian changes in urinary excretion of water, electrolytes (Na, K), urea and creatinine, and total proteins.[19-51] Similar data have been reported in man.[27] A fair correlation seems to exist with renal hemodynamics (e.g., GFR and RPF). In man, Wesson[30] reported a coincidence between GFR and Na, K excretion in recumbent subjects but not in patients changing posture.

In renal physiology, excretory function is assumed to result from glomerular and tubular function, as glomerular ultrafiltration and urinary excretion temporal changes.

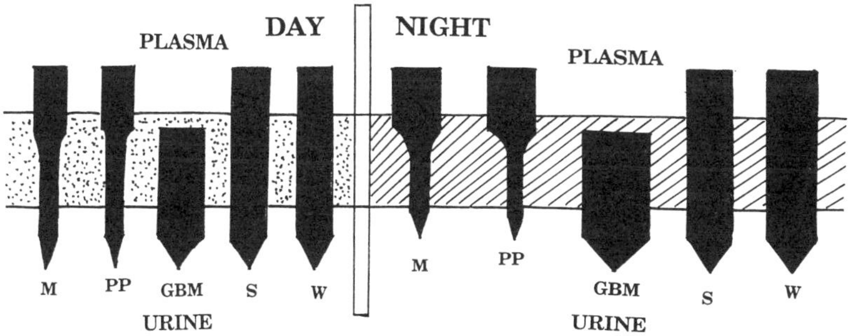

FIGURE 5. Day/night difference (with a nightly increase) in qualitative and quantitative pathway of water and numerous solutes through the glomerular barrier in the rat. GBM = glomerular basement membrane components; M = exogenous macromolecules; PP = plasma proteins; S = solutes; W = water.

Nevertheless, the difference in their amplitude suggests the major part of the tubules changes in excretion changes in excretion rhythms. But information on tubular changes is scarce.

In patients with renal disorders (renal insufficiency, renal failure, nephrotic syndrome, glomerulonephritis, etc.), renal hemodynamics and urinary excretion can be significantly changed. We know the clinical consequences of such renal disturbances for drug elimination.

From a chronobiologic view, most patients with nephrotic syndrome demonstrate a circadian rhythm for albumin, transferrin, and IGG excretion in urine with a maximum protein excretion at 4 PM and a minimum at 3 AM, even during a regimen of complete bedrest or defined identical meals.

Moreover, circadian rhythm in the selectivity index (IGG clearance/transferrin clearance) was seen in many patients, also with a nighttime minimum. Recently, similar circadian variations in urinary β_2-microglobulin excretion were found in most patients (75%) with the nephrotic syndrome.[48]

In parallel, numerous urinary components such as H_2O, Na, K, or proteins show circadian rhythms that are blunted in amplitude, of zero amplitude, or even of inverted phase in proportion to the severity of the disorder.[26,27]

These effects are due to a progressive overriding of circadian effects by postural reflexes, particularly if fluid is distributed abnormally with edema.[30–52] These results link renal hemodynamics and renal function, but possible changes in systemic blood pressure might also play a role.[53,54] Daily changes in glomerular filtration rate and renal plasma flow can also be controlled by various other factors such as the sympathetic nervous system or hormones of the renin-angiotensin-aldosterone system.

Correlation between Circadian Changes in Glomerular Filtration Rate and in the Autonomic Nervous System

It is well known that the sympathetic nerves play a role in the control of renal hemodynamics. All of the foregoing data suggest the role of the autonomic nervous system in renal pathology, especially in renal neuropathy[55] or renal transplantation,[21] but the rhythms are not uniformly changed, possibly indicating the effects of hormones or posture.

In patients with normal movement, two peaks of plasma adrenaline can be observed in the middle of the afternoon and early in the morning. This second peak can be attributed to upright posture, because it disappears in the recumbent position and in patients with denervated grafts. Nevertheless, a recent paper reported that after kidney transplantation, the circadian rhythms in GFR remained unchanged in patients regarded as a model of renal denervation and chronically treated with cyclosporin a nephrotoxic agent, that could by itself also damage the renal rhythmicity.[21]

In conclusion, the role of the autonomic nervous system in renal hemodynamics or tubular function is difficult to appreciate now and seems to be multifactorial.

Correlation between Circadian Changes in Glomerular Filtration Rate and in the Renin-Angiotensin-Aldosterone System

As noted previously in this report, renal hemodynamics are regulated by many factors, especially the numerous hormones of the renin-angiotensin-aldosterone system (RAAS).

As we reviewed,[56] all components of the RAAS present temporal changes with fair similarity to changes in plasma renin, aldosterone, and corticosterone in man[57–59] as well as in the rat[35,60,61] with a peak occurring in the activity period of both.

As angiotensin II can alter glomerular ultrafiltration via its influence on the ultrafiltration coefficient (Kf) through action upon mesangial cells,[62] its temporal plasma change can explain in part temporal variations in renal hemodynamics. Although plasma renin activity has a definite circadian rhythm in man, this rhythm has a different phase in subjects living a normal lifestyle than in subjects having constant bedrest.[27]

Others hormones such as prostaglandins, atrial natriuretic peptide, vasopressin, and cortisol are reported to influence renal hemodynamics and to present a circadian rhythmicity.[49,63–65]

In conclusion, it is not clear whether and how rhythmic changes in the activity of these hormones have a direct influence on the circadian rhythms of GFR and RPF.

We have also pointed out the existence of numerous local tissue RAAS that can regulate locally, perhaps independently, the vascular resistance or blood flow.[11] The existence of a temporal structure in the circulating RAAS suggests similar investigations at different tissue levels. Considerable chronobiologic studies may be required to identify such local chronostructures and to discover, by comparison, the eventual correlation between all of them.

CONCLUSION

Whether mediated by neural, humoral or intrarenal physiologic factors, regulation of the renal hemodynamics ultimately depends on resistance changes resulting from constriction or relaxation of vascular smooth muscle. Indeed, all renal vessels (arcuate and interlobular arteries, afferent or efferent arterioles) contain smooth muscle. Moreover, the glomerular mesangium, which contains contractile proteins, is also the site of action and production of many such substances.

A variety of hormonal and vasoactive substances, including prostaglandins, kinins, adenosine, endothelium-dependent relaxing factor, and mainly the renin-angiotensin-aldosterone-atrial natriuretic factor system have been implicated in the autoregulation of renal plasma flow and glomerular filtration rate.

These substances can regulate the tone of preglomerular and postglomerular vessels showing a permanent balance between the vasodilating and vasoconstrictor agents. It is obvious that temporal changes in the plasma level of these vasoactive substances induce temporal changes in renal hemodynamics. Such circadian changes in RPF and GFR in rat as well as in man have been well demonstrated.

The anatomic arrangement of the various components of the glomerular capillaries is well adapted to ultrafiltration and plays a crucial role in determining glomerular hydraulic conductivity. If we consider the equation GFR = GFP × Kf, it is clear that an increase in GFR can be due to an increase in GFP or Kf or both.

The GFR increase reported during the active period in mammals is the result of increased transglomerular fluid and solute flow and can also be due to an increase in vessel resistance or solute permselectivity.[66]

In conclusion, renal hemodynamics are controlled by many factors we have discussed from a physiologic point of view. But all of these factors present circadian changes, explaining why it is possible to report circadian changes in GFR and RPF. The fair correlation between renal hemodynamics and systemic hemodynamics, glomerular basement membrane structure and permselectivity, urinary excretion, autonomic nervous system and renin angiotensin system are discussed. All these data permit us to understand the role of temporal changes in renal hemodynamics on blood pressure rhythmicity.

REFERENCES

1. LEMMER, B. 1989. Circadian rhythms in the cardiovascular system. *In* Biological Rhythm in Clinical Practices. J. Arendt, J. Waterhouse & D. Minors, Eds.. 51–70. Butterworths Publishing. London.
2. CAMBAR, J., J. C. CAL & J. TRANCHOT. 1992. Renal excretion: rhythms in physiology and pathology. *In* Biologic Rhythms in Clinical and Laboratory Medicine. Y. Touitou & E. Haus, Eds.: 470–482. Springer Verlag. Berlin.
3. ORLOFF, J. & R. W. BERLINER. Eds. 1973. Handbook of Physiology. Sect 8: Renal Physiology. American Physiological Society Washington.

4. SELDIN, D. W. & G. GIEBISCH, Ed. 1985. Physiology and Pathophysiology of the Kidney. Raven Press. New York.
5. BRENNER, B. M., Ed. 1996. The Kidney, 5th Ed. Saunders Co. Philadelphia.
6. SPIRO, R. G. & P. S. MOHAN. 1987. Biochemical exploration of the macromolecular organization of the glomerular basement membrane: Interrelationship of collagen, proteoglycan and glycoprotein components. *In* Renal Basement Membranes in Health and Disease. R. G. Price & B. G. Hudson, Eds.: 11–23. Academic Press. London.
7. FARQUHAR, M. G. 1981. The glomerular basement membrane. A selective macromolecular filter. *In* Cell Biology of Extracellular Matrix. E. D. Hay, Ed.: 335–378. Plenum Publishing.
8. KANWAR, Y. S., L. J. ROSENZWEIG & D. I. KERJASCHKI. 1981. Glycosaminoglycans of the glomerular basement membrane in normal and nephrotic states. Renal Physiol. **4:** 121–130.
9. TAUGNER, R. & E. HACKENTHAL. 1989. The Juxtaglomerular Apparatus. Springer Verlag. Berlin.
10. KRIZ, W. 1987. A parietal pathway for intrarenal distribution of renin. Kidney Int. **31:** 51–56.
11. CAMPBELL, D. J. 1987. Circulating and tissue angiotensin systems. J. Clin. Invest. **79:** 1–6.
12. PIPERNO, E. Detection of drug induced nephrotoxicity with urinalysis and enzymuria assessment. 1981. *In* Toxicology of the Kidney. J. B. Book, Ed.: 31–55 Raven Press. New York.
13. TOTO, R. D. 1995. Conventional measurement of renal function utilizing serum creatinine, creatinine clearance, inulin and para-aminohippuric acid clearance. Curr. Opin. Nephrol. Hypertens. **4:** 505–509.
14. DONDI, M. & S. FANTI. 1995. Determination of individual renal function through noninvasive methodologies. Curr. Opin. Nephrol. Hypertens. **4:** 520–524.
15. RAFF, H. & K. D. FAGIN. 1984. Measurement of hormones and blood gases during hypoxia in conscious cannulated rats. Appl. Physiol. **56:** 1426–1430.
16. CONRAD, K. P., T. BRINK-JOHNSEN, M. GELLAI & H. VALTIN. 1984. Renal autoregulation of chronically catheterized conscious rats. Renal Fluid Electr. Physiol. **16:** F229–F233.
17. SMITH, T. L., T. G. COLEMAN, K. A. STANKEK & W. R. MURPHY. 1987. Hemodynamic monitoring for 24 hours in unanesthetized rats. Am. J. Physiol. **253:** H1335–H1341.
18. PONS, M., J. TRANCHOT, B. L'AZOU & J. CAMBAR. 1994. Circadian rhythms of renal hemodynamics in unanaesthetized, unrestrained rats. Chronobiol. Int. **11:** 301–308.
19. CAMBAR, J., F. LEMOIGNE & C. TOUSSAINT. 1979. Etude des variations nycthémérales de la filtration glomérulaire chez le rat. Experentia **35:** 1607–1609.
20. LUKE, D. R., K. M. WASSAN & K. VADIEI. 1991. Circadian variations in renal function of the obese rat. Renal Physiol. Biochem. **14:** 71–80.
21. BUIJSEN, J. G. M., B. A. C. VAN ACKER, G. C. M. KOOMEN, M. G. KOOPMAN & L. ARISZ. 1994. Circadian rhythm of glomerular filtration rate in patients after kidney transplantation. Nephrol. Dial. Transplant. **9:** 1330–1333.
22. WESSON, L. G. & D. P. LAULER. 1961. Diurnal cycle of GFR and sodium and chloride excretion during responses to altered salt and water balance in man. J. Clin. Invest. **40:** 1967–1973.
23. WESSON, L. G. 1964. Electrolyte excretion in relation to diurnal cycles of renal function. Plasma electrolyte concentrations and aldosterone secretion before and during salt and water balance changes in normotensive subjects. Medicine **43:** 547–592.
24. VAGNUCCI, A. I. & L. G. WESSON. 1964. Diurnal cycles of renal hemodynamics and excretion of chloride and potassium in hypertensive subjects. J. Clin. Invest. **43:** 522–531.
25. WESSON, L. G. 1979. Diurnal circadian rhythms of renal function and electrolytes excretion in heart failure. Int. J. Chronobiol. **6:** 109–117.
26. KOOPMAN, M. G., R. T. KREDIET & L. ARISZ. 1985. Circadian rhythms and the kidney. A review. Neth. J. Med. **28:** 416–423.
27. KOOPMAN, M. G., D. S. MINORS & J. M. WATERHOUSE. 1989. Urinary and renal circadian rhythms. *In* Biological Rhythms in Clinical Practice. J. Arendt, D. S. Minors & J. M. Waterhouse, Eds.: 83–98. Wright Edition. London.

28. KOOPMAN, M. G., G. C. M. KOOMEN, R. T. KREDIET, E. A. M. DEMOOR, F. J. HOEK & L. ARISZ. 1989. Circadian rhythm of glomerular filtration rate in normal individuals. 1989. Clin. Sci. **77:** 105–111.
29. KOOPMAN, M. G., R. T. KREDIET, F. M. J. ZUYDERHOUDT, E. A. M. DE MOOR & L. ARISZ. 1985. A circadian rhythm of proteinuria in patients with a nephrotic syndrome. Clin. Sci. **69:** 395–401.
30. WESSON, L. G. 1980. Diurnal circadian rhythms of electrolyte excretion and filtration rate in end stage renal disease. Nephron **26:** 211–214.
31. KOOPMAN, M. G., G. C. M. KOOMEN, B. A. C. VAN ACKER & L. ARISZ. 1990. Circadian rhythm in glomerular filtration rate (GFR) and effective renal plasma flow (ERPF) in patients with nephrotic syndrome. Nephrol. Dial. Transplant. **5:** 649–650.
32. VAN ACKER, B. A. C., G. C. M. KOOMEN, M. G. KOOPMAN, R. T. KREDIET & L. ARISZ. 1992. Discrepancy between circadian rhythms of inulin and creatinine clearance. J. Lab. Clin. Med. **120:** 400–407.
33. LABRECQUE, G., P. M. BELANGER, F. DORE & M. LALANDE. 1988. 24 hours variations in the distribution of labelled microspheres to the intestine liver and kidneys. Ann. Rev. Chronopharmacol. **5:** 445–448.
34. MILLAR-CRAIG, M. W., C. N. BISHOP & E. B. RAFTERY. 1978. Circadian variation of blood pressure. Lancet 795–797.
35. HILFENHAUS, M. 1976. Circadian rhythm of the renin angiotensin aldosterone system in the rat. Arch. Toxicol. **36:** 305–316.
36. BARTTER, F. C., J. C. CHAN & J. W. SIMPSON. 1979. Chronobiological aspects of plasma renin activity, plasma aldosterone and urinary electroltyes. *In* Endocrine Rhythms. D. T. Krieger, Ed.: 693–699. Raven Press. New York.
37. BEILIN, L. J., J. DEACON, C. MICHAEL, R. VANDONGEN, C. LALOR, A. BARDEN & L. DAVIDSON. 1982. Circadian rhythms of blood pressure and pressor hormones in normal and hypertensive pregnancy. Clin. Exp. Pharmacol. Physiol. **9:** 321.
38. BULTASOVA, H., A. VESELKOVA, V. BRODAN & P. PINSKER. 1986. Circadian rhythms of urinary sodium, potassium and some agents influencing their excretion in young borderline hypertensives. Endocrinol. Exp. **20:** 359–369.
39. PICKERING, T. G. & G. D. JAMES. 1993. Determinants and consequences of the diurnal rhythm of blood pressure. Am. J. Hypertens. **6:** 166S–169S.
40. COCA, A. 1994. Circadian rhythm and blood pressure control: Physiological and pathophysiological factors. J. Hypertens. **12:** S13–S21.
41. MANCIA, G., M. DI RIENZO & G. PARATI. 1993. Ambulatory blood pressure monitoring use in hypertension research and clinical practice. Hypertension **21:** 510–524.
42. O'BRIEN, E., J. SHERIDAN & K. O'MALLY. 1988. Dippers and non dippers. Lancet **11:** 397.
43. SPINGATE, J. E., L. G. FELD & D. GANTEN. 1993. Renal function in hypertensive rats transgenic for mouse renin gene. Am. J. Physiol. **266:** F731–F737.
44. LEMMER, B., A. MATTES, M. BOHM & D. GANTEN. 1993. Circadian blood pressure variation in transgenic hypertensive rats. Hypertension **22:** 97–101.
45. LEMMER, B., K. WITTE, D. MINORS & J. M. WATERHOUSE. 1995. Circadian rhythms of heart rate and blood pressure in four strains of rat: Difference due to, and separate from, locomotor activity. Biol. Rhythm Res. In press.
46. BUZIO, C. A., L. ARISI, F. CAPANI, R. BARANI & P. QUARETTI. 1987. Circadian rhythm of proteinuria in normal subjects but not in patients with glomerulonephritis. Ann. Clin. Res. **19:** 30–33.
47. CAPRANI, F., A. D'EIMILIO, R. MARCHIOLI, A. NICOLUCCI, E. VITACOLONNA, P. MONDUNI & S. SENSI. 1988. Circadian rhythm of albumin excretion rate in healthy subjects. Chronobiologia **15:** 213–216.
48. KOOPMAN, M. G., R. T. KREDIET, F. J. M. ZUYDERHANDT, E. A. M. DEMOOR & L. ARISZ. 1987. Circadian rhythm of urinary β2-microglobulin excretion in patients with a nephrotic syndrome. Nephron **45:** 140–146.
49. REDON, J., Y. LIAO, J. V. LOZANO, A. MIRALLES, J. M. PASQUAL & R. S. COOPE. 1994. Ambulatory blood pressure and microalbuminuria in essential hypertension: Role of circadian variability. Am. J. Hypertens. **12:** 947–953.
50. BIANCHI, S., X. BIGAZZI, G. BALDARI, G. SGHERRI & V. M. CAMPESE. 1991. Diurnal

variations of blood pressure and microalbuminuria in essential hypertension. Am. J. Hypertens. **7:** 23–29.
51. CAMBAR, J., CH. TOUSSAINT, F. LEMOIGNE, P. CALES & R. CROCKETT. 1981. Variations circadiennes de l'excrétion des électrolytes et des substances azotées dans l'urine de souris et de rat. J. Physiol. **77:** 887–890.
52. KOOPMAN, M. G., G. C. M. KOOMEN, B. A. C. VAN ACKER & L. ARISZ. 1994. Urinary sodium excretion in patients with nephrotic syndrome and its circadian variation. Q. J. Med. **87:** 109–117.
53. PORTALUPPI, F., L. MONTANARI, M. FERLINI & P. GILLI. 1990. Altered circadian rhythms of blood pressure and heart rate in nonhemodialysis chronic renal failure. Chronobiol. Int. **7:** 321–327.
54. KORZETS, Z., A. ERDBERG, E. GOLAN & J. BERNHEIM. 1994. Does diurnal variations in blood pressure exist in CAPD patients? Nephrol. Dial. Transplant. **9:** 274–276.
55. BELL, G. M., W. REID, D. J. EWING, A. DOIG, A. D. CUMMING, M. L. WATSON & B. F. CLARKE. 1987. Abnormal diurnal urinary sodium and water excretion in diabetic autonomic neuropathy. Clin. Sci. **73:** 259–265.
56. REDON, P., M. LAVAL & J. CAMBAR. 1988. Chronobiological aspects of the renin angiotensin system: Physiology and physiopathology. Ann. Rev. Chronopharmacol. **6:** 183–223.
57. MODLINGER, R. S., K. SHARIF-ZADEH, N. H. ERTEL & M. GUTKIN. 1976. The circadian rhythm of renin. J. Clin. Endocrinol. Metab. **43:** 1276–1282.
58. CUGINI, P., D. SCAVO, G. CORNELISSEN, J. Y. LEE, T. NEUCCI & F. HALBERG. 1981. Circadian rhythms of plasma renin, aldosterone and cortisol on habitual and low dietary sodium intake. Hormone Res. **15:** 7–27.
59. RICHARDS, A. M., M. G. NICHOLLS, E. A. ESPINER, H. IKRAM, M. CULLENS & D. HINTON. 1986. Diurnal patterns of blood pressure, heart rate and vasoreactive hormones in normal man. Clin. Exp. Ther. Pract. **A8:** 153–166.
60. RODRIGUEZ-SARGENT, C., J. L. CANGIANO & M. MARTINEZ MALDONADO. 1982. Circadian rhythms of plasma renin levels in the anesthetized rat. Renal. Physiol. **5:** 53–56.
61. LAVAL, M. 1988. Etude expérimentale et clinique de la structure temporelle du système rénine angiotensine. Doctorat d'Etat de Pharm., Univ. Bordeaux II, n°2.
62. BAYLIS, C. 1993. Renal responses to acute angiotensin II inhibition and administered angiotensin II in the aging conscious, chronically catheterized rat. Am. J. Kidney Dis. **22:** 842.
63. GEORGE, C. P. L., F. H. MESSERLI, J. GENEST, W. NOWACYNSKI, R. BOUCHER, O. KUCHEL & M. ROJO-ORTEGA. Diurnal variations of plasma vasopressin in man. J. Clin. Endocrinol. Metab. **41:** 332–338.
64. BOWDEN, R. E., J. H. WARE, D. L. DE METS, & H. R. KEISER. 1977. Urinary excretion of immunoreactive prostaglandin E: A circadian rhythm and the effect of posture. Prostaglandins **14:** 151–161.
65. DONCKIER, J., J. V. ANDERSON, T. YEO & S. R. BLOOM. 1986. Diurnal rhythm in the plasma concentrations of atrial natriuretic peptide. New Engl. J. Med. **315:** 710–711.
66. HANSEN, P. M. P. P. M. GODDIJN, A. KOFOED-ENEVOLDSEN, K. M. VAN TOL, H. J. G. BILO & T. DECKERT. 1995. Diurnal variation in glomerular charge selectivity, urinary albumin excretion and blood pressure in insulin dependent diabetic patients. Kidney Int. **48:** 1559–1562.

Endocrine Mechanisms of Blood Pressure Rhythms

FRANCESCO PORTALUPPI, LUCIANA VERGNANI,
ROBERTO MANFREDINI, AND CARMELO FERSINI

Hypertension Unit
First and Second Institute of Internal Medicine
University of Ferrara
Ferrara, Italy

The cardiovascular system is mainly controlled by neurohumoral influences. Not surprisingly, its temporal organization is also centered on the same neurohumoral mechanisms. The autonomic and generally the monoaminergic systems appear to integrate the major driving factors of temporal variability. On the other hand, many clinical manifestations of cardiovascular disease reflect the temporal structure of the underlying pathophysiologic mechanisms. The main sources of temporal variability in cardiovascular physiology and pathophysiology are external interfering stimuli such as physical and mental activity or stress, the sleep-wake cycle, and endogenous rhythmicity, even though it is often impossible to clearly separate the relative influence of each factor.

The sleep-wake cycle is a coordinated process involving rhythmic changes in sensory, motor, autonomic, hormonal, and cerebral processes, acting through the alternating prevalence of activating and deactivating systems.[1] Physiologic substances known to determine sleep induction or to activate arousal also have established actions on the cardiovascular system. Examples are melatonin, arginine vasopressin, vasoactive intestinal peptide, somatotropin, insulin, steroid hormones and their metabolites, and serotonin among the sleep factors and corticotropin-releasing factor (CRF), corticotropin (ACTH), TRH, endogenous opioids, and prostaglandin E_2 among the waking factors. Their temporal pattern of variation is bound to be reflected in phasic oscillations of the cardiovascular function. Although secretion of these substances is typically episodic, the probability of secretory episodes is "gated" by mechanisms that are coupled either to sleep itself or to an endogenous pacemaker that usually is predominantly (though not only) circadian.

Interactions between feedback responses to occasional stimuli and endogenous circadian rhythmicity appear to be biunivocal. For instance, changes induced by external stimuli on monoaminergic neurotransmitter synthesis, release, and activity, particularly autonomic activity, modulate endogenous rhythmicity.[2–4] These influences are not simply superimposed on, but are modulated in turn by rhythms endogenously generated by biologic pacemakers.[5–7] Cyclic variations in autonomic nervous system activity play an important role not only in determining the sleep-wake process, but also in mediating the influence of sleep and wakefulness on cardiovascular function.[1] Feedback responses to environmental stimuli may synchronize or mask the endogenous rhythmicity of various components of the cardiovascular system. On the other hand, the immediate adaptation of the cardiovascular system to the demands of the environment is modulated by the circadian-time–dependent responsiveness of the biologic oscillators. In mammals, the major circadian oscillator is located in the suprachiasmatic nucleus of the hypothalamus,[8] and its main hormonal effector is melatonin. Selective lesion of the suprachiasmatic nucleus in rats abolishes circadian rhythmicity in blood pressure (BP) and heart rate (HR) without affecting the sleep-wake and motor activity cycles.[9,10] Pineal melatonin production and secretion directly affect circadian rhythmicity of cardiac function, namely, the diurnal changes in (Ca^{2+}-Mg^{2+})-dependent ATPase activity in the myocardial sarcolemma.[11]

These notions clearly indicate that knowledge of the temporal organization of neurohumoral controlling mechanisms is essential for understanding the temporal variability of cardiovascular function, including BP. This review analyzes what is known about neuroendocrine integration of the biologic rhythms that may affect cardiovascular function in general and BP in particular.

BIOGENIC AMINES AND AUTONOMOUS NERVOUS SYSTEM ACTIVITY

By far the most studied in detail, circadian rhythms in autonomous nervous system activity are well known, with a diurnal sympathetic and a nocturnal vagal prevalence.[12–29] In man, both norepinephrine and epinephrine plasma levels peak in the morning and reach their lowest values at night during sleep.[15,30–34] Urinary catecholamine excretion also shows a circadian rhythm.[35–37] These patterns are also present in constantly recumbent subjects, who show a strong relationship between circadian changes in plasma dopamine levels and those in norepinephrine and epinephrine throughout a 24-hour period, indicating dopaminergic modulation of the circadian variations in sympathetic nervous system activity.[38]

In rats, turnover of central dopamine and serotonin and clearance of their acid metabolites from the brain exhibit significant variations, with increased turnover rates in the second half of the resting and first half of the active period.[39,40] Circadian variations are also found in the extracellular concentration of dopamine at quantitative microdialysis.[41]

Norepinephrine precursor tyrosine in brain and its flow into brain (calculated from plasma amino acid concentrations) show circadian fluctuations in rats.[42] Circadian variations in biogenic amines are seen in the median eminence and in the intermediate and posterior lobes of the pituitary gland of the rat.[43,44] Parallel fluctuations are seen in the concentration of the norepinephrine metabolites 3-methoxy-4-hydroxyphenylglycol in the hippocampus and 3,4-dihydroxyphenylglycol in the hypothalamus and occipital and parietal cortex of rat brain.[45] In the hypothalamus of the mouse, concentrations of norepinephrine, dopamine, and serotonin vary in a rhythmic pattern by as much as twofold. Daily alterations in parent biogenic amines are reflected by concurrent changes in their metabolites.[46] Also, brain stem (but not adrenal) phenylethanolamine *N*-methyltransferase (the enzyme that converts norepinephrine to epinephrine) activity shows a distinct diurnal fluctuation, with a nadir at 7 AM and a peak at 3 PM.[47]

Epinephrine in the peripheral organs of mammals is mostly derived from the adrenal gland, and circulating epinephrine is taken up by sympathetic nerve endings in the organs. In rats, 24-hour changes in nucleolar diameters, an ultrastructural feature of cellular activation, occur in adrenomedullary chromaffin cells with a peak in the late dark phase and a minimum in the late light phase.[5] The epinephrine content of the peripheral organs of the rat increases after electric foot-shock in a circadian-time–dependent way,[48] suggesting that epinephrine in body organs plays some role in the responses of the sympathetic nervous system to daily activities and stressful conditions.

In humans, catecholamine sulfates, biologically inactive, have a circadian pattern opposite those of free catecholamines, biologically active, in normotensive recumbent subjects.[49] The initial part of sleep is associated with a peak in plasma noradrenaline and adrenaline sulfates, whereas before awakening both free noradrenaline and adrenaline suddenly increase while their sulfoconjugate counterparts decrease. This suggests that the nocturnal decrease in biologically active catecholamines may be due to activation of an endogenous system of sulfoconjugation; conversely, deactivation of the same system may contribute to the sudden morning increase in biologically active catecholamines. Genetic control of the process of sulfoconjugation and deconjugation of catecholamines is well established and is regulated mainly by two enzymes, phenolsulfotransferase and aryl sulfatase.[50] Modulation of the gene expression of these enzymes may contribute to circadian variations in sympathetic activity.

In fact, environmental factors such as sleep/activity and light/dark changes as well as endogenous rhythmicity may contribute to the determination of circadian monoaminergic patterns. Norepinephrine (but not epinephrine) levels show a direct relationship to sleep and posture.[15] Rhythm in the plasma norepinephrine metabolite 3-methoxy-4-hydroxyphenylglycol is evoked by changes in physical activity and posture, whereas major components of the diurnal plasma variation of the dopamine metabolites homovanillic acid and 3,4-dihydroxyphenylacetic acid are regulated by a circadian oscillator that is independent of sleep or activity.[45,51] In healthy volunteers required to eat and walk around every hour to prevent sleep for 24 hours, persistence of a nocturnal decrease in norepinephrine was documented.[52] This suggests that norepinephrine secretion may be influenced by a biologic clock whose activity becomes visible when the influence of sleep, posture, and activity is removed.

Convincing evidence suggests that circadian variations in autonomous nervous system activity and biogenic amines are not explained by a single controlling influence. A significant component of the circadian rhythmicity of norepinephrine, dopamine, and serotonin is under the control of a circadian oscillator, whereas epinephrine seems more dependent on external stimuli.

The same is true at the receptor level. Radioligand binding studies have revealed the existence of circadian rhythms in several receptors of the rat forebrain, particularly α- and β-adrenergic, muscarinic cholinergic, and striatal dopamine receptors.[53–56] The endogenous nature of the rhythms in α- and β-adrenergic receptors of the rat forebrain and hypothalamus is strongly suggested by their persistence in the absence of time cues.[54,56] Moreover, ablation of the suprachiasmatic nuclei abolishes these circadian rhythms,[57] whereas sleep deprivation has almost no effect.[53,57] In the cerebral cortex the circadian rhythm in norepinephrine-stimulated cyclic AMP production is a biologic response to the circadian rhythms in α- and β-adrenergic receptors. Circadian rhythms in the postsynaptic responsiveness of hippocampal pyramidal cells to serotonin, norepinephrine, gamma-aminobutyric acid, and acetylcholine have been demonstrated in the rat brain.[58] Short-term exposure to constant light or darkness produced a phase shift in serotoninergic and cholinergic rhythms, suggesting their endogenous nature and their synchronization to clock time by the light/dark cycle. Diurnal rhythms in responsiveness to serotonin and norepinephrine show phase shifts from the December–April to May–August period in rats entrained to the 12:12-hour light/dark cycle, suggesting seasonal modulation of these rhythms.[58]

Circadian variations in the blood-brain barrier of rats acutely subjected to adrenaline-induced hypertension are also demonstrated with lower albumin leakage into the brain at night than during the day, possibly related to lower β-adrenoceptor sensitivity.[59,60]

Finally, circadian rhythms are exhibited by putative second messenger response system cGMP (in all brain regions except the parietal cortex) and cAMP (in piriform, temporal, occipital, cingulate, and parietal cortex, amygdala, and nucleus accumbens). When these rhythms are correlated with those of norepinephrine and dopamine metabolites, only partial parallelism is found.[45] Therefore, the rhythmicity measured in adrenergic receptors is not likely to be a specific response to rhythmic changes in adrenergic transmitter release. Instead, a stronger correlation is found with putative second messenger response systems, especially cGMP. Rhythms in cGMP are parallel in form and region to those in the α_1-adrenergic receptor and may act as a second messenger for that receptor.[45]

The driving role of the sympathetic system on the circadian rhythm of BP is well established. Circadian rhythms of BP and sympathoadrenergic activity (catecholamine concentrations in plasma and urine, cAMP concentrations in plasma or lymphocytes, and β-adrenoceptor density and affinity on lymphocytes) are synchronous not only in normotensive subjects, but also in those with essential hypertension and in states of abnormal catecholamine secretion with an abnormal circadian BP curve.[61] Intraarterial studies on the effect of a variety of antihypertensive agents indicate that the morning rise (both before awakening and upon arousal) is due to increased α-adrenoceptor activity.[62] Propranolol reduces the pressor range (difference between basal and maximum BP readings) in most patients, and an inverse relationship can be demonstrated between changes in pressor range and noradrenaline.[63] Circadian variations in sympathoadrenal and pressor reactivity to exercise are strongly correlated.[64] Spectral analysis of BP and interbeat interval recordings provides markers of autonomic activity and arterial baroreflex sensitivity. Using this technique, a recent study demonstrated that a clear 24-hour rhythm in sympathetic and vagal tone (but not in arterial baroreflex sensitivity) exists independent of changes in activity and position.[29]

Plasma and urinary norepinephrine[32,65–69] strikingly correlate with recumbent 24-hour mean arterial pressure in both normal and hypertensive subjects, even though elderly subjects appear to lose or invert this correlation.[31,67] Since bromocriptine treatment is associated with depression in mean arterial pressure throughout the 24-hour cycle and a lessening of the relationship with plasma norepinephrine, circadian variations in sympathetic nervous system activity and in patients with essential hypertension appear to be modulated by a central and/or peripheral dopaminergic mechanism.[65,70]

Ultradian oscillations of autonomic function are also a common property of mammalian systems. Brain stem networks are inherently capable of generating rhythms in sympathetic nerve discharge at 2–6 cycles per second.[71] In rats, spectrum analysis reveals ultradian norepinephrine rhythms with spectral peaks around 90 and 40 minutes. These rhythms can be synchronized by some environmental stimuli.[72] In the hypothalamus of the conscious rat, release rates of dopamine, noradrenaline, and adrenaline fluctuate according to two ultradian rhythms with different frequencies (dopamine, 92 minutes/cycle and 8 hours/cycle; noradrenaline, 92 minutes/cycle and 12 hours/cycle; adrenaline, 99 minutes/cycle and 12 hours/cycle). Release rates of dopamine and adrenaline are similar during light and dark periods, whereas the release rate of noradrenaline is slightly lower in the dark period than in the light.[73] In the rabbit, the resting release of hypothalamic catecholamines rhythmically varies with time at approximately 70-minute cycles.[74] Also in humans, plasma catecholamines have significant ultradian variations that persist after bilateral adrenalectomy.[75] Circaoctohoran rhythm of urine epinephrine excretion has also been reported in man.[76] Finally, endogenous 12-hour pulses of receptor escape from atropine blocking of acetylcholine have been demonstrated.[77]

Overall, the ultradian organization of biogenic amines and autonomous nervous system activity is less studied than is its circadian counterpart. Methodologic difficulties contribute to this fact, considering that ultradian fluctuations are not readily discernible from wider circadian and exogenous fluctuations. The ultradian organization of BP and its autonomic integration are analyzed in detail in another chapter of this book.

RENIN-ANGIOTENSIN-ALDOSTERONE SYSTEM

Prorenin, plasma renin activity (PRA), aldosterone, and angiotensin II are secreted with circadian rhythms in healthy and hypertensive subjects, and their temporal variations are independent of recumbency.[66,78–87] Some studies failed to demonstrate circadian rhythm in PRA,[88–90] showing only ultradian, nocturnal oscillations strongly related to a sleep pattern, namely, the rapid-eye-movement phase.[89,91] On the other hand, others documented circadian secretory patterns of renin and aldosterone with peaks in the morning, persisting to a certain extent in subjects who remain awake during the night.[92] This contradiction is more apparent than real considering that the rapid-eye-movement phase occurs preferentially in the second part of nocturnal sleep and may thus drive an early morning increase in PRA.

The sympathetic nervous system influences renin-aldosterone secretion.[93] In addition, in the rat pineal gland, norepinephrine exerts a negative effect on angiotensin-converting enzyme circadian activity.[94] Again, dopaminergic pathways modulate circadian variations in renin, angiotensin II, and aldosterone.[32,95,96] ACTH plays the main role in regulating plasma aldosterone oscillations, stimulating aldosterone secretion at night, whereas the renin-angiotensin system and dopaminergic pathways are additional modulators during the day.[32,93,97,98]

In normal- and high-renin hypertensive subjects, circadian cyclicity of aldosterone has a parallel acrophase with renin oscillations, which may modulate circadian secretion of cortisol. On the other hand, in low-renin hypertensive patients the renin-angiotensin system does not regulate the temporal organization of aldosterone circadian rhythm.[99] Moreover, the renin-angiotensin system may interact with the adrenal axis to modulate cortisol concentrations.[100]

Altogether, convincing evidence exists of a temporal organization with endogenous and exogenous components in circulating levels of renin, angiotensin, and aldosterone, whereas no data are available on the circadian rhythmicity of the tissue renin system. The temporal organization of the renin-angiotensin-aldosterone system is apparently interconnected with that of the autonomic, dopaminergic, and hypothalamic-pituitary-adrenal systems.

HYPOTHALAMIC-PITUITARY-ADRENAL SYSTEM

Glucocorticoids influence the cardiovascular system and the electrolyte balance acting directly on epithelial cells, smooth muscle cells, and myocytes of the heart. They also indirectly enhance vascular reactivity to other factors.[93,101] In humans, exogenous ACTH administration stimulates glucocorticoid secretion and induces a significant increase in BP, which is associated with elevated cardiac output and no variations in peripheral resistance[102,103] as well as elevated plasma volume but no significant increase in extracellular fluid.[104] In addition, ACTH-induced hypertension is influenced by variations in sodium intake.

Exogenous glucocorticoid administration[105] eliminates or reverses a physiologic nocturnal fall in BP, a phenomenon that is also found in patients with Cushing's syndrome.[106,107] ACTH-dependent adrenal steroids were demonstrated as major determinants of hypertension in rat strain TGR(mREN2)27, which has an additional mouse *Ren2* gene.[108] In these rats, BP abnormally peaks during the resting phase coincident with a twofold increase in plasma corticosterone levels, whereas aldosterone shows no rhythm.[109] In other words, changes in the circadian pattern of ACTH-dependent steroids are paralleled by changes in the circadian BP pattern of these transgenic rats. These findings indicate that the hypothalamic-pituitary-adrenal system plays a major role in modulating the circadian rhythm of BP.

ACTH and cortisol have a parallel temporal secretion profile with a peak in the early morning, elevation during late sleep, and minimal secretory concentration around midnight. Evidence on the adaptation of endogenous circadian rhythmicity of the adrenal system to variations in the sleep-wake cycle indicates that the morning peak and nocturnal nadir are modulated by different mechanisms.[110] Timing of the nocturnal secretory minimum seems to be genetically determined.[111] Furthermore, the early morning peak takes less time to readjust after shifts in the sleep-wake cycle, and carbohydrate intake may help to accelerate it.

Circadian ACTH rhythm drives the 24-hour profile of adrenal secretion, which is also influenced by a time-dependent variation in adrenal responsiveness to ACTH. Moreover, periodic fluctuations in CRH concentrations modulate the ACTH secretory pattern.[112]

Secretion of the adrenal steroids 11-hydroxyandrostenedione, dehydroepiandrosterone, and androstenedione is synchronized with the circadian profile of cortisol.[112] Normotensive and hypertensive subjects have a similar circadian rhythm of 18-hydroxycorticosterone.[113]

Catecholamines may modulate corticotropin function and rhythmicity.[47] Pharmacologic destruction of the pathway that conveys most catecholaminergic fibers from the locus coeruleus and the A1 and A2 medullary areas to the paraventricular nuclei causes obliteration of the circadian rhythmicity of ACTH and cortisol and the appearance in its place of ultradian oscillations of reduced amplitude.[114]

Evidence also suggests a regulatory role of serotoninergic patterns on the circadian profile of glucocorticoids.[115] On the other hand, circadian rhythmicity of corticosteroid levels may also occur independent of important norepinephrine and serotonin depletion in the central nervous system.[116] ACTH plays a major modulatory role in the pattern of aldosterone secretion.[93,98] Finally, glucocorticoids induce liver production of renin substrate, leading to increased circulating levels of angiotensin II.

HYPOTHALAMIC-PITUITARY-THYROID SYSTEM

Thyroid-stimulating hormone (TSH) secretion has a circadian pattern, independent of sleep, with a nocturnal rise, which begins in late afternoon and peaks after midnight.[8] Thyroid-stimulating hormone secretory oscillations result from a series of pulses throughout 24 hours with greater magnitude and frequency at night. Peak secretory concentrations are paralleled by a maximum norepinephrine content in the thyroid gland.[117,118] Moreover, pharmacologic studies with selective adrenergic agonists and antagonists confirm that catecholaminergic pathways influence TSH rhythmicity through α_2- and β-receptors.[117]

Dopamine modulates the TSH secretory pattern, blunting the pulse amplitude, and its effect is not decreased during the night. Hence, the nocturnal increase in TSH secretion is not regulated by reduced dopaminergic inhibition,[119–121] and the neuroendocrine mechanisms that drive the nocturnal peaks are still unknown.

Finally, the hypothalamic-pituitary-thyroid system acts on the cardiovascular system at multiple levels: direct positive inotropic and chronotropic effects on the heart, increased metabolic rate, and positive regulation of the sensitivity of β-adrenergic receptors of the myocardium. Based on the multiple influences exerted on the cardiovascular system by these hormones, a possible influence on BP levels and rhythmicity should be expected. Although hypertension is a well known feature of both hyper-[122,123] and hypothyroidism,[124–126] no data are available on the possible changes in BP rhythmicity in these conditions.

OPIOID SYSTEM

There are multiple types of both opioid peptides and receptors. Endogenous opioids derive from proteolytic cleavage of at least three separate precursors, proopiomelanocortin (POMC), preproopioenkephalin, and preproenkephalin B.[127–129] The bioactive opioid peptides are produced by specialized cells distributed throughout the central nervous system (in forebrain and hindbrain nuclei involved in baroregulation, sympathoadrenal activation, and several other vital autonomic functions) and in peripheral neural elements (in autonomic ganglia, adrenal gland, heart, vascular tissue, and kidneys). Whether functioning as hormones, neuromodulators, or neurotransmitters, opioids influence immune, endocrine, and cardiovascular systems.

In the heart of both normotensive and hypertensive rats, metenkephalins exhibit a circadian rhythm.[130] Moreover, a temporal secretory pattern exists in β-endorphin concentrations.[131,132] Involvement of both sympathetic and parasympathetic systems in the cardiovascular actions of these peptides is established. Opioids, administered centrally or systemically, produce both stimulatory and inhibitory effects on HR as well as hypotension and hypertension depending on dose, subtypes involved, specificity of receptor binding, animal species, and site of administration.

Many cardiovascular actions of opioid peptides and opiate receptors are supported by their effects on central nervous system areas and pathways involved in the regulation of cardiovascular homeostasis and profile. In man, endogenous opioids are involved in determination of the nocturnal decrease in BP.[133,134] First, injection of synthetic opioid peptides into the cisterna magna of anesthetized dogs induces hypotension, and this action is reversed by naloxone.[133] Moreover, the nucleus of the solitary tract, which shows a high concentration of opioid receptors, plays a major role in blood pressure regulation.[135] Evidence exists that morphine-like peptides can modulate the baroreceptor reflex at the level of the nucleus of the solitary tract as well as does sleep. Prolactin and growth hormone concentrations are elevated during the nocturnal BP fall, and endogenous opioids stimulate the secretion of these hormones.[136,137] ∂-Opioid receptors may play a role in nocturnal suppression of the sympathetic nervous system and hypothalamic-pituitary-adrenal axis, considering their known inhibitory effects[138,139] and the relationships of the respective circadian rhythms.

Physical exercise activates the inhibitory effect of the central opioid system on blood pressure and determines postexercise hypotension. Naloxone pretreatment at high doses blunts the BP decrease after running in hypertensive rats but not in normotensive animals.[140] On the other hand, infusion of naloxone attenuates the hemodynamic, sympathoneural, and hypothalamic pituitary responses after exercise in normotensive men, but it does not alter the aftereffects of exercise in hypertensive subjects, suggesting a non-opioidergic pathway for postexercise hypotension in hypertensive humans. Furthermore, these observations suggest an impaired endogenous opioid system response to exercise in mild hypertension.[141]

Most cardiac effects of opioid peptides have been ascribed to their presynaptic modulation of acetylcholine and catecholamine release at the level of the pathways supplying heart muscle.[142] On the other hand, the cardiac sarcolemma possesses both δ- and κ-opioid receptors, suggesting a direct action of opioids on myocardial cells.[143]

To date, however, little is known about the mechanisms of this involvement, especially in humans. Two main difficulties have slowed down the progress in this field, marked species specificity in the anatomic and functional organization of the opioid system, so that animal models are only partially representative of the human situation, and poor receptor selectivity of most agonist and antagonist substances that have been used until recently to investigate the opioid system.

VASOACTIVE PEPTIDES

The family of natriuretic peptides include A-type natriuretic peptide (ANP) and B-type natriuretic peptide (BNP), both synthesized by atrial cells, and C-type natriuretic peptide (CNP) isolated from brain tissue. Among these three peptides, the CNP gene is expressed at low levels in cardiac cells. The major sources of ANP are atrial granules, but evidence of ANP gene transcription has been found outside the heart, in lung, kidney, aorta, pituitary, adrenal medulla, and central nervous system.

The most important stimulus to ANP secretion is increased atrial transmural pressure.[144] In addition, adrenergic agonists, acetylcholine, glucocorticoids, endothelin, vasopressin, calcitonin gene-related peptide, and angiotensin II may regulate ANP secretion.[145]

ANP at physiologic concentrations blunts angiotensin II and aldosterone secretion, plasma renin activity, and catecholamine concentrations, whereas it increases sodium excretion and plasma and urinary cGMP concentrations. The ANP final effect on the cardiovascular system leads to lower BP and total peripheral resistance. Moreover, the brain pool of CNP has been identified in specific brain areas involved in cardiovascular regulation, suggesting interactions between circulating ANP and BNP and central CNP with the autonomic nervous system. Finally, ANP modulates nitric oxide production in humans.[146]

A significant circadian rhythm in plasma concentrations of ANP, unrelated to postural changes, with fluctuations of ~10 pmol/l in magnitude and peaking between 11 PM and 4 AM when plasma renin activity and aldosterone levels are still in the lower part of their circadian rhythms, is established in normal and hypertensive subjects.[32,147–154] Fluctuation in plasma ANP concentrations over 24 hours is paralleled by a circadian variation in numeric densities of secretory granules of myoendocrine cells in rat heart[155] and by a circadian rhythm in plasma concentration of second messenger cGMP in humans.[156] Alterations in the circadian rhythm of BP are paralleled by consistent changes in the circadian rhythm of ANP in chronic renal failure[157] and congestive heart failure[158] both before and after treatment.[159] An inverse circadian relationship between ANP and BP rhythms was reported in both normotensive[149] and hypertensive[154] men. Recently, this relationship was confirmed in normotensive individuals studied 5 years apart; each year the circadian parameters of ANP (mean 24-hour levels, 24-hour amplitude of variation, and time of occurrence of peaks and troughs) showed a significant negative correlation with corresponding BP parameters.[160] The negative correlation of ANP and BP throughout the 24 hours in normal subjects is not lost in patients with essential hypertension.[154] These observations suggest that natriuretic peptides contribute to the circadian regulation of BP.

Another family of peptides has been recognized as a physiologic regulator of vasomotor tone. Endothelins (ET), 21 amino acid peptides, were identified, from animal and human endothelial cell cultures in three isoforms, ET-1, ET-2, and ET-3. Evidence suggests that the major sources of ET-1 are endothelial cells, vascular smooth muscle cells, and neurons of the paraventricular and supraoptic nuclei of the hypothalamus.[161] Although ET-2 and ET-3 are widely distributed in different tissues, the precise source is still unknown. ET-1 shows a positive inotropic effect through direct action on the heart. Moreover, it produces an increase in BP and sympathetic activity, suggesting a role in the brain's modulation of cardiovascular function.

Other vasoactive factors, such as epinephrine, angiotensin II, and arginine vasopressin, regulate ET-1 gene transcription and increase peptide secretion. Nitric oxide seems to inhibit ET-1 release, suggesting an intracellular modulatory mechanism.[162] Higher ET-1 levels have been found in hypertensive subjects.[163] Endothelins appear to stimulate the synthesis of the ANP prohormone (i.e., increase ANP messenger RNA) as well as the release of ANP.[164]

In normal subjects, ET-1 shows a temporal secretion profile with a peak occurring in early morning.[165] In subjects with elevated cardiovascular risk factors, circadian rhythm is still present but with a peak shift towards late morning or afternoon.[165]

Calcitonin gene-related peptide (CGRP) is a 37 amino-acid peptide produced by tissue-specific alternative RNA processing of the calcitonin gene.[166–168] CGRP binding sites are distributed in different parts of the central and peripheral nervous system of animals and humans,[169–173] and CGRP-containing fibers have been found throughout the cardiovascular system, in the heart, especially within the coronary arteries, in the sinoatrial and atrioventricular nodes, and in papillary muscles.[174,175] CGRP concentration is high in the adventitia of arteries and veins,[174] and perivascular CGRP fibers were found in all vascular beds.[14] Evidence indicates that blood levels of CGRP represent a spillover phenomenon related to the release of the peptide from nerve terminals to promote vasodilation.[176,177] Hence, CGRP appears to have a role in the regulation of peripheral vascular tone and blood flow.[178–180] Studies both *in vivo* and *in vitro* suggested that CGRP is a potent vasodilator in human and animal vessels.[181–189] The injection of CGRP into the coronary system of rat, pig, and man results in dose-related increments in coronary flow.[183,184] Systemic infusion of CGRP and perfusion of isolated rat kidney decrease vascular resistance and enhance both renal blood flow and glomerular filtration.[188]

Plasma CGRP levels increase in healthy subjects after assumption of the upright position and low-dose infusion of angiotensin II.[190] Hence, the renin-angiotensin system may modulate plasma CGRP secretion either directly through a vasopressor effect on the peripheral blood vessels or indirectly by involving neurohormonal mechanisms that modulate vascular tone and BP. In fact, acute activation of cholinergic transmission by the acetylcholinesterase inhibitor pyridostigmine bromide results in an increase in basal plasma levels of CGRP, whereas the muscarinic receptor antagonist pirenzepine fails to exert any significant effect.[191] This implicates the cholinergic system in modulating the circulating levels of CGRP in the absence of a tonic cholinergic contribution to CGRP secretion under physiologic basal conditions.

Circadian rhythm in plasma concentrations of CGRP is demonstrated in both normotensive[192,193] and hypertensive subjects.[194] An inverse circadian relationship between CGRP and BP rhythms was reported in both normotensive[192] and hypertensive[194] persons, suggesting that CGRP contributes to the circadian regulation of BP in healthy normotensive subjects and in those with uncomplicated essential hypertension.

Other vasoactive hormones are known about whose temporal variations little or nothing is known. For instance, circulating concentrations of plasma vasopressin show circadian rhythm,[195] but nothing is known of their temporal relationships with other vasoactive peptides and BP.

Temporal relationships in the circadian patterns of atrial natriuretic peptides, endothelins, CGRP, and BP imply that vasodilating and vasoconstrictive peptides participate in a finely tuned mechanism of 24-hour BP control in normotensive individuals. In uncomplicated essential hypertension the balance between dilating and constrictive factors seems preserved, as indicated by normal temporal relationships between the respective circadian patterns. On the other hand, the balance seems lost in complicated and secondary forms of hypertension, where the circadian patterns of vasoactive peptides have either shifted in phase or are lost. The ability of these peptides to modulate their antagonists further suggests that an imbalance in their complex interrelationships may represent a pathogenetic mechanism in hypertension. Further investigations of these relationships are warranted.

SUMMARY

The temporal organization of blood pressure is mainly controlled by neuroendocrine mechanisms. The monoaminergic systems appear to integrate the major driving factors of temporal variability, but evidence also indicates a role of the hypothalamic-pituitary-adrenal, hypothalamic-pituitary-thyroid, opioid, renin-angiotensin-aldosterone, and endothelial systems as well as other vasoactive peptides. Although their hormonal secretions are typically episodic, the probability of secretory episodes is "gated" by mechanisms that are coupled either to sleep or to an endogenous pacemaker which usually is predominantly (though not only) circadian. Many hormones with established actions on the cardiovascular system (arginine vasopressin, vasoactive intestinal peptide, melatonin, somatotropin, insulin, steroids, serotonin, CRF, ACTH, TRH, endogenous opioids, and prostaglandin E_2) are also involved in sleep induction or arousal. Hence, physical, mental, and pathologic stimuli, which may drive activation or inhibition of these neuroendocrine effectors of biologic rhythmicity, may also interfere with the temporal blood pressure structure. On the other hand, the immediate adaptation of the exogenous components of blood pressure rhythms to the demands of the environment are modulated by the circadian-time–dependent responsiveness of the biologic oscillators and their neuroendocrine effectors. These notions may contribute to a better understanding of the pathophysiology and therapeutics of changes in blood pressure.

REFERENCES

1. McGinty, D. & R. Szymusiak. 1994. Neurobiology of sleep. *In* Sleep and Breathing. N. A. Saunders & C. E. Sullivan, Eds., **71:** 1–26. Marcel Dekker, Inc. New York, Basel, Hong Kong.
2. Axelrod, J. 1974. The pineal gland: A neurochemical transducer. Science **184:** 1341–1348.
3. Zatz, M. & D. A. Mullen. 1988. Norepinephrine, acting via adenylate cyclase, inhibits melatonin output but does not phase-shift the pacemaker in cultured chick pineal cells. Brain Res. **450:** 137–143.
4. Stehle, J. H., N. S. Foulkes, C. A. Molina, V. Simonneaux, P. Pevet & P. Sassone-Corsi. 1993. Adrenergic signals direct rhythmic expression of transcriptional repressor CREM in the pineal gland. Nature **365:** 314–320.

5. KACHI, T., T. K. BANERJI & W. B. QUAY. 1984. Quantitative cytological analysis of functional changes in adrenomedullary chromaffin cells in normal, sham-operated, and pinealectomized rats in relation to time of day. I. Nucleolar size. J. Pineal Res. **1:** 31–49.
6. YOSHIDA, T. & G. A. BRAY. 1988. Effects of food and light on norepinephrine turnover. Am. J. Physiol. **254:** R821–R827.
7. CARNEIRO, R. C., J. CIPOLLA-NETO & R. P. MARKUS. 1991. Diurnal variation of the rat vas deferens contraction induced by stimulation of presynaptic nicotinic receptors and pineal function. J. Pharmacol. Exp. Ther. **259:** 614–619.
8. ASCHOFF, J., S. DAAN & G. A. GROOS. 1982. Vertebrate Circadian Systems. Springer-Verlag. Berlin, Heidelberg & New York.
9. JANSSEN, B. J., C. M. TYSSEN, H. DUINDAM & W. J. RIETVELD. 1994. Suprachiasmatic lesions eliminate 24-h blood pressure variability in rats. Physiol. Behav. **55:** 307–311.
10. WITTE, K., A. SCHNECKO, R. BUIJS & B. LEMMER. 1995. Circadian rhythms in blood pressure and heart rate in SCN-lesioned and unlesioned transgenic hypertensive rats (abstract). Biol. Rhythm Res. **26:** 458–459.
11. CHEN, L. D., D. X. TAN, R. J. REITER *et al.* 1993. In vivo and in vitro effects of the pineal gland and melatonin on [$Ca^{(2+)}$ + Mg^{2+}]-dependent ATPase in cardiac sarcolemma. J. Pineal Res. **14:** 178–183.
12. COWLEY, A. W. JR., J. F. LIARD & A. C. GUYTON. 1973. Role of baroreceptor reflex in daily control of arterial blood pressure and other variables in dogs. Circ. Res. **32:** 564–576.
13. EWING, D. J., D. Q. BORSEY, P. TRAVIS, F. BELLAVERE, J. M. NEILSON & B. F. CLARKE. 1983. Abnormalities of ambulatory 24-hour heart rate in diabetes mellitus. Diabetes **32:** 101–105.
14. CLEMENT, D. L., N. DE PUE, L. J. JORDAENS & L. PACKET. 1985. Adrenergic and vagal influences on blood pressure variability. Clin. Exp. Hypertens. [A] **7:** 159–166.
15. LINSELL, C. R., S. L. LIGHTMAN, P. E. MULLEN, M. J. BROWN & R. C. CAUSON. 1985. Circadian rhythms of epinephrine and norepinephrine in man. J. Clin. Endocrinol. Metab. **60:** 1210–1215.
16. BEXTON, R. S., H. O. VALLIN & A. J. CAMM. 1986. Diurnal variation of the QT interval—influence of the autonomic nervous system. Br. Heart J. **55:** 253–258.
17. MANCIA, G., G. PARATI, G. POMIDOSSI, R. CASADEI, M. DI RIENZO & A. ZANCHETTI. 1986. Arterial baroreflexes and blood pressure and heart rate variabilities in humans. Hypertension **8:** 147–153.
18. KASTING, G. A., D. L. ECKBERG, J. M. FRITSCH & C. L. BIRKETT. 1987. Continuous resetting of the human carotid baroreceptor-cardiac reflex. Am. J. Physiol. **252:** R732–R736.
19. PARATI, G., M. DI RIENZO, G. BERTINIERI *et al.* 1988. Evaluation of the baroreceptor-heart rate reflex by 24-hour intra-arterial blood pressure monitoring in humans. Hypertension **12:** 214–222.
20. BELL, L. B., P. K. DORWARD & C. D. RUDD. 1990. Influence of cardiac afferents on time-dependent changes in the renal sympathetic baroreflex of conscious rabbits. Clin. Exp. Pharmacol. Physiol. **17:** 545–555.
21. FURLAN, R., S. GUZZETTI, W. CRIVELLARO *et al.* 1990. Continuous 24-hour assessment of the neural regulation of systemic arterial pressure and RR variabilities in ambulant subjects. Circulation **81:** 537–547.
22. HAYANO, J., Y. SAKAKIBARA, M. YAMADA *et al.* 1990. Diurnal variations in vagal and sympathetic cardiac control. Am. J. Physiol. **258:** H642–H646.
23. KAMATH, M. V. & E. L. FALLEN. 1991. Diurnal variations of neurocardiac rhythms in acute myocardial infarction. Am. J. Cardiol **68:** 155–160.
24. MOLGAARD, H., P. D. CHRISTENSEN, K. E. SORENSEN, C. K. CHRISTENSEN & C. E. MOGENSEN. 1992. Association of 24-h cardiac parasympathetic activity and degree of nephropathy in IDDM patients. Diabetes **41:** 812–817.
25. GOLDSMITH, R. L., J. T. BIGGER, R. C. STEINMAN & J. L. FLEISS. 1992. Comparison of 24-hour parasympathetic activity in endurance-trained and untrained young men. J. Am. Coll. Cardiol. **20:** 552–558.

26. DETOLLENAERE, M. S., D. A. DUPREZ, M. L. DE BUYZERE, H. J. VANDEKERCKHOVE, G. G. DE BACKER & D. L. CLEMENT. 1993. 24 hour ambulatory blood pressure variability and cardiac parasympathetic function 2 and 6 weeks after acute myocardial infarction. Clin. Auton. Res. **3:** 255–259.
27. HARTIKAINEN, J., I. TARKIAINEN, K. TAHVANAINEN, M. MANTYSAARI, E. LANSIMIES & K. PYORALA. 1993. Circadian variation of cardiac autonomic regulation during 24-h bed rest. Clin. Physiol. **13:** 185–196.
28. SOMERS, V. K., D. PHIL, M. E. DYKEN, A. L. MARK & F. M. ABBOUD. 1993. Sympathetic-nerve activity during sleep in normal subjects. N. Engl. J. Med. **328:** 303–307.
29. VAN DE BORNE, P., H. NGUYEN, P. BISTON, P. LINKOWSKI & J. P. DEGAUTE. 1994. Effects of wake and sleep stages on the 24-h autonomic control of blood pressure and heart rate in recumbent men. Am. J. Physiol. **266:** H548–H554.
30. TURTON, M. B. & T. DEEGAN. 1974. Circadian variations of plasma catecholamine, cortisol and immunoreactive insulin concentrations in supine subjects. Clin. Chim. Acta **55:** 389–397.
31. PRINZ, P. N., J. HALTER, C. BENEDETTI & M. RASKIND. 1979. Circadian variation of plasma catecholamines in young and old men: Relation to rapid eye movement and slow wave sleep. J. Clin. Endocrinol. Metab. **49:** 300–304.
32. RICHARDS, A. M., M. G. NICHOLLS, E. A. ESPINER, H. IKRAM, M. CULLENS & D. HINTON. 1986. Diurnal patterns of blood pressure, heart rate and vasoactive hormones in normal man. Clin. Exp. Hypertens. [A] **8:** 153–166.
33. KAWANO, Y., T. KAWASAKI, N. KAWAZOE *et al.* 1990. Circadian variations of urinary dopamine, norepinephrine, epinephrine and sodium in normotensive and hypertensive subjects. Nephron **55:** 277–282.
34. KAWANO, Y., O. TOCHIKUBO, K. MINAMISAWA, E. MIYAJIMA & M. ISHII. 1994. Circadian variation of hemodynamics in patients with essential hypertension: Comparison between early morning and evening. J. Hypertens. **12:** 1405–1412.
35. LAKATUA, D. J., E. HAUS, F. HALBERG *et al.* 1986. Circadian characteristics of urinary epinephrine and norepinephrine from healthy young women in Japan and U.S.A. Chronobiol. Int. **3:** 189–195.
36. WESTERINK, B. H. & N. TEN KATE. 1986. 24 h excretion patterns of free, conjugated and methylated catecholamines in man. J. Clin. Chem. Clin. Biochem. **24:** 513–519.
37. LAKATUA, D. J., G. Y. NICOLAU, C. BOGDAN *et al.* 1987. Chronobiology of catecholamine excretion in different age groups. Prog. Clin. Biol. Res. **227:** 31–50.
38. SOWERS, J. R. & N. VLACHAKIS. 1984. Circadian variation in plasma dopamine levels in man. J. Endocrinol. Invest. **7:** 341–345.
39. LEMMER, B. & T. BERGER. 1978. Diurnal rhythm in the central dopamine turnover in the rat. Naunyn-Schmiedeberg's Arch. Pharmacol. **303:** 257–261.
40. HUTSON, P. H., G. S. SARNA & G. CURZON. 1984. Determination of daily variations of brain 5-hydroxytryptamine and dopamine turnovers and of the clearance of their acidic metabolites in conscious rats by repeated sampling of cerebrospinal fluid. J. Neurochem. **43:** 291–293.
41. SMITH, A. D., R. J. OLSON & J. B. J. JUSTICE. 1992. Quantitative microdialysis of dopamine in the striatum: Effect of circadian variation. J. Neurosci. Methods **44:** 33–41.
42. SCHWEIGER, U., M. WARNHOFF & K. M. PIRKE. 1985. Norepinephrine turnover in the hypothalamus of adult male rats: Alteration of circadian patterns by semistarvation. J. Neurochem. **45:** 706–709.
43. SCHEVING, L. E., W. H. HARRISON, P. GORDON & J. E. PAULY. 1968. Daily fluctuation (circadian and ultradian) in biogenic amines of the rat brain. Am. J. Physiol. **214:** 166–173.
44. KOULU, M., N. BJELOGRLIC, H. AGREN, J. M. SAAVEDRA, W. Z. POTTER & M. LINNOILA. 1989. Diurnal variation in the concentrations of catecholamines and indoleamines in the median eminence and in the intermediate and posterior lobes of the pituitary gland of the male rat. Brain Res. **503:** 246–252.
45. KAFKA, M. S., M. A. BENEDITO, R. H. ROTH, L. K. STEELE, W. W. WOLFE & G. N.

CATRAVAS. 1986. Circadian rhythms in catecholamine metabolites and cyclic nucleotide production. Chronobiol. Int. **3:** 101–115.

46. HUIE, J. M., R. P. SHARMA & R. A. J. COULOMBE. 1989. Diurnal alterations of catecholamines, indoleamines and their metabolites in specific brain regions of the mouse. Comp. Biochem. Physiol. [C] **94:** 575–579.
47. TURNER, B. B., T. E. WILENS, K. A. SCHROEDER, R. J. KATZ & B. J. CARROLL. 1981. Comparison of brainstem and adrenal circadian patterns of epinephrine synthesis. Neuroendocrinology **32:** 257–261.
48. SUDO, A. 1987. Adrenaline in various organs of the rat: Its origin, location and diurnal fluctuation. Life Sci. **41:** 2477–2484.
49. KUCHEL, O. & N. T. BUU. 1985. Circadian variations of free and sulfoconjugated catecholamines in normal subjects. Endocr. Res. **11:** 17–25.
50. KUCHEL, O. 1994. Clinical implications of genetic and acquired defects in catecholamine synthesis and metabolism. Clin. Invest. Med. **17:** 354–373.
51. SACK, D. A., S. P. JAMES, A. R. DORAN, M. A. SHERER, M. LINNOILA & T. A. WEHR. 1988. The diurnal variation in plasma homovanillic acid level persists but the variation in 3-methoxy-4-hydroxyphenylglycol level is abolished under constant conditions. Arch. Gen. Psychiatry **45:** 162–166.
52. CANDITO, M., D. PRINGUEY, Y. JACOMET *et al.* 1992. Circadian rhythm in plasma noradrenaline of healthy sleep-deprived subjects. Chronobiol. Int. **9:** 444–447.
53. WIRZ-JUSTICE, A., I. TOBLER, M. S. KAFKA *et al.* 1981. Sleep deprivation: Effects on circadian rhythms of rat brain neurotransmitter receptors. Psychiatry Res. **5:** 67–76.
54. KAFKA, M. S., A. WIRZ-JUSTICE & D. NABER. 1981. Circadian and seasonal rhythms in alpha- and beta-adrenergic receptors in the rat brain. Brain Res. **207:** 409–419.
55. LEMMER, B. & P. H. LANG. 1984. Evidence for a circadian-phase-dependency in ^{3}H-dihydroalprenolol binding and cyclic AMP content in heart ventricles of L-D synchronized rats. Pol. J. Pharmacol. Pharm. **36:** 275–280.
56. MEHTA, J., M. MALLOY, D. LAWSON & L. LOPEZ. 1989. Circadian variation in platelet alpha 2-adrenoceptor affinity in normal subjects. Am. J. Cardiol. **63:** 1002–1005.
57. KAFKA, M. S., A. WIRZ-JUSTICE, D. NABER, R. Y. MOORE & M. A. BENEDITO. 1983. Circadian rhythms in rat brain neurotransmitter receptors. Fed. Proc. **42:** 2796–2801.
58. BRUNEL, S. & C. DE MONTIGNY. 1987. Diurnal rhythms in the responsiveness of hippocampal pyramidal neurons to serotonin, norepinephrine, gamma-aminobutyric acid and acetylcholine. Brain Res. Bull. **18:** 205–212.
59. JOHANSSON, B. B. & L. MARTINSSON. 1980. The blood-brain barrier in adrenaline-induced hypertension. Circadian variations and modification by beta-adrenoreceptor antagonists. Acta Neurol. Scand. **62:** 96–102.
60. MATO, M., S. OOKAWARA, K. TOOYAMA & T. ISHIZAKI. 1981. Chronobiological studies on the blood-brain barrier. Experientia **37:** 1013–1015.
61. MIDDEKE, M. 1992. Synchronizität von zirkadianer Blutdruckrhythmik und sympathoadrenerger Aktivität. Z. Kardiol. **81** (Suppl. 2): 55–58.
62. GOULD, B. A. & E. B. RAFTERY. 1991. Twenty-four-hour blood pressure control: An intraarterial review. Chronobiol. Int. **8:** 495–505.
63. DE LEEUW, P. W., H. E. FALKE, T. L. KHO, R. VANDONGEN, A. WESTER & W. H. BIRKENHAGER. 1977. Effects of beta-adrenergic blockade on diurnal variability of blood pressure and plasma noradrenaline levels. Acta Med. Scand. **202:** 389–392.
64. HICKEY, M. S., D. L. COSTILL, M. D. VUKOVICH, K. KRYZMENSKI & J. J. WIDRICK. 1993. Time of day effects on sympathoadrenal and pressor reactivity to exercise in healthy men. Eur. J. Appl. Physiol. **67:** 159–163.
65. SOWERS, J. R. 1981. Dopaminergic control of circadian norepinephrine levels in patients with essential hypertension. J. Clin. Endocrinol. Metab. **53:** 1133–1137.
66. TUCK, M. L., N. STERN & J. R. SOWERS. 1985. Enhanced 24-hour norepinephrine and renin secretion in young patients with essential hypertension: Relation with the circadian pattern of arterial blood pressure. Am. J. Cardiol. **55:** 112–115.
67. THOMPSON, M. E., G. Y. NICOLAU, D. J. LAKATUA *et al.* 1987. Endocrine factors of blood pressure regulation in different age groups. Prog. Clin. Biol. Res. **227B:** 79–95.
68. BERNARDI, M., F. TREVISANI, R. DE PALMA *et al.* 1987. Chronobiological evaluation of

sympathoadrenergic function in cirrhosis. Relationship with arterial pressure and heart rate. Gastroenterology **93:** 1178–1186.

69. Brickman, A. S., N. Stern & J. R. Sowers. 1990. Circadian variations of catecholamines and blood pressure in patients with pseudohypoparathyroidism and hypertension. Chronobiologia **17:** 37–44.
70. Sowers, J. R., M. Nyby & K. Jasberg. 1982. Dopaminergic control of prolactin and blood pressure: Altered control in essential hypertension. Hypertension **4:** 431–437.
71. Gebber, G. L. & S. M. Barman. 1980. Rhythmogenesis in the sympathetic nervous system. Fed. Proc. **39:** 2526–2530.
72. Tapp, W. N., B. E. Levin & B. H. Natelson. 1981. Ultradian rhythm of plasma norepinephrine in rats. Endocrinology **109:** 1781–1783.
73. Dietl, H., H. Prast & A. Philippu. 1993. Pulsatile release of catecholamines in the hypothalamus of conscious rats. Naunyn-Schmiedeberg's Arch. Pharmacol. **347:** 28–33.
74. Robinson, R. L., H. Dietl, M. Bald, A. Kraus & A. Philippu. 1983. Effects of short-lasting and long-lasting blood pressure changes on the release of endogenous catecholamines in the hypothalamus of the conscious, freely moving rabbit. Naunyn-Schmiedeberg's Arch. Pharmacol. **322:** 203–209.
75. Takiyyuddin, M. A., H. P. Neumann, J. H. Cervenka *et al.* 1991. Ultradian variations of chromogranin A in humans. Am. J. Physiol. **261:** R939–R944.
76. Simpson, H. W., F. C. Bartter, F. Halberg *et al.* 1990. A circaoctohoran rhythm of urine epinephrine excretion in man. Prog. Clin. Biol. Res. **341A:** 157–160.
77. Rounds, H. D. 1984. A cholinergic receptor 12-hour interval-timer displaying "circa" periodicities. Comp. Biochem. Physiol. [C] **77:** 59–63.
78. Gordon, R. D., L. K. Wolfe, D. P. Island & G. W. Liddle. 1966. A diurnal rhythm in plasma renin activity in man. J. Clin. Invest. **45:** 1587–1592.
79. Katz, F. H., P. Romfh & J. A. Smith. 1972. Episodic secretion of aldosterone in supine man; relationship to cortisol. J. Clin. Endocrinol. Metab. **35:** 178–181.
80. Armbruster, H., W. Vetter, R. Beckerhoff, J. Nussberger, H. Vetter & W. Siegenthaler. 1975. Diurnal variations of plasma aldosterone in supine man: Relationship to plasma renin activity and plasma cortisol. Acta Endocrinol. (Copenh.) **80:** 95–103.
81. Katz, F. H., P. Romfh & J. A. Smith. 1975. Diurnal variation of plasma aldosterone, cortisol and renin activity in supine man. J. Clin. Endocrinol. Metab. **40:** 125–134.
82. Modlinger, R. S., K. Sharif-Zadeh, N. H. Ertel & M. Gutkin. 1976. The circadian rhythm of renin. J. Clin. Endocrinol. Metab. **43:** 1276–1282.
83. Liebau, H. & J. Manitius. 1982. Diurnal and daily variations of PRA, plasma catecholamines and blood pressure in normotensive and hypertensive man. Contrib. Nephrol. **30:** 57–63.
84. Beilin, L. J., J. Deacon, C. A. Michael *et al.* 1982. Circadian rhythms of blood pressure and pressor hormones in normal and hypertensive pregnancy. Clin. Exp. Pharmacol. Physiol. **9:** 321–326.
85. Freeman, R. H., J. O. Davis, G. M. Williams & A. A. Seymour. 1982. Circadian changes in plasma renin activity and plasma aldosterone concentration in one-kidney hypertensive rats. Proc. Soc. Exp. Biol. Med. **169:** 86–89.
86. Shiraki, K., N. Konda, S. Sagawa, J. R. Claybaugh & S. K. Hong. 1986. Cardiorenal-endocrine responses to head-out immersion at night. J. Appl. Physiol. **60:** 176–183.
87. Kool, M. J., J. A. Wijnen, F. H. Derkx, H. A. Struijker Boudier & L. M. Van Bortel. 1994. Diurnal variation in prorenin in relation to other humoral factors and hemodynamics. Am. J. Hypertens. **7:** 723–730.
88. Mullen, P. E., V. H. James, S. L. Lightman, C. Linsell & W. S. Peart. 1980. A relationship between plasma renin activity and the rapid eye movement phase of sleep in man. J. Clin. Endocrinol. Metab. **50:** 466–469.
89. Lightman, S. L., V. H. James, C. Linsell, P. E. Mullen, W. S. Peart & P. S. Sever. 1981. Studies of diurnal changes in plasma renin activity, and plasma noradrenaline, aldosterone and cortisol concentrations in man. Clin. Endocrinol. (Oxf.) **14:** 213–223.
90. Brandenberger, G., M. Follenius, A. Muzet, J. Ehrhart & J. P. Schieber. 1985.

Ultradian oscillations in plasma renin activity: Their relationships to meals and sleep stages. J. Clin. Endocrinol. Metab. **61:** 280–284.

91. BRANDENBERGER, G., M. FOLLENIUS, C. SIMON, J. EHRHART & J. P. LIBERT. 1988. Nocturnal oscillations in plasma renin activity and REM-NREM sleep cycles in humans: A common regulatory mechanism? Sleep **11:** 242–250.
92. STUMPE, K. O., R. KOLLOCH, H. VETTER *et al.* 1976. Acute and long-term studies of the mechanisms of action of beta-blocking drugs in lowering blood pressure. Am. J. Med. **60:** 853–865.
93. NICHOLLS, M. G., E. A. ESPINER, H. IKRAM, A. H. MASLOWSKI, E. J. HAMILTON & P. J. BONES. 1984. Hormone and blood pressure relationships in primary aldosteronism. Clin. Exp. Hypertens. [A] **6:** 1441–1458.
94. NAHMOD, V. E., M. S. BALDA, C. J. PIROLA, S. FINKIELMAN, P. V. GEJMAN & D. P. CARDINALI. 1982. Circadian rhythm and neural regulation of rat pineal angiotensin converting enzyme. Brain Res. **236:** 216–220.
95. VAGNUCCI, A. H., R. H. J. R. MCDONALD, A. L. DRASH & A. K. WONG. 1974. Intradiem changes of plasma aldosterone, cortisol, corticosterone and growth hormone in sodium restriction. J. Clin. Endocrinol. Metab. **38:** 761–776.
96. SOWERS, J. R., N. STERN, M. D. NYBY & K. A. JASBERG. 1982. Dopaminergic regulation of circadian rhythms of blood pressure, renin and aldosterone in essential hypertension. Cardiovasc. Res. **16:** 317–323.
97. TAKEDA, R., I. MIYAMORI, M. IKEDA *et al.* 1984. Circadian rhythm of plasma aldosterone and time dependent alterations of aldosterone regulators. J. Steroid Biochem. **20:** 321–323.
98. FOLLENIUS, M., M. O. KRAUTH, J. SAINI & G. BRANDENBERGER. 1992. Effect of awakening on aldosterone. J. Endocrinol. Invest. **15:** 475–478.
99. CUGINI, P., R. MANCONI, R. SERDOZ, A. MANCINI, T. MEUCCI & D. SCAVO. 1980. Rhythm characteristics of plasma renin, aldosterone and cortisol in five subtypes of mesor-hypertension. J. Endocrinol. Invest. **3:** 143–147.
100. ATLAS, S. A., D. B. CASE, J. E. SEALEY & J. H. LARAGH. 1981. Relationship between plasma renin and cortisol in hypertensive patients. Clin. Sci. **61:** S265s–S268s.
101. CAMPBELL, P. J., G. B. BOLLI, P. E. CRYER & J. E. GERICH. 1985. Sequence of events during development of the dawn phenomenon in insulin-dependent diabetes mellitus. Metabolism **34:** 1100–1104.
102. WITWORTH, J. A., D. SAINES, R. THATCHER, A. BURTKUS & B. A. SCOGGINS. 1983. Blood pressure and metabolic effects of ACTH in normotensive and hypertensive man. Clin. Exp. Hypertens. A **5:** 501–522.
103. SPENCE, C. D., A. BROWN, J. P. COGHLAN, D. G. PAERKES, J. A. WITWORTH & B. A. SCOGGINS. 1989. Rapid haemodynamic response to adrenocorticotropin and the role of peripheral resistance in adrenoorticotropin-induced hypertension in conscious sheep. J. Hypertens. **7:** 277–285.
104. SCOGGINS, B. A., J. P. COGHLAN, D. A. DENTON *et al.* 1975. The role of adrenocortical hormones in ACTH-induced hypertension. Clin. Exp. Pharmacol. Physiol. Suppl. **2:** 119–122.
105. IMAI, Y., K. ABE, S. SASAKI *et al.* 1989. Exogenous glucocorticoid eliminates or reverses circadian blood pressure variations. J. Hypertens. **7:** 113–120.
106. IMAI, Y., K. ABE, S. SASAKI *et al.* 1988. Altered circadian blood pressure rhythm in patients with Cushing's syndrome. Hypertension **12:** 11–19.
107. PANARELLI, M., M. TERZOLO, A. PIOVESAN *et al.* 1990. A 24-hour profiles of blood pressure and heart rate in Cushing's syndrome. Evidence for a differential control of cardiovascular variables by glucocorticoids. Ann. Ital. Med. Int. **5:** 18–25.
108. DJAVIDANI, B., M. SANDER, R. KREUTZ *et al.* 1995. Chronic dexamethasone treatment suppresses hypertension development in the transgenic rat TGR(mRen2)27. J. Hypertens. **13:** 637–645.
109. SCHNECKO, A., K. WITTE, A. SCHANZER & B. LEMMER. 1995. Corticosterone and aldosterone concentration in normotensive Sprague Dawley and transgenic hypertensive rats (abstract). Biol. Rhythm Res. **26:** 442.

110. VAN CAUTER, E. & S. REFETEFF. 1985. Multifactorial control of the 24-hour secretory profiles of pituitary hormones. J. Endocrinol. Invest. **8:** 381–391.
111. VAN DE BORNE, P., H. NGUYEN, P. LINKOWSKI & J.-P. DEGAUTE. 1993. Sleep quality and continuous, non-ivasive beat-to-beat blood pressure recording. J. Hypertens. **11:** 1423–1427.
112. LEJEUNE-LENAIN, C., E. VAN CAUTER & D. DESIR. 1987. Control of circadian and episodic variations of adrenal androgen secretion in man. J. Endocrinol. Invest. **10:** 267–276.
113. SOWERS, J. R. & F. W. BECK. 1983. Dopaminergic regulation of 18-hydroxycorticosterone and aldosterone secretion in man. Acta Endocrinol. (Copenh.) **102:** 258–264.
114. SZAFARCZYK, A., G. ALONSO, G. IXART, F. MALAVAL & I. ASSENMACHER. 1985. Diurnal-stimulated and stress-induced ACTH release in rats is mediated by ventral noradrenergic bundle. Am. J. Physiol. **249:** E219 E226.
115. JOHNSON, R. F. & W. RANDALL. 1983. Mediation of the photoperiodic effect on grooming reflexes by glucocorticoid hormones and serotonin. Behav. Neurosci. **97:** 195–209.
116. KRIEGER, D. T. 1975. Effect of intraventricular neonatal 6-OH dopamine or 5,6-dihydroxytryptamine administration on the circadian periodicity of plasma corticosteroid levels in the rat. Neuroendocrinology **17:** 62–74.
117. JAFFER, A., V. A. RUSSELL & J. J. TALJAARD. 1987. Noradrenergic and dopaminergic modulation of thyrotropin secretion in the rat. Brain Res. **404:** 267–272.
118. BARONTINI, M., H. E. ROMEO, I. ARMANDO & D. P. CARDINALI. 1988. 24 hour changes in catecholamine content of rat thyroid and submaxillary glands. J. Neural. Transm. **71:** 189–194.
119. SCANLON, M. F., A. P. WEETMAN, M. LEWIS *et al.* 1980. Dopaminergic modulation of circadian thyrotropin rhythms and thyroid hormone levels in euthyroid subjects. J. Clin. Endocrinol. Metab. **51:** 1251–1256.
120. PEREZ-LOPEZ, F. R., C. M. GONZALEZ-MORENO, M. D. ABOS, J. A. ANDONEGUI & R. H. CORVO. 1982. Pituitary response to a dopamine antagonist at different times of the day in normal women. Acta Endocrinol. (Copenh.) **100:** 481–485.
121. ROSSMANITH, W. G., J. F. MORTOLA, G. A. LAUGHLIN & S. S. YEN. 1988. Dopaminergic control of circadian and pulsatile pituitary thyrotropin release in women. J. Clin. Endocrinol. Metab. **67:** 560–564.
122. HURXTHAL, L. M. 1931. Blood pressure before and after operation in hyperthyroidism. Arch. Intern. Med. **47:**
123. MERILLON, J. P., P. PASSA, J. CHASTRE, A. WOLF & R. GOURGON. 1981. Left ventricular function and hyperthyroidism. Br. Heart J. **46:** 137–143.
124. GRAETTINGER, J. S., J. J. MUENSTER, C. S. CHECCHIA, R. L. GRISSOM & J. A. CAMPBELL. 1957. A correlation of clinical hemodynamic studies in patients with hypothyroidism. J. Clin. Invest. **1957:** 502–512.
125. SAITO, I., K. ITO & T. SARUTA. 1983. Hypothyroidism as a cause of hypertension. Hypertension **5:** 112–115.
126. STREETEN, D. H., G. H. ANDERSON, JR., T. HOWLAND, R. CHIANG & H. SMULYAN. 1988. Effects of thyroid function on blood pressure. Recognition of hypothyroid hypertension. Hypertension **11:** 78–83.
127. CARMODY, J. J. 1987. Opiate receptors: An introduction. Anaesth. Intensive Care **15:** 27–37.
128. HERZ, A. 1983. Multiple opiate receptors and their functional significance. J. Neural Transm. **18**(Suppl.): 227–230.
129. PASTERNACK, G. W., A. R. GINTZLER, R. A. HOUGHTON *et al.* 1983. Biochemical and pharmacological evidence for opioid receptor multiplicity in the central nervous system. Life Sci. **33**(Suppl. 1): 167–173.
130. DUMONT, M., M. OUELLETTE, L. BRAKIER-GINGRAS & S. LEMAIRE. 1991. Circadian regulation of the biosynthesis of cardiac Met-enkephalin and precursors in normotensive and spontaneously hypertensive rats. Life Sci. **48:** 1895–1902.
131. SHANKS, M. F., V. CLEMENT-JONES, C. J. LINSELL, P. E. MULLEN, L. H. REES & G. M. BESSER. 1981. A study of 24-hour profiles of plasma met-enkephalin in man. Brain Res. **212:** 403–409.

132. McIntosh, T. K. 1987. Prolonged disruption of plasma beta-endorphin dynamics after trauma in the nonhuman primate. Endocrinology **120:** 1734–1741.
133. Laubie, M., H. Schmitt, M. Vincent & G. Remond. 1977. Central cardiovascular effects of morphinomimetic peptides in dogs. Eur. J. Pharmacol. **46:** 67–71.
134. Rubin, P., T. F. Blaschke & C. Guilleminault. 1981. Effect of naloxone, a specific opioid inhibitor, on blood pressure fall during sleep. Circulation **63:** 117–121.
135. Doba, N. & D. J. Reis. 1974. Role of central and peripheral adrenergic mechanisms in neurogenic hypertension produced by brainstem lesions in rat. Circ. Res. **34:** 293–301.
136. Grandison, L. & A. Guidotti. 1977. Regulation of prolactin release by endogenous opiates. Nature **270:** 357–359.
137. Dupont, A., L. Cusan, M. Garon, F. Labrie & C. H. Li. 1977. Beta-endorphin: Stimulation of growth hormone release *in vivo.* Proc. Natl. Acad. Sci. USA **74:** 358–359.
138. degli Uberti, E. C., S. Salvadori, G. Trasforini *et al.* 1992. Effect of deltorphin on pituitary-adrenal response to insulin-induced hypoglycemia and ovine corticotropin-releasing hormone in healthy man. J. Clin. Endocrinol. Metab. **75:** 370–374.
139. degli Uberti, E. C., M. R. Ambrosio, L. Vergnani *et al.* 1993. Stress-induced activation of sympathetic nervous system is attenuated by the selective ∂-opioid receptor agonist deltorphin in healthy man. J. Clin. Endocrinol. Metab. **77:** 1490–1494.
140. Thoren, P., J. S. Floras, P. Hoffman & D. R. Seals. 1990. Endorphins and exercise: Physiological mechanisms and clinical implications. Med. Sci. Sports Exerc. **22:** 417–428.
141. Hara, K. & J. S. Floras. 1995. Influence of naloxone on muscle sympathetic nerve activity, systemic and calf haemodynamics and ambulatory blood pressure after exercise in mild essential hypertension. J. Hypertens. **13:** 447–461.
142. Lang, R. E., H. K., R. Dietz *et al.* 1983. Evidence for the presence of enkephalins in the heart. Life Sci. **32:** 399–406.
143. Ventura, C., L. Bastagli, P. Bernardi, C. M. Caldarera & C. Guarnieri. 1989. Opioid receptors in rat cardiac sarcolemma: Effect of phenylephrine and isoproterenolo. Biochem. Biophys. Acta **987:** 69–74.
144. Portaluppi, F., A. Pradella, L. Montanari *et al.* 1990. Atrial strain is the main determinant of release of atrial natriuretic peptide. Int. J. Cardiol. **29:** 297–303.
145. Portaluppi, F., L. Vergnani & E. C. degli Uberti. 1993. Atrial natriuretic peptide and circadian blood pressure regulation: Clues from a chronobiological approach. Chronobiol. Int. **10:** 176–189.
146. McLay, J. S., P. K. Chatterjee, A. G. Jardine & G. M. Hawksworth. 1995. Atrial natriuretic factor modulates nitric oxide production: An ANF-C receptor-mediated effect. J. Hypertens. **13:** 625–630.
147. Donckier, J., J. V. Anderson, T. Yeo & S. R. Bloom. 1986. Diurnal rhythm in the plasma concentration of atrial natriuretic peptide (letter). N. Engl. J. Med. **315:** 710–711.
148. Winters, C. J., A. L. Sallman & D. L. Vesely. 1988. Circadian rhythm of prohormone atrial natriuretic peptides 1–30, 31–67 and 99–126 in man. Chronobiol. Int. **5:** 403–409.
149. Portaluppi, F., L. Montanari, B. Bagni, E. degli Uberti, G. Trasforini & A. Margutti. 1989. Circadian rhythms of atrial natriuretic peptide, blood pressure and heart rate in normal subjects. Cardiology **76:** 428–432.
150. McCance, D. R., G. Roberts, B. Sheridan *et al.* 1989. Variations in the plasma concentration of atrial natriuretic factor across 24 hours. Acta Endocrinol. (Copenh.) **120:** 266–270.
151. Colantonio, D., P. Pasqualetti, R. Casale & G. Natali. 1989. Is atrial natriuretic peptide important in the circadian rhythm of arterial blood pressure? (letter). Am. J. Cardiol. **63:** 1166.
152. Miyamoto, S., H. Shimokawa, H. Sumioki & H. Nakano. 1989. Physiologic role of endogenous human atrial natriuretic peptide in preeclamptic pregnancies. Am. J. Obstet. Gynecol. **160:** 155–159.

153. SUMIOKI, H., H. SHIMOKAWA, S. MIYAMOTO, K. UEZONO, T. UTSUNOMIYA & H. NAKANO. 1989. Circadian variations of plasma atrial natriuretic peptide in four types of hypertensive disorder during pregnancy. Br. J. Obstet. Gynaecol. **96:** 922–927.
154. PORTALUPPI, F., B. BAGNI, E. DEGLI UBERTI *et al.* 1990. Circadian rhythms of atrial natriuretic peptide, renin, aldosterone, cortisol, blood pressure and heart rate in normal and hypertensive subjects. J. Hypertens. **8:** 85–95.
155. WATANABE, M. & Y. UCHIYAMA. 1988. Twenty-four hour variations in subcellular structures of rat pancreatic islet B-, A- and D-cells, and of portal plasma glucose and insulin levels. Cell Tissue Res. **253:** 337–345.
156. WEIL, J., T. M. STROM, J. M. HEIM *et al.* 1988. Influence of diurnal rhythm, posture and right atrial size on plasma atrial natriuretic peptide levels. Z. Kardiol. **77**(Suppl. 2): 36–40.
157. PORTALUPPI, F., L. MONTANARI, L. VERGNANI *et al.* 1992. Loss of nocturnal increase in plasma concentration of atrial natriuretic peptide in hypertensive chronic renal failure. Cardiology **80:** 312–323.
158. PORTALUPPI, F., L. MONTANARI, M. FERLINI *et al.* 1991. Consistent changes in the circadian rhythms of blood pressure and atrial natriuretic peptide in congestive heart failure. Chronobiol. Int. **8:** 432–439.
159. PORTALUPPI, F., L. MONTANARI, M. FERLINI, L. VERGNANI, B. BAGNI & E. C. DEGLI UBERTI. 1992. Differences in blood pressure regulation of congestive heart failure, before and after treatment, correlate with changes in the circulating pattern of atrial natriuretic peptide. Eur. Heart J. **13:** 990–996.
160. SOTHERN, R. B., D. L. VESELY, E. L. KANABROCKI *et al.* 1994. Temporal (circadian) and functional relationship between atrial natriuretic peptides and blood pressure. Chronobiol. Int. **12:** 106–120.
161. YOSHIZAWA, T., O. SHIMMI & A. GIAID. 1990. Endothelin: A novel peptide in the posterior pituitary system. Science **247:** 462–464.
162. BOULANGER, C. & T. F. LUSCHER. 1990. Release of endothelin from porcine aorta: Inhibition by endothelium-derived nitric oxide. J. Clin. Invest. **85:** 587–590.
163. SAITO, Y., K. NAKAO, M. MUKOYAMA & H. IMURA. 1990. Increased plasma endothelin level in patients with essential hypertension (letter). N. Engl. J. Med. **322:** 205.
164. FUKUDA, Y., Y. HIRATA, S. TAKETANI *et al.* 1989. Endothelin stimulates accumulations of cellular atrial natriuretic peptide and its messenger RNA in rat cardiocytes. Biochem. Biophys. Res. Commun. **164:** 1431–1436.
165. PERFETTO, F., R. TARQUINI, A. PILUSO *et al.* 1995. Sex difference, age and circadian rhythmometry of circulating endothelin-1 (abstract). Biol. Rhythm Res. **26:** 432.
166. AMARA, S. G., V. JONAS, M. G. ROSENFELD, E. S. ONG & R. M. EVANS. 1982. Alternative RNA processing in calcitonin gene expression generates mRNAs encoding different polypeptide products. Nature **298:** 240–244.
167. KAWAI, Y., K. TAKAMI, S. SHIOSAKA *et al.* 1985. Topographic localization of calcitonin gene-related peptide in the rat brain: An immunohistochemical analysis. Neuroscience **15:** 747–763.
168. ROSENFELD, M. G., S. G. AMARA & R. M. EVANS. 1984. Alternative RNA processing: Determining neuronal phenotype. Science **225:** 1315–1320.
169. TSCHOPP, F. A., P. H. TOBLER & J. A. FISCHER. 1984. Calcitonin gene-related peptide in the human thyroid, pituitary and brain. Mol. Cell Endocrinol. **36:** 53–57.
170. TSCHOPP, F. A., H. HENKE, J. B. PETERMANN *et al.* 1985. Calcitonin gene-related peptide and its binding sites in the human central nervous system and pituitary. Proc. Natl. Acad. Sci. USA **82:** 248–252.
171. SKOFITSCH, G. & D. M. JACOBOWITZ. 1985. Autoradiographic distribution of 125I calcitonin gene-related peptide binding sites in the rat central nervous system. Peptides **6:** 975–986.
172. SKOFITSCH, G. & D. M. JACOBOWITZ. 1985. Calcitonin gene-related peptide: Detailed immunohistochemical distribution in the central nervous system. Peptides **6:** 721–745.
173. SKOFITSCH, G. & D. M. JACOBOWITZ. 1985. Quantitative distribution of calcitonin gene-related peptide in the rat central nervous system. Peptides **6:** 1069–1073.

174. MULDERRY, P. K., M. A. GHATEI, J. RODRIGO *et al.* 1985. Calcitonin gene-related peptide in cardiovascular tissues of the rat. Neuroscience **14:** 947–954.
175. WIMALAWANSA, S. J. & I. MACINTYRE. 1988. Calcitonin gene-related peptide and its specific binding sites in the cardiovascular system of rat. Int. J. Cardiol. **20:** 29–37.
176. UDDMAN, R., L. EDVINSSON, E. EKBLAD, R. HAKANSON & F. SUNDLER. 1986. Calcitonin gene-related peptide (CGRP): Perivascular distribution and vasodilatory effects. Regul. Pept. **15:** 1–23.
177. BENDTSEN, F., S. SCHIFTER & J. H. HENRIKSEN. 1991. Increased circulating calcitonin gene-related peptide (CGRP) in cirrhosis. J. Hepatol. **12:** 118–123.
178. GIRGIS, S. I., D. W. MACDONALD, J. C. STEVENSON *et al.* 1985. Calcitonin gene-related peptide: Potent vasodilator and major product of calcitonin gene. Lancet **2:** 14–16.
179. BRAIN, S. D., T. J. WILLIAMS, J. R. TIPPINS, H. R. MORRIS & I. MACINTYRE. 1985. Calcitonin gene-related peptide is a potent vasodilator. Nature **313:** 54–56.
180. GOODMAN, E. C. & L. L. IVERSEN. 1986. Calcitonin gene-related peptide: Novel neuropeptide. Life Sci. **38:** 2169–2178.
181. TIPPINS, J. R. 1986. CGRP: a novel neuropeptide from the calcitonin gene is the most potent vasodilator known. J. Hypertens. **4:** S102–S105.
182. STRUTHERS, A. D., M. J. BROWN, D. W. MACDONALD *et al.* 1986. Human calcitonin gene related peptide: A potent endogenous vasodilator in man. Clin. Sci. **70:** 389–393.
183. MCEWAN, J., S. LARKIN, G. DAVIES *et al.* 1986. Calcitonin gene-related peptide: A potent dilator of human epicardial coronary arteries. Circulation **74:** 1243–1247.
184. EZRA, D., F. R. LAURINDO, D. S. GOLDSTEIN, R. E. GOLDSTEIN & G. FEUERSTEIN. 1987. Calcitonin gene-related peptide: A potent modulator of coronary flow. Eur. J. Pharmacol. **137:** 101–105.
185. DIPETTE, D. J., K. SCHWARZENBERGER, N. KERR & O. B. HOLLAND. 1987. Systemic and regional hemodynamic effects of calcitonin gene-related peptide. Hypertension. **9:** III142–III146.
186. LAPPE, R. W., M. J. SLIVJAK, J. A. TODT & R. L. WENDT. 1987. Hemodynamic effects of calcitonin gene-related peptide in conscious rats. Regul. Pept. **19:** 307–312.
187. WANG, B. C., P. BIE, R. J. J. LEADLEY & K. L. GOETZ. 1989. Cardiovascular effects of calcitonin gene-related peptide in conscious dogs. Am. J. Physiol. **257:** R726–R731.
188. KURTZ, A., R. MUFF & J. A. FISCHER. 1989. Calcitonin gene products and the kidney. Klin. Wochenschr. **67:** 870–875.
189. GENNARI, C. & J. A. FISCHER. 1985. Cardiovascular action of calcitonin gene-related peptide in humans. Calcif. Tissue Int. **37:** 581–584.
190. PORTALUPPI, F., L. VERGNANI, A. MARGUTTI *et al.* 1993. Modulatory effect of the renin-angiotensin system on the plasma levels of calcitonin gene-related peptide in normal man. J. Clin. Endocrinol. Metab. **77:** 816–820.
191. TRASFORINI, G., A. MARGUTTI, L. VERGNANI *et al.* 1994. Evidence that enhancement of cholinergic tone increases basal plasma levels of calcitonin gene-related peptide in normal man. J. Clin. Endocrinol. Metab. **78:** 763–766.
192. TRASFORINI, G., A. MARGUTTI, F. PORTALUPPI *et al.* 1991. Circadian profile of plasma calcitonin gene-related peptide in healthy man. J. Clin. Endocrinol. Metab. **73:** 945–951.
193. DE LOS SANTOS, E. T. & E. L. MAZZAFERRI. 1991. Calcitonin gene-related peptide: 24-hour profile and responses to volume contraction and expansion in normal men. J. Clin. Endocrinol. Metab. **72:** 1031–1035.
194. PORTALUPPI, F., G. TRASFORINI, A. MARGUTTI *et al.* 1992. Circadian rhythm of calcitonin gene-related peptide in uncomplicated essential hypertension. J. Hypertens. **10:** 1227–1234.
195. KATZ, F. H., J. A. SMITH, J. P. LOCK & D. E. LOEFFEL. 1979. Plasma vasopressin variation and renin activity in normal active humans. Horm. Res. **10:** 289–302.

Behavioral Influences on Diurnal Blood Pressure Rhythms

THOMAS G. PICKERING, JOSEPH E. SCHWARTZ, AND ARTHUR STONE

Hypertension Center
New York Hospital-Cornell University Medical Center
New York, New York 10021

Department of Psychiatry
State University of New York at Stonybrook
Stonybrook, New York

There is no dispute that blood pressure is subject to a pronounced diurnal rhythm, one of the first descriptions of which was made by Janeway[1] in 1904 using the Riva-Rocci technique. This rhythm constitutes the major source of blood pressure variability. The introduction of portable blood pressure monitors has enabled the characteristics and determinants of the diurnal rhythm to be described in detail. The first ambulatory monitor to be introduced was the Remler, a noninvasive device that required manual inflation of the cuff. Although it provided much useful information about the influences of behavioral factors on blood pressure, it was impossible to study the diurnal rhythm of pressure.[2] This subsequently became possible with the introduction of the Oxford Medilog device, with which the entire 24-hour profile of blood pressure could be studied in freely moving subjects.[3] One of the first articles to be published on the diurnal rhythm of blood pressure using this technique showed a smooth curve of blood pressure, with the readings highest in the morning, lower in the afternoon, and lowest during sleep.[4] This type of analysis led to the mistaken idea that blood pressure shows a smooth decline at night, followed by a smooth increase in the early morning. It was subsequently pointed out that this pattern is an artifact arising from the integration of records of subjects with different times of waking and sleeping, as will be described.

DIURNAL PATTERNING OF ACTIVITIES AND MOODS

Most people follow a consistent pattern of activities during a normal working day. In a study of ambulatory blood pressure and its determinants we examined the diaries of 94 subjects, who recorded their position, activity, and mood every 15 minutes during the day.[5] The timing of different activities was for the most part predictable; eating showed peaks at 12 noon and 7 PM, but somewhat less expected was the finding that social discourse was markedly less after 5 PM. Of 10 moods tested, Anxious, Happy, Rushed, and Tired were predicted by the time of day, whereas Annoyed, Angry, and Tense were not. Being Anxious was reported more in the early part of the day, while Rushed showed peaks at noon, 4 PM, and 6 PM, and Happy and Tired were more frequent in the evening. After removing the effects of activities and locations, three moods showed diurnal cycles. Rushed decreased throughout the day; Sad was commonest in the early morning and evening, and Tired increased progressively throughout the day.

We also examined the effects of these moods on blood pressure,[6] using diary entry data from 246 subjects who were part of the same study. Sixty-five percent of these readings were reported as "Neutral," 12% as Happy, 7% as Anxious, and less than 1% as Angry. After controlling for position and location, significant increases in both systolic and diastolic pressure relative to average waking pressure were observed for Angry (4/4 mm Hg), Excited (3/2 mm Hg), and Happy (1/1 mm Hg). A significant decrease in blood pressure was seen for Tired.

IS THE DIURNAL RHYTHM DETERMINED EXTRINSICALLY OR INTRINSICALLY?

The diurnal rhythm of blood pressure has been conceptualized in two different ways, the salient features of which are summarized in FIGURE 1. The first one, which

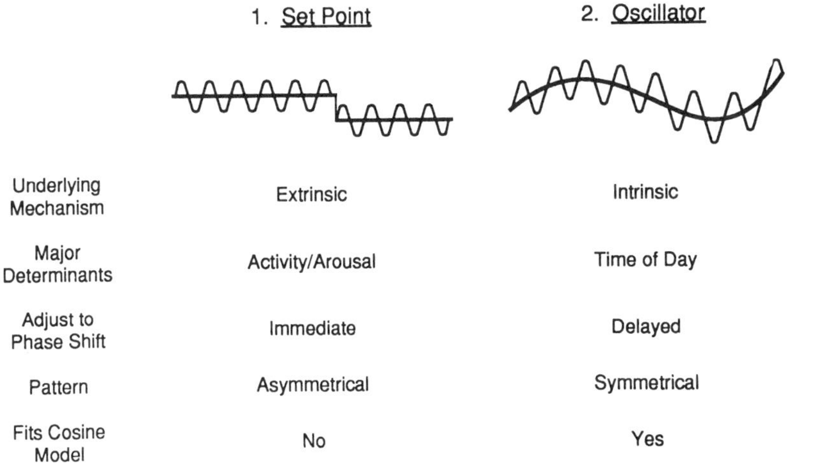

FIGURE 1. Two models for describing the diurnal rhythm of blood pressure.

we will call the "Set Point Model," conceives blood pressure as having a tonic component (or possibly two, corresponding to wakefulness and sleep) about which short-term oscillations occur (phasic component). The second, which we will term the "Oscillator Model," views blood pressure rhythm as an intrinsic oscillator with a periodicity of 24 hours. Short-term variations are again superimposed on the basic rhythm. The most important difference between the two models is that the Set Point Model is influenced primarily by the level of activity and arousal, whereas the Oscillator Model is more dependent on an intrinsic rhythm. We have argued elsewhere that most evidence favors the Set Point Model.[7]

The case for an intrinsic sinusoidal pattern of blood pressure variation was argued most forcefully by Halberg,[8] who found that such curves fit the blood pressure pattern of subjects kept in a confined environment for several days. It was also proposed by Raftery *et al.*[4] that a gradual increase in blood pressure occurs during the early morning hours (3–6 AM) before awakening[4] and that this increase could contribute to the high incidence of cerebral hemorrhage and myocardial infarction in the early morning.[9] Other workers, however, presented data which indicate that the apparently sinusoidal pattern of blood pressure is mostly an artifact attributable to the averaging of records from individuals who wake at different times; when the recordings are synchronized to the time of waking rather than the time of day, the gradual early morning rise of pressure is no longer seen or is much less pronounced.[10,11] Instead, blood pressure remains relatively stable during the hour before waking, but rises abruptly at the moment of waking. This pattern is consistent with the changes described in studies in which the electroencephalogram as well as blood pressure was recorded.[12] Evidence favoring the extrinsic or activity-dependent origin of the diurnal rhythm of blood pressure can be summarized as follows.

Correlations between Activity and Blood Pressure. Two studies, including one of our own, have shown that a major portion of blood pressure variability during ambulatory monitoring can be accounted for by changes in activity recorded in the subjects' diaries, and that time of day makes little additional contribution.[13,14] In addition, direct recording of physical activity using ambulatory activity monitors has shown a close correlation between changes in activity and blood pressure.[15] One study reported that activity accounts for 20% of systolic and 26% of diastolic blood pressure variation.[16]

The role of posture is somewhat controversial. Going from the lying to the upright position results in a modest increase in diastolic pressure and heart rate, without much change in systolic pressure.[9] Some analyses of ambulatory recordings[6,17] have concluded that posture can account for a significant part of the overall variance in blood pressure; in our own analysis this variance was 6.2% for systolic pressure, and 9.7% for diastolic pressure.[6] Both systolic and diastolic pressures were 3.4 mm Hg higher while standing than when sitting. A third study, which monitored paramedics on both a work day and a non-work day, found no effect of posture on systolic pressure, but diastolic pressure was 3.6 mm Hg higher when standing than when sitting.[18] The apparently large effect of posture on systolic pressure is most probably due to the effects of associated activities rather than to postural change per se.

We analyzed data from 315 subjects in our worksite study, all of whom were monitored on a work day. FIGURE 2 shows the aggregated rhythm of systolic pressure before and after controlling for the effects of posture (lying, sitting, or standing) and location (mostly at home or at work). Raw data show the typical asymmetric sine wave appearance noted previously. Most variance is eliminated by controlling simply for posture and location; what remains can probably be attributed to the effects of sleep and other activities.

Mancia *et al.*[23] examined average blood pressure levels over consecutive 30-minute periods throughout the 24 hours and also found no early morning increase preceding waking. All the half-hour averages during waking were substantially similar, leading them to conclude that there is no circadian rhythm of blood pressure. The relatively flat profile of blood pressure during the day observed in their study may be attributed to the fact that their patients were hospitalized.

Effects of Restricted Activity. If there is an endogenous circadian rhythm of blood pressure, it should still be apparent when the influence of external stimuli is removed. This was done in the study by Athanassiadis *et al.*,[20] in which blood pressure was monitored noninvasively for 24 hours in patients in an orthopedic ward, who were immobilized by plaster casts. In this situation, blood pressure showed no evidence of any sinusoidal change, being relatively constant during the day and decreasing during sleep. There were in effect two levels of pressure, one corresponding to waking and the other to sleeping. Similar observations were made in hospitalized patients kept on bedrest by Mann *et al.*[21] In the latter study, diurnal variations of pressure became more pronounced when subjects were studied a second time, but while being physically active during the day. When subjects stay in bed for several hours after waking up, blood pressure remains low until they get up.[22]

Shift Workers. Studies of shift workers also show a close linkage between activity and blood pressure. Sundberg *et al.*[24] monitored 24-hour pressure and heart rate in seven normotensive nurses on three different days. On the first day, they worked a normal daytime shift and showed a typical diurnal pattern, with the highest pressures during work and the lowest pressures at night. The second recording, made on the first day they worked a night shift, showed complete reversal of the diurnal blood pressure pattern, with the lowest blood pressure occurring while they were asleep during the day. The third recording, made a few days later, was similar to the second.

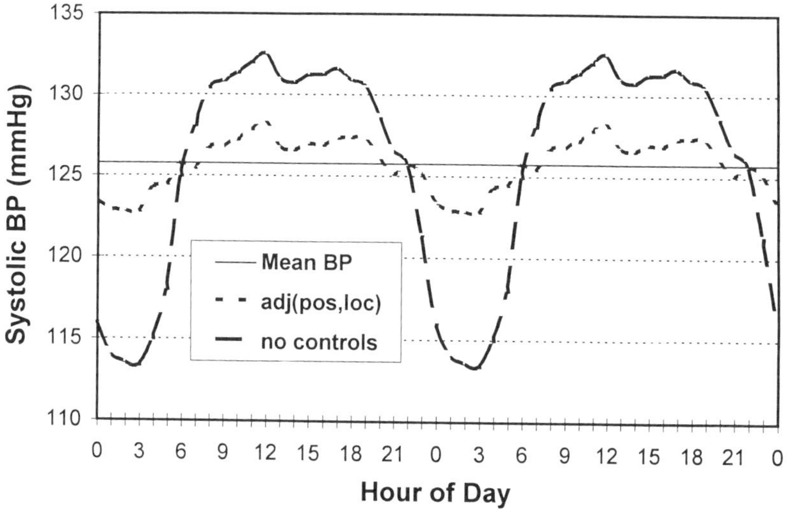

FIGURE 2. The diurnal rhythm of systolic pressure in 315 subjects monitored on a work day. The *horizontal line* shows the 24-hour mean blood pressure level, and the *darker dotted line* the raw (unadjusted) data. The *lighter dotted line* shows the data after controlling for position and location.

In contrast to the immediate reversal of blood pressure rhythm, the diurnal rhythm of heart rate was still not fully reversed by the third recording. Baumgart *et al.*,[25] who studied industrial workers on both a day and a night shift, also found that the diurnal pattern of blood pressure changed immediately when the workers switched their shifts. Furthermore, amplitudes of the diurnal pressure variations were the same on both shifts. One other study found that the range of blood pressure was greater in day workers than in night workers, which was attributed to the night workers getting less sleep.[26] In another study of shift workers, Chau *et al.*[27] monitored the 24-hour blood pressures of 15 workers in a chemical factory, who each worked three different shifts: morning (4 AM to noon); afternoon (noon to 8 PM), and night (8 PM to 4 AM). In all cases the highest pressures were seen during the hours of work, but the diurnal patterns of blood pressure were different for the three shifts. The periods of high blood pressure were longer when subjects worked during the morning or night than when they worked during the afternoon, although the average 24-hour pressures were not very different (123/83 mm Hg for the morning shifts, 120/79 for the afternoon, and 124/83 for the night shift). These differences can presumably be explained by differences in sleep patterns (which were not reported in this study).

The close linkage of the diurnal rhythm of blood pressure to the activity cycle is in marked contrast to some other bodily rhythms, which are more closely entrained to an "internal clock." Examples of the latter sort are body temperature, urinary electrolyte excretion, and cortisol secretion. All of these are less easily disrupted by sudden changes in the sleep-wakefulness cycle, but they can eventually be reversed.[28–30]

WHICH METHOD OF ANALYSIS GIVES THE BEST FIT TO THE DIURNAL RHYTHM OF BLOOD PRESSURE?

Many biologic variables show intrinsic rhythmicity within a period of approximately 24 hours, hence the term circadian (around the day) rhythms. Halberg *et al.*[8] have been a strong advocate of the use of "rhythmometry" to analyze human diurnal blood pressure patterns. The basis of this approach is to analyze the observed patterns as a series of cosine waves, within a basic period of 24 hours. A number of terms need to be defined for this analysis. The mesor[a] is the mean level, which is close (but not necessarily identical) to the average 24-hour blood pressure; the Amplitude is the distance from the peak (or trough) to the mesor, which in this case would be the difference between the average 24-hour pressure and the peak pressure.[48] On this schema, hypertension can be classified as "mesor hypertension" when the entire profile is shifted to a higher level, or "amplitude hypertension." Ideally, this type of analysis should be performed over periods of more than 24 hours, preferably 48 hours.

While this approach has mathematical appeal, it may be criticized on two grounds. First, available evidence overwhelmingly suggests that blood pressure is not determined by an endogenous rhythm to the same extent that, for example, heart rate or plasma cortisol are, and second, it imposes a symmetry that may be more artifactual than real. In this sense, it may be a procrustean bed. Thus, the peak and trough of blood pressure are assumed to be exactly 12 hours apart and also equidistant from the mesor. Neither of these criteria is likely to be met in reality (FIG. 2). The fit between the theoretic waveform and the observed data points can be improved by including as many as four harmonics in the Fourier analysis,[31] but the physiologic relevance of this is questionable. The study of shift workers by Chau *et al.*[27] just cited used an elaborate Fourier analysis procedure to describe diurnal patterns during the three shifts. High and low blood pressure spans were identified, and a Fourier equation was used to fit the data. The simple cosinor model (with one harmonic) gave a very poor fit, and in fact four harmonics were needed.

An analysis comparing a square wave model with a sinusoidal model of 24-hour blood pressure found that the former provided a better fit to the data.[32] This same analysis, which used 20-minute averages of blood pressure readings made from continuous intraarterial recording, also determined that the distribution of blood pressure over 24 hours was bimodal, which implies that a square wave model is appropriate. The transitions between the high and low pressure periods, which were

[a] The word mesor is derived from the Midline Estimating Statistic of Rhythm.

determined empirically, showed a surprisingly wide range, from 7:40 to 12:40 PM for the nocturnal decrease and from 5:40 to 9:40 AM for the morning increase. Whether these changes coincided with the transitions between sleep and wakefulness was not determined.

HOW SHOULD DAYTIME AND NIGHTTIME BE DEFINED?

In 1988 O'Brien *et al.*[33] suggested that people could be classified as "dippers" or "non-dippers" according to whether their blood pressure decreased during the night by more or less than 10%. Although most people would be classified as dippers, a variety of clinical situations are associated with a diminished nocturnal fall in blood pressure. Nondipping may be seen in healthy normotensive individuals, but is more commonly seen in hypertensive persons. A potentially interesting observation is that blacks tend to be nondippers more commonly than do whites, but this difference is probably environmental rather than genetic, because it is not seen in all populations. Nondipping has also been associated with several forms of secondary hypertension, including preeclampsia, chronic renal disease, some types of adrenocortical hypertension, such as Cushing's disease, and pheochromocytoma. Several cross-sectional studies have shown that target organ damage may be more pronounced in nondippers than in dippers.[34,35]

The subject is thus of clinical importance, but up to now there has been no consensus as to how "day" and "night" blood pressures should be defined, and as many as 12 different time schedules have been used by different authors.[36] In 1990 the Scientific Committee of an International Conference on ambulatory blood pressure monitoring recommended standardization of data analysis and proposed using 7 AM to 10 PM for the day and 10 PM to 7 AM for the night.[37] Physiologically this makes little sense, because people go to bed and get up at different times.

Three recent studies compared the consequences of expressing the day-night differences in blood pressure either by a fixed time of day (which was different in all three) or by the patients' reported times of going to sleep and waking up.[36,38,39] One concluded that the fixed time methods overestimated the level of blood pressure during sleep and hence underestimated the day-night differences.[39] The other two papers focused on the classification of dippers, and both found that the fixed time methods tended to give a higher prevalence of dippers than did the diary method; however, in one of them these differences were relatively small.[38]

It is clear that no single mechanism accounts for individual differences in the extent of dipping. There is a tendency to regard it as a measure of the fall in blood pressure occurring with sleep, but it should also be recognized that the increases occurring during the day are of equal importance. If we took sleeping pressure as the basal level, we might refer to peakers rather than to dippers.[40] As already described herein, the day-night differences are smaller in subjects who stay in bed for 24 hours than in those who are ambulatory during the day. We found that demographic factors such as smoking and habitual alcohol intake also influence the difference, but whether the effect is primarily on the daytime or nighttime pressure is not resolved.[41]

HYPERTENSION AS A DISTURBANCE OF THE "SET POINT" OF BLOOD PRESSURE

Numerous ambulatory monitoring studies comparing normotensive and hypertensive subjects have shown that the diurnal rhythm of blood pressure is preserved in most (but by no means all) hypertensive subjects but is reset at a higher level of blood pressure throughout the day and night. Thus, hypertension is primarily a disturbance of the tonic level of blood pressure. This is associated with an increase in the absolute level of the phasic component of blood pressure (measured as the standard deviation), although when expressed in relative terms (e.g., as the coefficient of variation), it is little altered.[42] Furthermore, studies of young normotensive subjects with a positive family history of hypertension have shown a subtle increase in blood pressure throughout the 24 hours, without any associated change in either short-term variability of blood pressure or reactivity to stressful stimuli.[43]

CONCLUSIONS

The diurnal rhythm of blood pressure is a major source of blood pressure variability, whose pathologic significance is only just beginning to be appreciated. Physicians should no longer think of blood pressure as characterized by a single pair of numbers, and ambulatory monitoring, which is the only available method for evaluating the diurnal changes in blood pressure, is likely to play an increasingly important role in the management of hypertensive patients. The normal pattern is not a smooth sinusoidal curve, but an asymmetrical pattern showing the biggest changes on going to sleep and getting up in the morning. Several lines of evidence point to the cycle of activity and arousal as the most important determinant of blood pressure rhythm in contrast to other biologic rhythms where intrinsic mechanisms are predominant.

REFERENCES

1. Janeway, T. C. 1904. The Clinical Study of Blood-Pressure. A Guide to the Use of the Sphygmomanometer in Medical, Surgical, and Obstetrical Practice, with a Summary of the Experimental and Clinical Facts Relating to the Blood-Pressure in Health and Disease. Appleton. New York.
2. Kain, H. K., A. T. Hinman & M. Sokolow. 1964. Arterial blood pressure measurements with a portable recorder in hypertensive patients. I. Variability and correlation with "casual" pressure. Circulation **30:** 882–892.
3. Bevan, A. T., A. J. Honour & F. H. Stott. 1969. Direct arterial pressure recording in unrestricted man. Clin. Sci. **36:** 329–344.
4. Millar-Craig, M. W., C. N. Bishop & E. B. Raftery. 1978. Circadian variation of blood-pressure. Lancet **1:** 795–797.
5. Stone, A. A., J. M. Smyth, T. G. Pickering & J. E. Schwartz. 1995. Daily mood variability: form of diurnal patterns and determinants of diurnal patterns. In press.
6. Schwartz, J. E., K. Warren & T. G. Pickering. 1994. Mood, location and physical position as predictors of ambulatory blood pressure and heart rate: Application of a multi-level random effects model. Ann. Behav. Med. **16:** 210–220.
7. Pickering, T. G. 1992. The ninth Sir George Pickering memorial lecture. Ambulatory monitoring and the definition of hypertension. J. Hypertens. **10:** 401–409.
8. Halberg, F., E. Halberg, J. Halberg & F. Halberg. 1984. Chronobiologic assessment of human blood pressure variation in health and disease. *In* Ambulatory Blood Pressure Monitoring. M. A. Weber & J. I. M. Drayer, Eds.: 137–156. Steinkopff-Darmstadt.
9. Pickering, T. G. 1991. Ambulatory Monitoring and Blood Pressure Variability. Science Press. London.
10. Littler, W. A. & R. D. S. Watson. 1978. Circadian variation in blood pressure. Lancet **1:** 995–996.
11. Floras, J. S., J. V. Jones, J. A. Johnston, D. E. Brooks, M. O. Hassan & P. Sleight.

1978. Arousal and the circadian rhythm of blood pressure. Clin. Sci. Mol. Med. **55:** 395S–397S.

12. Bristow, J. D., A. J. Honour, T. G. Pickering & P. Sleight. 1969. Cardiovascular and respiratory changes during sleep in normal and hypertensive subjects. Cardiovasc. Med. **3:** 476–486.
13. Clark, L. A., L. Denby, D. Pregibon, G. A. Harshfield, T. G. Pickering, S. Blank & J. H. Laragh. 1987. The effects of activity and time of day on the diurnal variations of blood pressure. J. Chron. Dis. **40:** 671–681.
14. Turjanmaa, V., M. Tuomisto, M. Fredrikson, S. Kalli & A. Uusitalo. 1991. Blood pressure and heart rate variability and reactivity as related to daily activities in normotensive men measured with 24-h intra-arterial recording. J. Hypertens. **9:** 665–673.
15. Van Egeren, L. F. 1991. Computer-based monitoring of physical activity. *In* Clinical Evaluation and Physiological Monitoring. L. Miles & R. Broughton, Eds. Raven Press. New York.
16. Stewart, M. J., H. Brown & P. L. Padfield. 1993. Can simultaneous ambulatory blood pressure and activity monitoring improve the definition of blood pressure? Am. J. Hypertens. **6:** 174S–178S.
17. Gellman, M., S. Spitzer, G. Ironson, M. Llabre, P. Saab, R. P. Decarlo, D. J. Weidler & N. Schneiderman. 1990. Posture, place and mood effects on ambulatory blood pressure. Psychophysiology, **27:** 544–551.
18. Goldstein, I. B., L. D. Jamner & D. Shapiro. 1992. Ambulatory blood pressure and heart rate in healthy male paramedics during a workday and a nonworkday. Health Psychol. **11:** 48–54.
19. Van Der Meiracker, A. H., A. J. Man In't Veld, H. G. Ritsema Van Eck, G. J. Wenting & M. A. D. H. Schalekamp. 1988. Determinants of short term blood pressure variability. Effects of bed rest and sensory deprivation in essential hypertension. Am. J. Hypertens. **1:** 22–26.
20. Athanassiadis, D., G. D. Draper, A. J. Honour & W. I. Cranston. 1969. Variability of automatic blood pressure measurements over 24 hour periods. Clin. Sci. **36:** 147–156.
21. Mann, S., M. W. Millar-Craig, D. I. Melville, V. Balasubramanian & E. B. Raftery. 1979. Physical activity and the circadian rhythm of blood pressure. Clin. Sci. **57:** 291S–294S.
22. Khoury, A. F., P. Sunderajan & N. M. Kaplan. 1992. The early morning rise in blood pressure is related mainly to ambulation. Am. J. Hypertens. **5:** 339–344.
23. Mancia, G., A. Ferrari, L. Gregorini, G. Parati, G. Pomidossi, G. Bertinieri, G. Grassi, M. Dirienzo, A. Pedottri & A. Zanchetti. 1983. Blood pressure and heart rate variability in normotensive and hypertensive human beings. Circ. Res. **53:** 96–104.
24. Sundberg, S., A. Kohvakka & A. Gordin. 1988. Rapid reversal of circadian blood pressure rhythm in shift workers. J. Hypertens. **6:** 393–396.
25. Baumgart, P., P. Walger, G. Fuchs, K. G. Dorst, H. Vetter & K. H. Rahm. 1989. Twenty-four-hour blood pressure is not dependent on endogenous circadian rhythm. J. Hypertens. **7:** 331–334.
26. Sternberg, H., T. Rosenthal, A. Shamisss & M. Green. 1995. Altered circadian rhythm of blood pressure in shift workers. J. Hum. Hypertens. **9:** 349–353.
27. Chau, H. P., J. P. Mallion, R. de Gaudemaris, E. Ruche, J. P. Siche, O. Pelen & G. Mathern. 1989. Twenty-four-hour ambulatory blood pressure in shift workers. Circulation **80:** 341–347.
28. Lewis, P. R. & M. C. Lobban. 1957. Dissociation of diurnal rhythms in human subjects on abnormal time routines. Quart. J. Exp. Physiol. **42:** 371–386.
29. Kleitman, N. & E. Kleitman. 1953. Effect of non-twenty-four-hour routine living on oral temperature and heart rate. J. Appl. Physiol. **6:** 283–291.
30. Mills, J. N. & S. W. Stanbury. 1952. Persistent 24 hr renal excretory rhythm on a 12 hr cycle of activity. J. Physiol. **117:** 22–37.
31. Thijs, L., J. Staessen & R. Fagard. 1992. Analysis of the diurnal pressure curve. High Blood Press. **1:** 17–28.
32. Idema, R. N., E. S. Gelsema, G.-J. Wenting, J. L. Grashuis, A. H. Van Der

MEIRACKER, R. M. L. BROUWER & A. J. MAN IN'T VELD. 1992. A new model for diurnal blood pressure profiling. Square wave fit compared with conventional methods. Hypertension **19:** 595–605.

33. O'BRIEN, E., J. SHERIDAN & K. O'MALLEY. 1988. Dippers and non-dippers (letter). Lancet **2:** 397.
34. KUWAJIMA, I., Y. SUZUKI, T. SHIMOSAWA, A. KANEMARU, S. HOSHINO & K. KURAMOTO. 1992. Diminished nocturnal decline in blood pressure in elderly hypertensive patients with left ventricular hypertrophy. Am. Heart J. **67:** 1307–1311.
35. SHIMADA, K., A. KAWAMOTO, K. MATSUBAYASHI, M. NISHINAGA, S. KIMURA & T. OZAWA. 1992. Diurnal blood pressure variations and silent cerebrovascular damage in elderly patients with hypertension. J. Hypertens. **10:** 875–878.
36. PEIXOTO, A. J., G. A. MANSOOR & W. B. WHITE. 1995. Effects of actual versus arbitrary awake and asleep times on analyses of 24-hour blood pressure. Am. J. Hypertens. **8:** 676–680.
37. THE SCIENTIFIC COMMITTEE. Consensus document on non-invasive ambulatory blood pressure monitoring. 1990. J. Hypertens. **8**(suppl. 6): 135–140.
38. ROSANSKY, S. J., S. J. MENACHERY, C. M. WAGNER & K. JACKSON. 1995. The effect of sleep intervals on analysis of 24-hour ambulatory blood pressure data. Am. J. Hypertens. **8:** 672–675.
39. GATZKA, C. D. & R. E. SCHMIEDER. 1995. Improved classification of dippers by individualized analysis of ambulatory blood pressure profiles. Am. J. Hypertens. **8:** 666–671.
40. PICKERING, T. G. & G. D. JAMES. 1993. Determinants and consequences of the diurnal rhythm of blood pressure. Am. J. Hypertens. **6:** 266s–269s.
41. JAMES, G. D., T. TOLEDANO & T. G. PICKERING. 1995. Factors influencing the awake-sleep difference in ambulatory blood pressure: Main effects and sex differences. J. Hum. Hypertens. In press.
42. MANCIA, G., A. FERRARI, L. GREGORINI, G. PARATI, G. POMIDOSSI, G. BERTINIERI, G. GRASSI, M. DI RIENZO, A. PEDOTTI & A. ZANCHETTI. 1983. Blood pressure and heart rate variabilities in normotensive and hypertensive human beings. Circ. Res. **53:** 96–104.
43. RAVOGLI, A., S. TRAZZI, A. VILLANI *et al.* 1990. Early 24-hour blood pressure elevation in normotensive subjects with parental hypertension. Hypertension **16:** 491–497.

Relationships of the Circadian Rhythms of Thrombotic, Ischemic, Hemorrhagic, and Arrhythmic Events to Blood Pressure Rhythms

ROBERTO MANFREDINI, MASSIMO GALLERANI, FRANCESCO PORTALUPPI, AND CARMELO FERSINI

First Institute of Internal Medicine;
Emergency Department
St. Anna Hospital; and
Hypertension Unit, First Institute of Internal Medicine
University of Ferrara
Ferrara, Italy

Many biologic functions oscillate around a mean state following roughly a 24-hour pattern (also called a circadian pattern). When reproducible and statistically significant, these fluctuations are called rhythms. In the last decade, it has become increasingly evident that these fluctuations and rhythms are potentially active as triggering factors that favor or precipitate a pathologic event. This finding has increased interest in a new research field called "chronoepidemiology."

Epidemiology enables us to define several conditions as "risk factors" because of their association with the development of a given disease. Accordingly, we usually compute the expectation and predictability of a certain disease on the basis of the presence or absence of one or more risk factors. This concept of risk, however, is totally dissociated from the possibility that the risk may have a temporal variability. The actual concept of risk in medicine, in fact, is quantistic: the higher the given risk factor, the more dangerous it becomes. However, the temporal concurrence of multiple oscillatory variables, sometimes not even individually altered to a significant extent, can facilitate the onset or manifestation of a given disease at a certain moment. Hence, "chronorisk" is defined as risk that has systematic periodicity in its time of occurrence.

On these premises, a growing body of evidence indicates that several unfavorable cardiovascular events show a temporal distribution in their occurrence.[1,2] This was shown to be true for myocardial angina and infarction, sudden cardiac death, pulmonary thromboembolism, arrhythmic events, and cerebrovascular diseases. At the same time, deeper knowledge of the underlying oscillatory mechanisms has opened up the potential for temporized optimization of the therapy of cardiovascular disease.[3–5]

MYOCARDIAL ISCHEMIA

Myocardial ischemia may occur at any time of the day or night.[6–8] However, the frequency of attacks is increased between 2 and 4 AM in patients with variant angina.[9] The extended availability of ambulatory electrocardiographic monitoring techniques allowed the demonstration that in the absence of antiischemic therapy, most patients exhibit a circadian pattern in the occurrence of episodes of transient myocardial ischemia with ST-segment depression.[10–18] These studies revealed that a two- or threefold increase in transient ischemic episodes occurs between 6 AM and noon. After data correction for individual variability in the time of awakening, events are most frequent during the first 2 hours after arousal.[19] An analogous pattern is also found in patients with silent myocardial ischemia.[20] Some studies have demonstrated a second increase in the frequency of transient ischemic episodes in late evening, between 6 and 8 PM,[21] while others have demonstrated a lowered ischemic threshold at night.[22] Using 24-hour Holter monitoring in patients with established coronary artery disease, both symptomatic and silent episodes of ST-segment elevations were recorded in 72% of patients between midnight and 9 AM.[23]

A study of chest pain in patients with variant angina performing an exercise treadmill test at different hours of the day (early morning and afternoon) showed that an increase in angina associated with ST-segment elevation occurred more frequently in the morning than in the afternoon.[24]

MYOCARDIAL INFARCTION

Myocardial infarction does not occur randomly throughout the day, but shows a peculiar diurnal pattern. Studies based on retrospective assessment of the time of symptom onset[25,26] had indicated a morning peak occurrence for myocardial infarction. Then two other studies, the Multicenter Investigation of Limitation of Infarct Size[27] and the Intravenous Streptokinase in Acute Myocardial Infarction study,[28] identified more objectively the onset of myocardial infarction using serial determinations of plasma creatine kinase levels. A circadian relation between clinical onset of myocardial infarction and time of appearance of creatine kinase in the plasma was demonstrated in both studies. On the basis of these and other reports,[29–33] it is well established that the frequency of onset of myocardial infarction is lowest during the first part of the night, increases in the second part of the night, and peaks between 6 AM and noon. When reference is made to the actual time of awakening, the peak morning incidence of myocardial infarction occurs in the first few hours after awakening.[34] Moreover, in a population of shift workers the temporal pattern of occurrence of cardiovascular events changed in relation to alterations in the sleep-wake cycle, its peak coinciding with the hours around awakening.[35]

Epidemiologic studies of myocardial infarction also found a secondary, smaller peak incidence in the evening,[29,30,33] sometimes accompanied by other smaller peaks during the day,[30,31] suggesting that subgroups of patients might present different temporal patterns in the onset of disease. Some investigators reported that age[29,30] and gender[30] affect the circadian onset pattern of myocardial infarction, but others found no such influence.[32,33] A history of either smoking, diabetes,[29,30] cardiovascular disease, stroke,[33] or previous myocardial infarction other than non-Q wave[30] as well as the presence of congestive heart failure or non-Q-wave type myocardial infarction[29,30,36] has all been associated with a minor evening peak or no diurnal variation in infarct onset. Altogether, these findings indicate that some clinical characteristics of the patients who undergo myocardial infarction, namely, a history or the presence of other cerebrocardiovascular diseases, may act as modifiers of the typical circadian onset pattern of myocardial infarction.

On the other hand, the circadian time of occurrence can affect the clinical course of myocardial infarction.[37] Symptom onset between 6 AM and noon were reported to be associated with greater infarct size, whereas onset time between 12 noon and 6 AM was associated with a significantly lower risk of circulatory arrests from ventricular arrhythmias.[38]

Finally, the circadian-time–dependent occurrence of myocardial infarction is also likely to have therapeutic implications. A lower success rate of thrombolysis was found in patients with morning onset compared to those with onset at other times of the day.[39]

SUDDEN CARDIAC DEATH

A circadian variation in sudden cardiac death was first observed by retrospective analysis of more than 2,200 mortality records from a Massachusetts population.[40] A prominent peak was found between 9 and 11 AM, while a secondary peak was present in the afternoon. To overcome the limited value of information derived from death certificates, a further study analyzed the occurrence of sudden death in 5,209 subjects between the ages of 30 and 62 from the Framingham Heart Study.[41] Time of death for both definite (n = 264) and possible (n = 165) sudden cardiac death showed a circadian variation, with a prominent morning peak between 6 and 9 AM and the lowest frequency during the night. Other confirmations of this temporal pattern came from the Beta-Blocker Heart Attack trial,[42] the prospective study on 1,009 cases of out-of-hospital cardiac arrest in Houston,[43] and the 362 cases of cardiac arrest in elderly patients in New York.[44] In a population study of out-of-hospital sudden deaths in which necroscopic confirmation was available, morning peak occurrence was associated with afternoon and evening secondary peaks. We analyzed subgroups of patients according to different pathophysiologic and clinicoanatomic characteristics. Deaths from fatal arrhythmias and myocardial infarction showed a morning and an afternoon peak incidence in the total population, whereas diabetic subjects had a peak in the afternoon-evening hours.[45,46]

PULMONARY THROMBOEMBOLISM

Pulmonary thromboembolism has a circadian pattern characterized by a morning peak. This pattern was first reported in 152 cases.[47] It was then confirmed in 13 hospitalized patients with acute spinal cord injury[48] and in 48 cases of fatal pulmonary thromboembolism among 507 sudden out-of-hospital deaths at autopsy over a 7-year period.[49] The same peak frequency between 8 AM and 3 PM was found in 90 hospitalized subjects whose cause of death, determined at autopsy, was pulmonary thromboembolism,[50] suggesting that this pattern is independent of possible alterations in daily activity caused by the constraints of hospital stay. Sex and age, however, may influence this circadian pattern, which seems more pronounced in females and elderly subjects.[51]

CEREBROVASCULAR DISEASES

Ischemic Disease

Although earlier reports showed a nocturnal peak occurrence of acute cerebrovascular accidents,[52,53] recent studies documented a higher frequency in the morning.[54–63] A second subsidiary peak during the evening was also described by some investigators,[56,59–61] but no change in this temporal pattern was found in subgroups of cases obtained by anatomic and pathophysiologic classification.[63]

Hemorrhagic Diseases

A circadian pattern in the occurrence of intracranial hemorrhage was first reported 20 years ago.[53] Several years later, a study of 194 patients with subarachnoid and 118 with intracerebral hemorrhage documented a bimodal pattern with a prevalent morning peak (between 10 AM and 12 noon) in both groups[54]; a minor peak was found in the early evening (between 6 and 8 PM) for subarachnoid hemorrhage and in the afternoon (between 4 and 6 PM) for intracerebral hemorrhage.

Since then, other reports have documented the circadian onset distribution of intracranial hemorrhage, confirming the existence of a prevalent morning peak,[62,64–66] sometimes associated with a secondary afternoon peak.[65] In one study, the primary peak for subarachnoid hemorrhage occurred in the afternoon,[64] whereas in another study, performed retrospectively on 159 cerebral hemorrhages over a 17-year period, no specific significant circadian pattern was found.[60]

A correlation between this circadian pattern and hypertension was detected,[65] but a later study did not confirm this finding.[66]

ARRHYTHMIC EVENTS

Relatively few data on hypokinetic arrhythmias are available. A progressive loss of day-night variability in heart rate was observed in patients with sick sinus syndrome,[67] in relation to the increasing severity of the disease. In complete atrioventricular block, both atrial and ventricular rates have a typical 24-hour sinusoidal pattern,[68] suggesting that both sinus node and ventricular pacemakers maintain their normal circadian variability in this arrhythmia.

Much more is known about hyperkynetic arrhythmias. A prominent circadian pattern in the ventricular response to chronic atrial fibrillation is described, with an estimated peak at 1 PM and a trough at 4 AM, with a fluctuation amplitude of 22%.[69] The same pattern is found in control subjects matched by sex and underlying heart condition, suggesting that both the sinus pacemaker and atrioventricular conduction maintain their responsiveness to the diurnal variations of neurovegetative tone in this arrhythmia.

Reports on temporal distribution in the onset of paroxysmal atrial fibrillation are discordant. In large populations, a higher frequency was found during the day than the night,[70,71] with morning and evening peaks. However, another large study of 726 subjects over a 10-year span led to contrasting evidence of a significantly higher frequency at night,[72] although two secondary peaks were also found in the morning and early afternoon. Finally, no circadian pattern was documented in another limited study of 37 patients.[73] Interestingly, administration of β-adrenoceptor antagonists eliminated the morning peak observed in untreated patients and shifted the peak in paroxysmal events to the evening and night hours.[71]

For paroxysmal supraventricular tachycardia, a midafternoon onset peak was unanimously recognized.[72–75]

A circadian pattern in ventricular ectopic beats, with a dominant morning peak and a possible accessory afternoon peak, is well established.[76–80] A morning and subsidiary afternoon peak for ventricular and sustained tachycardia was also described.[81–84] Obviously, it is difficult to obtain data on ventricular fibrillation. The only information available on the temporal distribution of this arrhythmia comes from the records of 294 patients collected by a first-aid interventional squad.[85] Again, a morning peak with a secondary one in the afternoon was found.

PATHOPHYSIOLOGIC MECHANISMS OF MYOCARDIAL ISCHEMIA

Increased Myocardial Oxygen Demand

Several physiologic determinants may participate in enhancing myocardial oxygen demand, with consequent pathologic and therapeutic implications. Until now, heart rate (HR) was not considered to play a leading role in the pathogenesis of myocardial ischemia.[8,86] Recent studies, however, focused on HR responsivity as a triggering factor. Results from the Angina and Silent Ischemia Study Group[87] showed that more than 80% of episodes of ambulatory ischemia were associated with substantial increases in HR. The likelihood of developing ischemia could be estimated by the magnitude and duration of the increase in HR and the preceding baseline HR. In those authors' opinion, time of day per se does not influence the likelihood of developing ischemia other than providing a temporal frame within which increases in HR occur. However, no doubt the circadian rhythmicity of HR[67,68,88] is intrinsic. A strong genetic influence on its rhythm was demonstrated.[89] Moreover, sleep accounts for only a minor part of the nocturnal decline in HR, as demonstrated by a persistent nocturnal bradycardia for months after sleep deprivation in fatal familial insomnia,[90] whereas autonomic function appears to play a major role.[91] In fact, the progressive alterations in 24-hour patterns of HR and catecholamines found in fatal familial insomnia closely parallel the clinical and instrumental signs of progressive sympathetic overactivity and are reflected in strikingly similar profiles throughout different stages of the disease.[90] Because sleep bradycardia is due more to parasympathetic activation than to sympathetic withdrawal, its preservation is justified in fatal familial insomnia patients, who have preserved parasympathetic activity. Huikuri *et al.*[92] studied the power spectral components of HR variability and their circadian pattern in 22 survivors of out-of-hospital cardiac arrest not associated with myocardial infarction and 22 control patients. They found a distinct reproducible circadian pattern of HR variability in both groups, with a morning peak before arousal. In survivors of cardiac arrest an abrupt decrease in HR variability after arousal occurred as a result of a very low vagal control, suggesting that an autonomic imbalance should be considered a potential factor in morning cardiovascular accidents. Benhorin *et al.*,[93] using various statistical analytic methods to evaluate the degree of HR changes, concluded that the morning increase in ischemic episodes is largely due to increased oxygen demand rather than reduced ischemic threshold. They demonstrated that both the frequency of ischemic episodes and the ischemic threshold might be modified by common factors such as mental arousal, postural changes, and physical activities. Another study by Barry *et al.*[94] found that waking and rising at night is highly associated with transient myocardial ischemia. The characteristic changes in HR and ECG signs of ischemia at night resemble the events occurring on waking and rising in the morning. Most ischemic events, in fact, are preceded by a modest increase in HR starting 5–30 minutes before the onset of ECG changes. Mulcahy *et al.*[95] demonstrated that potentially negative hemodynamic modifications observed after morning arousal can also be reproduced after an afternoon siesta, possibly carrying cardiovascular risks similar to those associated with morning waking hours. Those authors drew two conclusions: (1) sleep is a relatively protected period, and (2) the period soon after waking and getting up is particularly associated with cardiovascular risks.

Concerning sleep, however, some considerations have to be added. It is well known that ischemic events can occur during sleep, especially in patients with severe coronary artery disease and vasospastic angina. In particular, rapid eye movement (REM) sleep is associated with myocardial ischemia.[96,97] In the study by Barry *et al.*[94] of 30 patients who had nocturnal ischemia, 9 had episodes of ischemia at night only, during sleep, and did not report rising at night. They had increased HR which became significant 5 minutes before the onset of ischemia. Contemporaneous sleep recording was not performed, but periods of HR increase are likely to correspond to REM sleep. In normal subjects, sympathetic nerve activity, BP, and HR are lower during deep non-REM sleep than during wakefulness.[98] During REM sleep, sympathetic nerve activity increases above the levels of wakefulness, whereas BP and HR return to levels recorded in the awake state. Thus, the triggering period that is usually estimated to start from arousal is anticipated to also include the last part of sleep, when the REM phase preferentially occurs. The circadian rhythm in the occurrence of ischemic events may be related to the circadian variation in REM sleep propensity through phasic sympathetic, BP, and HR fluctuations which occur during this phase of sleep. This concept of REM sleep as a period of potential cardiovascular risk is supported by the demonstration of a concomitant reduction in coronary blood flow[99] and an increased occurrence of coronary spasm.[97] Incidentally, the duration and organization of human sleep are strictly circadian-time dependent.[100] REM sleep is closely coupled with the circadian rhythm of core temperature,[101] which in turn is controlled by the biologic clock independent of the sleep period.[102,103] Finally, recent data indicate that BP fluctuations during REM sleep do not originate in the induction area of REM sleep itself (pontine tegmental area) but in the forebrain and are likely to be influenced by factors independent of sleep.[104]

Arousal is undoubtedly a risk period for unfavorable cardiovascular events because of a series of concomitant factors.

Plasma norepinephrine levels[105,106] and plasma renin activity[107] are higher in the morning and may cause coronary vasoconstriction. Moreover, a circadian rhythm in basal vascular tone, due either partly or entirely to increased α-sympathetic vasoconstrictor activity, may also contribute to higher BP.[108] It is not known if the morning surge in BP begins before or after awakening. Millar-Craig *et al.*,[109] after invasive monitoring, noted that BP started to rise during sleep in the early morning hours. Other authors reported a rise in pressure immediately in relation to physical activity following arousal.[110,111] Through the years, the pre-waking rise was alternately denied[112,113] and confirmed.[114–116] It is difficult to determine in a single patient the precise timing of the arousal status, and this might explain different interpretations of BP curves. A recent study compared morning changes in BP in patients immediately active after arousing with that in patients remaining in bed for a while after arousing.[117] Blood pressure showed a significant morning rise in patients who were immediately active, suggesting a relation with physical activity. It has been postulated that the morning surge in blood pressure consists of two different patterns, a gradual one beginning before awakening (especially in younger subjects), and the other starting with physical activity, especially in elderly subjects.[118,119] Hence, the morning surge of blood pressure in elderly patients may play a particularly important role in the genesis of unfavorable cardiovascular events.

Myocardial Oxygen Supply

Studies in animals show that the resting coronary tone is highest in the morning and less intense at night.[120] A circadian rhythm in vascular basal tone was also demonstrated[108] in relation to increased α-sympathetic vasoconstrictor activity during the morning hours. In addition, ischemic threshold is lower during this part of the day,[121] suggesting that ischemia-induced coronary vascular resistance is increased during those hours. This finding is also supported by a similar variation in postischemic forearm vascular resistance. Thus, in the morning, a low ischemic threshold presumably reflects elevated coronary vascular resistances. However, ischemic episodes also occur at night. The most likely mechanism for these nocturnal episodes is increased vascular resistance, particularly the increased coronary tone that may be related to REM sleep, especially in the second part of the night (discussed in the section on "Increased Myocardial Oxygen Demand" in this chapter). In addition, the circadian pattern of plasma cortisol secretion, characterized by a morning rise,[122] may reduce coronary blood flow by increasing sensitivity of epicardial vessels to vasoconstrictor stimuli.[12]

Another factor contributing to decreased blood supply may be an imbalance between coagulation and fibrinolysis, whose circadian patterns are demonstrated.[123,124]

Abnormalities of the coagulation, fibrinolytic, and inhibitory systems are present in patients with myocardial angina and infarction.[125] Circadian fluctuations exist in several coagulation parameters, leading to a pro-coagulative tendency in the morning, when fibrinogen peaks are found, concurrent with low prothrombin and activated prothrombin time.[126–128] Moreover, platelet aggregability is higher in the morning in both healthy subjects[106,129] and those with atherosclerotic[130] or coronary artery disease.[131] Platelet activity is diminished when the subject is in the supine position[132] and rises with assumption of the upright posture[133] but not during exercise,[134] reflecting parallel changes in sympathetic activity. Moreover, higher aggregability of white blood cells combined with low levels of plasma scavengers, both predisposing to microcirculatory occlusion and thrombosis, are present in patients with ischemic heart disease.[135]

Concurrent with this prothrombotic tendency, which is also favored by the morning increase in hematocrit[136,137] and plasma viscosity,[138] endogenous fibrinolytic activity is at its lowest. A circadian rhythm in fibrinolytic activity is well established.[139–143] The antiphasic rhythms in tissue-type plasminogen activator (t-PA) and plasminogen activator-inhibitor 1 (PAI-1) activity are characterized by concentration excursions of up to 250%.[144] A higher frequency of cardiovascular accidents is found at the times of day when PAI-1 levels are higher,[143] and patients with coronary artery disease have significantly reduced fibrinolytic activity compared to that of healthy subjects, because of the presence of peaks of activity of PAI-1 and the absence of t-PA.[145–147]

PATHOPHYSIOLOGY OF CEREBROVASCULAR DISEASE AND ITS RELATIONS TO BLOOD PRESSURE

We reviewed in a previous section of this chapter the available evidence in favor of a morning prevalence of cerebrovascular disease. Based on that evidence, we conclude that the temporal pattern in the onset of stroke does not appear to be influenced by the different etiologies. Although tradition holds that thrombotic stroke is more likely to occur at night and embolic stroke during activity, this difference was not confirmed in the literature. Some of the mechanisms proposed to explain such patterns have been extensively discussed before. Circadian changes in physical activity, catecholamine levels, vascular resistance, blood coagulability, platelet aggregability,[148] blood viscosity, and arterial BP play a concurrent favoring role. High hematocrit and BP seem to play a synergistic effect on the risk of stroke, because hypertensive subjects with raised hematocrit levels have more than a twofold increase in risk compared to other hypertensive subjects and over a ninefold increase compared to normotensive subjects.[149] An additional determinant of the higher morning risk of stroke is a diminished vasodilator reserve in the morning.[150] Moreover, endogenous release of dopamine, which may promote cell damage in experimentally produced cerebral ischemia,[151] shows a circadian pattern in the central nervous system with enhanced metabolism or release in the morning.[152] Such a pattern, on the one hand, could benefit dopaminergic-dependent functions, but on the other might contribute to the increased risk or extent of cerebral infarction at this time.[153]

Blood pressure deserves some considerations alone. Apart from the well known long-term cerebrovascular consequences of hypertension,[154] several studies consistently demonstrate that it is positively, and apparently independently, associated with the primary incidence of stroke, particularly among those subjects whose diastolic BP was 90 mm Hg or greater at the start of the study. Diastolic BP, in fact, has generally been used to characterize the BP regimen of an individual irrespective of other indices, such as the systolic or mean BP. Prolonged differences in usual diastolic BP of 5, 7.5, and 10 mm Hg are respectively associated with at least 34%, 46%, and 56% less stroke.[155] Treating hypertension reduces considerably the frequency of cerebrovascular accidents within just a few years.[156] Reduction in the odds of stroke averages 42% (95% confidence intervals 33–50%).[156] Isolated systolic hypertension also has an established role in the determination of stroke, particularly in elderly people, and its treatment reduces fatal and nonfatal strokes by 36% after an average of 4.5 years.[157]

Epidemiologic studies showing a morning onset of stroke challenge the hypothesis that stroke coincides with low pressures, and in fact the onset of ischemic stroke is maximal when BP rises. Concomitant vascular, rheologic, and coagulative patterns facilitate ischemic processes. Intracerebral and subarachnoid hemorrhages show morning prevalences which are more obviously related to BP patterns. Their morning peak of occurrence, in fact, is very similar to that described for other hemorrhagic diseases of the gastrointestinal system.[158,159]

If the diurnal pattern of BP influences the occurrence of cerebrovascular events, the BP pattern is also modified both in the course of certain cerebral diseases and after an accident has occurred.

For example, although BP is lower at night than during the day in normal subjects, average blood pressure at night is slightly higher than that in the day in patients with lacunar type and Binswanger type dementia,[160] Shy-Drager syndrome,[161] and autonomic failure.[162]

On the other hand, different changes in the circadian BP pattern are observed after brain infarction, depending on the pathogenesis and location of the stroke, indicating a different disturbance in circadian blood pressure regulation. Thromboembolic brain stem infarctions result in blunted or reversed circadian BP patterns,[163–166] whereas hemodynamic brain infarctions give rise to a higher range of variation (20% systolic and 23% diastolic) between day and night BP values.[166] A prolonged disturbance of the blood-brain barrier is also found in these patients.[167]

RELATIONSHIPS BETWEEN BLOOD PRESSURE AND ARRHYTHMIAS

Hypertension and hypertensive heart disease are considered major risk factors in coronary artery disease.[168] Moreover, systemic hypertension, with and without left ventricular hypertrophy, favors an increased risk of sudden deaths from tachyarrhythmias.[169–171] However, left ventricular hypertrophy is also often related to ventricular arrhythmias,[172,173] and other conditions (such as hypokalemia) are debated to be predictive factors of the presence of ventricular arrhythmias as well.[174–176] Several mechanisms facilitate ventricular arrhythmias. Examples at the cellular level are myocyte hypertrophy associated with prolongation of the action potential, lowered conduction velocity and resting membrane potential, and dispersion of refractoriness.[177] Important at the tissue level are the reduced subendocardial coronary vascular reserve and subendocardial ischemia, both facilitating arrhythmias[178] through enhancement of differences in conduction velocity and recovery of excitability.[179,180] Moreover, a reduced metabolic tolerance to ischemia[181] lowers the fibrillation threshold.

A strict correlation between the occurrence of ventricular arrhythmias and BP rhythms exists.[182] In hypertensive patients, ventricular arrhythmias show a preferential morning pattern.[80] Ventricular arrhythmias and asymptomatic transient ST-segment alterations are common in hypertensive subjects, and the interaction of both these risk factors may result in fatal events.[183] Therapies that modify morning peaks in blood pressure together with heart rate, vasoconstriction, and blood coagulability are strongly recommended in reducing the development of ventricular arrhythmias and sudden death.

SUMMARY

Convincing evidence has recently accumulated that several unfavorable cardiovascular events show a well defined pattern in their occurrence throughout the day. Myocardial angina and infarction, sudden cardiac death, arrhythmias, fatal pulmonary thromboembolism, and ischemic and hemorrhagic cerebrovascular accidents occur more frequently in the morning, after awaking, until noon. Diurnal variations in multiple biologic functions, such as assumption of an upright posture associated with increased platelet aggregability, changes in blood clotting, fibrinolysis, and vascular tone and resistance, may be potentially active triggering factors. Moreover, variations in sympathetic tone, catecholamine secretion, and blood pressure have to be considered. The role of triggering factors and their relationships with blood pressure patterns is discussed in view of an optimized pharmacologic treatment.

REFERENCES

1. Detry, J. M. & M. Vincent. 1992. Circadian rhythms in cardiovascular disease: The crucial hours. J. Hum. Hypertens. **6**(Suppl. 1): S3–8.
2. Jafri, S. M. & S. Goldstein. 1993. Circadian variation of onset of ischemic heart disease. *In* Advances in Internal Medicine. J. J. Leonard, Ed. **38:** 107–122. Mosby-Year Book, Inc. Chicago.
3. Pepine, C. J. 1991. Circadian variations in myocardial ischemia. Implications for management. JAMA **265:** 386–390.
4. Manfredini, R., M. Gallerani, R. Salmi & C. Fersini. 1994. Circadian rhythms and the heart: Implications for chronotherapy of cardiovascular diseases. Clin. Pharmacol. Ther. **56:** 244–247.
5. Cooke, H. M. & A. Lynch. 1994. Biorhythms and chronotherapy in cardiovascular disease. Am. J. Hosp. Pharm. **51:** 2569–2580.
6. Stern, S. & D. Tzivoni. 1974. Early detection of silent ischaemic heart disease by 24-hour electrocardiographic monitoring of active subjects. Br. Heart J. **36:** 481–486.
7. Schang, S. J. & C. J. Pepine. 1977. Transient asymptomatic S-T segment depression during daily activity. Am. J. Cardiol. **39:** 396–402.
8. Deanfield, J. E., A. Maseri, A. P. Selwyn *et al.* 1983. Myocardial ischaemia during

daily life in patients with stable angina: Its relation to symptoms and heart rate changes. Lancet **2:** 753–758.
9. OGAWA, H., H. YASUE, S. OSHIMA, K. OKUMURA, K. MATSUYAMA & K. OBATA. 1989. Circadian variation of plasma fibrinopeptide A level in patients with variant angina. Circulation **80:** 1617–1626.
10. WATERS, D. D., D. D. MILLER, A. BOUCHARD, X. BOSCH & P. THEROUX. 1984. Circadian variation in variant angina. Am. J. Cardiol. **54:** 61–64.
11. QUYYUMI, A. A., L. MOCKUS, C. WRIGHT & K. M. FOX. 1985. Morphology of ambulatory ST segment changes in patients with varying severity of coronary artery disease. Investigation of the frequency of nocturnal ischaemia and coronary spasm. Br. Heart J. **53:** 186–193.
12. ROCCO, M. B., E. G. NABEL & A. P. SELWYN. 1987. Circadian rhythms and coronary artery disease. Am. J. Cardiol. **59:** 13C–17C.
13. NADEMANEE, K., V. INTARACHOT, M. A. JOSEPHSON & B. N. SINGH. 1987. Circadian variation in occurrence of transient overt and silent myocardial ischemia in chronic stable angina and comparison with Prinzmetal angina in men. Am. J. Cardiol. **60:** 494–498.
14. ASLANIAN, N. L. 1990. Chronobiological aspects of ischemic heart disease. Prog. Clin. Biol. Res. **341B:** 583–592.
15. REBUZZI, A. G., G. A. LANZA, M. LUCENTE *et al.* 1990. Circadian rhythm of ischemic episodes in patients with and without previous myocardial infarction. Prog. Clin. Biol. Res. **341A:** 419–422.
16. ROCCO, M. B. 1990. Timing and triggers of transient myocardial ischemia. Am. J. Cardiol. **66:** 18G–21G.
17. BEHAR, S., H. REICHER-REISS, U. GOLDBOURT & E. KAPLINSKY. 1991. Circadian variation in pain onset in unstable angina pectoris. Am. J. Cardiol. **67:** 91–93.
18. HAUSMANN, D., P. R. LICHTLEN, P. NIKUTTA, P. WENZLAFF & W. G. DANIEL. 1991. Circadian variation of myocardial ischemia in patients with stable coronary artery disease. Chronobiol. Int. **8:** 385–398.
19. ROCCO, M. B., J. BARRY, S. CAMPBELL *et al.* 1987. Circadian variation of transient myocardial ischemia in patients with coronary artery disease. Circulation **75:** 395–400.
20. FOX, K., D. MULCAHY, J. KEEGAN & C. WRIGHT. 1989. Circadian patterns of myocardial ischemia. Am. Heart J. **118:** 1084–1087.
21. MULCAHY, D., J. KEEGAN, D. CUNNINGHAM *et al.* 1988. Circadian variation of total ischaemic burden and its alteration with anti-anginal agents. Lancet **2:** 755–759.
22. FIGUERAS, J., J. CINCA, F. BALDA, A. MOYA & J. RIUS. 1986. Resting angina with fixed coronary artery stenosis: Nocturnal decline in ischemic threshold. Circulation **74:** 1248–1254.
23. ARAKI, H., Y. KOIWAYA, O. NAKAGAKI & M. NAKAMURA. 1983. Diurnal distribution of ST-segment elevation and related arrhythmias in patients with variant angina: A study by ambulatory ECG monitoring. Circulation **67:** 995–1000.
24. YASUE, H., S. OMOTE, A. TAKIZAWA, M. NAGAO, K. MIWA & S. TANAKA. 1979. Circadian variation of exercise capacity in patients with Prinzmetal's variant angina: Role of exercise-induced coronary arterial spasm. Circulation **59:** 938–948.
25. PELL, S. & C. A. D'ALONZO. 1963. Acute myocardial infarction in a large industrial population: Report of a 6-year study of 1,356 cases. JAMA **185:** 831–838.
26. THOMPSON, D. R., R. L. BLANDFORD, T. W. SUTTON & P. R. MARCHANT. 1985. Time of onset of chest pain in acute myocardial infarction. Int. J. Cardiol **7:** 139–148.
27. MULLER, J. E., P. H. STONE, Z. G. TURI *et al.* 1985. Circadian variation in the frequency of onset of acute myocardial infarction. N. Engl. J. Med. **313:** 1315–1322.
28. WILLICH, S. N., T. LINDERER, K. WEGSCHEIDER, A. LEIZOROVICZ, I. ALAMERCERY & R. SCHRODER. 1989. Increased morning incidence of myocardial infarction in the ISAM Study: Absence with prior beta-adrenergic blockade. ISAM Study Group. Circulation **80:** 853–858.
29. HJALMARSON, A., E. A. GILPIN, P. NICOD *et al.* 1989. Differing circadian patterns of symptom onset in subgroups of patients with acute myocardial infarction. Circulation **80:** 267–275.

30. GILPIN, E. A., A. HJALMARSON & J. ROSS. 1990. Subgroups of patients with atypical circadian patterns of symptom onset in acute myocardial infarction. Am. J. Cardiol. **66:** 7G–11G.
31. HANSEN, O., B. W. JOHANSSON & B. GULLBERG. 1992. Circadian distribution of onset of acute myocardial infarction in subgroups from analysis of 10,791 patients treated in a single center. Am. J. Cardiol. **69:** 1003–1008.
32. GALLERANI, M., R. MANFREDINI, L. RICCI, C. GOLDONI, A. COCURULLO & P. L. PARESCHI. 1993. Circadian variation in the onset of acute myocardial infarction: Lack of an effect due to age and sex. J. Int. Med. Res. **21:** 158–160.
33. BEHAR, S., M. HALABI, H. REICHER-REISS *et al.* 1993. Circadian variation and possible external triggers of onset of myocardial infarction. SPRINT Study Group. Am. J. Med. **94:** 395–400.
34. GOLDBERG, R. J., P. BRADY, J. E. MULLER *et al.* 1990. Time of onset of symptoms of acute myocardial infarction. Am. J. Cardiol. **66:** 140–144.
35. KNUTSSON, A. 1989. Shift work and coronary heart disease. Scand. J. Soc. Med. Suppl. **44**: 1–36.
36. KLEIMAN, N. S., K. B. SCHECHTMAN, P. M. YOUNG *et al.* 1990. Lack of diurnal variation in the onset of non-Q wave infarction. Circulation **81:** 548–555.
37. BARASH, D., R. A. SILVERMAN, P. GENNIS, N. BUDNER, M. MATOS & E. J. GALLAGHER. 1989. Circadian variation in the frequency of myocardial infarction and death associated with acute pulmonary edema. J. Emerg. Med. **7:** 119–121.
38. HANSEN, O., B. W. JOHANSSON & B. GULLBERG. 1993. The clinical outcome of acute myocardial infarction is related to the circadian rhythm of myocardial infarction onset. Angiology **44:** 509–516.
39. FUJITA, M., E. ARAIE, K. YAMANISHI, K. MIWA, M. KIDA & H. NAKAJIMA. 1993. Circadian variation in the success rate of intracoronary thrombolysis for acute myocardial infarction. Am. J. Cardiol. **71:** 1369–1371.
40. MULLER, J. E., P. L. LUDMER, S. N. WILLICH *et al.* 1987. Circadian variation in the frequency of sudden cardiac death. Circulation **75:** 131–138.
41. WILLICH, S. N., D. LEVY, M. B. ROCCO, G. H. TOFLER, P. H. STONE & J. E. MULLER. 1987. Circadian variation in the incidence of sudden cardiac death in the Framingham Heart Study population. Am. J. Cardiol. **60:** 801–806.
42. PETERS, R. W., J. E. MULLER, S. GOLDSTEIN, R. BYINGTON & L. M. FRIEDMAN. 1989. Propranolol and the morning increase in the frequency of sudden cardiac death (BHAT Study). Am. J. Cardiol. **63:** 1518–1520.
43. LEVINE, R. L., P. E. PEPE, R. E. FROMM, P. A. CURKA & P. A. CLARK. 1992. Prospective evidence of a circadian rhythm for out-of-hospital cardiac arrests. JAMA **267:** 2935–2937.
44. ARONOW, W. S. & C. AHN. 1993. Circadian variation of primary cardiac arrest or sudden cardiac death in patients aged 62 to 100 years (mean 82). Am. J. Cardiol. **71:** 1455–1456.
45. GALLERANI, M., R. MANFREDINI, L. RICCI *et al.* 1993. Sudden death may show a circadian time of risk depending on its anatomo-clinical causes and age. Jpn. Heart J. **34:** 729–739.
46. GALLERANI, M., R. MANFREDINI, R. GRAZIANI *et al.* 1994. Sudden death in diabetic subjects: Evidence for a peculiar circadian variation in occurrence. Panminerva Med. **36:** 134–137.
47. COLANTONIO, D., R. CASALE, B. P. ABRUZZO, G. LORENZETTI & P. PASQUALETTI. 1989. Circadian distribution in fatal pulmonary thromboembolism. Am. J. Cardiol. **64:** 403–404.
48. FRISBIE, J. H. & G. V. SHARMA. 1992. Circadian rhythm of pulmonary embolism in patients with acute spinal cord injury. Am. J. Cardiol. **70:** 827–828.
49. GALLERANI, M., R. MANFREDINI, L. RICCI *et al.* 1992. Sudden death from pulmonary thromboembolism: Chronobiological aspects. Eur. Heart J. **13:** 661–665.
50. MANFREDINI, R., M. GALLERANI, R. SALMI, E. GRANDI & C. FERSINI. 1993. Circadian rhythmicity in the onset of fatal pulmonary embolism in hospitalized subjects. Eur. J. Med. **2:** 183–185.

51. GALLERANI, M., R. MANFREDINI, F. PORTALUPPI *et al.* 1994. Circadian variation in the occurrence of fatal pulmonary embolism. Differences depending on sex and age. Jpn. Heart J. **35:** 765–770.
52. HOSSMANN, V. 1971. Circadian changes of blood pressure and stroke. *In* Cerebral Circulation and Stroke. K. J. Zulch, Ed.: 203–208. Springer-Verlag. Berlin, Heidelberg, New York.
53. MARSHALL, J. 1977. Diurnal variation in occurrence of strokes. Stroke **8:** 230–231.
54. TSEMENTZIS, S. A., J. S. GILL, E. R. HITCHCOCK, S. K. GILL & D. G. BEEVERS. 1985. Diurnal variation of and activity during the onset of stroke. Neurosurgery **17:** 901–904.
55. FERSINI, C., R. MANFREDINI, F. MANFREDINI, G. BALBONI & G. FERSINI. 1987. Chronobiologic aspects of recurrent transient ischemic attack. Prog. Clin. Biol. Res. **227B:** 167–171.
56. MARLER, J. R., T. R. PRICE, G. L. CLARK *et al.* 1989. Morning increase in onset of ischemic stroke. Stroke **20:** 473–476.
57. ARBOIX, A. & J. L. MARTI-VILALTA. 1990. Acute stroke and circadian rhythm [letter]. Stroke **21:** 826.
58. ARGENTINO, C., D. TONI, M. RASURA *et al.* 1990. Circadian variation in the frequency of ischemic stroke. Stroke **21:** 387–389.
59. MARSH, E. E., J. BILLER, H. P. ADAMS *et al.* 1990. Circadian variation in onset of acute ischemic stroke. Arch. Neurol. **47:** 1178–1180.
60. PASQUALETTI, P., G. NATALI, R. CASALE & D. COLANTONIO. 1990. Epidemiological chronorisk of stroke. Acta Neurol. Scand. **81:** 71–74.
61. TONI, D., C. ARGENTINO, M. GENTILE *et al.* 1991. Circadian variation in the onset of acute cerebral ischemia: Ethiopathogenetic correlates in 80 patients given angiography. Chronobiol. Int. **8:** 321–326.
62. RICCI, S., M. G. CELANI, R. VITALI, F. LA ROSA, E. RIGHETTI & E. DUCA. 1992. Diurnal and seasonal variations in the occurrence of stroke: A community-based study. Neuroepidemiology **11:** 59–64.
63. GALLERANI, M., R. MANFREDINI, L. RICCI *et al.* 1993. Chronobiological aspects of acute cerebrovascular diseases. Acta Neurol. Scand. **87:** 482–487.
64. JOHANSSON, B. B., B. NORRVING, H. WIDNER, J. Y. WU & F. HALBERG. 1990. Stroke incidence: Circadian and circaseptan (about weekly) variations in onset. Prog. Clin. Biol. Res. **341A:** 427–436.
65. SLOAN, M. A., T. R. PRICE, M. A. FOULKES *et al.* 1992. Circadian rhythmicity of stroke onset. Intracerebral and subarachnoid hemorrhage. Stroke **23:** 1420–1426.
66. GALLERANI, M., G. TRAPPELLA, R. MANFREDINI, M. PASIN, M. NAPOLITANO & A. MIGLIORE. 1994. Acute intracerebral haemorrhage: Circadian and circannual patterns of onset. Acta Neurol. Scand. **89:** 280–286.
67. ALBONI, P., E. DEGLI UBERTI, L. CODECA *et al.* 1982. Circadian variations of sinus rate in subjects with sinus node dysfunction. Chronobiologia **9:** 173–183.
68. CHRIST, J. E. & H. E. HOFF. 1975. An analysis of the circadian rhythmicity of atrial and ventricular rates in complete heart block. J. Electrocardiol. **8:** 69–72.
69. RAEDER, E. A. 1990. Circadian fluctuations in ventricular response to atrial fibrillation. Am. J. Cardiol. **66:** 1013–1016.
70. CUGINI, P., L. DI PALMA, P. BATTISTI *et al.* 1990. Ultradian, circadian and infradian periodicity of some cardiovascular emergencies. Am. J. Cardiol. **66:** 240–243.
71. KUPARI, M., P. KOSKINEN & H. LEINONEN. 1990. Double-peaking circadian variation in the occurrence of sustained supraventricular tachyarrhythmias. Am. Heart J. **120:** 1364–1369.
72. ROSTAGNO, C., T. TADDEI, B. PALADINI, P. A. MODESTI, P. UTARI & G. BERTINI. 1993. The onset of symptomatic atrial fibrillation and paroxysmal supraventricular tachycardia is characterized by different circadian rhythms. Am. J. Cardiol. **71:** 453–455.
73. CLAIR, W. K., W. E. WILKINSON, E. A. MCCARTHY, R. L. PAGE & E. L. PRITCHETT. 1993. Spontaneous occurrence of symptomatic paroxysmal atrial fibrillation and paroxysmal supraventricular tachycardia in untreated patients. Circulation **87:** 1114–1122.
74. IRWIN, J. M., E. A. MCCARTHY, W. E. WILKINSON & E. L. PRITCHETT. 1988. Circadian

occurrence of symptomatic paroxysmal supraventricular tachycardia in untreated patients. Circulation **77:** 298–300.
75. MANFREDINI, R., M. GALLERANI, F. PORTALUPPI, R. SALMI, F. CHIERICI & C. FERSINI. 1995. Circadian variation in the occurrence of paroxysmal supraventricular tachycardia in clinically healthy subjects. Chronobiol. Int. **12:** 55–61.
76. DE LEONARDIS, V., M. DE SCALZI, R. VERGASSOLA, S. ROMANO, A. BECUCCI & P. CINELLI. 1987. Circadian variations of heart rate and premature beats in healthy subjects and in patients with previous myocardial infarction. Chronobiol. Int. **4:** 283–289.
77. RAEDER, E. A., S. H. HOHNLOSER, T. B. GRABOYS, P. J. PODRID, S. LAMPERT & B. LOWN. 1988. Spontaneous variability and circadian distribution of ectopic activity in patients with malignant ventricular arrhythmia. J. Am. Coll. Cardiol. **12:** 656–661.
78. LANZA, G. A., M. C. CORTELLESSA, A. G. REBUZZI *et al.* 1990. Reproducibility in circadian rhythm of ventricular premature complexes. Am. J. Cardiol. **66:** 1099–1106.
79. GILLIS, A. M., R. W. PETERS, L. B. MITCHELL, H. J. DUFF, M. MCDONALD & D. G. WYSE. 1992. Effects of left ventricular dysfunction on the circadian variation of ventricular premature complexes in healed myocardial infarction. Am. J. Cardiol. **69:** 1009–1014.
80. SIEGEL, D., D. M. BLACK, D. G. SEELEY & S. B. HULLEY. 1992. Circadian variation in ventricular arrhythmias in hypertensive men. Am. J. Cardiol. **69:** 344–347.
81. LUCENTE, M., A. G. REBUZZI, G. A. LANZA *et al.* 1988. Circadian variation of ventricular tachycardia in acute myocardial infarction. Am. J. Cardiol. **62:** 670–674.
82. TWIDALE, N., S. TAYLOR, W. F. HEDDLE, B. F. AYRES & A. M. TONKIN. 1989. Morning increase in the time of onset of sustained ventricular tachycardia. Am. J. Cardiol. **64:** 1204–1206.
83. VALKAMA, J. O., H. V. HUIKURI, M. K. LINNALUOTO & J. T. TAKKUNEN. 1992. Circadian variation of ventricular tachycardia in patients with coronary arterial disease. Int. J. Cardiol. **34:** 173–178.
84. LAMPERT, R., L. ROSENFELD, W. BATSFORD, F. LEE & C. MCPHERSON. 1994. Circadian variation of sustained ventricular tachycardia in patients with coronary artery disease and implantable cardioverter-defibrillators. Circulation **90:** 241–247.
85. ARNTZ, H. R., S. N. WILLICH, M. OEFF *et al.* 1993. Circadian variation of sudden cardiac death reflects age-related variability in ventricular fibrillation. Circulation **88:** 2284–2289.
86. CHIERCHIA, S., A. GALLINO, G. SMITH *et al.* 1984. Role of heart rate in pathophysiology of chronic stable angina. Lancet **2:** 1353–1357.
87. ANDREWS, T. C., T. FENTON, N. TOYOSAKI *et al.* 1993. Subsets of ambulatory myocardial ischemia based on heart rate activity. Circadian distribution and response to anti-ischemic medication. The Angina and Silent Ischemia Study Group (ASIS). Circulation **88:** 92–100.
88. ASHKAR, E. 1979. Twenty-four-hour pattern of circulation by radiotelemetry in the unrestrained dog. Am. J. Physiol. **236:** R231–R236.
89. DEGAUTE, J. P., E. VAN CAUTER, P. VAN DE BORNE & P. LINKOWSKI. 1994. Twenty-four-hour blood pressure and heart rate profiles in humans. A twin study. Hypertension **23:** 244–253.
90. PORTALUPPI, F., P. CORTELLI, P. AVONI *et al.* 1994. Diurnal blood pressure variation and hormonal correlates in fatal familial insomnia. Hypertension **23:** 569–576.
91. FURLAN, R., S. GUZZETTI, W. CRIVELLARO *et al.* 1990. Continuous 24-hour assessment of the neural regulation of systemic arterial pressure and RR variabilities in ambulant subjects. Circulation **81:** 537–547.
92. HUIKURI, H. V., M. K. LINNALUOTO, T. SEPPANEN *et al.* 1992. Circadian rhythm of heart rate variability in survivors of cardiac arrest. Am. J. Cardiol. **70:** 610–615.
93. BENHORIN, J., S. BANAI, M. MORIEL *et al.* 1993. Circadian variations in ischemic threshold and their relation to the occurrence of ischemic episodes. Circulation **87:** 808–814.
94. BARRY, J., S. CAMPBELL, A. C. YEUNG, K. E. RABY & A. P. SELWYN. 1991. Waking and rising at night as a trigger of myocardial ischemia. Am. J. Cardiol. **67:** 1067–1072.
95. MULCAHY, D., C. WRIGHT, J. SPARROW *et al.* 1993. Heart rate and blood pressure

consequences of an afternoon SIESTA (snooze-induced excitation of sympathetic triggered activity). Am. J. Cardiol. **71:** 611–614.

96. NOWLIN, J. B., W. G. TROYER & W. S. COLLINS. 1965. The association of nocturnal angina pectoris with dreaming. Ann. Intern. Med. **63:** 1040–1046.
97. KING, M. J., L. M. ZIR, A. J. KALTMAN & A. C. FOX. 1973. Variant angina associated with angiographically demonstrated coronary artery spasm and REM sleep. Am. J. Med. Sci. **265:** 419–422.
98. SOMERS, V. K., M. E. DYKEN, A. L. MARK & F. M. ABBOUD. 1993. Sympathetic-nerve activity during sleep in normal subjects. N. Engl. J. Med. **328:** 303–307.
99. KIRBY, D. A. & R. L. VERRIER. 1989. Differential effects of sleep stage on coronary hemodynamic function during stenosis. Physiol. Behav. **45:** 1017–1020.
100. CZEISLER, C. A., E. D. WEITZMAN, M. C. MOORE-EDE, J. C. ZIMMERMAN & R. S. KNAUER. 1980. Human sleep: Its duration and organization depend on its circadian phase. Science **210:** 1264–1267.
101. CZEISLER, C. A., J. C. ZIMMERMAN, J. M. RONDA, M. C. MOORE-EDE & E. D. WEITZMAN. 1980. Timing of REM sleep is coupled to the circadian rhythm of body temperature in man. Sleep **2:** 329–346.
102. ASCHOFF, J. & R. WEVER. 1976. Human circadian rhythms: A multioscillatory system. Fed. Proc. **35:** 236–232.
103. REFINETTI, R. & M. MENAKER. 1992. The circadian rhythm of body temperature. Physiol. Behav. **51:** 613–637.
104. KANAMORI, N., K. SAKAI & H. SEI. 1995. Effects of decerebration on blood pressure during paradoxical sleep in cats. Brain Res. Bull. **37:** 545–549.
105. TURTON, M. B. & T. DEEGAN. 1974. Circadian variations of plasma catecholamine, cortisol and immunoreactive insulin concentrations in supine subjects. Clin. Chim. Acta **55:** 389–397.
106. TOFLER, G. H., D. BREZINSKI, A. I. SCHAFER *et al.* 1987. Concurrent morning increase in platelet aggregability and the risk of myocardial infarction and sudden cardiac death. N. Engl. J. Med. **316:** 1514–1518.
107. GORDON, R. D., L. K. WOLFE, D. P. ISLAND & G. W. LIDDLE. 1966. A diurnal rhythm in plasma renin activity in man. J. Clin. Invest. **45:** 1587–1592.
108. PANZA, J. A., S. E. EPSTEIN & A. A. QUYYUMI. 1991. Circadian variation in vascular tone and its relation to alpha-sympathetic vasoconstrictor activity. N. Engl. J. Med. **325:** 986–990.
109. MILLAR-CRAIG, M. W., C. N. BISHOP & E. B. RAFTERY. 1978. Circadian variation of blood-pressure. Lancet **1:** 795–797.
110. FLORAS, J. S., J. V. JONES, J. A. JOHNSTON, D. E. BROOKS, M. O. HASSAN & P. SLEIGHT. 1978. Arousal and the circadian rhythm of blood pressure. Clin. Sci. Mol. Med. **4**(Suppl. 1): 395s–397s.
111. LITTLER, W. A. 1979. Sleep and blood pressure: Further observations. Am. Heart J. **97:** 35–37.
112. ATHANASSIADIS, D., G. J. DRAPER, A. J. HONOUR & W. I. CRANSTON. 1969. Variability of automatic blood pressure measurements over 24 hour periods. Clin. Sci. **36:** 147–156.
113. MANCIA, G., A. FERRARI, L. GREGORINI *et al.* 1983. Blood pressure and heart rate variabilities in normotensive and hypertensive human beings. Circ. Res. **53:** 96–104.
114. BROADHURST, P., G. BRIGDEN, P. DASGUPTA, A. LAHIRI & E. B. RAFTERY. 1990. Ambulatory intra-arterial blood pressure in normal subjects. Am. Heart J. **120:** 160–166.
115. GOULD, B. A. & E. B. RAFTERY. 1991. Twenty-four-hour blood pressure control: An intraarterial review. Chronobiol. Int. **8:** 495–505.
116. DEGAUTE, J. P., P. VAN DE BORNE, P. LINKOWSKI & E. VAN CAUTER. 1991. Quantitative analysis of the 24-hour blood pressure and heart rate patterns in young men [see comments]. Hypertension **18:** 199–210.
117. KHOURY, A. F., P. SUNDERAJAN & N. M. KAPLAN. 1992. The early morning rise in blood pressure is related mainly to ambulation. Am. J. Hypertens. **5:** 339–344.
118. SUZUKI, Y., I. KUWAJIMA, K. MITANI & K. KURAMOTO. 1993. The relation of morning rise

in blood pressure to pressure response to exercise and mental stress. Jpn J. Geriatr. **30:** 841–848.
119. KUWAJIMA, I., K. MITANI, M. MIYAO, Y. SUZUKI, K. KURAMOTO & T. OZAWA. 1995. Cardiac implications of the morning surge in blood pressure in elderly hypertensive patients: Relation to arising time. Am. J. Hypertens. **8:** 29–33.
120. FUJITA, M. & D. FRANKLIN. 1987. Diurnal changes in coronary blood flow in conscious dogs. Circulation **76:** 488–491.
121. QUYYUMI, A. A., J. A. PANZA, J. G. DIODATI, E. LAKATOS & S. E. EPSTEIN. 1992. Circadian variation in ischemic threshold. A mechanism underlying the circadian variation in ischemic events. Circulation **86:** 22–28.
122. WEITZMAN, E. D., D. FUKUSHIMA, C. NOGEIRE, H. ROFFWARG, T. F. GALLAGHER & L. HELLMAN. 1971. Twenty-four hour pattern of the episodic secretion of cortisol in normal subjects. J. Clin. Endocrinol. Metab. **33:** 14–22.
123. LABRECQUE, G. & G. SOULBAN. 1991. Biological rhythms in the physiology and pharmacology of blood coagulation. Chronobiol. Int. **8:** 361–372.
124. ANDREOTTI, F. & C. KLUFT. 1991. Circadian variation of fibrinolytic activity in blood. Chronobiol. Int. **8:** 336–351.
125. VAZIRI, N. D., S. C. KENNEDY, D. KENNEDY & E. GONZALES. 1992. Coagulation, fibrinolytic, and inhibitory proteins in acute myocardial infarction and angina pectoris. Am. J. Med. **93:** 651–657.
126. PETRALITO, A., R. A. MANGIAFICO, S. GIBIINO, M. A. CUFFARI, M. F. MIANO & C. E. FIORE. 1982. Daily modifications of plasma fibrinogen platelets aggregation, Howell's time, PTT, TT, and antithrombin II in normal subjects and in patients with vascular disease. Chronobiologia **9:** 195–201.
127. CONCHONNET, P., H. DECOUSUS & B. PERPOINT. 1990. Morning hypercoagulability in man. Annu. Rev. Chronopharmacol. **7:** 165–168.
128. HAUS, E., M. CUSULOS, L. SACKETT-LUNDEEN & J. SWOYER. 1990. Circadian variations in blood coagulation parameters, alpha-antitrypsin antigen and platelet aggregation and retention in clinically healthy subjects. Chronobiol. Int. **7:** 203–216.
129. JOVICIC, A. & S. MANDIC. 1991. Circadian variations of platelet aggregability and fibrinolytic activity in healthy subjects. Thromb. Res. **62:** 65–74.
130. MUSUMECI, V., S. ROSA, A. CARUSO, C. ZUPPI, B. ZAPPACOSTA & F. TUTINELLI. 1986. Abnormal diurnal changes in in-vivo platelet activation in patients with atherosclerotic diseases. Atherosclerosis **60:** 231–236.
131. WILLICH, S. N., S. P. SINTONEN & S. S. BATHIA. 1988. Morning increase in platelet aggregability in patients with coronary artery disease (abstract). J. Am. Coll. Cardiol. **11:** 204.
132. JAFRI, S. M., M. VAN ROLLINS, T. OZAWA, E. F. MAMMEN, A. D. GOLDBERG & S. GOLDSTEIN. 1992. Circadian variation in platelet function in healthy volunteers. Am. J. Cardiol. **69:** 951–954.
133. BREZINSKI, D. A., G. H. TOFLER, J. E. MULLER *et al.* 1988. Morning increase in platelet aggregability. Association with assumption of the upright posture. Circulation **78:** 35–40.
134. WINTHER, K., W. HILLEGASS, G. H. TOFLER *et al.* 1992. Effects on platelet aggregation and fibrinolytic activity during upright posture and exercise in healthy men. Am. J. Cardiol. **70:** 1051–1055.
135. BRIDGES, A. B., N. A. SCOTT, G. P. MCNEILL, T. H. PRINGLE & J. J. BELCH. 1992. Circadian variation of white blood cell aggregation and free radical indices in men with ischaemic heart disease. Eur. Heart J. **13:** 1632–1636.
136. TOUITOU, Y., C. TOUITOU, A. BOGDAN *et al.* 1986. Differences between young and elderly subjects in seasonal and circadian variations of total plasma proteins and blood volume as reflected by hemoglobin, hematocrit, and erythrocyte counts. Clin. Chem. **32:** 801–804.
137. KUBOTA, K., T. SAKURAI, J. TAMURA & T. SHIRAKURA. 1987. Is the circadian change in hematocrit and blood viscosity a factor triggering cerebral and myocardial infarction? [letter]. Stroke **18:** 812–813.

138. EHRLY, A. M. & G. JUNG. 1973. Circadian rhythm of human blood viscosity. Biorheology **10:** 577–583.
139. FEARNLEY, G. R., G. BALMFORTH & E. FEARNLEY. 1957. Evidence of a diurnal fibrinolytic rhythm with a simple method of measuring natural fibrinolysis. Clin. Sci. **16:** 645–650.
140. GRIMAUDO, V., J. HAUERT, F. BACHMANN & E. K. KRUITHOF. 1988. Diurnal variation of the fibrinolytic system. Thromb. Haemost. **59:** 495–499.
141. KLUFT, C., A. F. JIE, D. C. RIJKEN & J. H. VERHEIJEN. 1988. Daytime fluctuations in blood of tissue-type plasminogen activator (t-PA) and its fast-acting inhibitor (PAI-1). Thromb. Haemost. **59:** 329–332.
142. ANGLETON, P., W. L. CHANDLER & G. SCHMER. 1989. Diurnal variation of tissue-type plasminogen activator and its rapid inhibitor (PAI-1). Circulation **79:** 101–106.
143. HUBER, K., D. ROSE, I. RESCH, D. H. GLOGAR, F. KAINDL & B. R. BINDER. 1989. Circadian fluctuations of plasma levels of tissue plasminogen activator antigen and plasminogen activator inhibitor activity. Fibrinolysis **3:** 41–43.
144. ANDREOTTI, F., G. J. DAVIES, D. R. HACKETT *et al.* 1988. Major circadian fluctuations in fibrinolytic factors and possible relevance to time of onset of myocardial infarction, sudden cardiac death and stroke. Am. J. Cardiol. **62:** 635–637.
145. SAKATA, K., T. HOSHINO, H. YOSHIDA *et al.* 1992. Circadian fluctuations of tissue plasminogen activator antigen and plasminogen activator inhibitor-1 antigens in vasospastic angina. Am. Heart J. **124:** 854–860.
146. BRIDGES, A. B., M. MCLAREN, N. A. SCOTT, T. H. PRINGLE, G. P. MCNEILL & J. J. BELCH. 1993. Circadian variation of tissue plasminogen activator and its inhibitor, von Willebrand factor antigen, and prostacyclin stimulating factor in men with ischaemic heart disease. Br. Heart J. **69:** 121–124.
147. MASUDA, T., H. OGAWA, Y. MIYAO *et al.* 1994. Circadian variation in fibrinolytic activity in patients with variant angina. Br. Heart J. **71:** 156–161.
148. JOVICIC, A., V. IVANISEVIC & R. NIKOLAJEVIC. 1991. Circadian variations of platelet aggregability and fibrinolytic activity in patients with ischemic stroke. Thromb. Res. **64:** 487–491.
149. WANNAMETHEE, G., I. J. PERRY & A. G. SHAPER. 1994. Haematocrit, hypertension and risk of stroke. J. Intern. Med. **235:** 163–168.
150. AMERISO, S. F., J. G. MOHLER, M. SUAREZ & M. FISHER. 1994. Morning reduction of cerebral vasomotor reactivity. Neurology **44:** 1907–1909.
151. GLOBUS, M. Y., M. D. GINSBERG, S. I. HARIK, R. BUSTO & W. D. DIETRICH. 1987. Role of dopamine in ischemic striatal injury: Metabolic evidence. Neurology **37:** 1712–1719.
152. FAHN, S. 1981. Fluctuations of disability in Parkinson's disease: Pathophysiological aspects. *In* Movement Disorders. C. D. Marsden & S. Fahn, Eds.: 123– 145. Butterworth Scientific. London.
153. STRUCK, L. K. & R. L. RODNITZKY. 1989. Periodicity of ischemic stroke. Stroke **20:** 1590.
154. STRANDGAARD, S. & O. B. PAULSON. 1994. Cerebrovascular consequences of hypertension. Lancet **344:** 519–521.
155. MAC MAHON, S., R. PETO, J. CUTLER *et al.* 1990. Blood pressure, stroke, and coronary heart disease. Part 1. Prolonged differences in blood pressure: Prospective observational studies corrected for the regression dilution bias. Lancet **335:** 765–774.
156. COLLINS, R., R. PETO, S. MAC MAHON *et al.* 1990. Blood pressure, stroke, and coronary heart disease. Part 2. Short-term reductions in blood pressure: Overview of randomised drug trials in their epidemiological context. Lancet **335:** 827–838.
157. SHEP COOPERATIVE RESEARCH GROUP. 1991. Prevention of stroke by antihypertensive drug treatment in older persons with isolated systolic hypertension. JAMA **265:** 3255–3264.
158. ERGUN, G. A. & B. RIGAS. 1990. Circadian periodicity of the time of onset of acute lower gastrointestinal bleeding. Prog. Clin. Biol. Res. **341B:** 221–228.
159. MANFREDINI, R., M. GALLERANI, R. SALMI *et al.* 1994. Circadian variation in the time of onset of acute gastrointestinal bleeding. J. Emerg. Med. **12:** 5–9.
160. TOHGI, H., K. CHIBA & M. KIMURA. 1991. Twenty-four-hour variation of blood pressure in vascular dementia of the Binswanger type. Stroke **22:** 603–608.

161. MARTINELLI, P., G. COCCAGNA, N. RIZZUTO & E. LUGARESI. 1981. Changes in systemic arterial pressure during sleep in Shy Drager syndrome. Sleep **4:** 139–146.
162. MANN, S., D. G. ALTMAN, E. B. RAFTERY & R. BANNISTER. 1983. Circadian variation of blood pressure in autonomic failure. Circulation **68:** 477–483.
163. STOICA, E. & O. ENULESCU. 1983. Inability to deactivate sympathetic nervous system in brainstem infarct patients. J. Neurol. Sci. **58:** 223–234.
164. SHIMADA, K., A. KAWAMOTO, K. MATSUBAYASHI, M. NISHINAGA, S. KIMURA & T. OZAWA. 1992. Diurnal blood pressure variations and silent cerebrovascular damage in elderly patients with hypertension. J. Hypertens. **10:** 875–878.
165. MATSUMURA, K., I. ABE, M. FUKUHARA *et al.* 1993. Attenuation of nocturnal BP fall in essential hypertensives with cerebral infarction (letter). J. Hum. Hypertens. **7:** 309–310.
166. SANDER, D. & J. KLINGELHÖFER. 1994. Changes of circadian blood pressure patterns after hemodynamic and thromboembolic brain infarction. Stroke **25:** 1730–1737.
167. SANDER, D. & J. KLINGELHOFER. 1993. Circadian blood pressure patterns in four cases with hemodynamic brain infarction and prolonged blood-brain barrier disturbance. Clin. Neurol. Neurosurg. **95:** 221–229.
168. GORDON, T. & W. B. KANNEL. 1971. Premature mortality from coronary heart disease. The Framingham study. JAMA **215:** 1617–1625.
169. MOSS, A. J. 1980. Clinical significance of ventricular arrhythmias in patients with and without coronary artery disease. Prog. Cardiovasc. Dis. **23:** 33–52.
170. MESSERLI, F. H., H. O. VENTURA, D. J. ELIZARDI, F. G. DUNN & E. D. FROHLICH. 1984. Hypertension and sudden death. Increased ventricular ectopic activity in left ventricular hypertrophy. Am. J. Med. **77:** 18–22.
171. KOREN, M. J., R. B. DEVEREUX, P. N. CASALE, D. D. SAVAGE & J. H. LARAGH. 1991. Relation of left ventricular mass and geometry to morbidity and mortality in uncomplicated essential hypertension. Ann. Intern. Med. **114:** 345–352.
172. LEVY, D., K. M. ANDERSON, D. D. SAVAGE, S. A. BALKUS, W. B. KANNEL & W. P. CASTELLI. 1987. Risk of ventricular arrhythmias in left ventricular hypertrophy: The Framingham Heart Study. Am. J. Cardiol. **60:** 560–565.
173. MCLENACHAN, J. M., E. HENDERSON, K. I. MORRIS & H. J. DARGIE. 1987. Ventricular arrhythmias in patients with hypertensive left ventricular hypertrophy. N. Engl. J. Med. **317:** 787–792.
174. WHELTON, P. K. 1985. Systemic hypertension, diuretic drugs, arrhythmias and death. Am. J. Cardiol. **55:** 221–222.
175. STEWART, D. E., H. IKRAM, E. A. ESPINER & M. G. NICHOLLS. 1985. Arrhythmogenic potential of diuretic induced hypokalaemia in patients with mild hypertension and ischaemic heart disease. Br. Heart J. **54:** 290–297.
176. JAMES, M. A. & J. V. JONES. 1989. Ventricular arrhythmia in untreated newly presenting hypertensive patients compared with matched normal population. J. Hypertens. **7:** 409–415.
177. HELLER, L. J. & E. K. STAUFFER. 1981. Membrane potentials and contractile events of hypertrophied rat cardiac muscle. Proc. Soc. Exp. Biol. Med. **166:** 141–147.
178. MARTINS, J. B., W. KIM & M. L. MARCUS. 1989. Chronic hypertension and left ventricular hypertrophy facilitate induction of sustained ventricular tachycardia in dogs 3 hours after left circumflex coronary artery occlusion. J. Am. Coll. Cardiol. **14:** 1365–1373.
179. KOYANAGI, S., C. EASTHAM & M. L. MARCUS. 1982. Effects of chronic hypertension and left ventricular hypertrophy on the incidence of sudden cardiac death after coronary artery occlusion in conscious dogs. Circulation **65:** 1192–1197.
180. INOU, T., W. C. LAMBERTH, S. KOYANAGI, D. G. HARRISON, C. L. EASTHAM & M. L. MARCUS. 1987. Relative importance of hypertension after coronary occlusion in chronic hypertensive dogs with LVH. Am. J. Physiol. **253:** H1148–1158.
181. GAASCH, W. H., M. R. ZILE, P. K. HOSHINO, E. O. WEINBERG, D. R. RHODES & C. S. APSTEIN. 1990. Tolerance of the hypertrophic heart to ischemia. Studies in compensated and failing dog hearts with pressure overload hypertrophy. Circulation **81:** 1644–1653.
182. NOVO, S., M. BARBAGALLO, M. G. ABRIGNANI *et al.* 1990. Cardiac arrhythmias as

correlated with the circadian rhythm of arterial pressure in hypertensive subjects with and without left ventricular hypertrophy. Eur. J. Clin. Pharmacol. **39**(Suppl 1): S49–S51.

183. ZEHENDER, M., T. MEINERTZ, S. HOHNLOSER *et al.* 1992. Prevalence of circadian variations and spontaneous variability of cardiac disorders and ECG changes suggestive of myocardial ischemia in systemic arterial hypertension. Circulation **85:** 1808–1815.

Relationships of Cardiac Function and Structure to Blood Pressure Rhythms

DAMIANO RIZZONI[a] AND ENRICO AGABITI-ROSEI

Chair of Semeiotica e Metodologia Medica
Department of Medical Sciences
University of Brescia
Brescia, Italy

The most important and consistent sources of blood pressure variation are the diurnal changes associated with the sleep/wake cycle.[1] It is widely accepted that blood pressure follows a circadian rhythm. Blood pressure levels are usually lowest during the night, and awakening is commonly associated with a rapid increase in both systolic and diastolic blood pressure values. A small reduction in blood pressure in the early afternoon is less constant and possibly associated with daily habits (e.g., afternoon sleep).

Deviations from the physiologic diurnal rhythm may have pathologic relevance. In fact, in recent years, the development of fully automated, portable devices for recording blood pressure has facilitated the monitoring of ambulatory blood pressure during a 24-hour period. The widespread clinical use of noninvasive ambulatory blood pressure monitoring (ABPM) has permitted an insight into the interrelationships between blood pressure circadian rhythms and cardiac structure and function. In this paper, the possible association between left ventricular hypertrophy and alterations of circadian blood pressure patterns are discussed.

CARDIAC FUNCTION AND BLOOD PRESSURE RHYTHMS

Increasing attention has focused on 24-hour rhythms of the cardiovascular system. Diurnal variations were demonstrated for blood pressure, but also for heart function.[2,3] In a study by Kool and colleagues,[2] during nighttime blood pressure, cardiac index, stroke index, and heart rate decreased in a classic day-night pattern (−4%, −19%, −10%, and −10%, respectively).

The behavior of systemic vascular resistance is more controversial, because both an increase in peripheral resistance,[2,4] no change,[3] or a decrease[3,5,6] was observed during nighttime. However, in the latter study, evaluation of vascular tone in the morning, afternoon, and evening (9 PM) was performed by means of a plethysmographic technique, with patients awake, and cannulation of a brachial artery.[6] Therefore, these conditions could hardly be considered physiologic, and no information about the sleeping period was provided by this study.

Blood pressure, cardiac output, and peripheral resistance are linked to each other by the well-known Poiseuille relation. Each term of this relation is under the control of autonomic nervous system activity. In fact, in a study performed by Schested *et al.*,[7] cardiac transplant recipients demonstrated that diurnal variations in

[a] Address for correspondence: Damiano Rizzoni, MD, Cattedra di Semeiotica e Metodologia Medica, Scienze Mediche, University of Brescia, c/o la Medicina, Spedali Civili, 25100 Brescia, Italy (tel. ++39 30 396044; fax ++ 39 30 3384348).

cardiac function or catecholamine levels are not responsible or sufficient to produce a nocturnal decrease in blood pressure. The presence of an innervated heart appears to be crucial in this situation.

Variation in the autonomic drive directed to the heart may therefore be important in the genesis of circadian changes in cardiac function. In fact, day-to-night modulation of sympathetic and parasympatetic nervous system activity was observed by spectral analysis of heart rate and blood pressure signals.[8–10] Furlan *et al.*[8] observed a clear-cut reduction in the power spectral density of the low frequency component (about 0.10 Hz) of systolic blood pressure during the night which was accompanied by a less constant increase in the high frequency component (about 0.25 Hz). Consensual reduction in the power spectral density of the low frequency component and increases in the high frequency component of the heart period during the night were observed by Guzzetti *et al.*[9]

Genesis of the power spectral density in the low frequency band, however, cannot be safely ascribed to the sole effect of the sympathetic nervous system, but also reflects the influence of other factors (spontaneous vasomotor activity, parasympathetic activity, mechanics of respiration, etc.).[11] In fact, with a different methodologic approach and using special care to avoid overmodeling of the data, it was demonstrated that circadian changes in the different indexes of autonomic nervous system activity are complex and somehow difficult to explain. The mid-frequency (index, at least in part, of sympathetic activity) and the high frequency (index, at least in part, of parasympathetic activity) powers of diastolic blood pressure and the mid-frequency power of systolic blood pressure were markedly reduced during the night than in the day, whereas the opposite occurred for the high frequency of the pulse interval.[10]

TABLE 1. Current Definitions

Dippers: Subjects with a nocturnal fall of systolic or diastolic blood pressure.
Non-dippers: Subjects with a reduced or absent nocturnal fall of blood pressure.
Peakers: Subjects with a daytime rise in blood pressure.
Non-peakers: Subjects without a daytime rise in blood pressure.
Night-peakers: Subjects with a rise of blood pressure during nighttime.
Over-dippers: Subjects with a fall in blood pressure during nighttime greater than 20% of daytime blood pressure.

Taken together, these data indicated that circadian cardiovascular variability in humans has its origin in the central nervous system. However, other factors (mechanical, humoral, etc.) may be involved in the control of cardiac function. A classic day-night pattern was demonstrated for epinephrine, norepinephrine, and prorenin.[2] Circulating catecholamines seem to be particularly important for the morning rise in blood pressure.[12] The heart itself plays an important role in the establishment of diurnal variation in blood pressure independently of variations in heart rate, because a normally innervated heart is a prerequisite for the generation of a circadian rhythm in blood pressure.

LEFT VENTRICULAR HYPERTROPHY AND BLOOD PRESSURE RHYTHMS

Ambulatory blood pressure readings obtained using automated, portable recorders are better predictors of cardiac structural changes in hypertension. It was demonstrated that the left ventricular mass correlates more closely with 24-hour blood pressure or daytime and nighttime blood pressure than with clinic blood pressure.[13]

Moreover, many empiric observations suggested that in hypertensive patients with left ventricular hypertrophy there was a lack of the physiologic reduction of blood pressure during the night, especially when left ventricular hypertrophy was severe. In the last years, some clinical studies demonstrated an association between cardiac structural alterations and circadian blood pressure patterns.[13–15] Higher values of left ventricular mass were observed in those patients in whom the nocturnal decline in blood pressure was reduced or absent[13,15]; these patients were defined as "non-dippers" (TABLE 1). A statistically significant inverse correlation between left ventricular mass index and percentage of nocturnal reduction in daytime ambulatory systolic and diastolic blood pressure was also observed.[15]

These data support the hypothesis that cardiac structural changes are associated with impairment in the circadian modulation of blood pressure. However, there are some methodologic problems in this area. First, the criterion for the definition of a "non-dipper" has not yet been established. In one study it was defined as a subject with less than 10% or a reduction of both systolic and diastolic blood pressure during the night than the day.[15] In other studies the cut-off point was set at 10% reduction in mean blood pressure[13] or at a reduction of 5 and 10 mm Hg for systolic and diastolic blood pressure, respectively.[16]

All the proposed definitions are entirely arbitrary, because the distinction between the physiologic and the pathologic value of the nocturnal reduction in blood pressure is not clear. In addition, some healthy subjects may also present a reduced or absent circadian rhythm of blood pressure.

The prevalence of dippers and non-dippers, defined according to Verdecchia *et al.*,[15] was 35.7% in hypertensive patients and 34.7% in normotensive subjects.[17]

A second problem regards the definition of nighttime; the time intervals used to define daytime and nighttime vary between studies. In our study,[13] we considered nighttime hours as 10 PM to 6 AM, Verdecchia *et al.*[15] defined nighttime as 8 PM to 6 AM, while O'Brien *et al.*[16] restricted the period to 1 AM to 8 AM. The timing of the sleep/inactivity periods and their duration show a high intrasubject and intersubject variability; because sleeping habits may affect the circadian pattern of blood pressure substantially, the present uncertainty in the definition of nighttime may have affected the previously mentioned results. When hypertensive dippers and non-dippers are matched for daytime ambulatory blood pressure, the latter may have greater target organ damage in relation to higher mean blood pressure levels over 24 hours and not necessarily as a result of altered blood pressure rhythm per se. In fact, in men, the difference in left ventricular mass between dippers and non-dippers was not significant after adjustment for daytime or 24-hour blood pressure,[18] whereas after adjustment for nighttime blood pressure, left ventricular mass was greater in dippers (more properly "peakers"). In women, left ventricular mass was greater in non-dippers than in dippers before and after adjustment for 24-hour blood pressure.[18,19] Therefore, at least in men, the previously observed increase in left ventricular mass in dippers could be an artifact of matching patients for daytime blood pressure, by resulting in a higher 24-hour blood pressure in non-dippers.

Further problems arise from the repeatability in time of the definition of dippers and non-dippers. Changes in ambulatory blood pressure from day to night show remarkable intersubject variability.[20] In a reproducibility study of 63 untreated hypertensive patients who underwent 24-hour ambulatory blood pressure monitoring twice (3–5 days apart), 73% of patients remained in the same dipper/non-dipper category for both sessions, whereas the other 27% moved from one group to the other.[17,21] The long-term repeatability of this classification is unknown; therefore, its clinical value is partially limited.

Other important confounding factors may affect the categorization of patients as dippers and non-dippers, such as the quality of sleep,[22] antihypertensive drugs,[23,24] daily activities, and sleep habits. In particular, an excessive high frequency of noninvasive blood pressure measurements at night might disturb sleep sufficiently to increase the nighttime blood pressure, thereby reducing the day-night difference, but this possibility has been excluded, at least for a measurement frequency of two per hour[25] and three per hour.[26]

In a study performed by Fagard *et al.*,[27] no difference in left ventricular mass between dippers and non-dippers was observed. The multicenter SAMPLE study (Study on Ambulatory Monitoring of Pressure and Lisinopril Evaluation), performed in 206 hypertensive patients with left ventricular hypertrophy, was recently completed. Also in this study the association between circadian rhythms and cardiac structure was not confirmed, because a statistically significant nocturnal reduction in blood pressure was observed (-14.3 ± 9.7 mm Hg for systolic blood pressure, -12.3 ± 7.2 mm Hg for diastolic blood pressure).[28] Metaanalysis of comparative studies on the relationships between left ventricular mass and daytime and nighttime blood pressure, recently published,[29] concluded that an association between cardiac structure and impaired nocturnal reduction in blood pressure was not proven beyond doubt.

AUTONOMIC NERVOUS SYSTEM ACTIVITY, LEFT VENTRICULAR HYPERTROPHY, AND CIRCADIAN PATTERNS OF BLOOD PRESSURE

Notwithstanding the methodologic limitations of the available studies, it is theoretically probable and it seems reasonable that patients with left ventricular hypertrophy may have an altered circadian pattern of blood pressure. These patients had increased sympathetic activity during the day compared with hypertensive patients without left ventricular hypertrophy[30] (FIG. 1). Moreover, in patients with left ventricular hypertrophy the circadian modulation of autonomic nervous system activity is absent, and persistently elevated sympathetic activity may be observed during the night[30] (FIG. 1). In effect, also in diabetic patients (who frequently show autonomic neuropathy), the nocturnal decrease in blood pressure is reduced.[31]

Patients with left ventricular hypertrophy show impaired cardiopulmonary low pressure baroreceptor function.[32–34] During loading and unloading of cardiopulmonary low pressure receptors, a reduced response in terms of circulating levels of catecholamines, of plasma renin activity,[31,32] and of changes in sympathetic nervous system activity was observed.[33] Because these receptors are activated by increased venous return to the heart (through an increased stretch of cardiac and vascular walls), a reduction in sympathetic nervous system activity and/or an increase in vagal activity should be expected during their loading (for example, by applying positive pressure to the lower body or by passive leg raising), whereas changes in the opposite direction should be expected during their unloading, by applying negative pressure to the lower body. These modifications in sympathetic nervous system activity and in circulating levels of several humoral factors are blunted in hypertensive patients with

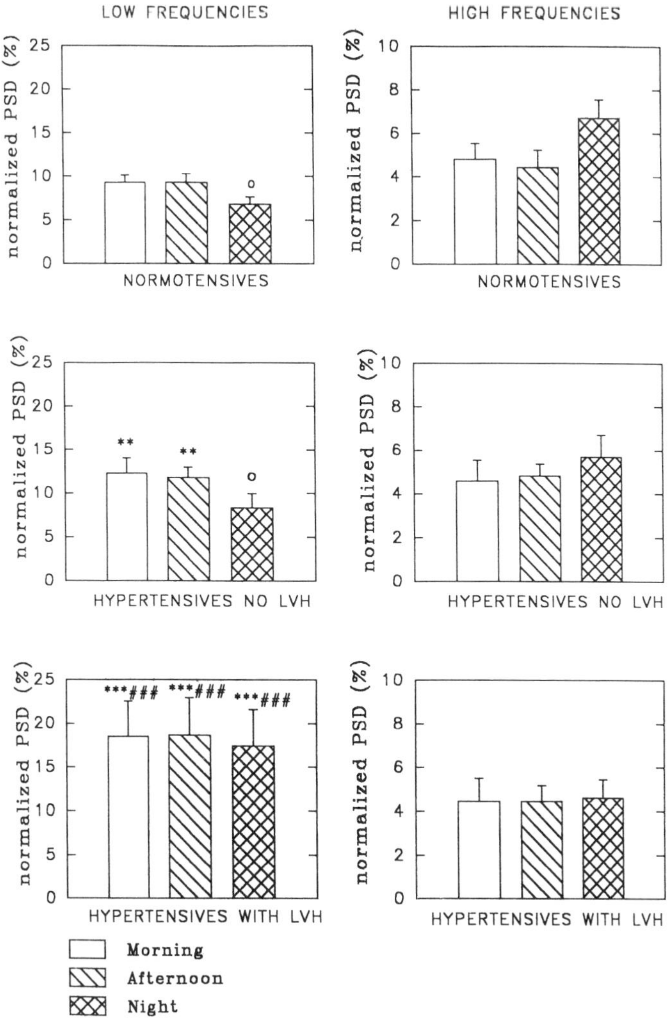

FIGURE 1. Normalized power spectral densities of the low frequency (partially reflecting sympathetic activity) and high frequency (partially reflecting parasympathetic activity) components of heart rate variability in normotensive subjects (normotensives, $n = 5$) and hypertensive patients with (hypertensives with LVH, $n = 10$) and without (hypertensives no LVH, $n = 7$) left ventricular hypertrophy. $^{\circ}p < 0.05$ at least compared with morning and afternoon; $^{**}p < 0.01$; $^{***}p < 0.001$ compared with normotensive subjects; $^{\#\#\#}p < 0.001$ compared with hypertensive patients without left ventricular hypertrophy.

left ventricular hypertrophy.[32-36] The impairment of low pressure baroreceptor function is not due to the presence of hypertension per se, because it is also present in normotensive athletes with an increased cardiac mass.[37]

This finding may have important consequences in controlling the circadian

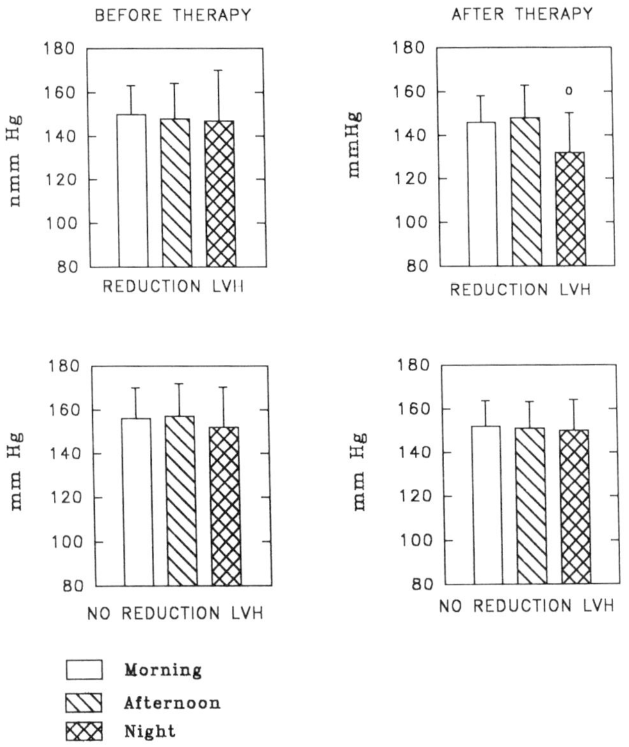

FIGURE 2. Systolic blood pressure values in hypertensive patients with left ventricular hypertrophy, with a reduced left ventricular mass (reduction LVH, $n = 5$), and without a reduction in left ventricular mass (no reduction LVH, $n = 5$) before and after antihypertensive therapy. $^{\circ}p < 0.05$ at least compared with morning and afternoon.

pattern of blood pressure. Cardiac structural alterations (hypertrophy, remodeling, and fibrosis) could mechanically impair the responsiveness of cardiopulmonary low pressure receptors to an increased venous return, which usually occurs when the subject lies supine during the night. The negative feedback on adrenergic activity during the night may be impaired, and therefore the physiologic reduction in blood pressure may also be reduced or absent.

Also, impairment of the response of several vasoactive humoral factors to the loading and unloading of cardiopulmonary baroreceptors may be important and, at least in part, may help explain the reduction of circadian changes in blood pressure.[32,33,35,36,38]

The interrelationships between left ventricular hypertrophy and circadian rhythms of blood pressure, however, may be mutual. Left ventricular hypertrophy may impair the nocturnal fall in blood pressure through impairment of cardiopulmonary low pressure baroreceptor function or through a humoral or an unknown mechanism. On the other hand, an impaired nocturnal fall in blood pressure may increase total 24-hour hemodynamic load and ultimately induce an increase in left ventricular mass. It is well known that the afterload is one of the most important determinants of cardiac structure.

REGRESSION OF LEFT VENTRICULAR HYPERTROPHY AND CIRCADIAN PATTERN OF BLOOD PRESSURE

If the *primum movens* in the relation between left ventricular hypertrophy and impairment of a nocturnal fall in blood pressure is represented by the former term, a beneficial effect of the reduction in left ventricular mass by antihypertensive therapy on the circadian pattern of blood pressure would be expected. In a preliminary study,[39] regression of left ventricular hypertrophy is associated with a greater reduction in nighttime arterial pressure values as measured by noninvasive ABPM compared to pretreatment evaluation. Moreover, restoration of the circadian pattern in blood pressure (FIGS. 2 and 3) and autonomic nervous system activity (FIGS. 4 and 5) was observed only in patients with regression of left ventricular hypertrophy,

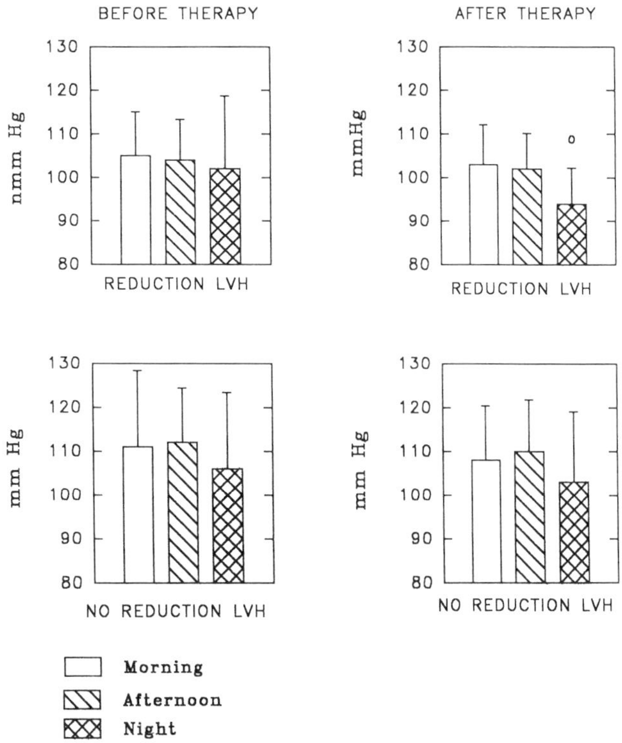

FIGURE 3. Diastolic blood pressure values in hypertensive patients with left ventricular hypertrophy, with a reduced left ventricular mass (reduction LVH, $n = 5$), and without a reduction in left ventricular mass (no reduction LVH, $n = 5$) before and after antihypertensive therapy. °$p < 0.05$ at least compared with morning and afternoon.

whereas no change was observed in those patients with no reduction in left ventricular mass despite satisfactory control of blood pressure. The two groups of patients had comparable pretreatment levels of blood pressure. To avoid the confounding effect of blood pressure reduction, the effect of antihypertensive treatment was evaluated after prolonged treatment withdrawal, when hypertension had almost completely redeveloped.

These preliminary data support the hypothesis that left ventricular hypertrophy directly affects blood pressure patterns and that the reduction of left ventricular mass by effective antihypertensive therapy is associated with a restoration of the physiologic day/night modulation of blood pressure and autonomic nervous system activity.

These results may be extremely important, because alterations in the circadian pattern of blood pressure are associated not only with cardiac structural alterations, but also with a worse prognosis, independently for 24-hour blood pressure and left ventricular mass values.[16,40,41]

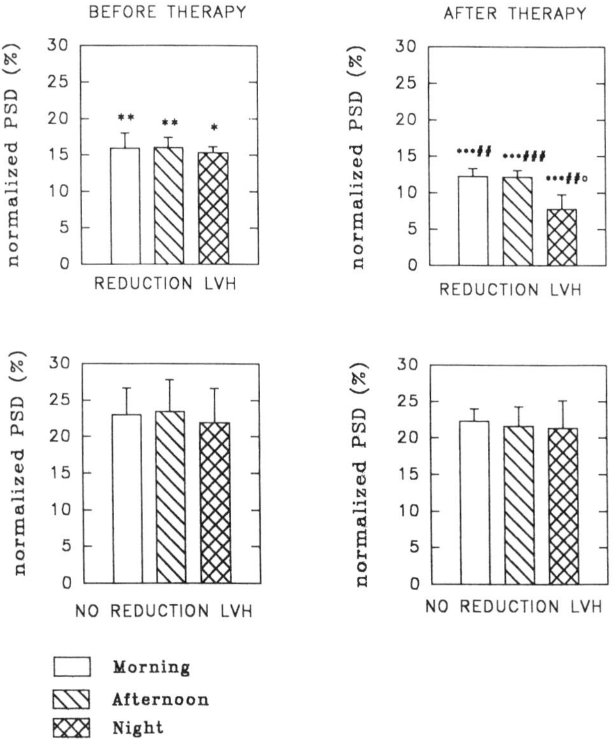

FIGURE 4. Normalized power spectral densities of the low frequency (partially reflecting sympathetic activity) component of heart rate variability in hypertensive patients with left ventricular hypertrophy, with a reduced left ventricular mass (reduction LVH, $n = 5$), and without a reduced left ventricular mass (no reduction LVH, $n = 5$) before and after antihypertensive therapy. $^{\circ}p < 0.05$ at least compared with morning and afternoon; $^{*}p < 0.05$, $^{**}p < 0.01$, $^{***}p < 0.001$ compared with patients with no reduction in left ventricular hypertrophy; $^{\#\#}p < 0.01$, $^{\#\#\#}p < 0.001$ compared before therapy.

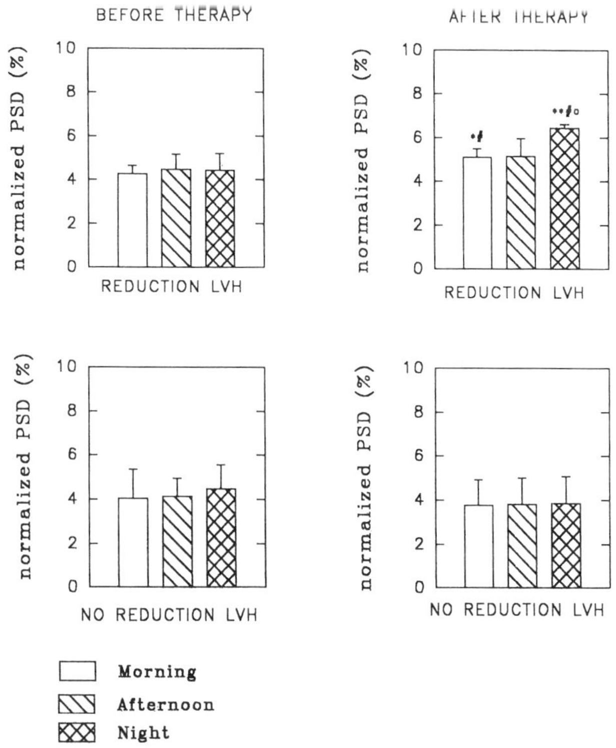

FIGURE 5. Normalized power spectral densities of the high frequency (partially reflecting parasympathetic activity) component of heart rate variability in hypertensive patients with left ventricular hypertrophy, with a reduced left ventricular mass (reduction LVH, $n = 5$), and without a reduced left ventricular mass (no reduction LVH, $n = 5$) before and after antihypertensive therapy. $^{\circ}p < 0.05$ at least compared with morning and afternoon; $^{*}p < 0.05$; $^{**}p < 0.01$ compared with patients with no reduction of left ventricular hypertrophy; $^{\#}\ p < 0.05$ compared before therapy.

NEW PERSPECTIVES AND CONCLUSIONS

Calculations of the pressure difference[16,42] or of the percentage pressure difference[13,15,42] between average daytime and nighttime blood pressure assume fixed sleep/wake timing and are subject to the previously mentioned limitations. Therefore, new methods or new clinical applications of old methods for evaluating and quantifying the circadian rhythm of blood pressure are needed. Among the presently available methods, the following are of particular interest: the cumulative sums method,[44] the spectral methods (fast Fourier transform, etc.), Halberg's cosinor method,[45,46] periodic spline models,[47] and other statistical analyses, such as separate linear models fitted to the data of each subject.[48] However, there is currently no universally accepted method of quantifying the circadian pattern of blood pressure.

In conclusion, patients with left ventricular hypertrophy probably have an altered circadian pattern of blood pressure, and several clinical and experimental studies support this hypothesis. However, the information provided by the available clinical studies on this topic are limited, because they are biased due to several methodologic problems. High degrees of left ventricular hypertrophy are commonly associated with impaired autonomic nervous system modulation and a circadian pattern of blood pressure. Preliminary data favor a beneficial effect of left ventricular mass reduction on the circadian rhythms of blood pressure. Further studies are needed to elucidate the clinical impact of all these observations in patients with mild left ventricular hypertrophy. Improved methods of quantifying the circadian pattern of blood pressure are welcome.

SUMMARY

It is well known that hemodynamic load is one of the most important determinants of cardiac structure and function. Circadian variations in blood pressure (BP) values are usually accompanied by consensual changes in peripheral resistance and/or cardiac output. In recent years, in hypertensive patients with left ventricular hypertrophy (LVH), a reduction in the circadian variations of BP and, in particular, a lack of nocturnal decline were observed; patients with only a small reduction in BP or none at all during the night were considered "non-dippers." In patients in whom a regression of LVH was obtained after prolonged antihypertensive therapy, restoration of the circadian rhythm of BP was also observed. However, the division of patients into "dippers" and "non-dippers" is arbitrary and poorly standardized and repeatable, and in the recent SAMPLE study, most hypertensive patients with LVH were dippers. Therefore, we should be particularly cautious about the conclusions drawn using this index. On the other hand, in patients with LVH, reduced activity of the low pressure cardiopulmonary baroreceptors and impaired day-to-night modulation of autonomic nervous system activity were observed. Therefore, cardiac structural alterations may possibly impair BP modulation; on the other hand, the opposite could also be true: a primarily altered BP modulation, through a persistently elevated afterload, could increase cardiac mass. Therefore, the interrelationships between cardiac structure and BP modulation are complex, and as a result, new and more specific methods of evaluation circadian changes in BP are needed to better clarify their reciprocal influences.

REFERENCES

1. Pickering, T. G. 1991. Diurnal, ultradian and seasonal rhythms of blood pressure. *In* Ambulatory Monitoring and Blood Pressure Variability. T. G. Pickering, C. Pieper & C. B. Schechter, Eds. 5.1–6.1. Science Press Ltd. London.
2. Kool, M. J., J. A. Wijnen, F. H. Derkx, H. A. Struijker Boudier & L. M. Van Bortel. 1994. Diurnal variation in prorenin in relation to other humoral factors and hemodynamics. Am. J. Hypertens. **7:** 723–730.
3. Minamisawa, K., O. Tochikubo & M. Ishii. 1994. Systemic hemodynamics during sleep in young or middle-aged and elderly patients with essential hypertension. Hypertension **23:** 167–173.
4. Veerman, D. P., B. P. M. Imholz, W. Wieling, K. H. Wesseling & G. A. Van Montfrans. 1995. Circadian profile of systemic hemodynamics. Hypertension **26:** 55–59.
5. Kawano, Y., O. Tochikubo, K. Minamisawa, E. Miyajima & M. Ishii. 1994. Circadian variation of hemodynamics in patients with essential hypertension: Comparison between early morning and evening. J. Hypertens. **12:** 1405–1412.
6. Panza, J. A., S. E. Epstein & A. A. Quyyumi. 1991. Circadian variation in vascular tone and its relation to α-sympathetic vasocontrictor activity. N. Engl. J. Med. **325:** 986–990.

7. SEHESTED, J., E. HAPPE, K. ISHINO, R. HETZER, U. SCHIESSLER & S. SCHIFTER. 1994. Diurnal variations in blood pressure in patients with biventricular assist devices and retained, nonpumping native hearts. Circulation **89:** 2601–2604.
8. FURLAN, R., S. GUZZETTI, W. CRIVELLARO, S. DASSI, M. TINELLI, G. BASELLI, S. CERUTTI, F. LOMBARDI, M. PAGANI & A. MALLIANI. 1990. Continuous 24-hour assessment of the neural regulation of systemic arterial pressure an RR variabilities in ambulant subjects. Circulation **81:** 537–547.
9. GUZZETTI, S., S. DASSI, M. PECIS, R. CASATI, A. M. MASU, P. LONGONI, M. TINELLI, S. CERUTTI, M. PAGANI & A. MALLIANI. 1991. Altered pattern of circadian neural control of heart period in mild hypertension. J. Hypertens. **9:** 831–838.
10. PARATI, G., P. CASTIGLIONI, M. DI RIENZO, S. OMBONI, A. PEDOTTI & G. MANCIA. 1990. Sequential spectral analysis of 24-hour blood pressure and pulse interval in humans. Hypertension **16:** 414–421.
11. PARATI, G., J. P. SAUL, M. DI RIENZO & G. MANCIA. 1995. Spectral analysis of blood pressure and heart rate variability in evaluating cardiovascular regulation. A critical appraisal. Hypertension **25:** 1276–1286.
12. KUCHEL, O. 1995. A new potential contributor to circadian sympathetic variations. J. Hypertens. **13:** 569–570.
13. RIZZONI, D., M. L. MUIESAN, G. MONTANI, R. ZULLI, S. CALEBICH & E. AGABITI ROSEI. 1992. Relationship between initial cardiovascular structural changes and daytime and nighttime blood pressure monitoring. Am. J. Hypertens. **5:** 180–186.
14. LANG, R. M., K. M. BOROW & L. WEINERT. 1988. Circadian blood pressure patterns in hypertensive patients: Importance of nocturnal blood pressure. Circulation **78**(Suppl. II): II517.
15. VERDECCHIA, P., G. SCHILLACI, M. GUERRIERI, C. GATTESCHI, G. BENEMIO, F. BOLDRINI & C. PORCELLATI. 1990. Circadian blood pressure changes and left ventricular hypertrophy in essential hypertension. Circulation **81:** 528–536.
16. O'BRIEN, E., J. SHERIDAN & K. O'MALLEY. 1988. Dippers and non-dippers. Lancet **ii:** 397.
17. VERDECCHIA, P., G. SCHILLACI & C. PORCELLATI. 1991. Dippers versus non-dippers. J. Hypertens. **9**(Suppl. 8): S42– S44.
18. VERDECCHIA, P., G. SCHILLACI, C. BORGIONI, A. CIUCCI, N. SACCHI, M. BATTISTELLI, M. GUERRIERI, E. COMPARATO & C. PORCELLATI. 1995. Gender, day-night blood pressure changes, and left ventricular mass in essential hypertension. Dipper and non-dippers. Am. J. Hypertens. **8:** 193–196.
19. VERDECCHIA, P., G. SCHILLACI, F. BOLDRINI, M. GUERRIERI & C. PORCELLATI. 1992. Sex, cardiac hypertrophy and diurnal blood pressure in essential hypertension. J. Hypertens. **10:** 683–692.
20. MANCIA, G. 1990. Presidential lecture: Ambulatory blood pressure monitoring. J. Hypertens. **8**(Suppl. 7): S1–S13.
21. VERDECCHIA, P., G. SCHILLACI, F. BOLDRINI, M. GUERRIERI, I. ZAMPI & C. PORCELLATI. 1991. Quantitative assessment of day-to-day spontaneous variability in non-invasive ambulatory blood pressure measurements in essential hypertension. J. Hypertens. **9**(Suppl. 6): S322–S323.
22. PALATINI, P., M. PENZO, A. RACIOPPA, E. ZUGNO, G. GUZZARDI, M. ANACLERIO & A. C. PESSINA. 1992. Clinical relevance of nighttime blood pressure and of daytime blood pressure variability. Arch. Intern. Med. **152:** 1855–1860.
23. JOHNSON, J. A., W. S. ALKERS, S. T. MILLER & W. B. APPLEGATE. 1995. Metoprolol minimizes night-time blood pressure dip in hypertensive black males. Am. J. Hypertens. **8:** 254–259.
24. RIZZONI, D., M. L. MUIESAN, R. ZULLI, M. CASTELLANO, G. BETTONI, E. PORTERI & E. AGABITI ROSEI. 1993. Effect of antihypertensive treatment on daytime and nighttime power spectral analysis of heart rate. Am. J. Hypertens. **6:** 204–208.
25. PARATI, G., G. POMIDOSSI & R. CASADEI. 1985. Ambulatory blood pressure monitoring does not interfere with the hemodynamic effects of sleep. J. Hypertens. **3**(Suppl. 2): S107– S109.
26. BRIGDEN, G., P. BROADHUST, P. CASHMAN & E. RAFTERY. 1991. Effect of noninvasive

ambulatory blood pressure measuring devices on blood pressure. Am. J. Cardiol. **66:** 1396–1398.

27. FAGARD, R., E. BIELEN & A. AMERY. 1992. Automated versus observer blood pressure as determinants of left ventricular structure. Eur. Heart J. **13:** 1373–1379.
28. MANCIA, G. & A. ZANCHETTI on behalf of the Sample Study Group. 1996. Prognostic value of 24 h ambulatory blood pressure monitoring. The SAMPLE Study Abstracts of the Seventh European Meeting on Hypertension. Milan, Italy, June 9–12, 1995. p. 115.
29. FAGARD, R., J. A. STAESSEN & L. THIJS. 1995. The relationships between left ventricular mass and daytime and night-time blood pressure: A meta-analysis of comparative trials. J. Hypertens. **13:** 823–830.
30. AGABITI ROSEI, E., D. RIZZONI, M. CASTELLANO, G. BETTONI, M. L. MUIESAN, M. MALERBA & E. PORTERI. 1992. Power spectral analysis of heart rate and arterial pressure in hypertensive patients with and without left ventricular hypertrophy. *In* Blood Pressure and Heart Rate Variability. M. Di Rienzo, G. Mancia, G. Parati, A. Pedotti & A. Zanchetti, Eds.: 230–241. IOS Press. Amsterdam.
31. FOGARI, R., A. ZOPPI, G. D. MALAMANI, P. LAZZARI, M. DESTRO & L. CORRADI. 1993. Ambulatory blood pressure monitoring in normotensive and hypertensive type 2 diabetics. Prevalence of impaired diurnal blood pressure patterns. Am. J. Hypertens. **6:** 1–7.
32. GRASSI, G., C. GIANNATTASIO, J. CLEROUX, C. CUSPIDI, L. SAMPIERI, G. B. BOLLA & G. MANCIA. 1988. Cardiopulmonary reflex before and after regression of left ventricular hypertrophy in essential hypertension. Hypertension **12:** 227–237.
33. GRASSI, G., C. GIANNATTASIO, A. SAINO, E. SABADINI, A. CAPOZI, L. SAMPIERI, C. CUSPIDI & G. MANCIA. 1988. Cardiopulmonary receptor modulation of plasma renin activity in normotensive and hypertensive subjects. Hypertension **11:** 92–99.
34. RIZZONI, D., M. CASTELLANO, M. BESCHI, M. L. MUIESAN, G. BETTONI, E. PORTERI & E. AGABITI ROSEI. 1992. Autonomic nervous system activity as evaluated by spectral analysis of heart rate and plasma norepinephrine during cardiopulmonary receptor deactivation in normal and hypertensive subjects. High Blood Press. **1:** 57–64.
35. TRIMARCO, B., N. DE LUCA, A. DE SIMONE, M. VOLPE, B. RICCIARDELLI, G. LEMBO & M. CONDORELLI. 1987. Impaired control of vasopressin release in hypertensive subjects with cardiac hypertrophy. Hypertension **10:** 595–602.
36. TRIMARCO, B., G. LEMBO, N. DE LUCA, M. VOLPE, B. RICCIARDELLI, G. CONDORELLI, G. ROSIELLO & M. CONDORELLI. 1989. Blunted sympathetic response to cardiopulmonary receptor unloading in hypertensive patients with left ventricular hypertrophy. A compensatory possible role of Atrial Natriuretic Factor. Circulation **80:** 883–892.
37. GIANNATTASIO, C., G. SERAVALLE, G. BOLLA, B. CATTANEO, J. CLEROUX, C. CUSPIDI, L. SAMPIERI, G. GRASSI & G. MANCIA. 1990. Cardiopulmonary reflexes in normotensive athletes with cardiac hypertrophy. Circulation **82:** 1222–1229.
38. RIZZONI, D., M. CASTELLANO, M. L. MUIESAN, M. BESCHI, G. MONTANI, G. PIZZOCCOLO, C. POIESI, A. RODELLA & E. AGABITI ROSEI. 1992. Antidiuretic hormone and atrial natriuretic peptide during lower body negative or positive pressure in hypertensive patients with and without left ventricular hypertrophy. Clin. Exp. Hypertens. Part A Theory Pract. **14:** 717–732.
39. MUIESAN, M. L., D. RIZZONI, R. ZULLI, S. CALEBICH, E. PORTERI & E. AGABITI ROSEI. 1993. Regression of cardiac hypertrophy and nighttime blood pressure reduction. J. Hypertens. **11**(Suppl. 5): S300–S301.
40. VERDECCHIA, P., G. SCHILLACI, C. GATTESCHI, I. ZAMPI, M. BATTISTELLI, C. BARTOCCINI & C. PORCELLATI. 1993. Blunted nocturnal fall in blood pressure in hypertensive women with future cardiovascular morbid events. Circulation **88:** 896–992.
41. VERDECCHIA, P., C. PORCELLATI, G. SCHILLACI, C. BORGIONI, A. CIUCCI, M. BATTISTELLI, M. GUERRIERI, C. GATTESCHI, I. ZAMPI, A. SANTUCCI, C. SANTUCCI & G. REBOLDI. 1994. Ambulatory blood pressure. An independent predictor of prognosis in essential hypertension. Hypertension **24:** 793–801.
42. KOBRIN, I., W. OIGMAN, A. KUMAN, H. O. VENTURA, F. H. MESSERLI, E. D. FROHLICH & F. G. DUNN. 1984. Diurnal variation of blood pressure in elderly patients with essential hypertension. Am. J. Geriatr. Soc. **32:** 896–899.

43. Valty, J., J. Y. Le Heuzey, A. Buoncuone, J. Y. Sagnes, A. Cohen & P. J. Scala. 1989. Etude des variation mycthemerales de la pression arterielle chez les sujets hospitalises. Arch. Mal Coeur **82:** 1039–1042.
44. Stanton, A., J. Cox, N. Atkins, K. O'Malley & E. O'Brian. 1992. Cumulative sums in quantifying circadian blood pressure patterns. Hypertension **19:** 93–101.
45. Halberg, F. 1069. Chronobiology. Annu. Rev. Physiol. **31:** 675–725.
46. Smolensky, M., S. E. Tatan, S. Bergman, J. Losman, C. N. Barbard, C. C. Dacso & I. A. Kraft. 1975. Circadian rhythmic aspects of human cardiovascular function: A review by chronobiological statistical methods. Chronobiologia **3:** 337–371.
47. Streitberg, B., W. Meyer-Sabellek & P. Baumgart. 1989. Statistical analysis of circadian blood pressure recordings in controlled clinical trials. J. Hypertens. **7**(Suppl. 3): S11–S17.
48. Marler, M. R., R. G. Jacob, J. P. Lehoczky & A. P. Shapiro. 1988. The statistical analysis of treatment effects in 24-hour ambulatory blood pressure recordings. Stat. Med. **7:** 697–716.

Circadian Blood Pressure Variation Related to Morbidity and Mortality from Cerebrovascular and Cardiovascular Diseases[a]

YUTAKA IMAI,[b] ICHIRO TSUJI,[c] KENICHI NAGAI,[d]
NORIKO WATANABE, TAKAYOSHI OHKUBO,[c]
MARIKO SAKUMA,[e] JUNICHIRO HASHIMOTO,
OSAMU ITOH, HIROSHI SATOH,[e]
SHIGERU HISAMICHI,[c] AND KEISHI ABE

[b]The Second Department of Medicine
[c]The Department of Public Health and
[e]The Department of Environmental Health Sciences
Tohoku University School of Medicine
Sendai, Japan

[d]Ohasma Hospital
Iwate, Japan

The mortality rate from stroke has decreased dramatically over the last 30 years in most industrialized countries,[1] particularly Japan, where the mortality rate from stroke is one fourth that in the 1960s (FIG. 1).[1] This decreased mortality from stroke in Japan as well as in the United States and European countries is attributable mainly to the decrease in cerebral hemorrhage.[2] However, the prevalence of cerebral infarction has increased recently in Japan[3] (FIG. 2) for several reasons. A decrease in fatal cerebral hemorrhage and an increase in nonfatal cerebral infarction are primarily responsible for the increased prevalence of cerebral infarction. Recent advances in medical technology such as computed tomography scan and magnetic resonance imaging have enhanced our ability to detect smaller cerebral infarctions and asymptomatic cerebrovascular lesions. Although improved antihypertensive drugs have dramatically decreased the incidence of cerebral hemorrhage, these drugs have not decreased the incidence of nonfatal cerebral infarction, especially among the elderly. We postulate that antihypertensive treatment relates to an increased prevalence of mild cerebral infarction in the elderly population, which interferes with daily activity and affects the quality of life. The question arises as to whether antihypertensive treatment for elderly people should be equal to that for the

[a]This work was supported in part by a grant from the Miyagi Prefectural Kidney Association, by a grant from Takeda Medical Research Foundation, by grants (No. 4C-3 and 5C-2) from the Ministry of Health and Welfare, by a grant entitled "Evaluation of the effect of drug treatment on hypertension and other chronic disease conditions in the elderly" from the Ministry of Health and Welfare, and by grants (07670420 and 07670646) from the Ministry of Education, Science and Culture, Japan.

[b]Address for correspondence: Yutaka Imai, MD, The Second Department of Medicine, Tohoku University School of Medicine, 1-1 Seiryomachi, Aobaku, Sendai 980-77, Japan.

younger population and whether blood pressure (BP) levels achieved with treatment should be the same regardless of age.

RELATIONS BETWEEN MORBIDITY AND MORTALITY RATES FROM CEREBROVASCULAR AND CARDIOVASCULAR DISEASE

Several intervention trials have demonstrated a J-shaped relation between the morbidity or mortality rate of coronary heart disease[4,5] and BP levels after antihypertensive treatment,[6] especially in patients with coronary arterial disease. However, the relation between the morbidity or mortality rate of stroke and BP levels remains unclear.[4,5] A recent report published by the Japanese Cardiovascular Center clearly demonstrated a J-shaped relation between recurrence rate of stroke and BP levels after antihypertensive treatment.[7] By contrast, metaanalysis of long-term observational studies by MacMahon *et al.*[8] demonstrated a direct, continuous, and apparently independent association of BP levels and risks of cardiovascular and cerebrovascular diseases with no evidence of threshold level of diastolic BP (DBP) below which lower levels of DBP were not associated with lower risks of stroke and coronary heart disease. Several intervention trials demonstrated lower mortality or morbidity rates due to stroke and/or coronary heart disease following treatment in elderly patients.[9,10] However, it has also been reported that in very old persons the survival rate may be better at a higher level of BP.[11] The lowest mortality rates were noted in those with a systolic BP (SBP) of 140 to 169 and a DBP of 70 to 99.[12] These studies evaluated the relation between casual (office, screening) BP measurements and cerebrovascular and cardiovascular morbidity and mortality rates, but ambulatory and home measurements of BP were not taken into account.

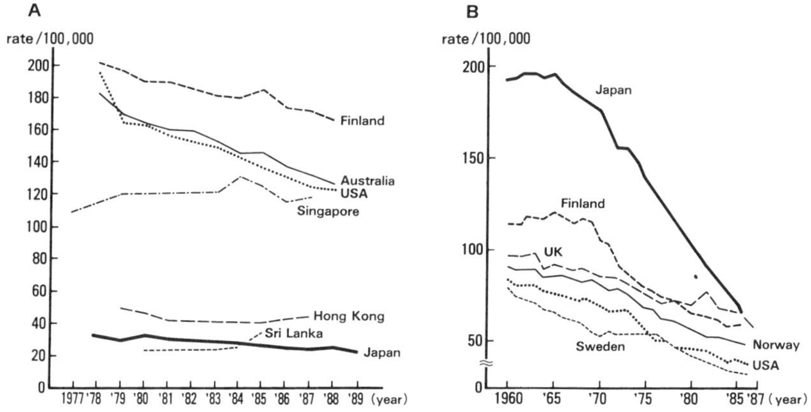

FIGURE 1. Trends in coronary heart disease mortality (**A**) and in stroke mortality (**B**) in major countries. Age was adjusted to the standard world population (WHO). (From Ueda & Omae[1] with permission from the authors and Churchill Livingstone, Japan, Kyoto.)

CASUAL, AMBULATORY, AND HOME MEASUREMENT OF BLOOD PRESSURE

Casual levels of BP reflect hypertensive organ damage as well as cardiovascular mortality and morbidity rates in the population at large. Almost all data on the risks of hypertension are based on only one or a few readings obtained in large study groups.[13,14] Although such data clearly have epidemiologic significance, there is controversy over whether a few casual readings constitute a reliable predictor of cardiovascular mortality or morbidity rates in an individual. Both ambulatory and home BP monitoring are known to be free from observer bias and the white coat effect. Especially with ambulatory BP monitoring, BP values, which are affected by several physical and mental conditions, can provide average BP for a certain period with different conditions, whereas the difference in daily activity makes questionable the reproducibility of ambulatory BP levels.[15,16]

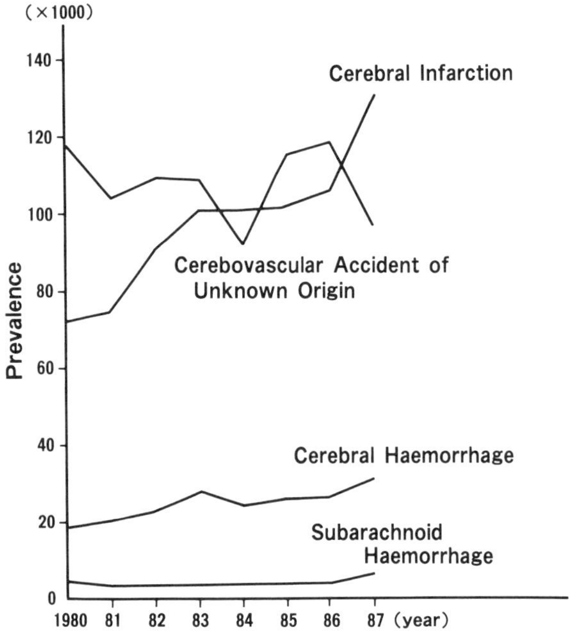

FIGURE 2. Trends in stroke prevalence in Japan. (From Kameyama[3] with permission from the author and Jikken Chiryo, Takeda Yakuhin Kogyo, Osaka.)

By contrast, home BP measurements are obtained under well-controlled conditions, that is, in the sitting position within 1 hour after awakening without observer bias. Thus, long-term average and variability of BP are available. In this context, the reproducibility of home BP is superior to that of ambulatory BP.[16] Both home BP and ambulatory BP levels, even the daytime average of ambulatory BP, are generally lower than casual office or screening BP levels,[17,18] mainly because of the absence of the white coat effect. Several reports have demonstrated the superiority of such home BP and ambulatory BP levels for reflecting hypertensive target organ damage as well as predicting cardiovascular mortality and morbidity rates.[19–26]

Despite the estimated 10,000 ambulatory BP monitoring devices and 20 million home BP measurement devices now being used in Japan,[16] the standard method of measuring BP in the diagnosis of hypertension uses a mercury sphygmomanometer based on the Korotkoff sound technique in the clinical setting, and ambulatory BP monitoring and home BP measurements are only supplementary. This is emphasized in the report of the 5th Joint National Committee on Detection, Evaluation and Treatment of High Blood Pressure[27] and the WHO/ISH Guidelines.[28] However, ambulatory BP and home BP monitoring would be used more frequently if their clinical significance was established and if the cost of the use of ambulatory BP monitoring or home BP measurements in the diagnosis and treatment of hypertension was covered by national or private health insurance.

Perloff *et al.*[25] studied the relation between clinical outcome and ambulatory BP in 1,706 patients followed an average of 5 years and observed that those with higher ambulatory BP than that predicted from the regression line between ambulatory BP and clinic BP exhibited a higher mortality rate and higher incidence of the first cardiovascular event than did those with lower ambulatory BP levels.[25] Pickering and James[26] recently demonstrated that ambulatory waking BP levels were better predictors than were BP measurements obtained in the clinical setting. They also reported that the variability of DBP (SD of ambulatory DBP) was a better predictor than any measurements of ambulatory or casual BP levels.[26] Relations between BP variability and target organ damage or prognosis of hypertension were reported by several researchers.[29,30] However, no reports demonstrated critical ambulatory BP levels as diagnostic criteria and therapeutic goals, and no information is available on the prognostic value of home BP measurements.

NOCTURNAL BP AND MORBIDITY AND MORTALITY RATES OF CEREBROVASCULAR AND CARDIOVASCULAR DISEASES

Recent studies of ambulatory BP monitoring indicate that hypertensive subjects with the least decrease in nocturnal BP had more severe left ventricular hypertrophy[31,32] than did those with a normal decrease in nocturnal BP. O'Brien *et al.*[33] referred to those subjects with the least reduction in nocturnal BP as "non-dippers" and those with an apparent fall in nocturnal BP as "dippers." A higher risk of stroke was found in non-dippers. Shimada *et al.*[34] demonstrated a correlation between silent cerebrovascular lesions (e.g., lacunar infarctions and periventricular hyperintensity [PVH]) and nocturnal levels of BP and reported that the non-dippers more frequently exhibited these lesions.[34] Investigators who studied the incidence of left ventricular hypertrophy and cerebrovascular lesions in non-dippers concluded that high nocturnal BP is the cause of target organ damage.[31–34] If the diurnal BP of non-dippers resembles that of dippers, the net BP load on the cardiovascular system is obviously greater in the former. Thus, more extensive cerebrovascular and cardiovascular lesions may occur in non-dippers. End-organ damage and cerebrovascular and cardiovascular complications may be prevented by inducing a decrease in nocturnal BP in non-dippers. However, other possibilities should also be considered. A decrease in nocturnal BP is generally observed in normotensive subjects as well as in those with mild to moderate essential hypertension regardless of the daytime BP level. The amplitude of the fall in nocturnal BP is decreased or even lost in men over the age of 80 years[35] as well as in hypertensive patients with advanced atherosclerotic or arteriosclerotic vascular lesions, such as silent cerebrovascular lesions, Binswanger type of vascular dementia, arteriosclerosis obliterans, and left ventricular hypertrophy.[36] Thus, it is possible that the circadian BP variation in patients with early-stage essential hypertension without end-organ damage is initiated in dippers, and hypertensive patients with accelerated end-organ damage experience the non-dipper type of circadian BP variation. It is unclear if the non-dipper represents the cause or the effect of end-organ damage and if the dipper condition is beneficial or harmful to patients with cerebrovascular and cardiovascular lesions.

Hypertension is a major risk factor for atherosclerosis and arteriosclerosis.[37–39] Regional cerebral blood flow is reduced by aging and hypertension. In normal subjects, decreases in cerebral blood flow and BP are observed in sleep stages II and III.[40] Regional cerebral blood flow decreases more markedly in elderly subjects with leukoaraiosis than in those without leukoaraiosis during sleep.[41] Elderly hypertensive patients with silent cerebrovascular lesions[34] and those with the Binswanger type of vascular dementia are reportedly non-dippers.[42] Patients with hypertensive, atherosclerotic, and arteriosclerotic end-organ damage also exhibit non-dipper type of circadian BP variation.[36] Patients with renovascular hypertension are reportedly non-dippers when this disorder is caused by atherosclerosis, whereas an ordinary fall in nocturnal BP is observed in young patients with renovascular hypertension when this disorder is caused by fibromuscular hyperplasia.[43] These reports suggest the existence of compensatory mechanisms that would diminish or eliminate the fall in nocturnal BP in patients with organ ischemia attributable to advanced atherosclerosis and/or arteriosclerosis, thus protecting against end-organ damage from a decrease in blood flow during sleep. We therefore hypothesize that a diminished reduction in nocturnal BP in patients with some types of hypertension is the result of hypertensive end-organ damage. Kario and Shimada[44] described an elderly hypertensive patient whose diurnal BP variability changed from that of an extreme dipper to that of a non-dipper following a small lacunar infarction. They concluded that non-dipper hypertension in elderly patients with ischemic cerebrovascular lesions may be secondary to minor cerebrovascular damage.

Blood flow is autoregulated in the brain, heart, and kidney within a certain range of perfusion pressure. Cerebral blood flow is reportedly reduced, and the lower limit of BP for autoregulation is shifted to higher levels in elderly hypertensive patients with stroke.[45,46] Cerebral blood flow may therefore easily be decreased in elderly hypertensive persons, especially in those with cerebrovascular lesions, at a perfusion pressure that does not normally reduce cerebral blood flow.[46] Hypertension and aging are the major risk factors for lacunar infarction and PVH. Hypertension causes atherosclerosis in the carotid and main cerebral arteries and hemodynamic and atherothrombotic infarctions in the cortical and external watershed regions.[47] Hypertension may also alter the deep penetrating arteries, leading to circulatory failure and lacunar infarction.[47] Prolonged hypertension thus produces hypertensive and atherosclerotic cerebrovascular changes. In elderly hypertensive patients with such cerebrovascular changes, a slight decrease in perfusion pressure can easily reduce cerebral blood flow. Studies of elderly hypertensive patients with PVH demonstrated decreased cerebral blood flow during sleep and disturbed cerebral blood flow autoregulation.[40,49] Thus, inappropriately low nocturnal BP or an excessive drop in nocturnal BP may cause silent cerebrovascular lesions.

FIGURE 3 shows age-dependent changes in the circadian BP variation exhibited by our cohort study in northern Japan. The amplitude of the nocturnal fall in BP decreased age-dependently in men but not in women, and elderly men exhibited a nondipper pattern of circadian BP variation, whereas some of the elderly women exhibited an extreme dipper pattern of circadian BP variation.

From this background, we performed two studies, a prospective cohort study on

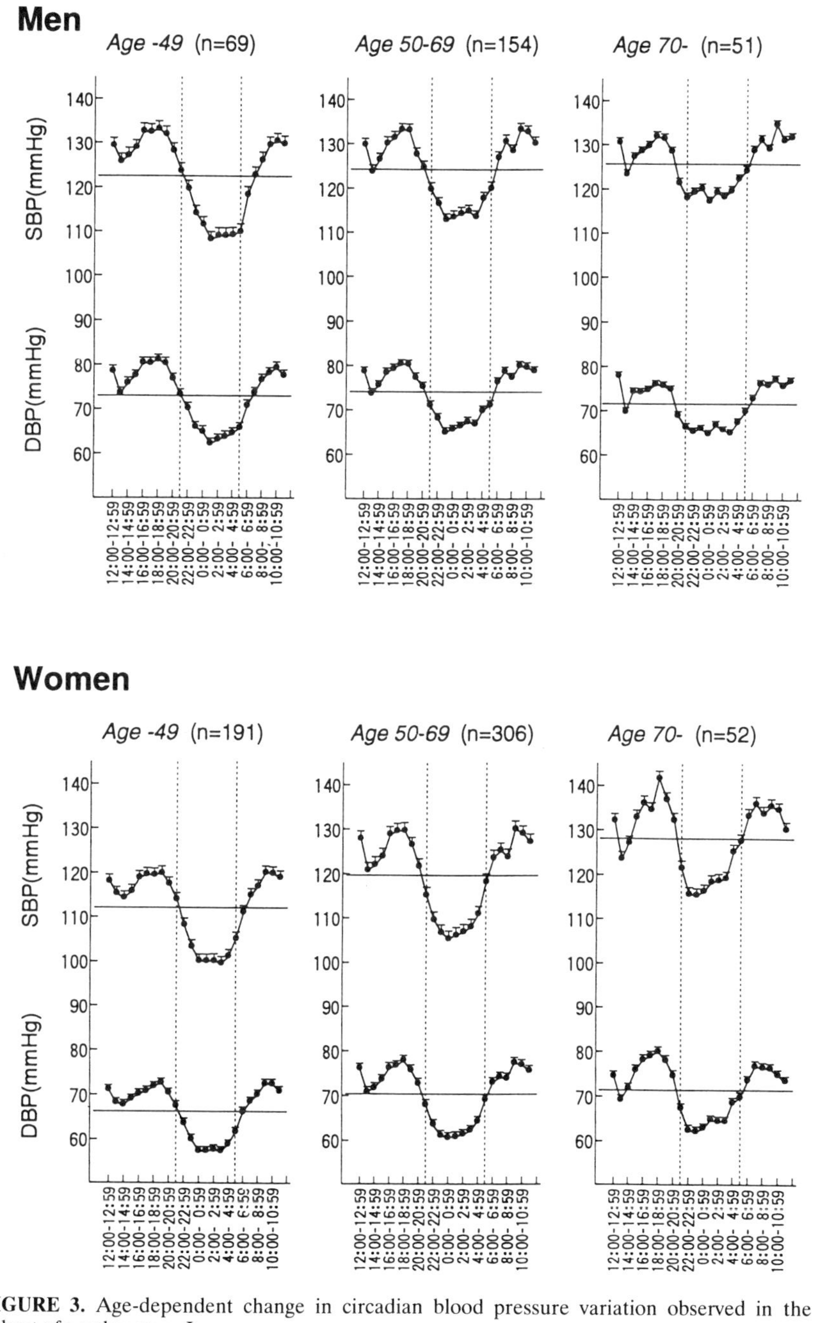

FIGURE 3. Age-dependent change in circadian blood pressure variation observed in the cohort of northeastern Japan.

mortality rate on the basis of ambulatory BP monitoring and home BP measurements, and a cross-sectional survey of the relationships between silent cerebrovascular lesions and circadian BP variation.[48]

PROGNOSTIC VALUE OF AMBULATORY BP MONITORING AND HOME BP MEASUREMENTS

We studied the prognostic value of ambulatory BP monitoring and home BP measurements in the cohort of Ohasama, northeastern Japan, which has a population of 8,500.[17,18] Most adults are employed as fruit farmers; the others work in offices or factories in Ohasama or neighboring cities. The mean per capita income in this area is approximately $15,000 (US). Ohasama was divided into four areas. All causes of mortality and cerebrovascular and cardiovascular mortality were studied since 1987 in two of these four areas, using screening BP measurements, ambulatory BP monitoring, and home BP measurements. People 20 years and older were asked to participate in the ambulatory BP monitoring program, and those 8 years and older were asked to record their home BP measurements. Ambulatory BP was measured using the Colin ABPM 630 (Komaki, Japan) every 30 minutes for 23.3 ± 1.6 hours in 1,192 subjects, and home BP was measured in the sitting position within 1 hour of awakening using automatic BP measuring devices (HEM 401C and 701C, Omron Life Science, Kyoto, Japan) in 1,962 subjects. More than three home BP measurements (3 days) were obtained for each subject, and the mean measurement frequency was 20.8 ± 8.3 times. TABLE 1 shows the number of subjects aged 50 years and older (regional population aged over 50 years and over = 1,556) who measured ambulatory BP, home BP, and screening BP, respectively, as well as the mean and maximum periods of observation. The incidence for all causes of mortality was obtained from Kaplan-Meier estimates of event probability. Statistical test of the relation between BP levels and the incidence of all causes of mortality was carried out using the Cox proportional hazard model, which involves comparison of the quintile with the lowest hazard (risk ratio: 1) to remaining quintiles of BP levels. Age and sex were selected as covariates in the multivariate analysis.

A Kaplan-Meier estimate was performed in all subjects including treated subjects as well as those with a history of stroke and coronary vascular disease. When screening BP levels were classified into quintiles, a significant decrease in survival

TABLE 1. Frequency of Antihypertensive Drug Treatment and Duration of Mean and Maximum Period of Observation in Subjects Aged 50 Years and Over Using Screening Casual Blood Pressure Measurements (CBP), Ambulatory Blood Pressure Monitoring (ABPM), and Home Blood Pressure Measurements (HBP)[a]

	CBP	ABP	HBP
Participants	1,057	893	1,226
Frequency of treatment (%)	32.5	37.6	30.9
Mean period of observation	4.98	4.53	5.19
Maximum period of observation	6.22	6.23	6.23

[a]Regional population aged 50 years and over = 1,556.

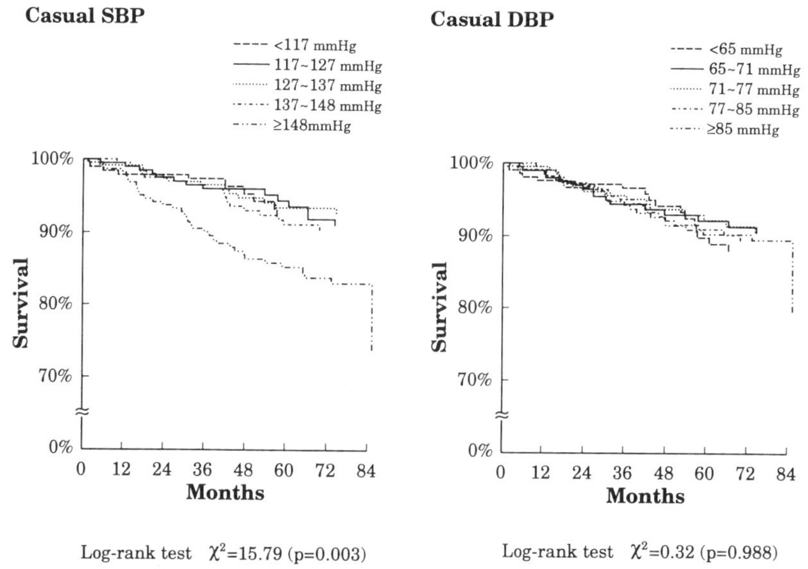

FIGURE 4. Kaplan-Meier estimate of event probability (all causes of mortality) based on screening blood pressure (BP) levels. Screening BP levels were classified into quintile, and survival rate is illustrated for each quintile.

rate was observed in the highest quintile of the SBP level, whereas no difference in survival rate was noted among the quintiles of the DBP level (FIG. 4). When ambulatory BP levels were classified into quintiles, a significant decrease in survival rate was observed in the highest quintile of the SBP level. The decrease in survival rate was observed in the lowest quintile of the ambulatory DBP level but not in the highest quintile, although this tendency was not statistically significant. A decrease in survival rate was clearly demonstrated in the highest quintile of the home SBP level during the 5-year period; a decrease in survival rate was observed in both the highest and the lowest quintiles of home DBP (FIG. 5).

The relative risk of all causes of mortality was analyzed by the Cox-proportional hazard model in subjects without antihypertensive treatment at least in the entry period of the prospective analysis. No significant risk ratio was observed in any quintile of the casual BP level (FIG. 6A). A significantly high risk ratio was observed in the lowest quintile of the ambulatory SBP and DBP levels but not in the highest quintiles (FIG. 6B). A significantly high risk ratio was observed in the highest and lowest quintiles of home DBP levels (J-shaped relationship) (FIG. 6C).

Cerebrovascular mortality was confirmed by death certificates, medical records, and computed tomography scan as well as magnetic resonance imaging. Cardiovascular mortality was confirmed by death certificates. As shown in FIGURE 7, cerebrovascular and cardiovascular mortality was higher in the highest and lowest quintiles of home DBP levels than in the remaining quintiles. This trend was not observed with casual BP levels. The incidence of cerebral hemorrhage tended to be greater in the highest quintile of home DBP, whereas that in cerebral infarction tended to be greater in the lowest and highest quintiles (J-shaped relationship).

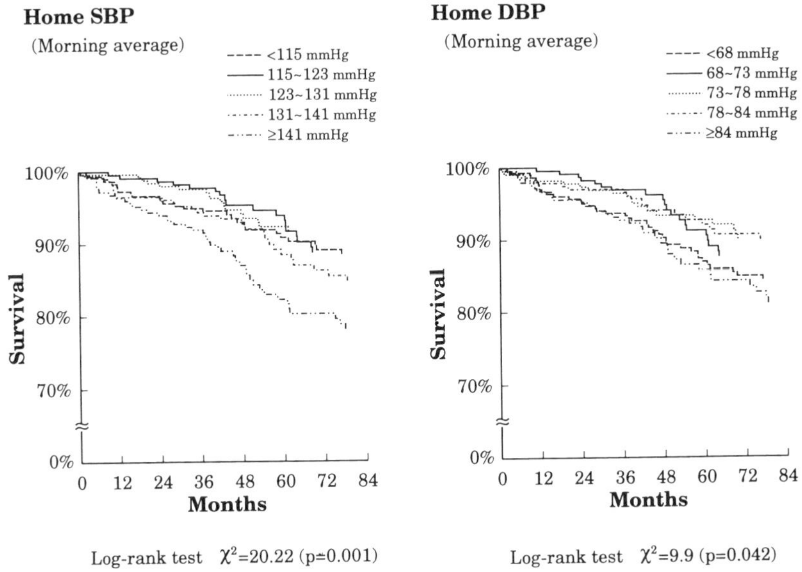

FIGURE 5. Kaplan-Meier estimate of event probability (all causes of mortality) based on home blood pressure (BP) levels. Home BP levels were classified into quintiles, and survival rate is illustrated for each quintile.

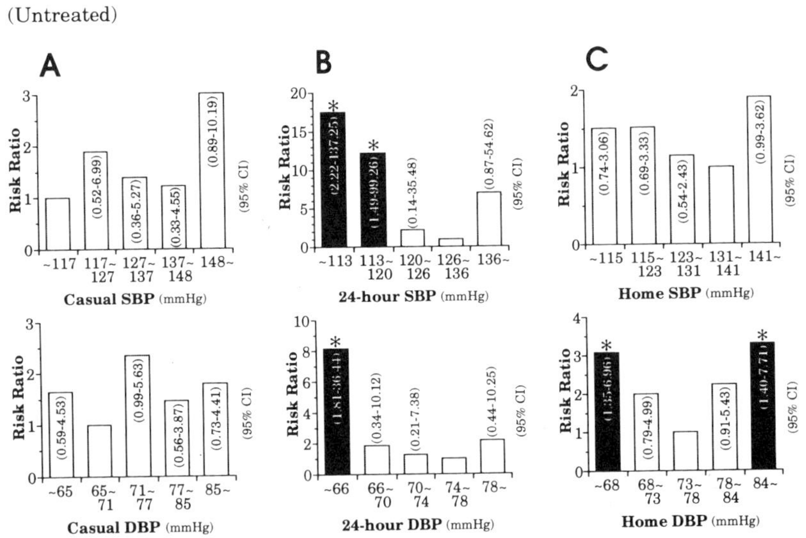

FIGURE 6. Relative risk of all causes of mortality analyzed by Cox proportional hazard model corrected for age and sex. Screening blood pressure (BP) levels (**A**), ambulatory BP levels (**B**), and home BP levels (**C**) were classified into the quintiles, respectively, and quintiles with the lowest hazard (risk ratio 1) were compared to the remaining quintiles. Numbers in columns represent confidence intervals.

NOCTURNAL BLOOD PRESSURE AND SILENT CEREBROVASCULAR LESIONS

We conducted a cross-sectional epidemiologic survey using ambulatory BP monitoring and brain magnetic resonance imaging (MRI) in the Ohasama cohort to determine if an inappropriately low nocturnal BP, or excess fall in nocturnal BP, may cause silent cerebrovascular lesions in the elderly.[48] Untreated subjects between the ages of 65 and 75 (29 men and 52 women, 91% of untreated regional population) participated in this survey. We evaluated the relationship between the amplitude (daytime average-nighttime average) of the fall in nocturnal BP and the incidence of silent cerebrovascular lesions.

Magnetic resonance images were obtained to evaluate the number of lacunae and the extent of periventricular signal abnormalities. Lacunae were defined as low signal intensity areas (3–10 mm) on T1-weighted images, which were visible as

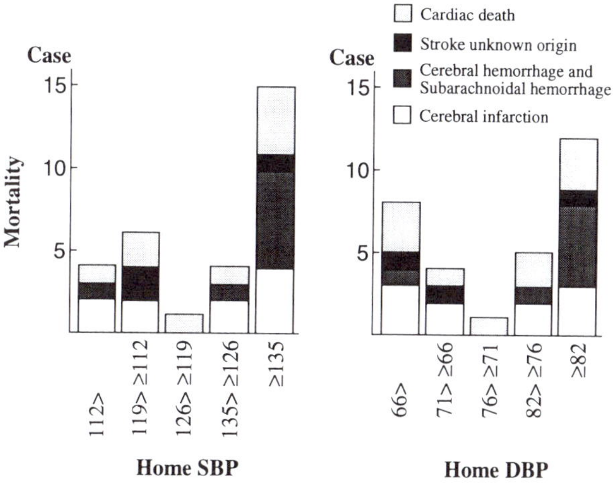

FIGURE 7. Cerebrovascular and cardiovascular mortality in each quintile of home blood pressure (BP) levels.

hyperintense lesions on the T2-weighted images. The number of lacunae in the supratentorial region was counted in each subject for classification into one of the following three groups: no lacunar infarction (Group A), 1–2 lacunar infarctions (Group B), and ≥ 3 lacunar infarctions (Group C).

Periventricular hyperintensities observed on the T2-weighted images were classified into five groups based on grade. When PVH was observed only in the region adjoining the lateral ventricle, it was scored as one point; when PVH was observed in the entire region from the lateral ventricle to the cortex, it was scored as 2 points. Frontal and occipital scores were added to give a score in the range of Grade 0 to 4.

Of the 81 elderly subjects evaluated, 41 had at least one lacunar infarction. The screening BP levels did not differ among subgroups that were classified according to the number of lacunar infarctions. The amplitude of the fall in nocturnal SBP in women with one to two lacunar infarctions (Group B) significantly exceeded that of women without lacunar infarction (Group A). The amplitude of the fall in nocturnal

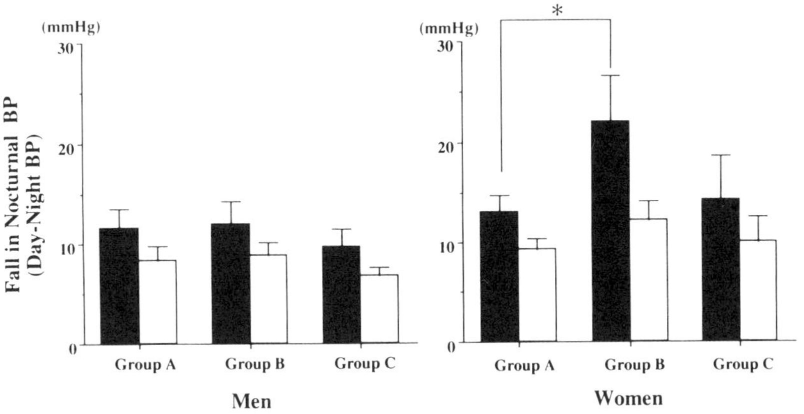

FIGURE 8. Relationship between number of lacunar infarctions and the amplitude of the reduction in nocturnal blood pressure (BP) in elderly subjects. Group A: no lacunar infarction; Group B: 1 to 2 lacunar infarctions; Group C: more than 3 lacunar infarctions. *Solid columns* represent SBP and *white columns* signify DBP.

BP in elderly women with one to two lacunar infarctions tended to be higher than that in women with more than three lacunar infarctions (Group C), but the difference was not statistically significant (FIG. 8).

The amplitude of the fall in nocturnal DBP did not differ among the subgroups of elderly women (FIG. 8) nor did the amplitude or rate of the fall in nocturnal BP differ among subgroups of elderly men (FIG. 8).

Of the 81 elderly subjects, 59 had PVH above Grade 1. The screening BP levels did not differ among the five PVH subgroups (ANOVA). The amplitude of the fall in nocturnal SBP in the elderly women increased as the grade of PVH increased (FIG.

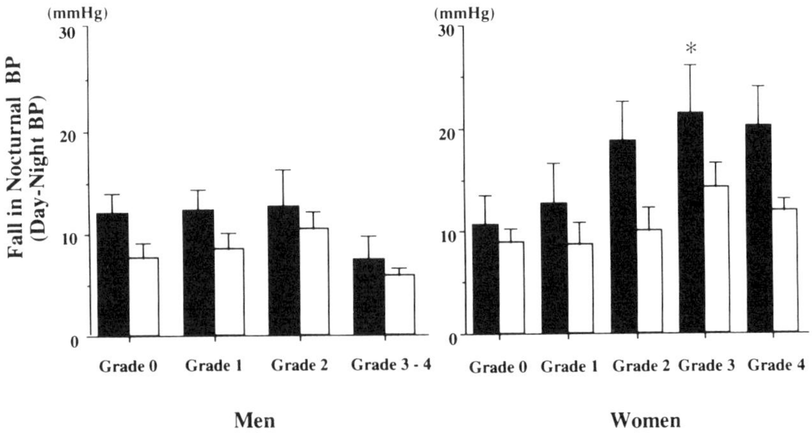

FIGURE 9. Relationship between periventricular hyperintensity (PVH) grade and amplitude of the reduction in nocturnal blood pressure (BP) in elderly subjects. Extent of PVH is expressed as Grade 0–4. *Solid columns* signify SBP and *white columns* represent DBP. $p < 0.05$ vs Grade 0.

9, $Y = 2.48X + 10.8, r = 0.32, p < 0.05$, where Y is the amplitude of the reduction in nocturnal BP and X is PVH grade). However, there was no correlation between the grade of PVH and the amplitude of the fall in nocturnal DBP.

The amplitude of the fall in nocturnal DBP in the elderly men with PVH greater than Grade 3 was smaller than that observed in the elderly men with Grade 2 PVH ($p < 0.05$) (FIG. 9). There was no difference in the amplitude of the fall in nocturnal SBP among subgroups of men with differing grades of PVH.

CONCLUSION

Our results clearly demonstrate that ambulatory BP monitoring and home BP measurements can predict the risk of low BP levels as well as high BP levels during a 5-year period of follow-up in a relatively small cohort, whereas casual screening BP cannot. Casual BP levels, which are subject to several biases including white coat effect,[17,18] tend to be higher than home BP levels or 24-hour ambulatory BP levels. However, these biased casual BP measurements reflected the prognosis in the population at large, suggesting that a reactive elevation of BP and/or lability of BP contribute to cardiovascular morbidity and mortality. However, casual BP measurements cannot determine true BP levels. Our findings which indicated a significantly higher cardiovascular risk in the lowest quintile of home and ambulatory BP levels may indicate a broader clinical significance of home and ambulatory BP monitoring than offered by casual BP measurements.

The relationship between a fall in nocturnal BP and silent vascular lesions in the elderly women in our study differs from that reported previously.[34,42] Although the present study per se is a cross-sectional one, and therefore the cause-and-effect relationship between a fall in nocturnal BP and silent cerebrovascular lesions is still uncertain, it should be emphasized that silent cerebrovascular lesions are more severe in elderly women with an extreme dipper pattern of circadian BP variation. The reason for this trend in elderly women is unknown. However, as shown in FIGURE 3, the circadian BP variation in elderly women differed from that in elderly men a priori. Some of the elderly women exhibited an extreme dipper pattern, whereas the elderly men exhibited non-dipper patterns (FIG. 3). The causal relationship between the reduction in nocturnal BP and silent cerebrovascular lesions will remain unclear until the cause of the difference in circadian BP variation between elderly men and women is clarified. The fall in nocturnal BP is generally thought to be mediated by a decrease in cardiac output[36] as well as by a potentiation of baroreflex function during sleep.[49] The baroreflex function acts to maintain BP during a decrease in cardiac output. Rapidly progressive atherosclerosis may disturb the baroreflex function in women with the disregulation of cardiac output during sleep, leading to a severe fall in nocturnal BP.

However, the Framingham study demonstrated lower rates of cardiovascular disease at all levels of SBP and DBP in women aged 35 to 94 years compared with men, which suggests that women tolerate hypertension better than men.[50] The risks of high BP levels may be masked in women, while the risks of low BP levels are more apparent.

In conclusion, results of our epidemiologic survey using ambulatory BP monitoring and home BP measurements suggest that low as well as high BP levels are associated with increased risk in the elderly population. Thus, ambulatory BP monitoring and home BP measurements are clinically significant predictors of the risk of low BP levels in the elderly. The risk of a low BP level should be determined in the general population as well as in patients receiving antihypertensive treatment.

SUMMARY

The mortality rate of stroke has decreased dramatically over the past 30 years in Japan, but the prevalence of cerebral infarction has increased, mainly because of the higher incidence of nonfatal cerebral infarction in the elderly. Significant advances in the development of antihypertensive drugs over the last 30 years have contributed to the decrease in stroke mortality, but antihypertensive treatment appears to relate to an increase in the prevalence of cerebral infarction in the elderly. We investigated the risks of low BP levels determined by ambulatory BP monitoring and home BP measurements to clarify the previously reported J-shaped relationships between the incidence of ischemic cardiovascular disease and BP levels after treatment. A long-term prospective study of all causes of mortality and cerebrovascular and cardiovascular mortality has been conducted in Ohasama, Japan, since 1987. Cox proportional hazard model after adjusting for age and sex demonstrated that during a 5-year follow-up period involving 893 of the subjects aged 50 years and over in this cohort, those with the lowest quintile of ambulatory BP levels exhibited a significantly high hazard ratio of cardiovascular as well as all causes of mortality. During the same follow-up period in 1,226 subjects aged 50 years and over, those with the lowest and highest quintiles of home BP levels demonstrated a significantly high hazard ratio of cardiovascular as well as all causes of mortality (i.e., J-shaped relationship). The amplitude of the fall in nocturnal BP obtained by ambulatory BP monitoring correlated positively with the severity of silent cerebrovascular lesions in elderly women, but not in elderly men. These results suggest the significantly high risk associated with low BP levels, which can be determined only by ambulatory and home BP measurements, but not by casual BP measurements.

REFERENCES

1. UEDA, K. & T. OMAE. 1992. The 12th Symposium of Hypertension, Clinical and Experimental. T. Omae, Ed.: 1–11. Shindan-to-Chiryosha. Tokyo.
2. OMAE, T. & K. UEDA. 1988. J. Hypertens. **6:** 343–349.
3. KAMEYAMA, M. 1992. Jikken-Chiryo **621:** 127–130 (in Japanese).
4. FARNETT, L., C. D. MULROW, W. D. LINN *et al.* 1991. JAMA **265:** 489–495.
5. FLETCHER, A. E. & C. J. BULPITT. 1992. N. Engl. J. Med. **326:** 251–254.
6. CRUICKSHANK, J. M. 1988. Br. Med. J. **297:** 1227–1230.
7. IRIE, K., T. YAMAGUCHI, K. MINEMATSU & T. OMAE. 1993. Stroke **24:** 1844–1849.
8. MACMAHON, S., R. PETO, J. CUTLER *et al.* 1990. Lancet **335:** 765–774.
9. SHEP COOPERATIVE RESEARCH GROUP. 1991. JAMA **266:** 3255–3264.
10. THIJS, L., R. FAGARD, P. LIJNEN *et al.* 1992. J. Hypertens. **10:** 1103–1109.
11. BULPITT, C. J. & A. E. FLETCHER. 1992. J. Hypertens. **10**(Suppl. 7): 45–49.
12. HEIKINHEIMO, R. J., M. V. HAAVISTO, R. H. KAARELA *et al.* 1990. J. Hypertens. **8:** 361–367.
13. KANNEL, W. B., T. R. DAWBER & D. L. MCGEE. 1980. Circulation **61:** 1179–1182.
14. THE WORKING GROUP ON RISK AND HIGH BLOOD PRESSURE. 1985. Hypertension **7:** 641–651.
15. STAESSEN, J. A., L. THIJS, D. CLEMENT *et al.* 1994. J. Hypertens. **12:** 1035–1039.
16. IMAI, Y. 1995. Tohoku J. Exp. Med. **176:** 1–15.
17. IMAI, Y., K. NAGAI, M. SAKUMA *et al.* 1993. Hypertension **22:** 900–912.
18. IMAI, Y., H. SATOH, K. NAGAI *et al.* 1993. J. Hypertens. **11:** 1441–1449.
19. JULIUS, S., C. N. ELLIS, A. V. PASCUAL *et al.* 1974. JAMA **229:** 663–666.
20. CONWAY, J. 1986. Clin. Exp. Hypertens. [A] **8:** 1247–1294.
21. SOKOLOW, M., D. WERDEGER, H. K. KAIN & A. T. HINMAN. 1966. Circulation **34:** 279–307.
22. DEVEREUX, R. B., T. G. PICKERING, G. A. HARSHFIELD *et al.* 1983. Circulation **68:** 470–476.

23. Parati, G., G. Pomidossi, F. Albini *et al.* 1987. J. Hypertens. **5:** 93–98.
24. White, W. B., P. Schulman, E. J. McCabe & H. M. Dey. 1989. JAMA **261:** 873–877.
25. Perloff, D., M. Sokolow & R. Cowan. 1983. JAMA **249:** 2792–2798.
26. Pickering, T. G. & G. D. James. 1994. J. Hypertens. **12**(Suppl. 8): 29–33.
27. The Joint National Committee on Detection, Evaluation, and Treatment of High Blood Pressure. 1993. Arch. Intern. Med. **153:** 154–183.
28. Zanchettii, A., J. Chalmers, K. Arakawa *et al.* 1993. ISH Hypertension News, Special Edition. 3–16.
29. Parati, G., G. Pomidossi, F. Albini *et al.* 1987. J. Hypertens. **5:** 93–98.
30. Frattola, A., G. Parati, C. Cuspidi *et al.* 1993. J. Hypertens. **11:** 1133–1137.
31. Verdecchia, P., G. Schillaci, M. Guerrieri *et al.* 1990. Circulation **81:** 528–536.
32. Kuwajima, I., Y. Suzuki, T. Shimosawa *et al.* 1992. Am. Heart J. **123:** 1307–1311.
33. O'Brien, E., J. Sheridan & K. O'Malley. 1988. Lancet **2:** 397.
34. Shimada, K., A. Kawamoto, K. Matsubayashi *et al.* 1990. Hypertension **16:** 692–699.
35. Imai, Y., M. Munakata, J. Hashimoto *et al.* 1993. Am. J. Hypertens. **6:** S179–S183.
36. Imai, Y., K. Abe, M. Munakata *et al.* 1990. J. Hypertens. **8:**(Suppl. 7): 125–132.
37. Pantano, P., J. C. Baron, P. Lebrun-Grandie *et al.* 1984. Stroke **15:** 635–641.
38. Shaw, T. G., K. F. Mortel, J. S. Meyer *et al.* 1984. Neurology **34:** 855–862.
39. Fujii, K., S. Sadoshima, Y. Okada *et al.* 1990. Stroke **21:** 283–290.
40. Madsen, P. L. & S. Vorstrup. 1991. Cerebrovasc. Brain Metab. Rev. **3:** 281–296.
41. Sakai, F. 1990. Rinsho-Shinkei (Clin. Neurol.) **30:** 1342–1344 (in Japanese with English abstract).
42. Tohgi, H., K. Chiba & M. Kimura. 1991. Stroke **22:** 603–608.
43. Imai, Y., K. Abe, S. Sasaki *et al.* 1992. Clin. Exp. Hypertens. **A14:** 1141–1167.
44. Kario, K. & K. Shimada. 1994. Lancet **344:** 200.
45. Strandgaard, S., J. Olesen, E. Skinhoj & N. A. Lassen. 1973. Br. Med. J. **1:** 507–510.
46. Fujishima, M., S. Ibayashi, K. Fujii & S. Mori. 1995. Hypertens. Res. **18:** 111–117.
47. The National Institute of Neurological Disorders and Stroke. 1990. Stroke **21:** 637–676.
48. Watanabe, N., Y. Imai, M. Sakuma *et al.* 1994. Hypertens. Res. **17:** 269 (abstr.).
49. Conway, J., N. Boon, J. Vann Jones & P. Sleight. 1983. Hypertension **5:** 746–748.
50. Levy, D. & W. B. Kannel. 1988. Am. Heart J. **116:** 266–272.

Aging and Blood Pressure Rhythms[a]

EOIN O'BRIEN

The Blood Pressure Unit
Beaumont Hospital
Dublin, Ireland

Elderly patients with hypertension constitute a group within the hypertensive population deserving of special attention. Not alone are the elderly susceptible to the adverse effects of blood pressure lowering drugs,[1] but blood pressure measurement in the elderly presents particular problems which, if not recognized, lead inevitably to incorrect diagnosis and unnecessary treatment.[2]

The prevalence of hypertension increases with age,[3] and hypertension is a major risk for the development of cardiovascular disease in elderly patients.[4] Furthermore, the European Working Party on Hypertension in the Elderly has shown that antihypertensive therapy is beneficial in reducing the cardiovascular complications of hypertension in elderly patients.[5] It has been estimated that more than 40% of elderly hypertensive patients receive regular antihypertensive medication,[6] and recently the accuracy of diagnosis which destines so many elderly people to life-long medication has been questioned.[2,7]

After nearly 20 years in the research laboratory, ambulatory blood pressure measurement (ABPM) has at last passed from research to clinical practice and is now accepted as a valuable procedure in the clinical management of hypertension.[8,9] ABPM permits more accurate characterization of blood pressure rhythms with aging and has a special role in the management of hypertension in the elderly. Firstly, the technique may permit more accurate diagnosis of hypertension,[10–12] especially in patients with isolated systolic hypertension.[13–16] ABPM also provides a better assessment of antihypertensive drug efficacy in elderly patients,[17] thereby avoiding overtreatment[1,8] which may have devastating consequences in older hypertensive subjects.[19] ABPM should prove to be a better technique than conventional measurement in predicting cardiovascular outcome in the elderly,[12,20] and one study is presently being conducted to determine this issue.[2,16,21] Finally, ABPM is also proving helpful in evaluating elderly patients with hypotension.[22]

TECHNICAL ASPECTS OF AMBULATORY BLOOD PRESSURE MEASUREMENT IN THE ELDERLY

This subject has been reviewed for ambulatory blood pressure measuring devices in general[23] and in the elderly in particular.[24,25] Since the introduction of 24-hour ambulatory blood pressure measurement into clinical practice, a large market for devices to measure ambulatory blood pressure was created. Forty-three such devices from 31 manufacturers or suppliers are now available to satisfy a market demand which is likely to increase. Of the 43 available devices, 18 have been validated according to the protocols of either the Association for the Advancement of Medical Instrumentation (AAMI) and/or the British Hypertension Society (BHS) in 25

[a]This work was supported by the Charitable Infirmary Charitable Trust, Royal College of Surgeons in Ireland, and Clinical Investigation Committee of Beaumont Hospital.

reported studies.[23] From this review of 43 devices on the market, it may be concluded that at the time of this writing, there is published evidence for only nine devices meeting the generally accepted criteria for accuracy and performance; these are: the A & D TM 2420 Models 6 and 7 and TM 2421, CH-Druck, Nissei ABPM DS-240, Profilomat, QuietTrak, SpaceLabs SL-90202, and SpaceLabs SL-90207 (TABLE 1).[23] However, these considerations apply to the general population, and evidence indicates that ABPM devices may perform less accurately in the elderly. In fact, the BHS, in its revised protocol for validating blood pressure measuring systems, has devoted a subsection to validation of devices in the elderly.[26] This recommendation is based on evidence that age may affect device accuracy,[27] and we have shown that device accuracy worsens with increasing blood pressure levels.[28]

Miller and his colleagues[27] showed a significant discrepancy between ABPM and a mercury standard which was systematically related to age. However, it would be unwise to apply their findings to all ambulatory systems, particularly as the device used in the study, the Accutracker II, has failed to fulfill the AAMI criteria.[27,29]

TABLE 1. Nine ABPM Devices That Have Fulfilled BHS[a] and AAMI[b] Accuracy Criteria[23] (Assessment in January 1995)

Device	AAMI Criteria	BHS[b] Criteria
CH-Druck	Passed	A/A[77]
Profilomat	Passed	B/A[78]
Nissei DS-240	Passed	B/A[79]
QuietTrak	Passed	B/B[81,82]
SpaceLabs 90202	Passed	B/B[73]
SpaceLabs 90207	Passed	B/B[74,83,84]
TM-2420 Model 6	Passed	B/B[49]
TM-2420 Model 7	Passed	B/B[70]
TM-2421	Passed	B/A[71]

Abbreviations: AAMI = Association for the Advancement of Medical Instrumentation; BHS = British Hypertension Society.

[a]Criteria for fulfillment of the BHS protocol: devices must achieve at least grade B/B.

[b]Criteria for fulfillment of the AAMI standard: mean difference ≤ 5 mm Hg/SD ≤ 8 mm Hg.

NORMAL AMBULATORY BLOOD PRESSURE MEASUREMENT IN THE ELDERLY

Discussion of abnormality is not a reasonable proposition until normality has been defined, yet how often in clinical science we attempt to grapple with the abnormal and so gain insight into normality. Such has been the case until recently with ABPM. However, the process of defining normality has come under careful scrutiny in recent years, and studies are being conducted to define normality in terms of outcome. In the context of the elderly, attention should be drawn to an important issue in considering normal versus abnormal pressure, namely, the definition of "elderly." Conventional definitions are based on a chronologic division of the nonelderly from the elderly at 60 or 65 years, and recently the very elderly were defined as those above the age of 80 years. With the conventional technique of measuring blood pressure with a mercury sphygmomanometer and stethoscope, chronologic definitions did not present a problem because the circumstance of blood pressure measurement was always static with the subject sitting or lying at rest.

However, with ambulatory measurement the subject is active and the amount of activity performed by elderly subjects may vary greatly. In considering ABPM in the elderly, it may be necessary, therefore, to broaden the chronologic classification to include a functional classification, such as the active healthy elderly, the less active ailing elderly, and finally the disabled immobile elderly. ABPM will not be the same across all three categories, and consideration must be given to this eventuality in future research and discussion of 24-hour blood pressure profiles in the elderly. There are three basic approaches to the issue of ABPM normality in any age group.[30]

Relationship of ABPM to Morbidity, Mortality, and Target Organ Involvement

Longitudinal Outcome Studies

Establishing the relationship of ABPM to the risk of heart attack, stroke, and death in longitudinal studies is the classic epidemiologic approach used in studies, such as the Framingham[31] and MRFIT[32] studies, in which clinic blood pressure levels were related to risk. A longitudinal study comparing ABPM with office blood pressure, which was first reported by Sokolow's group[33] in 1964 and subsequently in the intervening years[34–36] showed that the relation between risk and ABPM is stronger for ABPM than for clinic pressure in keeping with the prediction that such would be the case for true blood pressure as distinct from clinic blood pressure.[37]

The Office versus Ambulatory (OvA) study which was established as a prospective investigation of the correlation between ambulatory blood pressure measurement and prognosis in patients with primary hypertension undergoing treatment should determine if ABPM is a more sensitive predictor of outcome than is office blood pressure.[38]

In the Syst-Eur trial, another multicenter longitudinal study being conducted in Europe to determine the value of treating isolated systolic hypertension in the elderly, there is a side project devoted to determining the predictive value of 24-hour ABPM in elderly patients with isolated systolic hypertension.[39] As with the OvA study, this study should yield important data on the predictive characteristics of ABPM and the influence it might have on modifying therapeutic decisions in elderly patients with isolated systolic hypertension.

The Ambulatory Blood Pressure and Treatment of Hypertension (APTH) study was established to test the hypothesis that antihypertensive treatment based on ABPM may be more beneficial than treatment guided by conventional sphygmomanometry.[40]

In the Italian Pamela study in Monza, 2,400 subjects randomly selected from the population will have conventional blood pressure measurement, home blood pressure measurement, ABPM, and echocardiography performed, among other clinical and investigatory procedures.[41]

Verdecchia and his colleagues[43] prospectively followed up 1,187 subjects with hypertension for up to 7.5 years and showed that ABPM stratified cardiovascular risk independently of other traditional risk markers including echocardiographic left ventricular hypertrophy.

Target Organ Studies

Target organ involvement in hypertensive patients, especially echocardiographically demonstrated left ventricular hypertrophy, carries a poor prognosis. Conventional blood pressure measurement correlates with target organ damage and mortality. Growing evidence indicates that 24-hour ABPM is a better predictor of target organ involvement than is conventional blood pressure measurement. Moreover, ABPM permits the detection of white coat hypertension which, more than any other factor, makes conventional measurement so unreliable.[44]

From a review of the relationship of ABPM to target organ involvement in relation to the heart, kidney, brain, and arterial tree, little doubt exists that ABPM is superior to conventional measurement in predicting target organ involvement. How readily this data can be applied to the overall outcome of hypertension is for the moment debatable.[44]

Relationship of ABPM to Clinic Blood Pressure

A further approach in establishing the normal values of ABPM is to relate 24-hour pressures to clinic blood pressures and, as the risk for clinic blood pressure is known, to extrapolate from this relationship how ABPM relates to risk. This approach has been adopted in three studies.[45–47] From the evidence available to date, a continuous relationship appears to exist between ambulatory pressure levels and the incidence of cardiovascular complications. Thus, defining an operational threshold for ambulatory monitoring will necessitate consensus on arbitrary pressure limits in clinical practice, exactly as it has been the case for conventional sphygmomanometry.[8]

ABPM Levels in Normal Populations

An alternative option to the foregoing approaches is to determine the distribution of 24-hour blood pressures in normal population or community samples defining levels according to centiles or standard deviations, above or below the mean values for a given population. This approach has now been widely utilized for different nationalities, and reference tables showing normal levels for different ages and gender have been produced.[9,47,49,50–56] In considering such studies, it is necessary to categorize subjects in special groupings, *viz.,* adults and the elderly.

Adults. The first major study of ABPM in a population sample was the Allied Irish Bank (AIB) study[50] of 815 healthy bank employees (399 men and 416 women), 17 through 79 years old. The results of the AIB study are summarized in TABLE 2. Quite definite differences in relation to age and sex were observed. For all age groups men had higher 24-hour daytime and nighttime pressures.

In Belgium, a random sample of 328 healthy subjects aged 20–81 years were recruited for ABPM. The 24-hour blood pressures in this sample were almost identical to those obtained in the AIB study.[51,54] These findings were unchanged in a later follow-up study of a larger sample of 718 subjects.[47,56]

Metaanalysis of data from 19 studies performed in Europe, the United States, and the Far East and Australia was performed by Staessen and his colleagues.[53] The largest population analysis yet undertaken is that comprising data from 24 research groups contributing 24-hour values from 7,069 subjects in 12 countries.[49] The results are summarized in TABLE 3. Many experts would agree that ambulatory measurements more closely reflect a patient's true pressure than do conventional readings,[57,58] and the large international analysis suggests that using conventional blood pressure measurement to make clinical decisions may lead to misclassification of patients with hypertension. From the large amount of data available in these

TABLE 2. The AIB Study: Mean (±SD) Office and 24-Hour Ambulatory Blood Pressure in 815 People According to Age and Sex[50]

	Men				All	Women				All	Both
Age range	17–29	30–39	40–49	50–79	17–79	17–29	30–39	40–49	50–79	17–79	17–79
n	107	123	109	60	399	174	149	55	38	416	815
Office Measurements											
SBP	121 ± 12	122 ± 11	125 ± 16	133 ± 15	124 ± 14	110 ± 11	113 ± 10	121 ± 17	130 ± 24	115 ± 15	119 ± 15
DBP	73 ± 9	77 ± 8	81 ± 10	85 ± 11	78 ± 10	71 ± 8	72 ± 8	78 ± 9	81 ± 12	73 ± 9	76 ± 10
Ambulatory Measurements											
Day											
SBP	129 ± 8	128 ± 9	129 ± 12	132 ± 12	129 ± 10	118 ± 8	117 ± 8	121 ± 12	126 ± 18	118 ± 10	124 ± 12
DBP	77 ± 7	80 ± 6	83 ± 9	84 ± 9	81 ± 8	74 ± 6	75 ± 7	76 ± 9	78 ± 9	75 ± 7	78 ± 8
Night											
SBP	110 ± 9	108 ± 8	109 ± 12	113 ± 13	110 ± 10	102 ± 9	101 ± 9	103 ± 11	108 ± 12	102 ± 9	106 ± 11
DBP	59 ± 6	62 ± 6	66 ± 10	68 ± 9	63 ± 8	57 ± 6	58 ± 7	61 ± 8	63 ± 7	58 ± 7	61 ± 8
24-Hour											
SBP	123 ± 8	121 ± 8	122 ± 11	126 ± 12	123 ± 10	112 ± 7	111 ± 8	114 ± 10	120 ± 15	113 ± 9	118 ± 11
DBP	71 ± 5	74 ± 5	77 ± 8	79 ± 9	75 ± 7	68 ± 5	69 ± 6	71 ± 8	73 ± 8	69 ± 6	72 ± 7

Abbreviations: SBP, DBP = systolic and diastolic pressures (mm Hg), respectively.

TABLE 3. Mean and Standard Deviation of Blood Pressure in Normotensive[a] Subjects in Various Studies[49]

	n	Conventional Pressure	Ambulatory Pressure 24 Hours	Day	Night
Systolic pressure (mm Hg)					
Belgian population	574	119 ± 10	116 ± 8	122 ± 9	105 ± 9
AIB Study	806	117 ± 11	116 ± 8	122 ± 10	105 ± 9
Metaanalysis	3,414	na	118 ± 11	123 ± 11	106 ± 10
International database	4,577	119 ± 12	116 ± 10	122 ± 11	106 ± 11
Diastolic pressure (mm Hg)					
Belgian population	574	74 ± 8	70 ± 6	75 ± 7	60 ± 7
AIB Study	806	75 ± 8	71 ± 6	77 ± 7	60 ± 7
Metaanalysis	3,414	na	72 ± 8	76 ± 8	64 ± 9
International database	4,577	73 ± 9	70 ± 7	75 ± 8	61 ± 8

Note: Values are means ± standard deviation. na = not available.
[a]Normotension was a conventional pressure not exceeding 140 mm Hg systolic and 90 mm Hg diastolic.

population studies, it is possible to draw up operational thresholds of normalcy for use in clinical practice (TABLE 4).[30]

The Elderly. In the AIB study, blood pressures were influenced by gender and age. For example, mean daytime blood pressure in women aged 17–29 years averaged 117 ± 8 mm Hg, whereas in women aged 50–79 years, mean daytime pressure averaged 126 ± 18 mm Hg and daytime pressure in men aged 50–79 years averaged 132 ± 12 mm Hg.[50] Most of the subjects in the AIB study were under the age of 65 years, and there is clearly a need to characterize 24-hour blood pressure levels in the elderly population. Longitudinal studies relating 24-hour pressures to outcome are urgently needed, therefore, in elderly subjects. Many of the ongoing longitudinal studies will provide information on this issue. The ambulatory side-project of the Syst-Eur study, in particular, should provide data for the elderly with isolated systolic hypertension.[39] Another study, the Hypertension in the Very Elderly (HYVET) study, also has an ambulatory side-project which should provide interesting data in relation to hypertensive patients above the age of 80 years.[59]

ABNORMAL AMBULATORY BLOOD PRESSURE MEASUREMENT PATTERNS IN THE ELDERLY

ABPM has a number of uses in clinical practice which include the diagnosis of borderline systodiastolic hypertension, the exclusion of white coat hypertension, and

TABLE 4. Suggested Operational Thresholds for 24-Hour Ambulatory Pressure Rounded for Ease of Use in Clinical Practice[30]

	Possible Hypertension	P95		Definitely Hypertensive
Daytime pressure	135/85	131/81		≥140/90
Nighttime pressure	120/70	123/74		≥125/75
24-Hour pressure	130/80	131/81		≥135/85
	0 or 5 mm Hg	Round down	Round up	0 or 5 mm Hg

the diagnosis of secondary hypertension and resistant hypertension. However, in the elderly ABPM has a number of indications that do not generally apply to younger patients.

Diagnosis of Hypertension

As in younger patients, ABPM confirms if elevated blood pressure, either systolic, diastolic, or both, is sustained during the day- or nighttime period. It is not possible to predict using conventional measurement those patients with the white coat reaction, and it becomes especially important in patients with borderline elevation of blood pressure to determine if the observed elevation on conventional measurement is sustained with ABPM.

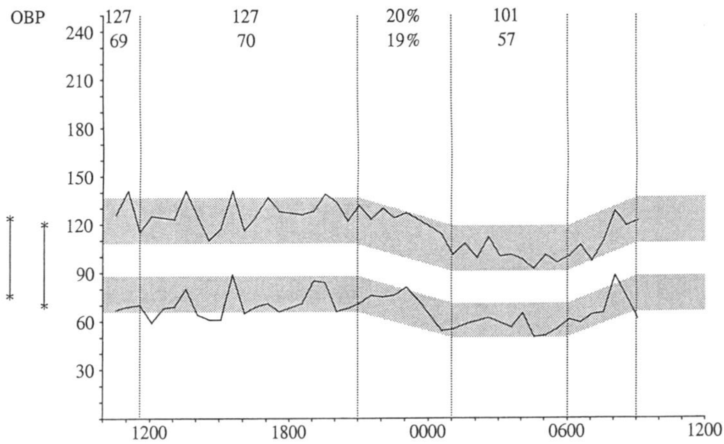

FIGURE 1. Scheme for presentation of data showing normal 24-hour pressures. Blood pressure levels are on the vertical axis, and time of day is plotted on the horizontal axis. The horizontal bands indicate the upper and lower limits of 24-hour pressures taken from the Allied Irish Bank study.[50] OBP = office blood pressure. The area between the first vertical lines is the "white coat window" in which any white coat effect may be carried into the ABPM recording; this is followed by the daytime window, the retiring window, the nocturnal window, and the awakening window. Mean blood pressures are indicated for each window on the *upper line* with the percentage nocturnal dip in blood pressure. The plot shown is clearly normal.

White Coat Hypertension. White coat hypertension, the phenomenon whereby office blood pressures are elevated but may fall to normal levels during ABPM, is common, affecting some 20% of patients with hypertension.[60] In discussing white coat hypertension, it is important to establish definitions so as to avoid confusion. We propose that the term *white coat hypertension* be reserved for patients who exhibit elevated office blood pressures which settle within a short time to normal daytime mean blood pressure (FIG. 2) and that the term *white coat effect* be reserved for patients who exhibit elevated office blood pressures which reduce with ABPM but do not lower the mean daytime blood pressure to normal (FIG. 3).

Blood pressure variability is increased in the elderly,[2] and some evidence indicates that both white coat hypertension and the white coat effect are more

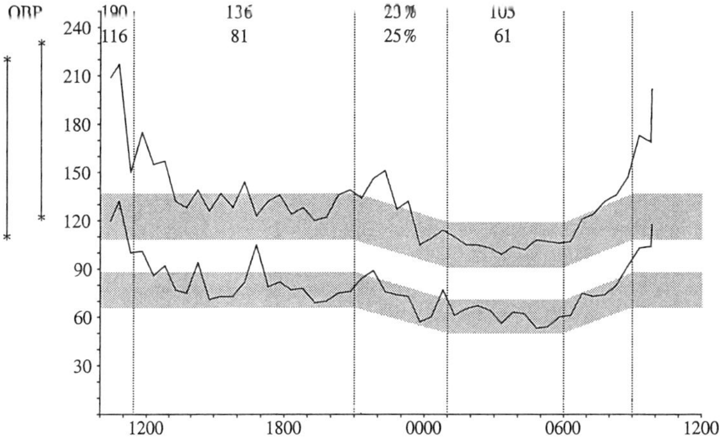

FIGURE 2. White coat hypertension. This patient has classic white coat hypertension with office pressure of 220/120 mm Hg and mean pressure in the white coat window of 190/116 mm Hg, but these pressures later settle to give mean daytime pressure of 136/81 mm Hg and nighttime pressure of 105/61 mm Hg.

prevalent in older subjects,[14,15,61,62] thus raising the important question as to how much reliance can be placed on conventional office blood pressure measurement in the diagnosis of hypertension in the elderly.[63] In fact, if we accept that the most effective method for diagnosing white coat hypertension and the white coat effect in younger patients with hypertension is by ABPM, then the argument in favor of applying the technique to the elderly becomes all the more compelling if the

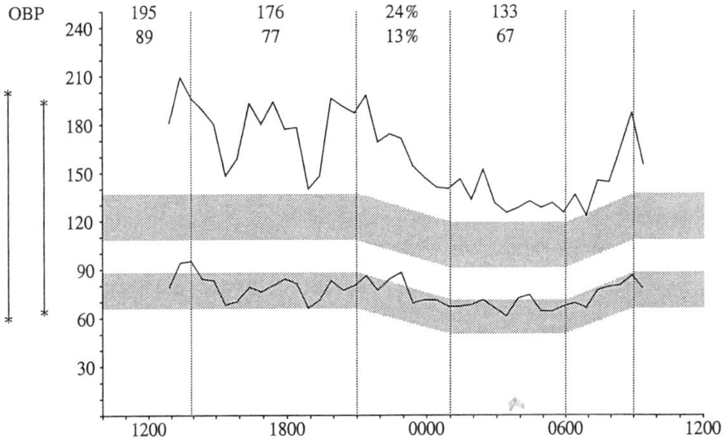

FIGURE 3. White coat effect. This patient has office pressure as high as 200/60 mm Hg with mean pressure in the white coat window of 195/89 mm Hg. During the day pressure settles to a mean level of 176/77 mm Hg and there is a 24/13% nocturnal dip in pressure to give mean nighttime pressure of 13/67 mm Hg. This is an example of the white coat effect in a patient with isolated systolic hypertension.

prevalence of the white coat phenomenon is greater in this age group. The case becomes even more persuasive when consideration is given to the issue of protecting elderly patients from unnecessary or excessive antihypertensive medication while being conscious of not withholding treatment that may have its greatest benefit in this age group. It is our practice to perform 24-hour ABPM in all elderly patients to exclude white coat responders, while accepting that the white coat response may not be a normal phenomenon but one that merits continuing observation but not necessarily antihypertensive medication.

Isolated Systolic Hypertension. The vascular, structural, and functional factors that lead to a tendency for elderly patients to develop predominantly systolic hypertension were reviewed by Messerli and colleagues.[2] The results of the SHEP study indicate considerable benefit in reducing isolated systolic hypertension in the elderly[64] but these results may not be applicable to all populations and the Syst-Eur study, which embodies in its design a side-project on 24-hour ABPM, is presently being conducted in Europe.[21] Until recently, therefore, the information on which we base therapeutic and management decisions in elderly patients with hypertension has been derived from conventional blood pressure measurement which may be influenced by factors common to all age groups with hypertension, such as the white coat effect and hypotension, and certain factors peculiar to the elderly, such as isolated systolic hypertension and pseudohypertension.[2]

Isolated systolic hypertension is a common condition of the elderly, occurring in 8% of people in the age stratum 60–69 years and rising to 22% in those aged 80 years or older.[8] These percentage estimates translate into a daunting reality that nearly 4 million Americans aged 60 and older have isolated systolic hypertension[64] and are likely to be considered candidates for antihypertensive medication. However, it must be borne in mind that all such statistics derive from data based on conventional blood pressure measurement. Although there is no denying the risk of isolated systolic hypertension,[3] a case can perhaps be made along the lines that while it is likely that benefit will be derived from the treatment of isolated hypertension, not all who receive such treatment need it. Is there any evidence to suggest that such an assertion can be upheld?

In 1990, Silagy and colleagues[14] demonstrated that in a small number of elderly subjects with isolated systolic hypertension, mean ambulatory blood pressures were consistent with that diagnosis for only 8% of the daytime period. In another study Cox and colleagues[15] also showed a marked discrepancy between clinic and daytime pressures in patients with isolated systolic hypertension, with the former being 27 mm Hg higher than ambulant pressures. These results were confirmed in a preliminary analysis of the ambulatory side-project of the Syst-Eur Study.[39] These findings suggest that isolated systolic hypertension is not always a sustained phenomenon and that in many patients the observed rise in pressure in the office is a manifestation of an alerting reaction.[65] We therefore have two forms of isolated systolic hypertension on ABPM, "sustained" isolated systolic hypertension and "transient" isolated systolic hypertension (FIGS. 4 and 5). It may be postulated that antihypertensive medication could be targeted particularly at those elderly patients with sustained isolated hypertension on ABPM. However, before making such a recommendation, it would be necessary to know if cardiovascular risk was predicted more accurately by ABPM than by conventional measurement.

An increasing number of studies have demonstrated that ambulatory pressures correlate more closely than clinic pressures with several different indices of target organ damage.[66–68] In a preliminary analysis of ambulatory measurements of blood pressure carried out on 46 patients in the run-in phase of the ambulatory monitoring side-project of the Syst-Eur study, ECG voltage criteria for left ventricular hypertro-

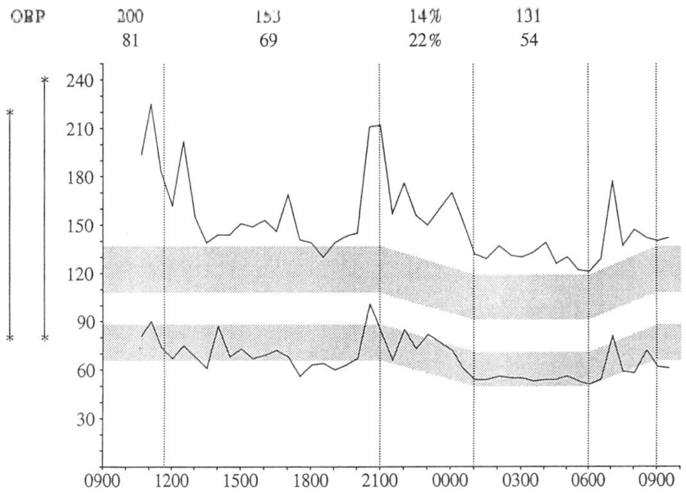

FIGURE 4. Transient isolated systolic hypertension. This appears to show severe isolated systolic hypertension on office pressures which reach 240/80 mm Hg, and this is carried into the white coat window where pressure averages 200/81 mm Hg. However, once the patient is away from the medical environment, pressure falls to give daytime pressures averaging 153/69 mm Hg and there is a 14/22% nocturnal dip in blood pressure to give nighttime pressure averaging 131/54 mm Hg.

phy correlated more with ABPM than with office blood pressure, suggesting that ambulatory measurement is a better predictor of left ventricular hypertrophy than is clinic measurement in elderly patients with hypertension.[69,70] It is anticipated that the Syst-Eur study should provide the answer to this important question before long.

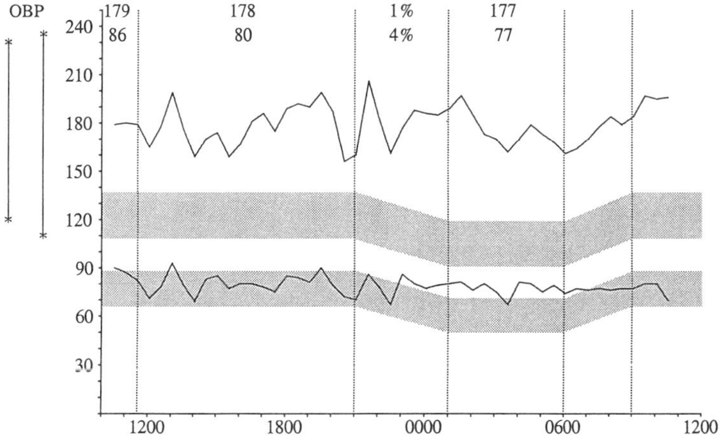

FIGURE 5. Sustained isolated systolic hypertension. In this example, daytime systolic pressures are persistently elevated, averaging 178/80 mm Hg and the patient may also be classified as a "non-dipper" in that nighttime pressures only fall by 1/4% to give average nocturnal pressures of 177/77 mm Hg.

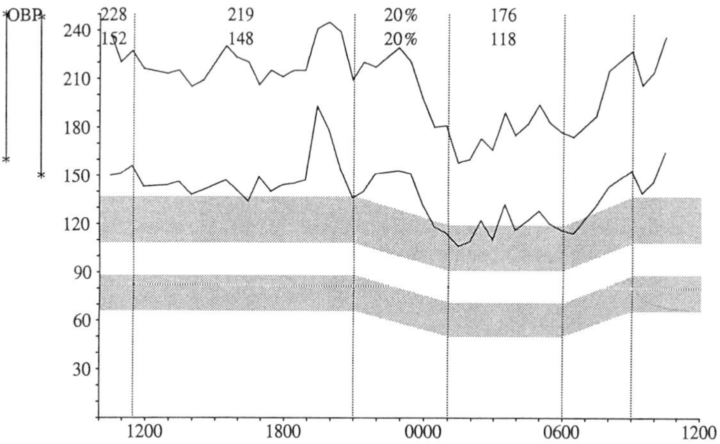

FIGURE 6. Hypertensive "dipper." This patient has severe systodiastolic hypertension with a mean daytime pressure of 219/148 mm Hg with a 20% nocturnal dip, giving mean nighttime pressures of 176/118 mm Hg.

Nocturnal Dipper Status. There is now growing evidence that hypertensive patients whose nocturnal blood pressure declines—so-called "dippers" (FIG. 6)—have less cardiovascular damage and consequent risk than do the minority of patients whose blood pressure fails to reduce at night—"non-dippers" (FIG. 7).[66,67,71,72] Yet little evidence suggests that blood pressure patterns in elderly patients during sleep differ greatly from those in younger age groups,[18] but as the quality of sleep alters with advancing years it may not be surprising if differences between age groups emerge as more data become available. The only means of determining a patient's nocturnal dipping status is ABPM.

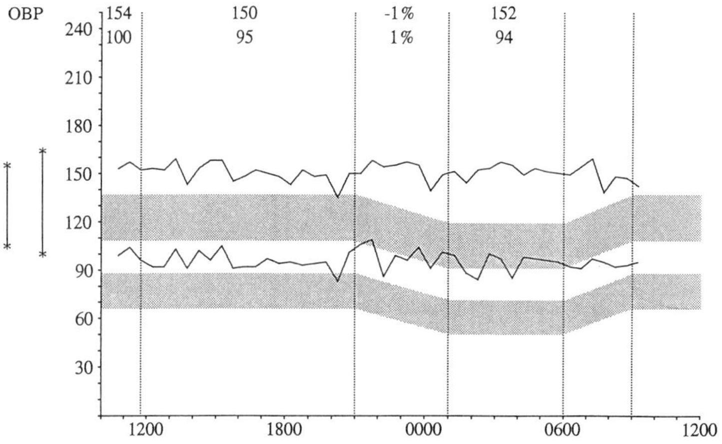

FIGURE 7. Hypertensive "non-dipper." This patient has moderate systodiastolic hypertension with daytime pressure averaging 150/95 mm Hg and is a "non-dipper" in that nighttime pressures show no decline, averaging 152/94 mm Hg.

Other Diagnostic Uses. In the elderly ambulatory measurement may also be helpful in identifying episodic hypertension in pheochromocytoma and in characterizing blood pressure behavior in patients with secondary hypertension in whom the diurnal dipping pattern may be lost, such as renal artery stenosis which becomes increasingly common with advancing years.

AMBULATORY BLOOD PRESSURE MEASUREMENT AND TREATMENT OF HYPERTENSION IN THE ELDERLY

ABPM may prove to be a valuable technique in guiding the prescribing physician in the choice of antihypertensive drug most suited for an individual patient regardless of age, although this role for ABPM in clinical practice awaits further study.[73] In this regard, however, ABPM may prove to be especially helpful in the elderly.[7,74] The technique permits determination of those elderly patients with elevated office blood pressure due to white coat hypertension or the white coat effect, and treatment can be selected for patients with sustained hypertension on ABPM. Moreover, ABPM provides an assessment of the entire 24-hour profile, thereby allowing determination of dipper status and selection of a short-acting drug for elderly patients with marked nocturnal reduction of blood pressure, or alternatively, selection of a longer acting drug for those with a non-dipping nocturnal pattern. Patients with autonomic failure and postural hypotension can be identified and the symptoms of hypotension avoided by judicious prescribing. Finally, with ABPM it is possible to determine if symptoms attributed to the blood pressure lowering effect of drugs are due to hypotension.

Some evidence indicates that different groups of antihypertensive drugs may perturb the circadian pattern of blood pressure in different ways.[75] Hypertensive individuals on angiotensin-converting enzyme inhibitors were shown to have had markedly accentuated systolic and diastolic dipping patterns when compared with untreated hypertensive subjects and patients on beta-blockers; however, hypertensive patients treated with beta-blockers, calcium antagonists, or diuretics had diastolic and systolic dipping patterns similar to those of the untreated groups. Whatever the explanation for these varying effects of different groups of antihypertensive drugs, their occurrence raises important questions in assessing antihypertensive drug effects and in choosing a drug for an individual patient.

HYPOTENSION IN THE ELDERLY

ABPM is proving useful in diagnosing symptoms due to low blood pressure, especially in elderly patients with autonomic failure, and the technique is permitting a revaluation of this troublesome condition through the association of symptoms with the level of blood pressure.

Postural Hypotension

Hypotension, particularly related to posture, may affect all age groups[76] but is particularly common in the elderly, occurring in 10% of otherwise healthy elderly hypertensive subjects.[77] It is possible for postural hypotension to coexist with raised supine and sitting blood pressure, and it is important that blood pressure is assessed in these positions as well as standing on initial assessment and from time to time if drugs known to cause postural hypotension are being taken. These include not only blood pressure lowering drugs, such as diuretics, but also noncardiovascular drugs, such as neuroleptics and tricyclic antidepressants. Symptomatic hypotension is particularly prevalent among the elderly in whom autonomic dysfunction is common.[78] Using conventional measurement these patients may appear hypertensive, but 24-hour ABPM may show periods of sustained hypotension (FIG. 8). Clearly the administration of antihypertensive medication to such patients could be disadvantageous, and ABPM is proving a valuable technique for assessing such patients.

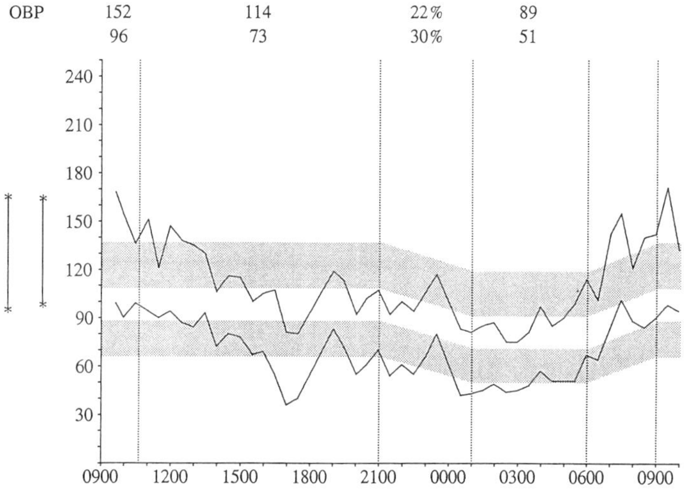

FIGURE 8. Autonomic failure. This elderly untreated patient exhibits many of the features of autonomic failure. Office blood pressure and pressure in the white coat window are clearly elevated, averaging 152/96 mm Hg, but these later fall to give daytime pressures averaging 114/73 mm Hg. However, there is a marked hypotensive period in the afternoon after eating when pressures reach as low as 80/33 mm Hg, indicating postprandial hypotension. Remaining pressures during the day and night are low but rise in the early morning.

Nocturnal Hypotension

Perhaps of equal importance in the aged is the possibility that excessive lowering of nocturnal blood pressure by critically altering perfusion of vital organs may predispose to cardiac and cerebral ischemic events.[79,80] It is reasonable to assume that the elderly, in whom cardiovascular hemodynamics are often compromised, are more susceptible to cerebrovascular events consequent upon excessive reduction of nocturnal blood pressure resulting from too much antihypertensive medication. Such an occurrence would be especially likely in elderly hypertensive patients who exhibit a pronounced nocturnal dipping pattern. Nocturnal blood pressure can only be characterized by 24-hour ABPM which should be performed in elderly patients so as to avoid excessive blood pressure lowering at night.

Postprandial Hypotension

Some elderly patients get a marked blood pressure fall after eating and this may be symptomatic. It is often not recognized how common this phenomenon is in the elderly.[2] Blood pressure may fall by as much as 25 mm Hg in response to a meal in elderly institutionalized patients.[81] The phenomenon of postprandial hypotension can be demonstrated by ABPM, thus identifying an important subgroup of elderly hypertensive patients who require special management.[78] Again this can only be diagnosed definitively by measuring blood pressure when standing after a meal.

CONCLUSION

The diversity of blood pressure patterns in the elderly presents a challenge to the diagnostic and management skills of their physician. On the one hand, there is the important issue of diagnosis. Does the patient have hypertension or not? The white coat phenomenon and the fact that isolated systolic hypertension is only an office phenomenon in many patients make ABPM an essential investigation for the accurate diagnosis of hypertension in the elderly. There are idiosyncratic aspects of blood pressure behavior in the elderly, among which are the phenomenon of isolated systolic hypertension, increased variability of blood pressure, and autonomic dysfunction, each of which can be categorized more accurately with ABPM than with conventional measurement. When it comes to treating the elderly patient with hypertension, ABPM can prove a useful adjunct to conventional measurement. It permits selection of the most appropriate blood pressure lowering drug for the individual patient, taking into account the pattern of blood pressure throughout the day and night and avoiding excessive drug administration during periods of endogenous blood pressure lowering, as during sleep. In assessing the common manifestations of hypotension in the elderly, ABPM permits the association of symptoms with actual blood pressure levels. In concluding that ABPM should now be used in the assessment of elderly patients with apparent hypertension on conventional measurement, the need to select carefully devices of proven accuracy is emphasized.

Finally, the relationship of ABPM to end organ damage is the subject of ongoing studies and should be a reminder that blood pressure, whether measured by conventional techniques or ABPM, is only one aspect of the elderly patient's cardiovascular profile. Assessing target organ disease by echocardiography and an increasing range of investigations permits assessment of the impact of blood pressure on the cardiovascular system which is likely to determine the ultimate outcome.

REFERENCES

1. Lamy, P. 1988. Potential adverse effects of antihypertensive drugs in the elderly. J. Hypertens. **6**(suppl 1): S81–S85.
2. Messerli, H. & R. E. Schmeider. 1991. Blood pressure measurement in the elderly. *In* Blood Pressure Measurement. E. O'Brien & K. O'Malley, Eds.: 148–154. Handbook of Hypertension. W. H. Birkenhager and J. L. Reid, Eds. Elsevier. Amsterdam.
3. Kannel, W. B. & F. N. Brand. 1985. Cardiovascular risk factors in the elderly. *In* Principles of Geriatric Medicine. R. Andreas, E. G. Bier-man & W. R. Hazzard, Eds.: 104–119. McGraw-Hill, New York.
4. Amery, A., L. Hansson, L. Andren, T. Gudbrandsson, R. Sivertsson & A. Svensson. 1981. Hypertension in the elderly. Acta Med. Scand. **210:** 221–229.
5. Amery, A., W. H. Birkenhager, P. Brixko *et al.* 1986. Efficacy of antihypertensive drug

treatment according to age, sex, blood pressure and previous cardiovascular disease in patients over the age of 60. Lancet **2:** 589–592.

6. FURBERG, C. D. & D. M. BLACK. 1988. The systolic hypertension in the elderly pilot program: Methodological issues. Eur. Heart J. **9:** 223–227.
7. O'BRIEN, E., G. BEEVERS & D. MARSHALL. 1995. ABC of Hypertension. BMJ Publications. London.
8. PICKERING, T., E. O'BRIEN & K. O'MALLEY. 1991. Second international consensus meeting on twenty-four-hour ambulatory blood pressure measurement: Consensus and conclusions. J. Hypertens. **9**(Suppl 8): S2–S6.
9. O'BRIEN, E., J. COX & K. O'MALLEY. 1991. The role of twenty-four-hour ambulatory blood pressure measurement in clinical practice. J. Hypertens. **9**(Suppl 8): S63–S65.
10. WITTENBERG, C., J. R. ZABLUDOWSKI & J. B. ROSENFELD. 1992. Overdiagnosis of hypertension in the elderly. J. Hum. Hypertens. **6:** 349–351.
11. SHIMADA, K., H. OGURA, A. KAWAMOTO, K. MATSUBAYASHI, H. ISHIDA & T. OZAWA. 1990. Noninvasive ambulatory blood pressure monitoring during clinic visit in elderly hypertensive patients. Clin. Exp. Hypertens. Theory Pract. **A12:** 151–170.
12. COX, J., E. O'BRIEN & K. O'MALLEY. 1991. Ambulatory blood pressure measurement in the elderly. J. Hypertens. **9**(Suppl. 3): S73–S77.
13. PRAGER, G., R. PRAGER & P. KLEIN. 1991. Comparison of twenty-four-hour blood pressure profile in elderly subjects with isolated systolic hypertension, essential hypertension and normotension. J. Hypertens. **9**(Suppl. 6): S102–S103.
14. SILAGY, C. A., J. J. MCNEIL & B. P. MCGRATH. 1990. Isolated systolic hypertension: Does it really exist on ambulatory blood pressure monitoring? Clin. Exp. Pharmacol. & Physiol. **17:** 203–206.
15. COX, J. P., N. ATKINS, K. O'MALLEY & E. O'BRIEN. 1991. Does isolated systolic hypertension occur with ambulatory blood pressure measurement? J. Hypertens. **9**(Suppl 6): S100–S101.
16. THIJS, L., J. COX, A. AMERY, D. CLEMENT, J. COX, P. DE CORT, G. FOWLER, C. CUO, R. FAGARD, G. MANCIA, R. MARM, P. PALATINI, E. O'BRIEN, K. O'MALLEY, G. MANCIA, G. PARATI, J. PETRIE, A. RAVOGH, J. ROSENFELD, J. STAESSEN & J. WEBSTER. 1992. Ambulatory blood pressure monitoring in elderly patients with isolated systolic hypertension. J. Hypertens. **10:** 693–699.
17. TORRIANI, S., B. WAEBER, A. PETRILLO, R. DI STEFANO, V. MOOSER, U. SCHERRER, J. NUSSBERGER, J.-R. HOFSTETTER & H. BRUNNER. 1988. Ambulatory blood pressure in the elderly hypertensive patient. J. Hypertens. **6**(Suppl. 1): S25–S27.
18. WAEBER, B., M. NIEDERBERGER, J. NUSSBERGER & H. R. BRUNNER. 1991. Ambulatory blood pressure monitoring in children, adolescents and elderly people. J. Hypertens. **9**(Suppl. 8): S72–S74.
19. TRAUB, Y. M., M. M. AYGEN & J. B. ROSENFELD. 1976. Hazards in treatment of systolic hypertension. Am. Heart J. **97:** 174–177.
20. KOBRIN, I., W. OIGMAN, A. KUMAR, H. O. VENTURA, F. H. MESSERLI, E. D. FROHLICH & F. G. DUNN. 1984. Diurnal variation of blood pressure in elderly patients with essential hypertension. Am. J. Geriatr. Soc. **32:** 896–899.
21. STAESSEN, J., A. AMERY, W. BIRKENHAGER, C. J. BULPITT, J. COX, P. DE LEEUW, P. DE CORT, R. FAGARD, A. FLETCHER, F. FORETTE, G. FOWLER, X. GIRRERD, G. LEONETTI, E. T. O'BRIEN, K. O'MALLEY, J. C. PETRIE, J. L. RODRICO, J. ROSEFIELD, W. SLOVIC, L. TEROZOLI, L. THIJS, J. TUOMILEHTO, J. WEBSTER & X. YODDFET. 1992. Syst-Eur, a multicentre trial for the treatment of isolated systolic hypertension in the elderly. First interim report. J. Cardiovasc. Pharmacol. **9:** 120–125.
22. IMHOLZ, B. P. M., H. A. DAMBRINK, J. M. KAREMAKER & W. WIELING. 1990. Orthostatic circulatory control in the elderly evaluated by non-invasive continuous blood pressure measurement. Clin. Sci. **79:** 73–79.
23. O'BRIEN, E., N. ATKINS & J. STAESSEN. 1995. State of the market: A review of ambulatory blood pressure monitoring devices. Hypertension **26:** 835–842.
24. O'BRIEN, E., F. MEE, N. ATKINS & K. O'MALLEY. 1993. Technical aspects of ambulatory blood pressure monitoring devices in the elderly. Cardiol. Elderly **1:** 464–469.
25. O'BRIEN, E. & K. O'MALLEY. 1994. Blood pressure measurement in the elderly with

special reference to ambulatory blood pressure measurement. *In* Hypertension in the Elderly. G. Leonetti & Cuspedi, Eds. Kluwer.: 13–25. Holland.
26. O'Brien, E., J. Petrie, W. A. Littler, M. de Swiet, P. L. Padfield, D. Altman, M. Bland, A. Coats & N. Atkins. 1993. The British Hypertension Society Protocol for the evaluation of blood pressure measuring devices. J. Hypertens. **11**(suppl. 2): S43–S63.
27. Miller, S. T., J. T. Elam, M. J. Graney & B. Applegate. 1992. Discrepancies in recording systolic blood pressure of elderly persons by ambulatory blood pressure monitor. Am. J. Hypertens. **5:** 16–21.
28. O'Brien, E., N. Atkins, F. Mee & K. O'Malley. 1993. Comparative accuracy of six ambulatory devices according to blood pressure levels. J. Hypertens. **11:** 673–675.
29. American National Standard for Electronic or Automated Sphygmomanometers. 1987. Association for the Advancement of Medical Instrumentation. Virginia, USA.
30. O'Brien, E. & J. Staessen. 1995. Normotension and hypertension as defined by 24-hour ambulatory blood pressure monitoring. Blood Pressure **4:** 266–282.
31. Kannel, W. B. 1974. Role of blood pressure in cardiovascular morbidity and mortality. Progr. Cardiovasc. Dis. **17:** 5–23.
32. Rutan, G. H., L. H. Kuller, J. D. Neaton, *et al.* 1988. Mortality associated with diastolic hypertension and isolated systolic hypertension among men screened for the Multiple Risk Factor Intervention Trial. Circulation **77:** 504–514.
33. Kain, H. R., A. T. Hinman & M. Sokolow. 1964. Arterial blood pressure measurements with a portable recorder in hypertensive patients. I. Variability and correlation with 'casual' pressures. Circulation **30:** 882–892.
34. Sokolow, M. D., H. K. Werdegar, H. K. Kain & A. T. Hinman. 1966. Relationship between level of blood pressure measured casually and by portable recorders and severity of complications in essential hypertension. Circulation **34:** 279–298.
35. Perloff, D., M. Sokolow & R. Cowan. 1983. The prognostic value of ambulatory blood pressures. JAMA **240:** 2792–2798.
36. Perloff, D. & M. Sokolow. 1991. Ambulatory blood pressure: Mortality and morbidity. J. Hypertens. **9**(suppl. 8): S31–S33.
37. MacMahon, S., R. Peto, J. Cutler, *et al.* 1990. Blood pressure, stroke, and coronary heart disease. Part I. Prolonged differences in blood pressure: Prospective observational studies corrected for the regression dilution bias. Lancet **335:** 765–774.
38. Clement, D., M. De Buyzere & D. D. Duprez. 1991. Ambulatory blood pressure and prognosis: Summary of ongoing studies. J. Hypertens. **9**(suppl. 8): S51–S53.
39. Staessen, J., A. Amery, D. Clement, *et al.* 1992. Twenty-four hour blood pressure monitoring in the Syst-Eur trial. Ageing Clin. Exp. Res. **4:** 85–91.
40. Staessen, J. & A. Amery on behalf of the Advisory Board. 1993. APTH—A trial on ambulatory blood pressure monitoring and treatment of hypertension: Objectives and protocol. Acta Cardiol. **48:** 25–42.
41. Cesana, G., G. De Vito, M. Ferrario, A. Libretti, G. Mancia, P. Mocarelli, R. Sega, F. Valagussa & A. Zanchetti. 1991. Ambulatory blood pressure normalcy: The Pamela Study. J. Hypertens. **9**(suppl. 3): S17–S23.
42. Sega, R., G. Cesana, R. Pagani, G. Bravi, G. De Vito, M. Ferrario, F. Valagussa, A. Libretti, G. Mancia & A. Zanchetti. 1993. Ambulatory and home blood pressure reference values: The Pamela Study. J. Hypertens. **11**(suppl. 5): S481–S482.
43. Verdecchia, P., C. Porcellati, G. Schillaci, C. Borgioni, A. Ciucci, M. Battistelli, M. Guerrieri, C. Gatteschi, I. Zampi, A. Santucci, C. Santucci & G. Reboldi. 1994. Ambulatory blood pressure: An independent predictor of prognosis in essential hypertension. Hypertension **24:** 793–801.
44. O'Brien, E. 1995. Ambulatory blood pressure measurements as a guide for predicting the occurrence of hypertensive organ involvement. *In* Handbook of Hypertension. L. Hansson & W. Birkenhager, Eds. Assessment of organ damage. Elsevier. Holland. In press.
45. Baumgart, P. 1990. Ambulatory blood pressure monitoring: Reference values and the standard blood pressure scale. J. Hypertens. **9**(suppl. 6): S65–S70.

46. PALATINI, P. & A. C. PESSINA. 1990. A new approach to define the upper normal limits of ambulatory blood pressure. J. Hypertens. **8**(suppl. 6): S65–S70.
47. STAESSEN, J., E. O'BRIEN, R. FAGARD, N. ATKINS, L. THIJS, G. VYNCKE & A. AMERY. 1994. A consistent reference frame for ambulatory blood pressure monitoring is found in different populations. J. Human Hypertens. **8:** 423–431.
48. THE WHO/ISH (INTERNATIONAL SOCIETY OF HYPERTENSION) MILD HYPERTENSION LIAISON COMMITTEE. 1989. 1988 guidelines for the management of mild hypertension: Memorandum from a WHO/ISH meeting. Bull WHO **67:** 493–498.
49. STAESSEN, J., E. T. O'BRIEN, A. AMERY, N. ATKINS, P. BAUMGART, P. DE CORT, J. P. DEGAUTE, P. DOLENC, R. DE GAUDEMARIS, I. ENSTRÖM, R. FAGARD, P. GOSSE, S. GOURLAY, H. HAYASHI, Y. IMAI, G. JAMES, T. KAWASAKI, E. KUSCHNIR, I. KUWAJIMA, L. LINDHOLM, L. LIU, F. MACOR, G. MANCIA, B. MCGRATH, M. MIDDEKE, J. MING, S. OMBONI, K. OTSUKA, P. PALATINI, G. PARATI, C. PIEPER, P. VERDECCHIA, P. ZACHARIAH & W. ZHANG. 1994. The ambulatory blood pressure in normotensive and hypertensive subjects: Results from an international database. J. Hypertens. **12**(suppl. 7): S1–S12.
50. O'BRIEN, E., J. MURPHY, A. TYNDALL, N. ATKINS, F. NEE, G. MCCARTHY, J. STAESSEN, J. COX & K. O'MALLEY. 1991. Twenty-four-hour ambulatory blood pressure in men and women aged 17 to 80 years: The Allied Irish Bank Study. J. Hypertens. **9:** 355–360.
51. STAESSEN, J., C. J. BULPITT, R. FAGARD, G. MANCIA, E. O'BRIEN, L. THIJS, G. VYNCKE & A. AMERY. 1991. Reference values for the ambulatory blood pressure and the blood pressure measured at home: A population study. J. Human Hypertens. **5:** 355–361.
52. STAESSEN, J., C. J. BULPITT, E. O'BRIEN, J. COX, R. FAGARD, A. STANTON, L. THIJS, S. VAN HULLE, G. VYNCKE & A. AMERY. 1992. The diurnal blood pressure profile: A population study. Am. J. Hypertens. **5:** 386–392.
53. STAESSEN, J., R. FAGARD, P. LIJNEN, L. THIJS, R. VAN HOOF & A. AMERY. 1994. Mean and range of the ambulatory pressure in normotensive subjects from a meta-analysis of 23 studies. Am. J. Cardiol. **67:** 723–737.
54. STAESSEN, J., C. J. BULPITT, R. FAGARD, G. MANCIA, E. T. O'BRIEN, L. THIJS, G. VYNCKE & A. AMERY. 1991. Reference values for ambulatory blood pressure: A population study. J. Hypertens. **9**(suppl. 6): S320–S321.
55. STAESSEN, J. A., N. ATKINS, R. FAGARD, E. T. O'BRIEN, L. THIJS, G. VYNCKE & A. AMERY. 1993. Correlates of the diurnal blood pressure profile in a population study. High Blood Press **2:** 271–282.
56. STAESSEN, J. A., R. FAGARD, P. LIJNEN, L. THIJS, S. VAN HULLE, G. VYNCKE & A. AMERY. 1994. Ambulatory blood pressure and blood pressure measured at home: Progress report on a population study. J. Cardiovasc. Pharmacol. **23**(suppl. 5): S5–S11.
57. THE SCIENTIFIC COMMITTEE. 1990. Consensus document on non-invasive ambulatory blood pressure monitoring. J. Hypertens. **8**(suppl. 6): 135–140.
58. PICKERING, T. G. 1992. The ninth Sir George Pickering memorial lecture. Ambulatory monitoring and definition of hypertension. J. Hypertens. **10:** 401–409.
59. BULPITT, C. J., A. E. FLETCHER, A. AMERY, J. COOPE, J. GRIMLEY EVANS, S. LIGHTOWLERS, K. O'MALLEY, A. PALMER, J. POTTER, P. SEVER, J. STAESSEN & C. SWIFT. 1994. The hypertension in the very elderly trial (HYVET). Drugs & Aging **5:** 171–183.
60. PICKERING, T. G., G. D. JAMES, C. BODDIE, G. A. HARSHFIELD, S. BLANK & J. H. LARAGH. 1988. How common is white coat hypertension? JAMA **259:** 225–228.
61. LERMAN, C. E., D. S. BRODY, T. HUI, C. LAZARO, D. G. SMITH & M. J. BLUM. 1989. The white-coat hypertension response: Prevelance and predictors. J. Gen. Intern. Med. **4:** 225–231.
62. RUDDY, M. C., G. B. BIALY, E. S. MALKA, C. R. LACY & J. B. KOSTIS. 1988. The relationship of plasma renin activity to clinic and ambulatory blood pressure in elderly people with isolated systolic hypertension. J. Hypertens. **6**(suppl. 4): S412–S415.
63. SHIMADA, K., H. OGURA, A. KAWAMOTO, K. MATSUBAYASHI, H. ISHIDA & T. OZAWA. 1990. Noninvasive ambulatory blood pressure monitoring during clinic visit in elderly hypertensive patients. Clin. & Exp. Hyper. Theory & Pract. **A12:** 151–170.
64. THE SYSTOLIC HYPERTENSION IN THE ELDERLY PROGRAM COOPERATIVE RESEARCH GROUP.

1993. Implications of the systolic hypertension in the elderly program. Hypertension **21:** 335–343.
65. SILAGY, C. A., J. J. MCNEIL, S. FARISH & B. P. MCGRATH. 1993. Comparison of repeated measurement of ambulatory and clinic blood pressure readings in isolated systolic hypertension. Clin & Exper Hypertens. **15:** 895–909.
66. O'BRIEN, E. & K. O'MALLEY. 1993. The relationship of 24-hour ambulatory blood pressure measurement to target organ damage. Ther. Res. **14:** 84–91.
67. O'BRIEN, E., N. ATKINS & K. O'MALLEY. 1993. Defining normal ambulatory blood pressure. Am. J. Hypertens. **6:** 201S–206S.
68. PICKERING, T. G. & R. B. DEVEREUX. 1987. Ambulatory monitoring of blood pressure as a predictor of cardiovascular risk. Am. Heart J. **114:** 925–928.
69. COX, J., A. AMERY, D. CLEMENT, P. DE CORT, R. FAGARD, G. FOWLER, R. M. IRANZO, G. MANCIA, E. O'BRIEN, K. O'MALLEY, P. PALATINI, G. PARATI, J. PETRIE, A. RAVOLGI, J. ROSENFELD, J. STAESSEN, L. THIJS & J. WEBSTER. 1993. Relationship between blood pressure measured in the clinic and by ambulatory monitoring and left ventricular size as measured by electrocardiogram in elderly patients with isolated systolic hypertension. J. Hypertens. **11:** 269–276.
70. COX, J., E. O'BRIEN & K. O'MALLEY. 1991. Ambulatory blood pressure measurement in the elderly. J. Hypertens. **9**(suppl. 3): S73–S77.
71. O'BRIEN, E., J. SHERIDAN & K. O'MALLEY. 1988. Dippers and non-dippers. Lancet **2:** 397.
72. PICKERING, T. G. 1990. The clinical significance of diurnal blood pressure variations; Dippers and non-dippers. Circulation **19:** 93–101.
73. O'BRIEN, E., K. O'MALLEY, J. COX & A. STANTON. 1991. Ambulatory blood pressure monitoring in the evaluation of drug efficacy. Am. Heart J. **121:** 999–1006.
74. WITTENBERG, C., J. R. ZABLUDOWSKI & J. B. ROSENFELD. 1992. Overdiagnosis of hypertension in the elderly. J. Human Hypertens. **6:** 349–351.
75. STANTON, A., N. ATKINS, E. O'BRIEN & K. O'MALLEY. 1993. Antihypertensive therapy and circadian blood pressure profiles: A retrospective analysis utilising cumulative sums. Blood Pressure **2:** 289–295.
76. IMHOLZ, B. P. M., J. H. A. DAMBRINK, J. M. KAREMAKER & W. WIELING. 1990. Orthostatic circulatory control in the elderly evaluated by non-invasive continuous blood pressure measurement. Clin. Sci. **79:** 73–79.
77. RODSTEIN, M. & F. D. ZEMAN. 1957. Postural blood pressure changes in the elderly. J. Chronic Dis. **6:** 581–588.
78. ZACHARIAH, P. K., J. KRIER & G. L. SCHWARTZ. 1991. Orthostatic hypotension and ambulatory blood pressure monitoring. J. Hypertens. **9**(suppl. 8): S78–S80.
79. FLORAS, J. S. 1988. Antihypertensive treatment, myocardial infarction and nocturnal ischaemia. Lancet **2:** 944–946.
80. STANTON, A. & E. O'BRIEN. 1993. Noninvasive 24 hour ambulatory blood pressure monitoring: Current status. Postgrad. med. J. **69:** 255–267.
81. LIPSITZ, L. A., R. P. NYQUIST, J. Y. WEI & J. W. ROWE. 1983. Postprandial reduction in blood pressure in the elderly. N. Engl. J. Med. **309:** 81–83.

Blood Pressure Rhythms in Sleep Disorders and Dysautonomia

P. CORTELLI, G. PIERANGELI, F. PROVINI,
G. PLAZZI, AND E. LUGARESI

Institute of Neurology
University of Bologna
Via Foscolo, 7
40123 Bologna, Italy

With the developments in polysomnographic techniques, sleep can be monitored during the night or throughout the whole nycthemeral cycle. This has led to remarkable advances in our understanding of mechanisms controlling blood pressure (BP), heart rate (HR), breathing, temperature, and the motor system.

This paper summarizes the pattern of BP changes during wakefulness and sleep in normal subjects, in sleep-related breathing disorders, and in neurodegenerative diseases with dysautonomia.

NORMAL HUMAN SLEEP

Sleep is a complex amalgam of physiologic and behavioral processes. Within sleep, two separate states have been defined on the basis of a constellation of physiologic parameters. These two states, nonrapid eye movement (NREM) and rapid eye movement (REM), exist in virtually all mammals and birds and are as distinct from one another as each of them is from wakefulness. NREM sleep is conventionally subdivided into four stages defined by electroencephalography (EEG) (FIG. 1). The EEG pattern in NREM sleep is described as synchronous, with such characteristic waveforms as sleep spindles, K complexes, and high voltage slow waves. NREM stages (stages 1–4) roughly parallel a "depth of sleep" continuum, with arousal thresholds lowest in stage 1 and highest in stage 4 sleep.

NREM sleep is usually associated with a dearth of mental activity, occurs mainly during the first third of the night, and usually constitutes 75–80% of sleep. REM sleep, by contrast, is defined by EEG activation, muscle atonia, and episodic bursts of REM. The mental activity of human REM sleep is associated with dreaming and is based on vivid dream recall reported after about 80% of arousals from this state of sleep. Inhibition of spinal motoneurons via brain stem mechanisms mediates suppression of postural motor tonus in REM sleep. REM sleep predominates in the last third of the night and usually comprises 20–25% of sleep occurring in four to six discrete episodes. The normal adult human enters sleep through NREM sleep. REM sleep does not occur until 80 minutes or longer, and NREM and REM sleep alternate cyclically throughout the night in periods of about 80 minutes (FIG. 1).

Respiratory and Cardiocirculatory Sleep Patterns

During sleep onset and light sleep (stages 1 and 2 of NREM sleep) breathing often becomes periodic, and sporadic central apneas are frequently observed.

Breathing oscillations occur periodically, every 20–30 seconds, associated with synchronous oscillations of cortical EEG activity, systemic BP, HR, and oxygen saturation (SaO_2) (FIG. 2). During deep sleep (stages 3 and 4 of NREM sleep) breathing becomes regular; BP and HR drop 20–30% below basal wakefulness values.

During REM sleep breathing and HR become irregular, and central apneas or hypopneas lasting 20–30 seconds occur sporadically, often in association with bursts of rapid eye movement. Hypertensive peaks, sometimes reaching 30–40 mm Hg, appear suddenly and continue throughout REM sleep (FIG. 2). Alveolar ventilation decreases, and pulmonary arterial pressure rises during sleep, but never reaches a pathologic level.[1–3] Changes in the sensitivity and function of homeostatic mechanisms account for the oscillatory phenomena that also occur, with similar periodicity, in other pathophysiologic situations[4,5] or at sleep onset.[3] The instability associated with the transition from mixed control of breathing (neurogenic and metabolic) typical of wakefulness to the purely metabolic state that operates during NREM

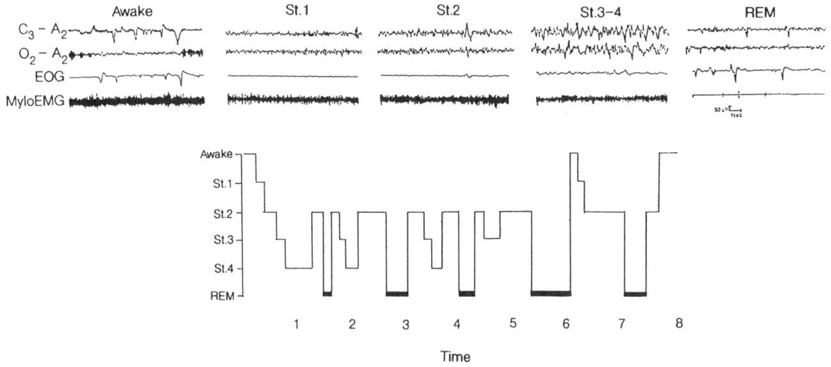

FIGURE 1. Polysomnographic example and sleep histogram show the typical repetition of the different stages of sleep (NREM = 1, 2, 3–4, and REM). C_3–A_2, O_2–A_2 = EEG channels; EOG = electrooculogram; MyloEMG = electromyography of submental muscles.

sleep is the underlying cause of periodic oscillations.[6] As periodic oscillations are simultaneously observed for breathing, systemic BP, HR, EEG, and limb muscular activity, the site responsible for this phenomenon was thought to be in the midbrain reticular formation,[7] but now this concept has to be expanded in view of the new wider notion of the central autonomic network where integration of these autonomic activities occurs.[8] During deep sleep (stages 3 and 4 of NREM sleep) breathing, BP, and HR are highly regular, indicating that metabolic control is well established and a new equilibrium is reached. On the other hand, the irregular respiratory and cardiocirculatory behavior during REM sleep suggests a lack of automatic control probably due to loss of homeostatic regulation by the hypothalamus.[9]

Circadian Rhythm and Blood Pressure Control

Blood pressure fluctuates continuously throughout day and night. The fluctuations in pressure may be of very short duration, from seconds to minutes, or of longer duration including day and night changes and seasonal variations. The most important and consistent sources of BP variation are associated with the sleep-wake cycle. In healthy normotensive subjects BP falls by about 20% from waking to sleeping hours[2,10–14] and further in women than in men.[15] The pressure tends to be higher in the early afternoon, decreasing gradually over the course of the day, and is lowest during the night. The nocturnal fall in BP is accompanied by a parallel fall in HR, total peripheral resistance, cardiac output, glomerular filtration rate, and electrolyte excretion. The fall in plasma catecholamine levels during sleep is consistent with diminished sympathetic activity. Antidiuretic hormone levels rise during the night and this may contribute to nocturnal oliguria.[16]

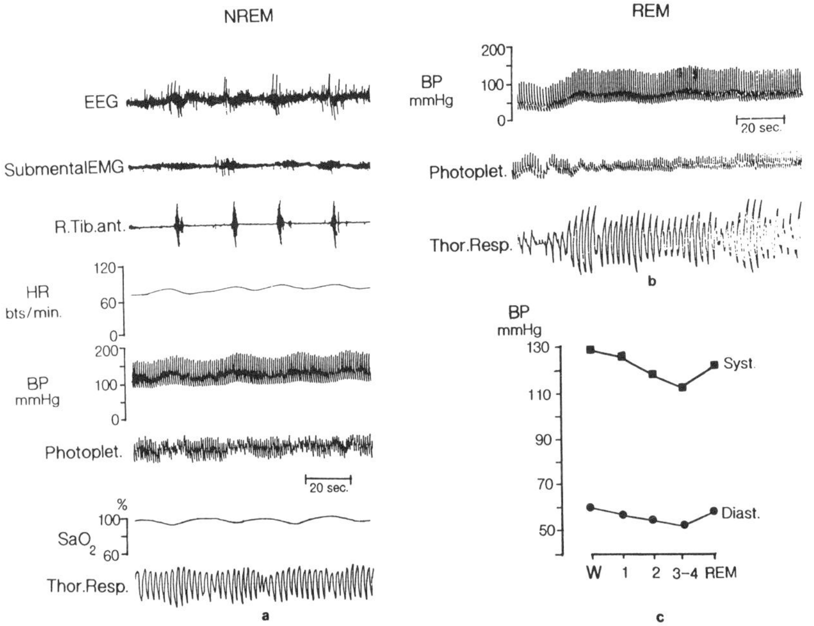

FIGURE 2. Cardiovascular functions during sleep. (**a**) Stages 1–2 of NREM sleep: synchronous oscillations of multiple systems with a common periodicity of about 30 seconds involving cortical (EEG), muscular (tonic submental EMG and phasic tibialis anterior jerks), autonomic (HR, BP, and vascular tone), and breathing (SaO_2 and thoracic respiration) activities. (**b**) Characteristic hypertensive peak during REM sleep. (**c**) Decrease in BP during NREM sleep with an increase during REM sleep in a control.

The variability in BP observed during the sleep-wake cycle depends largely on central cardiovascular control by the autonomic nervous system.[17–19] At the same time, the variability in BP is influenced by genetic and ethnic factors,[20–22] physiologic factors[16,23–28] (posture, exercise, and speech) or behavioral and environmental factors[29–34] (salt, caffeine, alcohol, and smoking), and pathologic conditions. In patients with essential hypertension, the day-night pattern of BP changes is generally similar to that of normotensive subjects with a significant nocturnal BP fall (dippers).[13,16,26,35] Nevertheless, in some cases of essential hypertension and in several forms of secondary hypertension, the nocturnal fall in BP is absent or reversed with a normal decrease in HR (non-dippers).[36–45] The nocturnal fall in BP may be clinically relevant because several studies have demonstrated that non-dippers are more likely to have morbid cardiovascular events than dippers[46–51] or to exhibit left ventricular hypertrophy.[48,52–54] The possibility of intrinsic circadian rhythm in BP was suggested by findings of a gradual increase in BP during the early morning hours before waking. However, when recordings are corrected to the waking time rather than the time of the day, it is clear that BP remains relatively stable during the hour before waking, rising abruptly on awaking.[55] Moreover, the BP decline during daytime sleep and nighttime sleep was similar,[56] suggesting a major effect of sleep itself probably through sympathetic withdrawal. These findings are in agreement with studies on shift workers[57–59] in whom rapid reversal of circadian BP rhythm was associated with a change in sleeping hours.

Thus, BP rhythm, closely following the sleep-wake cycle, is in marked contrast to some other rhythms closely linked with an internal clock such as HR, body temperature, or cortisol secretion rhythms.

METHODS

All-night or nycthemeral polygraphic recordings as well as physiologic, biochemical, and pharmacologic autonomic investigations were performed by standard methods, described in detail elsewhere.[60–63]

Snoring and Obstructive Sleep Apnea Syndrome or Heavy Snorers' Disease

Clinical Aspects. It is now well established that a continuum of clinical conditions exists between intense snoring and the severest forms of heavy snorers' disease (HSD), all of which originate in a narrowing of the upper airways.[64–68] The clinical features of this slowly evolving process include heavy and habitual snoring and the subsequent onset of intermittent snoring, related to obstructive sleep apneas and associated with impairment of daytime alertness. The progressive rise in the number and duration of sleep apneas aggravates daytime somnolence and induces slowly worsening cardiocirculatory changes (systemic and pulmonary hypertension, cardiac arrhythmias, and hypertrophy) and polycythemia. In more advanced stages of the disease, alveolar hypoventilation, previously confined to sleep, also occurs during wakefulness; death may supervene after some years due to heart or acute pulmonary failure.

Polysomnographic Aspects. During snoring each inspiration is characterized by increased activity of the intercostal and/or diaphragmatic muscles and by dramatic intrathoracic negative pressure, indicating that the inspiratory effort during snoring is greater than normal.[64] SaO_2 does not necessarily undergo marked variations, but in some heavy snorers we observed a 3–4% reduction in SaO_2 compared with wakefulness values. Pulmonary arterial pressure also exceeds the normal range for short periods, whereas systemic BP, instead of decreasing as it does in normal controls, remains at values comparable to or even slightly higher than those during wakefulness.[65] When apneas become persistent, alveolar hypoventilation worsens, and systemic and especially pulmonary arterial pressure reach high values,[69] particularly during REM sleep, when apneas become more prolonged. SaO_2 measured by ear oxymetry shows wide phasic oscillations closely correlated with each apnea. It is only

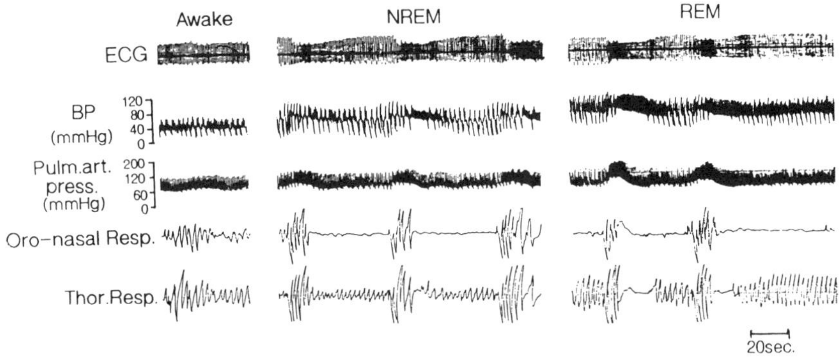

FIGURE 3. Polygraphic tracings of obstructive apneas in NREM and REM sleep. Channels (*from top to bottom*) are: electrocardiogram (ECG), systemic blood pressure (BP), pulmonary arterial pressure (pulm. art. press.), oro-nasal, and thoracic breathing. Characteristic cardiovascular changes associated with obstructive sleep apnea are shown.

in the more advanced stages of the disease, usually when hypoventilation is present during wakefulness, that phasic desaturations linked to the apneas are associated with a gradual fall in O_2 concentration throughout REM sleep. The cardiovascular changes that accompany obstructive sleep apnea events are now well recognized. Heart rate, systemic BP, and pulmonary arterial pressure initially fall with the onset of apnea, but they rise acutely with apnea termination (FIG. 3). The natural history of HSD has been divided into four evolutive stages[70] based on both clinical and polygraphic findings.

Cardiovascular Autonomic Function. In normotensive awake patients with obstructive sleep apnea syndrome (OSAS), cardiovascular reflex tests showed significantly higher HR and noradrenaline plasma levels at rest and a higher BP response to head-up tilt with a significantly lower respiratory arrhythmia, baroreflex sensitivity index, and Valsalva ratio associated with a greater decrease in HR induced by the cold face test.[62] Our data suggest sympathetic overactivity associated with a blunting of reflexes dependent on baroreceptor or pulmonary afferents and normal or increased cardiac vagal efferent activity.

Night-Day Blood Pressure Variation. Nine patients (7 males) with a previously established diagnosis of OSAS were examined. The mean age was 53 ± 9 years, with a mean duration of the reported onset of snoring and diurnal sleepiness of 22 ± 14 years. Mean body mass index (BMI) was 33 ± 7; six patients were obese (BMI >30). All patients were clinically stable on entering the study and drug free. Spirometry indicated that three subjects had airway obstruction. None had resting hypoxemia (arterial oxygen tension [PaO_2] <65 mm Hg) with mean PaO_2 of 90 ± 8, and one patient had hypercapnia (arterial carbon dioxide tension [$PaCO_2$] >45 mm Hg). The patients had moderate or severe OSAS with an apnea index (AI) of 57 ± 20. Most events were obstructive and mixed. The mean low SaO_2 was 62 ± 15%. No patients were hypertensive (systolic BP range 110–142 mm Hg, mean ± SD 128 ± 10 mm Hg; diastolic BP range 62–86 mm Hg, mean ± SD 73 ± 7 mm Hg). During the night systolic BP increased to 130 ± 18 mm Hg in NREM sleep and 138 ± 20 mm Hg in REM sleep. Diastolic BP was 70 ± 15 mm Hg during NREM sleep and 76 ± 18 mm Hg during REM sleep. The physiologic BP fall during sleep was absent in OSAS patients, and the variability of systemic BP was higher in sleep than in wakefulness (FIG. 4).

Relationship between Systemic Hypertension, Snoring, and Obstructive Sleep Apnea. Snoring, sleep apnea, and hypertension have been linked by epidemiologic studies.[67,71,72] Half or more of all patients with OSAS are hypertensive.[73–75] Conversely, several studies have shown an approximate 30% prevalence of sleep apnea in patients with primary hypertension.[76–79] Physiologically, sleep apnea is associated with distinctive patterns of nocturnal and diurnal BP variations; BP rises slowly throughout the night,[64,65] but reverts to normal if the abnormal breathing is eliminated.[80–86] Hypotheses have been advanced to explain how snoring and sleep apnea lead to sustained daytime elevation of BP and ultimately hypertension. Several factors associated with sleep apnea are candidate mechanisms for long-term diurnal BP elevation including the direct effects of episodic hypoxemia and hypercapnia on chemoreceptors and sympathetic activity,[87,88] the modification of the cardiovascular system (including fluid balance) in response to marked fluctuations in intrathoracic pressure during obstructed breathing,[89] the generalized stress from sleep disruption (the arousal effect),[90–92] genetic factors, and age. The relative contribution of each of these factors remains unclear, and particularly controversial is the role of recurrent hypoxia.[93–96] Conceivably, the concerted action of several factors (hypoxia, negative intrathoracic pressure, intermittent BP changes, arousal from sleep, and other metabolic or endocrine factors), possibly only in combination with a particular age and specific genetically determined characteristics, could bring about the development of hypertension. Whatever the causative factors, the common mechanism shared by hypertension and OSAS is activation of the sympathetic system[62,97–100] that by some mechanisms becomes chronic or acts as an intermediary, changing hormonal balance or causing vascular remodeling.

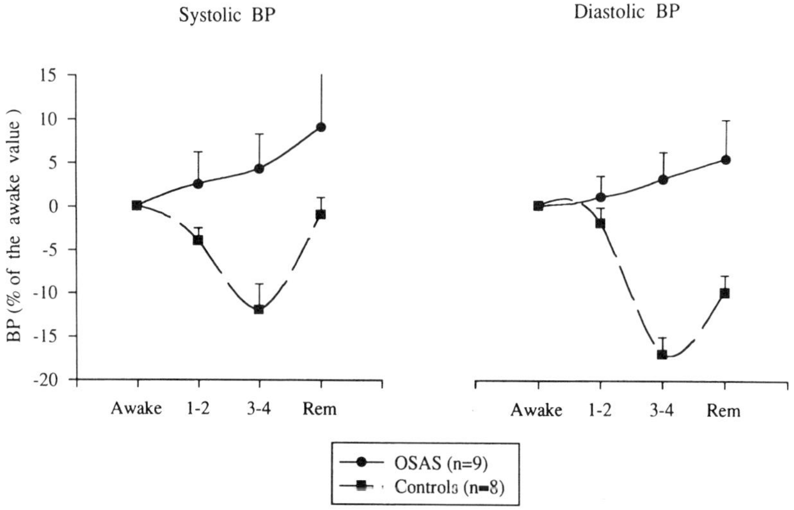

FIGURE 4. Systemic blood pressure changes (% of awake value) in patients with obstructive sleep apnea syndrome (OSAS) and controls in different stages of sleep (NREM = 1–2, 3–4, and REM). Blood pressure and its variability (standard deviation) increase during sleep in the OSAS group.

Multiple System Atrophy

Classification, Pathology, and Clinical Aspects. The term multiple system atrophy (MSA) first used by Graham and Oppenheimer[101] in 1969 refers to a sporadic adult-onset neurodegenerative disease that had previously been described variously as striatonigral degeneration, olivopontocerebellar atrophy, or Shy-Drager syndrome according to the clinical predominance of parkinsonism, cerebellar signs, or autonomic failure, respectively.[102–104] Autonomic failure may also occur in a "pure" form without any other neurologic signs, previously called "idiopathic orthostatic hypotension."[105] Neuropathologically, MSA is characterized by cell degeneration and gliosis in the striatum substantia nigra, Purkinje cell layer of the cerebellar cortex, pontine nuclei, and inferior olives. In almost all patients with autonomic failure whether they had pure autonomic failure or MSA there has been up to an 80% reduction in the number of sympathetic preganglionic neurons in the intermediolateral cell columns of the spinal cord. Brain stem abnormalities including those of the dorsal nucleus of the vagus, the locus coeruleus, and the nucleus tractus solitarii have also been noted.

The clinical presentation[106] in patients with MSA with autonomic failure is highly variable, at least at the onset of the disorder. Patients may present with parkinsonism poorly responding to levodopa, with a cerebellar syndrome, with progressive autonomic failure (postural hypotension, defective sweating, failure of erection in males, and constipation), or with a variable combination of these syndromes. The prognosis of MSA with autonomic failure is usually determined by the progress of neurologic features over the course of 6 years. The natural history of pure autonomic failure is of a slow progression over 10–15 years.

Patients and Cardiovascular Autonomic Function. We evaluated 14 patients with MSA and autonomic failure (5 females, 9 males, aged 61 ± 7 years), 7 patients with only MSA (4 females, 3 males, aged 54 ± 4 years), and 5 patients with pure autonomic failure (3 females, 2 males, aged 57 ± 11 years). Patients with probable MSA fulfilled the following clinical criteria: (a) progressive cerebellar ataxia and/or nonresponsive or poorly responsive levodopa parkinsonism and/or pyramidal signs without neuroradiologic evidence for symptomatic origin of the disorder; (b) absence of neurodegenerative disorders in relatives and no evidence of parental consanguinity; (c) absence of dementia or predominant downgaze supranuclear palsy; and (d) a clinical course of at least 3 years in patients with MSA and of at least 6 years for pure autonomic failure.

All patients with autonomic failure showed severe symptomatic postural hypotension, but subjects with MSA and autonomic failure had normal noradrenaline plasma levels at rest supine (200 ± 110 pg/ml; normal value 194 ± 106 pg/ml) without a significant increase on orthostatic stress (241 ± 133 pg/ml; normal value 374 ± 140 pg/ml); subjects with pure autonomic failure differed because of very low noradrenaline plasma levels at rest supine (40 ± 25 pg/ml) without an increase on orthostatic stress (50 ± 29 pg/ml).

The other physiologic tests (Valsalva maneuver, deep breathing, and isometric handgrip) in patients with either pure autonomic failure or MSA with autonomic failure showed a sympathetic efferent defect more pronounced in patients with pure autonomic failure, suggesting that the lesion is more completely distal in this group. The parasympathetic efferent lesion was less severe than the sympathetic one in both groups of patients. The pharmacologic tests of sympathetic function disclosed a greater increase in the sensitivity to infused noradrenaline in patients with pure autonomic failure than in those with MSA with autonomic failure, providing further evidence of more distal lesion in pure autonomic failure. Patients with only MSA had normal responses to the battery of physiologic and pharmacologic tests.

Polygraphic Aspects. The sleep-wake organization was abnormal in many patients studied except for the pure autonomic failure group. The most consistent abnormality was sleep fragmentation with increased sleep latency and frequent awakenings with 30–40% of the night spent awake along with reductions in REM sleep and stages 3 and 4 sleep. By contrast to quiescent sleep in normal persons, increased muscle tone and abnormal simple and complex movements are common during the sleep of patients with MSA with or without autonomic failure. During REM sleep, both groups presented prolonged elevation of muscle tone associated with more complex movements, vocalizations, and a fully developed REM sleep behavior disorder. No motor sleep abnormalities were found in the pure autonomic failure group. Analysis of the pattern of breathing revealed the existence of snoring and central, upper airway obstructive and mixed apneas in patients with MSA with or without autonomic failure (Apnea Hypopnea Index [AHI] = MSA: 18 ± 16; MSA + AF: 17 ± 15). Nocturnal stridor was present in six patients with MSA independently of the presence of autonomic failure. During obstructive sleep apneas no changes in BP or HR were noted in patients with autonomic failure in contrast to the usual brady-tachyarrhythmia and an increase in BP after a single episode of apnea in the MSA group without autonomic failure, suggesting that autonomic vasomotor reflexes are an important mediator of the hypertensive response to apnea. All patients with MSA with autonomic failure showed a consistent circadian trend in BP that was the inverse of the normal pattern, with the highest pressure at night and the lowest in the morning. This pattern was also present in the pure autonomic failure group but without nocturnal hypertension. Patients with only MSA had a normal circadian pattern of BP (FIG. 5).

Conclusions. Sleep abnormalities are common in MSA irrespective of the presence of autonomic failure and they are absent in pure autonomic failure patients, suggesting that the sleep disruption is mainly due to the central nervous system lesions. Whether this effect on sleep is primary or secondary to the discomfort caused by the neurologic signs remains a matter of debate.[107] Our findings showed a complete absence of the circadian changes of BP in pure autonomic failure patients, whereas the MSA with autonomic failure group exhibited an inverse of the normal circadian variation in pressure in contrast to the MSA group without autonomic failure who had a normal BP circadian variation. This strongly suggests that normal BP day-night changes depend on the proper functioning of the sympathetic nervous system. The low level of BP in the morning that corresponds to the peak incidence of postural symptoms could be the result of nocturnal polyuria and natriuresis as indicated by the improvement obtained with administration of desmopressin.[108]

FATAL FAMILIAL INSOMNIA

Fatal familial insomnia (FFI) is a recently identified prion disease[109] with distinctive clinical, pathologic, and molecular biology features which have already been reported in detail.[110–113] We shall focus on the autonomic abnormalities and BP patterns which together with the peculiar pathologic correlation allow interesting and novel conclusions to be drawn about the role of diencephalic structures in the regulation of these cyclic biologic activities.[114]

Clinical, Pathologic, and Genetic Aspects. FFI is an inherited disease transmitted in an autosomal dominant manner. Six patients from two unrelated Italian kindreds were followed until death, and the disease was verified postmortem in all of them.

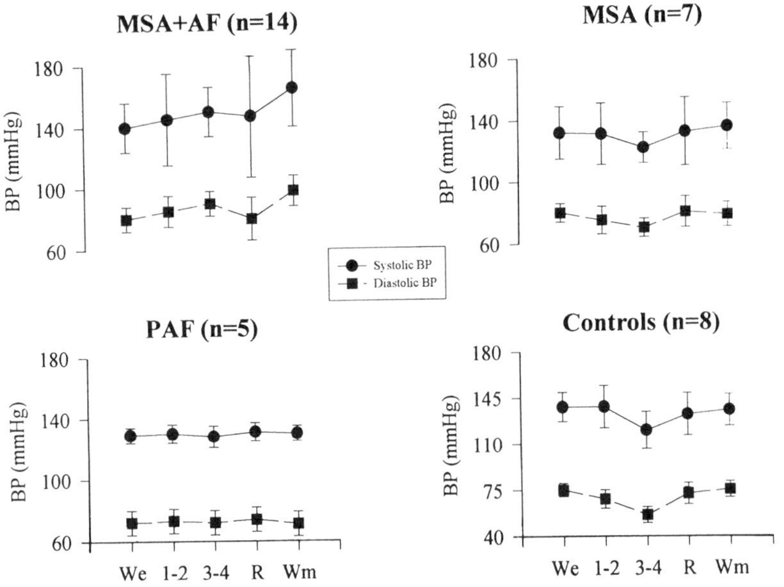

FIGURE 5. Day-night systemic blood pressure (BP) changes in patients with only multiple system atrophy (MSA) or MSA associated with autonomic failure (MSA + AF) or with "pure" autonomic failure (PAF) and controls. We = value corresponding to the awake state in the evening before sleep study; 1–2, 3–4 = stages of NREM sleep; Wm = value corresponding to the awake state on the morning after sleep study.

Age at onset varied between 36 and 62, with a mean of 52 ± 8 years. The disease was uniformly fatal, leading to death after only 7–9 months in acute cases or after 25–32 months when it assumed a chronic course. Insomnia was an early symptom, followed by peculiar and striking dream-like states, during which patients enact the content of a dream. Insomnia and oneiric stupor are particularly evident in the shorter-course cases, which also display early autonomic disturbances (sexual impotence, sphincter impairment, increased tearing, salivation and sweating, and increased core body temperature [T_{cb}], HR, and BP). In later stages of the disease motor abnormalities develop in the form of myoclonus, cerebellar and pyramidal involvement. Patients finally die suddenly (short-course cases) or after pulmonary infection. Postmortem brain studies disclosed severe loss of neurons and reactive astrogliosis in the anteroventral and dorsomedian thalamic nuclei; other thalamic nuclei were less severely and variously affected. The cerebral cortex displayed minimal astrogliosis, and in two cases spongiosis in the frontotemporal lobes and cingulate gyrus. Gliosis and loss of Purkinje cells were also common in the cerebellum, but basal ganglia, hypothalamus, and brain stem (except for olivary atrophy) remained remarkably unaffected. Recently, FFI was demonstrated to be a prion disease due to a point mutation at codon 178 of the prion-protein gene. A variant allele at codon 129 of the same gene, encoding for methionine, differentiates FFI from the inherited form of Creutzfeldt-Jakob disease that shares the 178 point mutation.[115]

Polysomnographic Aspects. In some patients 24-hour polysomnographic recordings demonstrated an almost total absence of sleep patterns detectable on the EEG from the early stage of the disease. Only short episodes of abnormal REM sleep occur without the physiologic loss of muscle tone lasting a few seconds or minutes. Patients with a more prolonged course of disease had progressively shortened sleep cycles, with progressive reduction in total sleep time, but the cyclic sleep organization and gradual transition from one stage to another was lost from the onset. Spindle activity and K complexes were totally absent or degraded from the onset of the disease in all patients. This was reflected in the inordinate and sudden passage from one phase of sleep to another, an expression of the loss of regulation of the intrinsic cyclic NREM and REM sleep activity. Polygraphic analysis showed consistent elevation of the values of BP, HR, and T_{cb} in FFI. Correlation with EEG tracings showed that in those cases in which short episodes of NREM still occurred, BP, HR, and T_{cb} fall abruptly, thus still displaying normal state-dependent behavior. In the first case recorded in which the patient displayed total agrypnia, no circadian oscillation of BP, HR, and T_{cb} was evident[109] (FIG. 6). In three other patients[116] studied at different stages of the disease in a temperature-controlled room with continuous recumbency and a controlled regimen for meals and dark-light cycle, nonlinear rhythm analysis identified significant time-dependent variations in systolic and diastolic BP as well as HR in all recording sessions except the preterminal one in one case. The 24-hour period predominated in all sessions. Amplitudes of the 24-hour components of BP were already reduced in the early phase of the disease and decreased further as the disease progressed, until the obliteration of BP and HR rhythms in the last stage of the disease. At night a marked reduction or complete loss in BP fall was apparent in all recording sessions. On the contrary, the 24-hour profile of HR was essentially unchanged although amplitude was reduced. Hence, in these patients the changes in BP did not correlate with those in HR but some 24-hour rhythmicity persisted even when the sleep-wake cycle had already completely disappeared.

Neuroendocrine Findings. In patients with FFI we performed appropriate studies to verify the integrity of the hypothalamic-pituitary axis and the functions of the target organs, and all tests gave normal results.[117] In addition, under standardized conditions and polysomnographic control, patients underwent repeated 24-hour

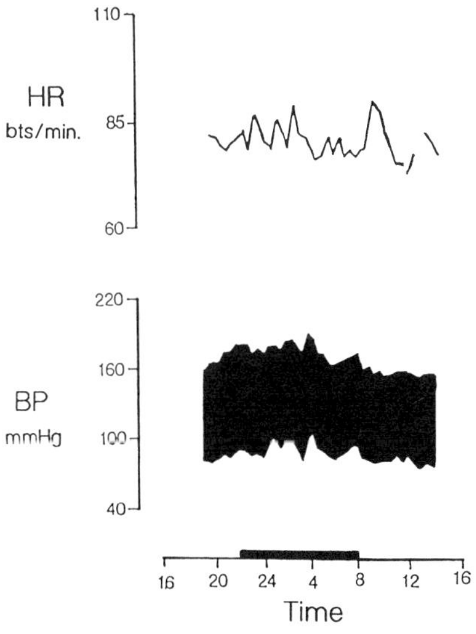

FIGURE 6. Circadian variations of heart rate (HR) and systemic blood pressure (BP, intra-arterial recording) in a patient with fatal familial insomnia.

study sessions covering the entire clinical course of the disease. Somatotropin, prolactin, cortisol, corticotropin (ACTH), plasma catecholamines (noradrenaline and adrenaline), and melatonin were assayed at 30-minute intervals. A dysfunction of the pituitary-adrenal axis was prominent in all cases with elevated serum cortisol and normal levels of ACTH.[116] Catecholamines displayed increased levels throughout the 24 hours considerably synchronized with HR. However, complete rhythm obliteration for these hormones occurred in the terminal stage of FFI. Plasma melatonin concentrations gradually decreased as the disease progressed. A significant circadian rhythm was detected in the earlier recordings, but disappeared in the most advanced stage.[118] The physiologic nocturnal elevation of somatotropin was absent simultaneously with sleep loss, whereas a significant 24-hour component of variations in serum prolactin levels was present for months after total disruption of the sleep/wake cycle with normally placed nocturnal acrophase.[119] Complete rhythm abolition was also observed later for prolactin. In summary, the advanced stages of the disease are invariably characterized by the disappearance of any circadian rhythmicity of the hormones evaluated.

Cardiovascular Autonomic Function. An imbalanced autonomic control with preserved parasympathetic activity and a higher background and stimulated sympathetic activity was the conclusion derived from the autonomic tests.[63] This sympathetic overactivity was evident in elevated noradrenaline plasma levels at rest, increasing further under orthostatic stress, the exaggerated BP responses to physiologic stimuli (postural change, Valsalva maneuver, isometric handgrip), and the absent BP response to noradrenaline infusion, the increased HR response to atropine, and the diminished depressor and sedative effects of clonidine. We confirmed that sympathetic hyperactivity is a hallmark of FFI by evaluating a larger number of patients at an earlier stage of the disease by means of power spectral analysis of HR variability.[120]

Conclusions. Fatal familial insomnia represents a model disease, emphasizing the correlations among the different sleep, autonomic, and neuroendocrine functions. Clinicopathologic correlations demonstrate the role of the thalamus as an integrative neural structure placed between the limbic system and the hypothalamus controlling the homeostatic balance of the organism. A lesion in the visceral part of the thalamus could release the hypothalamus from inhibitory cortical control, resulting in a permanently activated nervous system, a sort of brain unable to rest.[121] The loss of sleep, the persistent basal overactivation, and hyperactivity of the sympathetic nervous system all fit within the hypothesis that ergotrophic functions prevail over trophotropic functions.[122]

From the chronobiologic point of view, FFI offers a unique opportunity to observe in human subjects the daily patterns of hormones and cardiovascular parameters independently of sleep/wake cycles. The sympathetic overactivity may in itself be responsible for the hypercortisolism, and both represent the cause of the systemic hypertension typical of FFI, which can therefore be considered among the "secondary" types of human hypertension. Moreover, our data confirm that a rhythmic 24-hour component in the 24-hour pattern of BP variability may exist independently of sleep/wake and activity cycles. At a stage when the sleep deprivation of FFI was accompanied by a total loss of the nocturnal BP fall, it is noteworthy that nocturnal bradycardia was still preserved, suggesting that sleep accounts for only a minor part of the nocturnal decline in HR and that the 24-hour variations in HR reflect different modulatory influences from those controlling the BP pattern.

In conclusion, the mechanisms underlying the circadian variation of BP are not fully understood, but our data, showing that not only disorders with autonomic failure but also those with sympathetic hyperactivity are associated with the loss of a nocturnal BP fall, provide further evidence that a balanced autonomic function is essential for the normal circadian pattern of BP.

REFERENCES

1. BULOW, K. & D. H. INGVAR. 1961. Respiration and state of wakefulness in normal studied by spirography, capnography and E.E.G. A preliminary report. Acta Physiol. Scand. **51:** 230–238.
2. COCCAGNA, G., M. MANTOVANI, F. BRIGNANI, A. MANZINI & E. LUGARESI. 1971. Arterial pressure changes during spontaneous sleep in man. Electroencephalogr. Clin. Neurophysiol. **31:** 277–281.
3. LUGARESI, E., G. COCCAGNA, M. MANTOVANI & R. LEBRUN. 1972. Some periodic phenomena arising during drowsiness and sleep in man. Electroencephalogr. Clin. Neurophysiol. **32:** 701–705.
4. FERRETTI, R., N. S. CHERNIACK, G. LONGOBARDO, O. R. LEVINE, E. MORKIN, D. H. SINGER & A. P. FISHMAN. 1965. Systemic and pulmonary vasomotor waves. J. Physiol. **209:** 37–50.
5. CHERNIACK, N. S. & G. S. LONGOBARDO. 1973. Stokes breathing: An instability in physiologic control. N. Engl. J. Med. **288:** 952–957.
6. PHILLIPSON, E. A. 1978. Control of breathing during sleep. Am. Rev. Respir. Dis. **118:** 909–939.
7. DELL, P. 1963. Reticular homeostasis and critical reactivity. *In* Brain Mechanisms. G. Moruzzi *et al.,* Eds. Vol. 1. Elsevier. Amsterdam.
8. BENARROCH, E. E. 1993. The central autonomic network: Functional organization, dysfunction, and perspective. Mayo Clin. Proc. **68:** 988–1001.
9. PARMEGGIANI, P. L. 1979. Integrative aspects of hypothalamic influences and sleep. *In* Central Nervous Control Mechanisms in Breathing. C. von Euler & H. Lagercrantz, Eds. Pergamon. Oxford.
10. BEVAN, A. T., A. J. HONOUR & F. H. STOTT. 1969. Direct arterial pressure recording in unrestricted man. Br. Heart J. **31:** 387–388.
11. BRASH, H. M., J. B. IRVING, B. J. KIRBY & K. W. DONALD. 1973. Radiotelemetry of intrarterial pressure in man. Br. Heart J. **35:** 561.
12. MILLAR CRAIG, M. W., C. N. BISHOP & E. B. RAFTERY. 1978. Circadian variation of blood-pressure. Lancet **1**(8068): 795–797.
13. MANCIA, G., A. FERRARI, L. GREGORINI, G. PARATI, G. POMIDOSSI, G. BERTINIERI, G. GRASSI, M. DI RIENZA, A. PEDOTTI & A. ZANCHETTI. 1983. Blood pressure and heart rate variabilities in normotensive and hypertensive human beings. Circ. Res. **53:** 96–104.
14. STAESSEN, J. A., R. H. FAGARD, P. J. LIJNEN, L. THIJS, R. VAN HOOF & A. K. AMERY. 1991. Mean and range of the ambulatory pressure in normotensive subjects from a meta-analysis of 23 studies. Am. J. Cardiol. **67:** 723–727.
15. BROADHURST, P., G. BRIDGEN, P. DASGUPTA, A. LAHIRI & E. B. RAFTERY. 1990. Ambulatory intraarterial pressure in normal subjects. Am. Heart J. **120:** 160–166.
16. PICKERING, T. G. 1990. Diurnal rhythms and other sources of blood pressure in variability in normal and hypertensive subjects. *In* Hypertension: Pathophysiology, Diagnosis and Management. J. H. Laragh & B. M. Brenner, Eds.: 1397–1405. Raven Press. New York.
17. CONWAY, J., N. BOON, J. VANN JONES & P. SLEIGHT. 1985. Mechanisms concerned with blood pressure variability throughout the day. Clin. Exp. Hypertens. A. **7:** 153–157.
18. KRUM, H., W. J. LOUIS, D. J. BROWN, G. P. JACKMAN & L. G. HOWES. 1991. Diurnal blood pressure variation in quadriplegic chronic spinal cord injury patients. Clin. Sci. **80:** 271–276.
19. COCA, A. 1994. Circadian rhythm and blood pressure control: Physiological and pathophysiological factors. J. Hypertens. **12:** S13–S21.
20. HARSHFIELD, G. A., C. HWANG & C. E. GRIM. 1990. Circadian variation of blood pressure in blacks: Influence of age, gender and activity. J. Hum. Hypertens. **4:** 43–47.

21. Fumo, M. T., S. Teeger, R. M. Lang, J. Bednarz, P. Sareli & M. B. Murphy. 1992. Diurnal blood pressure variation and cardiac mass in American blacks and whites and South African blacks. Am. J. Hypertens. **5:** 111–116.
22. Gretler, D. D., M. T. Fumo, K. S. Nelson & M. B. Murphy. 1994. Ethnic differences in circadian hemodynamic profile. Am. J. Hypertens. **7:** 7–14.
23. Mann, S., M. W. Craig, D. I. Merville, V. Balasubramanian & E. B. Raftery. 1979. Physical activity and the circadian rhythm of blood pressure. Clin. Sci. **57:** 291s–294s.
24. Lynch, J. J., S. A. Thomas, J. M. Long, K. L. Malinow, G. Chickadonz & A. H. Katcher. 1980. Human speech and blood pressure. J. Nerv. Ment. Dis. **168:** 526–534.
25. Friedman, E., S. A. Thomas, D. Kulick-Ciuffo, J. J. Lynch & M. Suginahara. 1982. The effects of normal and rapid speech on blood pressure. Psychosom. Med. **44:** 545–553.
26. Pickering, T. G. 1990. Short-term variability of blood pressure, and the effects of physical and mental activity. *In* Ambulatory Monitoring and Blood Pressure Variability. T. G. Pickering, Ed. Part 1:4.1–4.17. Science Press. London.
27. Ghione, S., C. Rosa, L. Salerno, G. Ceccanti & D. Volterrani. 1993. Assessment of physical and mental activity using a standardized computer-assisted diary during ambulatory blood pressure monitoring. J. Hypertens. **11:** S284–S285.
28. Hartikainen, J., I. Tarkiainen, K. Tahvanainen, M. Mantysaari, E. Lansimies & K. Pyorala. 1993. Circadian variation of cardiac autonomic regulation during 24-h bed rest. Clin. Physiol. **13:** 185–196.
29. Cellina, G. U., A. J. Honour & W. A. Littler. 1975. Direct arterial pressure, heart rate, and electrocardiogram during cigarette smoking in unrestricted patients. Am. Heart J. **89:** 18–25.
30. Robertson, D., J. C. Frolich, R. K. Carr, J. T. Watson, J. W. Hollifield, D. G. Shand & J. A. Oates. 1978. Effects of caffeine on plasma renin activity, catecholamines and blood pressure. N. Engl. J. Med. **298:** 181–186.
31. Freestone, S. & L. E. Ramsey. 1982. Effect of coffee and cigarette smoking on the blood pressure of untreated and diuretic-treated hypertensive patients. Am. J. Med. **73:** 348–353.
32. Izzo, J. L., A. Ghosal, T. Kwong, R. B. Freeman & J. R. Jaenike. 1983. Age and prior caffeine use after the cardiovascular and adrenomedullary responses to oral caffeine. Am. J. Cardiol. **52:** 769–773.
33. Malhotra, H., D. Mathur, S. R. Mehta & P. D. Khandelwal. 1985. Pressor effects of alcohol in normotensive and hypertensive subjects. Lancet **ii:** 584–586.
34. Potter, J. F., A. Macdonald & D. G. Beevers. 1986. Alcohol raises blood pressure in hypertensive patients. J. Hypertens. **4:** 434–441.
35. Pickering, T. G. 1990. Diurnal, ultradiurnal and seasonal rhythms of blood pressure. *In* Ambulatory Monitoring and Blood Pressure Variability. T. G. Pickering, Ed. Part 1:5.1–5.15. Science Press. London.
36. Pickering, T. G. 1990. The clinical significance of diurnal blood pressure variations: Dippers and nondippers. Circulation **81:** 700–702.
37. Verdecchia, P., G. Schillaci & C. Porcellati. 1991. Dippers versus non-dippers. J. Hypertens. **9:** S42–S44.
38. Padfield, P. L. & M. J. Stewart. 1991. Ambulatory blood pressure monitoring in secondary hypertension. J. Hypertens. **9:** S69–S71.
39. White, W. B. 1992. Diurnal blood pressure and blood pressure variability in diabetic normotensive and hypertensive subjects. J. Hypertens. **10:** S35–S41.
40. Statius van Eps, R. G., A. H. van der Meiracker, F. Boomsma, A. J. Man in 't Veld & M. A. Schalekamp. 1993. Partial preservation of nocturnal fall in blood pressure in patients with catecholamine-producing tumors. J. Hypertens. **11:** S168–S169.
41. Kastrup, J., H. Wroblewski, J. Sindrup, H. Rolighed & N. Wiinberg. 1993. Diurnal blood pressure profile in patients with severe congestive heart failure: Dippers and non-dippers. Scand. J. Clin. Lab. Invest. **53:** 577–583.
42. Hassan, M. O., O. T. Shafie & W. J. Johnston. 1993. Loss of the nocturnal dip and increased variability of blood pressure in normotensive patients with noninsulin-dependent diabetes mellitus. Clin. Physiol. **13:** 519–523.

43. NODA, A., T. OKADA, H. HAYASHI, F. YASUMA & M. YOKOTA. 1993. 24-Hour ambulatory blood pressure variability in obstructive sleep apnea syndrome. Chest **103:** 1343–1347.
44. IDEMA, R. N., A. H. VAN DER MEIRACKER, A. J. MAN IN 'T VELD, A. BALK, E. BOS & M. A. D. H. SCHALEKAMP. 1993. Circadian hemodynamics in heart transplant recipients. J. Hypertens. **11:** S90–S91.
45. VAN DER MEIRACKER, A. H., J. H. M. TULEN, A. J. MAN IN 'T VELD, A. BALK, H. G. VAN STEENIS & M. A. D. H. SCHALEKAMP. 1993. Blood pressure and heart rate variability in clinical models of cardiac denervation, sympathicolysis and sympathico- and parasympathicolysis. J. Hypertens. **11:** S152–S153.
46. PARATI, G., G. POMIDOSSI, F. ALBINI, D. MALASPINA & G. MANCIA. 1987. Relationship of 24-hour blood pressure mean and variability to severity of target-organ damage in hypertension. J. Hypertens. **5:** 93–98.
47. O'BRIEN, E., J. SHERIDAN & K. O'MALLEY. 1988. Dippers and non-dippers. Lancet **ii:** 397.
48. PALATINI, P., M. PENZO, A. RACIOPPA, E. ZUGNO, G. GUZZARDI, M. ANACLERIO & A. C. PESSINA. 1992. Clinical relevance of nighttime blood pressure and of daytime blood pressure variability. Arch. Intern. Med. **52:** 1855–1860.
49. SHIMADA, K., A. KAWAMOTO, K. MATSUBAYASHI, M. NISHINAGA, S. KIMURA & T. OZAWA. 1992. Diurnal blood pressure variation and silent cerebrovascular change in elderly patients with hypertension. J. Hypertens. **10:** 875–878.
50. VERDECCHIA, P., G. SCHILLACI, C. GATTESCHI, I. ZAMPI, M. BATTISTELLI, C. BARTOCCINI & C. PORCELLATI. 1993. Blunted nocturnal fall in blood pressure in hypertensive women with future cardiovascular morbid events. Circulation **88:** 986–992.
51. FRATTOLA, A., G. PARATI, C. CUSPIDI, F. ALBINI & G. MANCIA. 1993. Prognostic value of 24-hour blood pressure variability. J. Hypertens. **11:** 1133–1137.
52. KOBRIN, I., W. OIGMAN, A. KUMAR, H. O. VENTURA, F. H. MESSERLI, E. D. FROHLICH & F. G. DUNN. 1984. Diurnal variation of blood pressure in elderly patients with essential hypertension. J. Am. Geriatr. Soc. **32:** 896–899.
53. VERDECCHIA, P., G. SCHILLACI, M. GUERRIERI, C. GATTESCHI, G. BENEMIO, F. BOLDRINI & C. PORCELLATI. 1990. Circadian blood pressure changes and left ventricular hypertrophy in essential hypertension. Circulation **81:** 526–528.
54. VERDECCHIA, P., G. SCHILLACI, F. BOLDRINI, M. GUERRIERI & C. PORCELLATI. 1992. Sex, cardiac hypertrophy and diurnal blood pressure variations in essential hypertension. J. Hypertens. **10:** 683–692.
55. FLORAS, J. S., J. V. JONES, J. A. JOHNSTON, D. E. BROOKS, M. O. HASSAN & P. SLEIGHT. 1978. Arousal and the circadian rhythm of blood pressure. Clin. Sci. Mol. Med. **55:** 395s–397s.
56. BURSZTYN, M., J. MEKLER, N. WACHTEL & D. BEN-ISHAY. 1994. Siesta and ambulatory blood pressure monitoring. Comparability of the afternoon nap and night sleep. Am. J. Hypertens. **7:** 217–221.
57. CHAU, N. P., J. M. MALLION, R. DE GANDEMERIS, E. RUCHIE, J. P. SICHE, O. PELEN & G. MATHERN. 1989. Twenty-four-hour ambulatory blood pressure in shift workers. Circulation **80:** 341–347.
58. SUNDBERG, S., A. KOLWAKKER & A. GORDIN. 1988. Rapid reveals of circadian blood pressure rhythm in shift workers. J. Hypertens. **6:** 393–396.
59. BAUMGART, P., P. WALGER, G. FUCHS, K. G. DORST, H. VETTER & K. H. RAHN. 1989. Twenty-four-hour blood pressure is not dependent on circadian rhythm. J. Hypertens. **7:** 331–334.
60. SFORZA, E., V. CAPECCHI & E. LUGARESI. 1992. Haemodynamic effects of short-term nasal continuous positive airway pressure therapy in sleep apnoea syndrome: Monitoring by a finger arterial pressure device. Eur. Respir. J. **5:** 858–863.
61. MATHIAS, C. J. & R. BANNISTER. 1992. Investigation of autonomic disorders. *In* Autonomic Failure. R. Bannister & C. J. Mathias, Eds. 3rd Ed.: 255–290. Oxford University Press. Oxford, New York, Tokyo.
62. CORTELLI, P., P. PARCHI, E. SFORZA, M. CONTIN, G. PIERANGELI, G. BARLETTA & E. LUGARESI. 1994. Cardiovascular autonomic dysfunction in normotensive awake subjects with obstructive sleep apnoea syndrome. Clin. Auton. Res. **4:** 57–62.

63. CORTELLI, P., P. PARCHI, M. CONTIN, G. PIERANGELI, P. AVONI, P. TINUPER, P. MONTAGNA, A. BARUZZI, P. GAMBETTI & E. LUGARESI. 1991. Cardiovascular dysautonomia in fatal familial insomnia. Clin. Auton. Res. **1:** 15–21.
64. LUGARESI, E., G. COCCAGNA, P. FARNETI, M. MANTOVANI & F. CIRIGNOTTA. 1975. Snoring. Electroencephalogr. Clin. Neurophysiol. **39:** 59–64.
65. LUGARESI, E., G. COCCAGNA, F. CIRIGNOTTA, P. FARNETI, R. GALLASSI, G. DI DONATO & P. VERUCCHI. 1978. Breathing during sleep in man in normal and pathological conditions. Adv. Exp. Med. Biol. **99:** 35–45.
66. LUGARESI, E., G. COCCAGNA & M. MANTOVANI. 1978. Hypersomnia with periodic apneas. *In* Advances in Sleep Research. E. Weitzman, Ed. Vol. 4. Spectrum. New York.
67. LUGARESI, E., F. CIRIGNOTTA, G. COCCAGNA & C. PIANA. 1980. Some epidemiological data on snoring and cardiocirculatory disturbances. Sleep **3:** 221–224.
68. LUGARESI, E., F. CIRIGNOTTA, G. COCCAGNA & P. MONTAGNA. 1984. Clinical significance of snoring. *In* Sleep and Breathing. N. A. Saunders & C. E. Sullivan, Eds. Marcel Dekker. New York.
69. COCCAGNA, G., M. MANTOVANI, F. BRIGNANI & C. PARCHI. 1972. Continuous recording of the pulmonary and systemic arterial pressure during sleep in syndrome of hypersomnia with periodic breathing. Bull. Physio-Path. Resp. **8:** 1087–1102.
70. LUGARESI, E., S. MONDINI, M. ZUCCONI, P. MONTAGNA & F. CIRIGNOTTA. 1983. Staging of heavy snorers disease. A proposal. Bull. Eur. Physiopathol. Respir. **19:** 590–594.
71. MONDINI, S., M. ZUCCONI, F. CIRIGNOTTA, U. AGUGLIA, P. L. LENZI, C. ZAULI & E. LUGARESI. 1983. Snoring as a risk factor for cardiac and circulatory problems: An epidemiological study. *In* Sleep/Wake Disorders Natural History, Epidemiology, and Long-Term Evolution. C. Guilleminault & E. Lugaresi, Eds. Raven, New York.
72. PARTINEN, M., J. KAPRIO, M. KOSKENVUO & H. LANGINVAINIO. 1983. Snoring and hypertension a cross-sectional study on 12808 Finns aged 24–65 years. Sleep Res. **12:** 273.
73. GUILLEMINAULT, C. A., TIKIAN & W. C. DEMENT. 1976. The sleep apnea syndromes. Annu. Rev. Med. **27:** 465–484.
74. TILKIAN, A., C. GUILLEMINAULT, J. S. SCHROEDER, K. L. LEHRMAN, F. B. SIMMONS & W. C. DEMENT. 1976. Hemodynamics in sleep-induced apnea: Studies during wakefulness and sleep. Ann. Intern. Med. **85:** 714–719.
75. KALES, A., R. J. CADIEUX, E. O. BIXLER, C. R. SOLDATOS, A. VELA-BUENO, C. A. MISOUL & T. W. LOCKE. 1985. Severe obstructive sleep apnea. I. Onset, clinical course, and characteristics. J. Chron. Dis. **38:** 419–425.
76. KALES, A., E. O. BIXLER, R. J. CADIEUX, D. W. SCHNECK, L. C. SHAW, T. W. LOCKE, A. VELABUENO & C. R. SOLDATOS. 1984. Sleep apnoea in a hypertensive population. Lancet **ii:** 1005–1008.
77. LAVIE, P., R. BEN-YOSEF & A. E. RUBIN. 1985. Prevalence of sleep apnea syndrome among patients with essential hypertension. Am. J. Cardiol. **55:** 1019–1022.
78. FLETCHER, E. C., R. D. DEBEHNKE, M. S. LOVOI & A. B. GORIN. 1985. Undiagnosed sleep apnea in patients with essential hypertension. Ann. Intern. Med. **103:** 190–195.
79. WILLIAMS, A. J., D. HOUSTON, S. FINBERG, C. LAM, J. L. KENNEY & S. SANTIAGO. 1985. Sleep apnea syndromes and essential hypertension. Am. J. Cardiol. **55:** 1019–1022.
80. LUGARESI, E., G. COCCAGNA, M. MANTOVANI & F. BRIGNANI. 1972. Effects of tracheostomy in two cases of hypersomnia with periodic breathing. J. Neurol. Neurosurg. Psychiatry **36:** 15–26.
81. GUILLEMINAULT, C., F. L. ELDRIDGE, F. B. SIMMONS & W. C. DEMENT. 1975. Sleep apnea syndrome: Can it induce hemodynamic changes? West J. Med. **123:** 7–16.
82. GUILLEMINAULT, C., F. B. SIMMONS, J. MOTTA, J. CUMMSKEY, M. ROSEKIND, J. S. SCHROEDER & W. C. DEMENT. 1981. Obstructive sleep apnea syndrome and tracheostomy. Arch. Intern. Med. **141:** 985–988.
83. JENNUM, P., G. WILDSCHIODTZ, N. J. CHRISTENSEN & T. SCHWARTZ. 1989. Blood pressure, catecholamines, and pancreatic polypeptide in obstructive sleep apnea with and without nasal continuous positive airway pressure (nCPAP) treatment. Am. J. Hypertens. **2:** 847–852.

84. MAYER, J., H. BECKER, U. BRANDENBURG, T. PENZEL, J. H. PETER & P. VON WICHERT. 1991. Blood pressure and sleep apnea: Results of long-term nasal continuous positive airway pressure therapy. Cardiology **79:** 84–92.
85. GUILLEMINAULT, C. & M. SUZUKI. 1992. Sleep-related hemodynamics and hypertension with partial or complete upper airway obstruction during sleep. Sleep **15:** S20–S24.
86. WILCOX, I., R. R. GRUNSTEIN, J. HEDNER, J. DOYLE, F. L. COLLINS, P. J. FLETCHER, D. T. KELLY & C. E. SULLIVAN. 1993. Effect of nasal continuous positive airway pressure during sleep on 24-hour blood pressure in obstructive sleep apnea. Sleep **16:** 539–544.
87. TRZEBSKI, A. 1992. Arterial chemoreceptor reflex and hypertension. Hypertension **19:** 562–566.
88. FLETCHER, E. C. 1995. An animal model of the relationship between systemic hypertension and repetitive episodic hypoxia as seen in sleep apnoea. J. Sleep Res. **4:** 71–77.
89. EHLENZ, K. 1995. Regulation of blood volume-implications for cardiovascular pathophysiology in sleep apnoea. J. Sleep Res. **4:** 30–33.
90. RINGLER, J., E. GARPESTAD, R. C. BASNER & J. WOODROW WEISS. 1994. Systemic blood pressure elevation after airway occlusion during NREM sleep. Am. J. Respir. Crit. Care Med. **150:** 1062–1066.
91. CALVERLEY, P. M. A. & K. REES. 1995. Systemic arterial blood pressure during obstructive sleep apnoea. J. Sleep Res. **4:** 93–96.
92. GUILLEMINAULT, C. & R. STOOHS. 1995. Arousal, increased respiratory efforts, blood pressure and obstructive sleep apnoea. J. Sleep Res. **4:** 117–124.
93. RINGLER, J., R. C. BASNER, R. SHANNON, R. SCHWARTZSTEIN, H. MANNING, S. E. WEINBERGER & J. W. WEISS. 1990. Hypoxemia alone does not explain blood pressure elevations after obstructive apneas. J. Appl. Physiol. **69:** 2143–2148.
94. MATEIKA, J. H., S. MATEIKA, A. S. SLUTSKY & V. HOFFSTEIN. 1992. The effect of snoring on mean arterial blood pressure during non-REM sleep. Am. Rev. Respir. Dis. **145:** 141–146.
95. HEDNER, J. A., I. WILCOX, L. LAKS, R. R. GRUNSTEIN & C. E. SULLIVAN. 1992. A specific and potent pressor effect of hypoxia in patients with sleep apnea. Am. Rev. Respir. Dis. **146:** 1240–1245.
96. WILCOX, I., F. L. COLLINS, R. R. GRUNSTEIN, J. HEDNER, D. T. KELLY & C. E. SULLIVAN. 1994. Relationship between chemosensitivity, obesity and blood pressure in obstructive sleep apnoea. Blood Press. **3:** 47–54.
97. FLETCHER, E. C., J. MILLER, J. W. SCHAAF & J. G. FLETCHER. 1987. Urinary catecholamines before and after tracheostomy in OSA and hypertension. Sleep **10:** 35–44.
98. HEDNER, J., H. EJNELL, J. SELLGREN, T. HEDNER & G. WALLIN. 1988. Is high and fluctuations muscle nerve sympathetic activity in the sleep apnea syndrome of pathogenetic importance for the development of hypertension? J. Hypertens. **6:** S529–S531.
99. BARUZZI, A., R. RIVA, F. CIRIGNOTTA, M. ZUCCONI, M. CAPPELLI & E. LUGARESI. 1991. Atrial natriuretic peptide and catecholamines in obstructive sleep apnea syndrome. Sleep **14:** 83–86.
100. MARRONE, O., L. RICCOBONO, A. SALVAGGIO, A. MIRABELLA, A. BONANNO & M. R. BONSIGNORE. 1993. Catecholamines and blood pressure in obstructive sleep apnea syndrome. Chest **103:** 722–727.
101. GRAHAM, J. G. & D. R. OPPENHEIMER. 1969. Orthostatic hypotension and nicotine sensitivity in a case of multiple system atrophy. J. Neurol. Neurosurg. Psychiatry **32:** 28–34.
102. ADAMS, R. D., L. VAN BOGAERT & H. VAN DER EECKEN. 1961. Dégénérescences nigro-striées et cérébello-nigro-striées. Psychiatr. Neurol. **142:** 219–259.
103. SHY, G. M. & G. A. DRAGER. 1960. A neurological syndrome associated with orthostatic hypotension. Arch. Neurol. **2:** 511–527.
104. DEJERINE, J. & A. A. THOMAS. 1900, L'atrophie olivo-ponto-cérébelleuse. Nouv. Iconog. Salpetrière **13:** 330–370.
105. BRADBURY, S. & C. EGGLESTON. 1925. Postural hypotension. A report of 3 cases. Am. Heart J. **1:** 73.
106. WENNING, G. K., Y. BEN SHLOMO, M. MAGALHAES, S. E. DANIEL & N. P. QUINN. 1994.

Clinical features and natural history of multiple system atrophy. An analysis of 100 cases. Brain **117:** 835–845.
107. Aldrich, M. S. 1994. Parkinsonism. *In* Principles and Practice of Sleep Medicine. M. H. Kryger, T. Roth & W. C. Dement, Eds. 2nd Ed. 783–789. W. B. Saunders Co. Philadelphia.
108. Mathias, C. J., P. Fosbraey, D. F. De Costa, A. Thorley & R. Bannister. 1986. Desmopressin reduces nocturnal polyuria, reverses overnight weight loss and improves morning postural hypotension in autonomic failure. Br. Med. J. **293:** 353–354.
109. Lugaresi, E., R. Medori, P. Montagna, A. Baruzzi, P. Cortelli, A. Lugaresi, P. Tinuper, M. Zucconi & P. Gambetti. 1986. Fatal familial insomnia and dysautonomia with selective degeneration of thalamic nuclei. N. Engl. J. Med. **315:** 997–1003.
110. Manetto, V., R. Medori, P. Cortelli, P. Montagna, P. Tinuper, A. Baruzzi, G. Rancurel, J. J. Hawn, J. J. Vanderhaeghen, P. Mailleux, O. Bugiani, F. Tagliavini, C. Bouras, N. Rizzuto, E. Lugaresi & P. Gambetti. 1992. Fatal familial insomnia: Clinical and pathologic study of five new cases. Neurology **42:** 312–319.
111. Medori, R., H. J. Tritschler, A. Leblanc, F. Villare, V. Manetto, H. Y. Chen, R. Xue, P. Montagna, P. Cortelli, P. Tinuper, P. Avoni, M. Mochi, A. Baruzzi, J. J. Hawn, E. Lugaresi, L. Autilio-Gambetti & P. Gambetti. 1992. Fatal familial insomnia: A prion disease with a mutation at codon 178 of the prion protein gene. N. Engl. J. Med. **326:** 444–487.
112. Medori, R., P. Montagna, H. J. Tritschler, A. Leblanc, P. Cortelli, P. Tinuper, E. Lugaresi & P. Gambetti. 1992. Fatal familial insomnia: A second kindred with mutation of prion protein gene at codon 178. Neurology **42:** 669–670.
113. Montagna, P., P. Cortelli, P. Tinuper, E. Sforza, P. Avoni, R. Gallassi, A. Morreale, I. Roiter, D. Perani, G. Lucignani, F. Fazio & E. Lugaresi. 1994. Fatal familial insomonia: A disease that emphasizes the role of the thalamus in the regulation of sleep and vegetative functions. *In* Fatal Familial Insomnia: Inherited Prion Diseases, Sleep, and the Thalamus. C. Guilleminault *et al.,* Eds. Chap. 1: 1–14. Raven Press. New York.
114. Montagna, P., P. Cortelli, P. Gambetti & E. Lugaresi. 1995. Fatal familial insomnia: Sleep, neuroendocrine and vegetative alterations. Adv. Neuroimmunol. **5:** 13–21.
115. Goldfarb, L. G., R. B. Peterson, M. Tabaton, P. Brown, A. C. Leblanc, P. Montagna, P. Cortelli, J. Julien, C. Vital, W. W. Pendelbury, M. Haltia, P. R. Wills, J. J. Hauw, P. E. McKeever, L. Monari, B. Schrank, G. D. Swergold, L. Autilio-Gambetti, D. C. Gajdusek, E. Lugaresi & P. Gambetti. 1992. Fatal familial insomnia and familial Creutzfeldt-Jakob disease: Disease phenotype determined by a DNA polymorphism. Science **258:** 806–809.
116. Portaluppi, F., P. Cortelli, P. Avoni, L. Vergnani, M. Contin, P. Maltoni, A. Pavani, E. Sforza, E. C. Degliuberti, P. Gambetti & E. Lugaresi. 1994. Diurnal blood pressure variation and hormonal correlates in fatal familial insomnia. Hypertension **23:** 569–576.
117. Lugaresi, A., A. Baruzzi, E. Cacciari, P. Cortelli, R. Medori, P. Montagna, P. Tinuper, M. Zucconi, I. Roiter & E. Lugaresi. 1987. Lack of vegetative and endocrine circadian rhythms in fatal familial thalamic degeneration. Clin. Endocrinol. **26:** 573–580.
118. Portaluppi, F., P. Cortelli, P. Avoni, L. Vergnani, P. Maltoni, A. Pavani, E. Sforza, E. C. Degliuberti, P. Gambetti & E. Lugaresi. 1994. Progressive disruption of the circadian rhythm of melatonin in fatal familial insomnia. J. Clin. Endocrinol. Metab. **78:** 1075–1078.
119. Portaluppi, F., P. Cortelli, P. Avoni, L. Vergnani, P. Maltoni, A. Pavani, E. Sforza, R. Manfredini, P. Montagna, I. Roiter, P. Gambetti, C. Fersini & E. Lugaresi. 1995. Dissociated 24-hour patterns of somatotropin and prolactin in fatal familial insomnia. Neuroendocrinology **61:** 731–737.
120. Cortelli, P., G. Pierangeli, P. Parchi, G. Barletta, M. Contin, P. Gambetti, P. Montagna & E. Lugaresi. 1994. Power spectral analysis reveals sympathetic hyper-

activity as an early feature of Fatal Familial Insomnia (FFI). Neurology **44**(Suppl. 2): A363.
121. LUGARESI, E. 1992. The thalamus and insomnia. Neurology **42**(Suppl. 6): 28–33.
122. HESS, W. R. 1969. Hypothalamus and Thalamus: Experimental Documentation. Georg Thieme. Stuttgart.

Long-Term and Short-Term Blood Pressure Variability in Transplanted Patients

P. VAN DE BORNE, M. LEEMAN, AND J. P. DEGAUTE

Hospital Erasme
Université Libre de Bruxelles
Brussels, Belgium

Hypertension is frequently encountered among patients with cardiac transplants (HTX)[1] and challenges the recent progress made in both surgical techniques and immunosuppression. Indeed, although the immunosuppressive drug cyclosporin has prolonged life expectancy,[1] its administration has increased the prevalence of hypertension among transplanted patients.[1]

As for essential hypertension, increased peripheral resistance is the hemodynamic hallmark of hypertensive HTX patients.[1] However, both entities differ in an important feature: while 24-hour variations in blood pressure and heart rate of both normotensive[2,3] and essential hypertensive patients are characterized by a diurnal increase and a nocturnal decrease,[4,5] HTX patients present abnormal nocturnal variations in both blood pressure and heart rate.[6–9] This finding has an important clinical implication, because both daytime and nighttime ambulatory blood pressure levels are predictive of cardiac organ damage,[10,11] and performing only daytime blood pressure measurements may underestimate the severity of hypertension in organ-transplanted patients.

The nocturnal decrease in both blood pressure and heart rate in normal subjects is thought to be due to a decrease in sympathetic activity during light and deep sleep[12] while vagal tone increases.[13,14] Blood pressure decrease then is a result of decreased cardiac output[15] and vascular resistance.[16] By contrast, the exact mechanism responsible for abnormal nocturnal variations in blood pressure in HTX patients remains unclear and may involve several mechanisms among which are: salt and water retention induced by cyclosporin therapy, the force-fed pump characteristics of the denervated heart, and the administration of exogenous glucocorticoids.

Although studies performed in our laboratory at the Free University of Brussels, Belgium do not provide conclusive evidence of the exact mechanism responsible for the nocturnal increase in blood pressure in cardiac transplanted patients, they reveal that: (1) salt and water retention per se is not responsible for the paradoxic nighttime increase in blood pressure; (2) cardiac denervation per se is not responsible for the paradoxic nighttime increase in BP in organ transplanted patients; (3) not all cardiac and liver transplanted patients present abnormal 24-hour BP profiles; and (4) attenuation of the nighttime decrease in BP is more pronounced during the first weeks after cardiac transplantation.

SALT AND WATER RETENTION PER SE IS NOT RESPONSIBLE FOR THE PARADOXIC NIGHTTIME INCREASE IN BLOOD PRESSURE

We recorded noninvasively 24-hour blood pressure and heart rate variations in 14 carefully selected chronic hemodialysis patients.[17] None had orthostatic hypotension, and all were free of any previously reported associated condition known to affect 24-hour blood pressure profiles. Five of these patients were anephric and five were anuric. Despite progressive salt and water retention, these chronic hemodialysis patients had normal 24-hour blood pressure profiles. Moreover, in 10 chronic hemodialysis patients, 48-hour ambulatory recordings showed that a mean increase of 1.7 ± 0.3 kg in body weight did not blunt the amplitude of two successive 24-hour systolic and diastolic blood pressure profiles.

CARDIAC DENERVATION PER SE IS NOT RESPONSIBLE FOR THE PARADOXIC NIGHTTIME INCREASE IN BLOOD PRESSURE IN ORGAN TRANSPLANTED PATIENTS

We explored the 24-hour blood pressure and heart rate patterns of patients with liver transplants (LTX) who underwent immunosuppressive therapy similar to that of cardiac transplanted patients but whose heart is not denervated by the surgical procedure.[18] The study revealed that almost 40% of these patients showed a paradoxic nighttime increase in systolic and diastolic blood pressure. By contrast, patients and control subjects had similar 24-hour variations in heart rate. Thus, paradoxic nighttime increases in blood pressure were observed in the presence of normal nighttime bradycardia. We reported similar independence between blood pressure and heart rate variability when we performed spectral analysis of short-term blood pressure and heart rate variabilities in cardiac transplanted patients.[19]

Previous fixed-rate pacemaker studies also showed the independence of 24-hour blood pressure and heart rate variations,[20] and our study suggests that the paradoxic increase in blood pressure in organ transplanted patients occurs in the presence of a preserved nocturnal decrease in cardiac sympathovagal balance. Moreover, the similar nighttime decrease in heart rate also suggests that patients without a nocturnal blood pressure decline were not poorer sleepers.[21]

NOT ALL CARDIAC AND LIVER TRANSPLANTED PATIENTS PRESENT ABNORMAL 24-HOUR BLOOD PRESSURE PROFILES

Analysis of blood pressure profiles revealed that cardiac and liver transplanted patients presented heterogeneous 24-hour patterns consisting of either normal daytime acrophases and nighttime nadirs or inverted profiles (daytime nadir and nighttime acrophase[9,18]). To elucidate possible factors responsible for the abnormal blood pressure patterns observed in some of the HTX and LTX patients, these patients were divided into two subgroups depending on whether they presented daytime acrophases and nighttime nadirs.

Both groups of HTX patients were similar in age, number of days after transplantation, left isotopic ejection fraction, serum creatinine, oral dose of azathioprine, and number taking antihypertensive medication. By contrast, patients presenting normal blood pressure profiles received a lower daily dose of prednisolone ($p < 0.001$) and cyclosporin ($p = 0.04$). Moreover, among the aforementioned parameters, only nighttime systolic and diastolic blood pressure decreases were inversely correlated with the daily oral dose of prednisolone ($r = -0.47$ and $r = -0.53$, $p < 0.05$). Similar results were found in the LTX patient groups which differed only in the daily dose of prednisolone ($p < 0.001$), and only the differences between daytime and nighttime systolic and diastolic blood pressure were inversely correlated with the daily oral dose of prednisolone ($r = -0.64$ and $r = -0.66, p < 0.01$).

ATTENUATION OF THE NIGHTTIME DECREASE IN BLOOD PRESSURE IS MORE PRONOUNCED DURING THE FIRST WEEKS AFTER CARDIAC TRANSPLANTATION

Our longitudinal study[9] showed that 60% of HTX patients had paradoxic nocturnal increases in blood pressure during the first month after cardiac transplantation but that a nighttime decrease in blood pressure reappeared 7 months after transplantation.

Variable levels of daytime physical activity and sleep quality could have contributed to these findings. However, all patients included in these studies were rehabilitated and fully ambulant. Furthermore, if environmental stimuli and changes in physical activity are minimized, a fall in blood pressure of approximately 20% still occurs during sleep.[22] Thus, a reduced level of physical activity cannot explain per se the increases observed in nighttime blood pressure.

We observed in this study that only a reduction in daily prednisolone dosage correlated with the greater nighttime decrease in systolic and diastolic blood pressure ($r = 0.61$ and $r = 0.54, p < 0.05$) observed 7 months after cardiac transplantation.

These results therefore suggest that exogenous glucocorticoid administration may play a role in the pathogenesis of the abnormal blood pressure profiles presented by some cardiac transplanted patients. These findings support those of Imai *et al.*[23] who reported abnormal blood pressure profiles and normal heart rate profiles in patients with Cushing's syndrome[23] or in those treated with exogenous glucocorticoids.[24] In particular, in a longitudinal study these authors reported the effects of the administration at 8 AM of a single mean dose of 45 mg of prednisolone (i.e., a dose similar to the one received by the HTX and LTX patients who presented abnormal 24-hour blood pressure profiles) for more than 1 week in six patients with lupus erythematosus or glomerulonephritis. All presented normal 24-hour blood pressure profiles before the administration of exogenous glucocorticoids. However, prednisolone induced an important paradoxic nighttime increase in systolic blood pressure and a disappearance of the nighttime decrease in diastolic blood pressure in these patients.

Glucocorticoids are reported to modulate the synthesis of sympathetic neurotransmitters[25] and the vascular response to catecholamines.[26–28] Moreover, under normal circumstances steroid hormones are thought to pass the blood-brain barrier[29] and glucocorticoids can act centrally to increase sympathetic nerve activity.[30]

The interactive effects of elevated systemic vascular resistance due to the vascular effects of cyclosporin[1] and the aforementioned effects of glucocorticoids on sympathetic nerve activity might account for the recent observation that the attenuated decrease in blood pressure in HTX patients is accompanied by a paradoxic 9 ± 9% increase in vascular resistance.[31]

REFERENCES

1. STARLING, R. C. & CODY, R. I. 1990. Cardiac transplant hypertension. Am. J. Cardiol. **65:** 106–111.
2. DEGAUTE, J. P., P. VAN DE BORNE, P. LINKOWSKI & E. VAN CAUTER. 1991. Quantitative analysis of the 24-hour blood pressure and heart rate patterns in young men. Hypertension **18:** 199–210.
3. DEGAUTE, J. P., E. VAN CAUTER, P. VAN DE BORNE & P. LINKOWSKI. 1994. Twenty-four-hour blood pressure and heart rate profiles in humans: A twin study. Hypertension **23:** 244–253.

4. MILLAR-CRAIG, M. W., C. N. BISHOP & E. B. RAFTERY. 1978. Circadian variation of blood pressure. Lancet **1:** 795–797.
5. PICKERING, T. G. 1990. Diurnal rhythms and other sources of blood pressure variability in normal and hypertensive subjects. *In* Hypertension, Pathophysiology, Diagnosis, and Management. J. H. Laragh & B. M. Brenner, Eds.: 1397–1405. Raven Press, New York.
6. REEVES, R. A., A. P. SHAPIRO, M. E. THOMPSON & A. M. JOHNSEN. 1986. Loss of nocturnal decline in blood pressure after cardiac transplantation. Circulation **73:** 401–408.
7. WENTING, G. J., A. H. VAN DEN MEIRACKER, H. J. RITSEMA VAN EEK, M. L. SIMOONS, E. BOS, W. WEIMAN, A. J. MAN IN'T VELT & M. A. D. H. SCHALEKAMP. 1986. Lack of circadian variation of blood pressure after heart transplantation. J. Hypertens. **4:** S78–S80.
8. VON POLNITZ, A., C. BRACHT, B. KEMKES & B. HÖFLING. 1990. Circadian pattern of blood pressure and heart rate in the longer term after heart transplantation. J. Cardiovasc. Pharmacol. **16:** S86–S89.
9. VAN DE BORNE, P., M. LEEMAN, G. PRIMO & J. P. DEGAUTE. 1992. Reappearance of a normal circadian blood pressure rhythm after cardiac transplantation. Am. J. Cardiol. **69:** 795–801.
10. SOKOLOW, M., D. WERDEGAR, H. K. KAIN & A. T. HIMMAN. 1966. Relationship between level of blood pressure measured casually and by portable recorders and severity of complications in essential hypertension. Circulation **34:** 279–298.
11. WHITE, W. B., H. M. DEY & P. SCHULMAN. 1989. Assessment of the daily blood pressure load as a determinant of cardiac function in patients with mild-to-moderate hypertension. Am. Heart J. **118:** 782–795.
12. SOMERS, V. K., M. E. DYKEN, A. L. MARK & F. M. ABBOUD. 1993. Sympathetic-nerve activity during sleep in normal subjects. N. Engl. J. Med. **328:** 303–307.
13. VAN DE BORNE, P., H. NGUYEN, P. BISTON, P. LINKOWSKI & J. P. DEGAUTE. 1994. Effects of wake and sleep stages to the 24-hour autonomic control of blood pressure and heart rate in recumbent men. Am. J. Physiol. **266:** H548–H554.
14. VAN DE BORNE, P., P. BISTON, M. PAÏVA, H. NGUYEN, P. LINKOWSKI & J. P. DEGAUTE. 1995. Cardiorespiratory transfer during sleep: A study in healthy young men. Am. J. Physiol.
15. KHATRI, I. M. & E. D. FREIS. 1967. Hemodynamic changes during sleep. J. Appl. Physiol. **22:** 867–873.
16. BRISTOW, J. D., A. J. HONOUR, T. G. PICKERING & P. SLEIGHT. Cardiovascular and respiratory changes during sleep in normal and hypertensive subjects. J. Cardiovasc. Med. **3:** 476–486.
17. VAN DE BORNE, P., C. TIELEMANS, F. COLLART, J. L. VANHERWEGHEM & J. P. DEGAUTE. 1993. Twenty-four-hour blood pressure and heart rate patterns in chronic hemodialysis patients. Am. J. Kidney Dis. **22:** 419–425.
18. VAN DE BORNE, P., M. GELIN, J. VAN DE STADT & J. P. DEGAUTE. 1993. Circadian rhythms of blood pressure after liver transplantation. Hypertension **21:** 398–405.
19. VAN DE BORNE, P., M. SCHINTGEN, G. NISET, P. SCHOENFELD, H. NGUYEN, S. DEGRE & J. P. DEGAUTE. 1994. Does cardiac denervation affect the short-term blood pressure variability in men? J. Hypertens. **12:** 1395–1403.
20. DAVIES, A. B., B. A. GOULD, P. M. M. CASHMAN & E. B. RAFFERTY. 1984. Circadian rhythm of blood pressure in patients dependent on ventricular demand pacemakers. Br. Heart J. **52:** 93–98.
21. PICKERING, T. G. 1991. Diurnal, ultradian and seasonal rhythms of blood pressure. *In* Ambulatory Monitoring and Blood Pressure Variability: 5.1–5.11. Science Press, Ltd. London.
22. PICKERING, T. G. 1991. Short-term variability of blood pressure and the effects of physical and mental activity. *In* Ambulatory Monitoring and Blood Pressure Variability: 4.1–4.15. Science Press, Ltd. London.
23. IMAI, Y., K. ABE, S. SASAKI, N. MINAMI, M. NIHEI, M. MUNAKATA, O. MURAKAMI, K. MATSUE, H. SEKINO, Y. MIURA & K. YOSHINAGA. 1988. Altered circadian blood pressure rhythm in patients with Cushing's syndrome. Hypertension **12:** 11–19.

24. IMAI, Y., K. ABE, S. SASAKI, N. MINAMI, M. MUNAKATA, M. NIHEI, H. SEKINO & K. YOSHINAGA. 1989. Exogenous glucocorticoid eliminates or reverses circadian blood pressure variations. J. Hypertens. **7:** 133–120.
25. HAMET, P. 1983. Endocrine hypertension: Cushing's syndrome, acromegaly, hyperparathyroidism, thyrotoxicosis, and hypothyroidism. *In* Hypertension: Physiopathology and Treatment. 2nd Ed. J. Genest, O. Kuchel, P. Hamet & M. Cantin, Eds: 964–976. McGraw-Hill. New York.
26. MENDLOWITZ, M., S. GITLOW & N. NAFTCHI. 1958. Work of digital vasoconstriction produced by infused norepinephrine in Cushing's syndrome. J. Appl. Physiol. **13:** 252–256.
27. KALSNER, S. 1969. Mechanism of hydrocortisone potentiation of responses to epinephrine and norepinephrine in rabbit aorta. Circ. Res. **24:** 383–395.
28. KRAKOFF, L., G. NICOLIS & B. AMSEL. 1975. Pathogenesis of hypertension in Cushing's syndrome. Am. J. Med. **58:** 216–220.
29. PFAFF, D. W. 1978. Peptide and steroid hormones and the neural mechanisms for female reproductive behavior. *In* The Hypothalamus. S. Reichlin, R. J. Baldessarini & J. B. Martin, Eds: 245–253. Raven Press, New York.
30. TAKAHASHI, H., K. TAKEDA, H. ASHIZAWA *ET AL.* 1983. Centrally induced cardiovascular and sympathetic responses to hydrocortisone in rats. Am. J. Physiol. **245:** H1013–H1018.
31. IDEMA, R. N., A. H. VAN DEN MEIRACKER, A. H. M. M. BALK, E. BOS, M. A. D. H. SCHALEKAMP & A. J. MAN IN'T VELD. 1994. Abnormal diurnal variation of blood pressure, cardiac output, and vascular resistance in cardiac transplant recipients. Circulation **90:** 2797–2803.

Temporal Blood Pressure Patterns in Children

RONALD J. PORTMAN AND ROBERT J. YETMAN

Department of Pediatrics
University of Texas–Houston, Medical School
and
Hermann Center for Chronobiology and Chronotherapeutics
Houston, Texas 77401

Blood pressure (BP) is measured in children to diagnose and treat hypertension and to identify, at the earliest possible time, those subjects at risk for hypertension later in life. Using current definitions, about 1% of children have hypertension,[1] whereas 2–65% of adults are given this diagnosis depending on age, race, and gender.[2] Accurate methods must be found to identify the potential for hypertension in children before well-known complications of this important disease process become evident. Molecular genetics may provide the best tools to predict individuals at risk[3]; however, it is critical to have an accurate picture of children's BP patterns so that the most reliable markers of the genetic component of BP can be ascertained. Achieving these goals will further the important aim of prevention that has been the hallmark of pediatrics since its inception as a specialty.

In this review, we briefly discuss the current status of BP measurement and diagnosis of hypertension in children, the development of biological rhythms in children, and the chronobiology of BP in children as determined by ambulatory blood pressure monitoring (ABPM).

MEASUREMENT OF BLOOD PRESSURE IN CHILDREN

Measurement of BP in children has been a routine part of the physical examination only in the last 20 years. The National Institutes of Health (NIH) published the first population, normative data of BP in children in 1977 [4]; this was later improved and replaced by current standards from the NIH Task Force of 1987.[1] Since that time, European studies[5] have reported similar data to that of the 1987 Task Force. All of these normative data are based on casual (often single) BP measurements and not on ABPM. The definitions of hypertension are based on epidemiologic data and not outcome measures. As we will discuss, we are unable not only to predict which child will become the hypertensive adult, but also to determine, except in cases of the most severe hypertension, which child will have end-organ damage.

DEVELOPMENT OF BLOOD PRESSURE IN CHILDREN

Blood pressure is lower in children than in adults, increasing throughout growth and development. In a full-term infant, BP is ~60/40 mm Hg systolic (SBP) and diastolic (DBP), respectively. An increase in SBP of 1 mm Hg per day for the first week of life occurs with a rather rapid rate of rise, albeit at a slower pace, for the first

2 months of life. After this initial rapid rise, BP rises slowly throughout the remainder of the first year of life. DBP follows a somewhat different pattern, increasing during the first week, then falling to a nadir at 3 months of age. DBP then increases parallel to SBP throughout the first year of life. From ages 1–13 years a linear rise in BP, based more on body size than age, is expected. Beginning at about age 5, SBP climbs 1–2 mm Hg per year until adolescence, with DBP rising less sharply than SBP. To determine the correct DBP, K4 (not K5) is used. Use of K4 rather than K5 has important implications for interpretation of ABPMs using auscultatory technology as these monitors cannot determine the muffling sound of K4 but rather identify K5 (loss of all sound). For children between the ages of 13 and 18 years, the SBP and DBP continue to rise at a constant rate.

The relationship of BP and body size was recognized by the NIH Task Force[1] and a caveat is included in their guidelines to diagnose hypertension only after consideration of the subject's height and weight. On these BP curves, the 90th percentiles for height and weight are noted. Thus, if BP is between the 90th and 95th percentile and the subject's height or weight is greater than the 90th percentile, the diagnosis of hypertension might not be appropriate.

FACTORS AFFECTING A CHILD'S BLOOD PRESSURE

Two main groups of factors affect BP in children: situational and inherent. Situational factors including body position, anxiety, wellness, ambient temperature, extremity selection, measuring device, and activity tend to be more important for casual measurements than for ABPM values. For both techniques the largest cuff that does not impinge on the antecubital fossa or axillary space and has an inflatable bladder large enough to encircle the arm completely should be used. Body position is clearly important for appropriate interpretation of intermittent measurements, whereas variable body positions are expected with ABPM. Anxiety, activity, wellness (vs illness), and ambient temperature affect a single, intermittent BP measurement more significantly than do a series of ABPM measurements. For intermittent measurements, more choices exist for extremity selection (arm vs leg) and monitor type (Doppler, Dinamapp, mercury, or aneroid). ABPM measures BP almost exclusively in the arm and by auscultatory and/or oscillometric technologies only.

Inherent factors of a child's BP include birth weight, gestational age, maternal BP, body mass, ethnicity, gender, family history, fitness, tracking, maturation, and mineral intake; all affect either mode of BP measurement. In the newborn period, birth weight, gestational age, and maternal BP are important. Outside the newborn period, factors operative in the determination of adult hypertension are also present in childhood.

IMPORTANCE OF FAMILY HISTORY

Much conflicting data exist on the role of family history in a child's BP; all of these data are based on casual and not ABPM. Sinaiko[6] suggests that despite a strong correlation of BP between parent and child, the presence of hypertension can only be used to identify risk and is not of value in predicting hypertension. Thus, offspring born to one or two hypertensive parents are at higher risk for hypertension than are those born to a family with no history of hypertension. It is impossible, however, to predict which of these children, if any, will develop hypertension later in life. In the future, molecular genetic studies may help to more accurately make this determination especially if used in conjunction with ABPM.

TRACKING OF BLOOD PRESSURE

Tracking is a general term that describes a child's tendency to maintain throughout time a particular percentile rank for a considered variable. The most commonly used tracking data are height and weight which are used with some degree of accuracy to predict final adult height and weight. Blood pressure, too, seems to track, but its predictive value for later hypertension remains controversial. Studies to date have been performed only with casual BP measurements which do not allow consideration of the white coat phenomenon; this factor alone might be enough to alter tracking data.

Gillman *et al.*[7] measured BP in 317 children 8–12 years apart and found evidence of tracking. The correlation coefficient for SBP tracking was 0.55 and for DBP 0.44. The positive predictive value was only 0.44. The specificity was good at 0.97, but the sensitivity was poor at 0.17. Thus, tracking is demonstrable, but the low sensitivity and positive predictive value for childhood BP as a screen for adult hypertension challenges the utility of this measure to identify children at high risk.

FOCUS ON PEDIATRIC HYPERTENSION?

A search of the *Index Medicus* from 1966 to the present elicits over 100,000 articles on hypertension in adults versus 2,000 in infants and children. The Fifth Report of the Joint National Committee Report on Detection, Evaluation, and Treatment of High Blood Pressure[8] is 30 pages long, yet only one page deals with pediatric issues. Why is there a lack of study of hypertension in children? As hypertension is unusual in children and presents few, if any, symptoms, no child is expected to have hypertension during a routine visit. Therefore, less emphasis is placed on accurate determination of BP. By contrast, the incidence is dramatically higher in adults, and all patients are candidates for hypertension. As with Type I diabetes mellitus, the sequelae of hypertension are rarely seen by the pediatrician, and the full impact of these diseases is seen many years later. If we are treating hypertension and its complications in the adult, are we too late?

DEFINITION OF HYPERTENSION

In children the incidence of hypertension after repeated casual measurements is generally considered to be 1%[1] based on Task Force data.[8] The definition of hypertension, however, is based on epidemiologic data only, that is, defining as abnormal a BP exceeding the 95th %ile in a population of presumably healthy subjects. In adults, more precise definition of hypertension has been developed with limits established by evidence of end-organ damage. Ideally, such a definition will be developed for children and would include subjects at risk for adverse outcomes (i.e., end-organ damage) when BPs exceed a determined threshold. Subjects with BPs lower than the definition would be at little risk for adverse outcomes.

The 1987 Task Force[1] offers several interpretations of casual BP measurements in children. "High normal" is defined as BP between the 90th and 95th percentiles for age. "Hypertension" is defined as BP greater than the 95th percentile for age but less than the 99th percentile. "Severe hypertension" includes BP greater than the 99th percentile. For none of these definitions is a relationship to end-organ damage included.

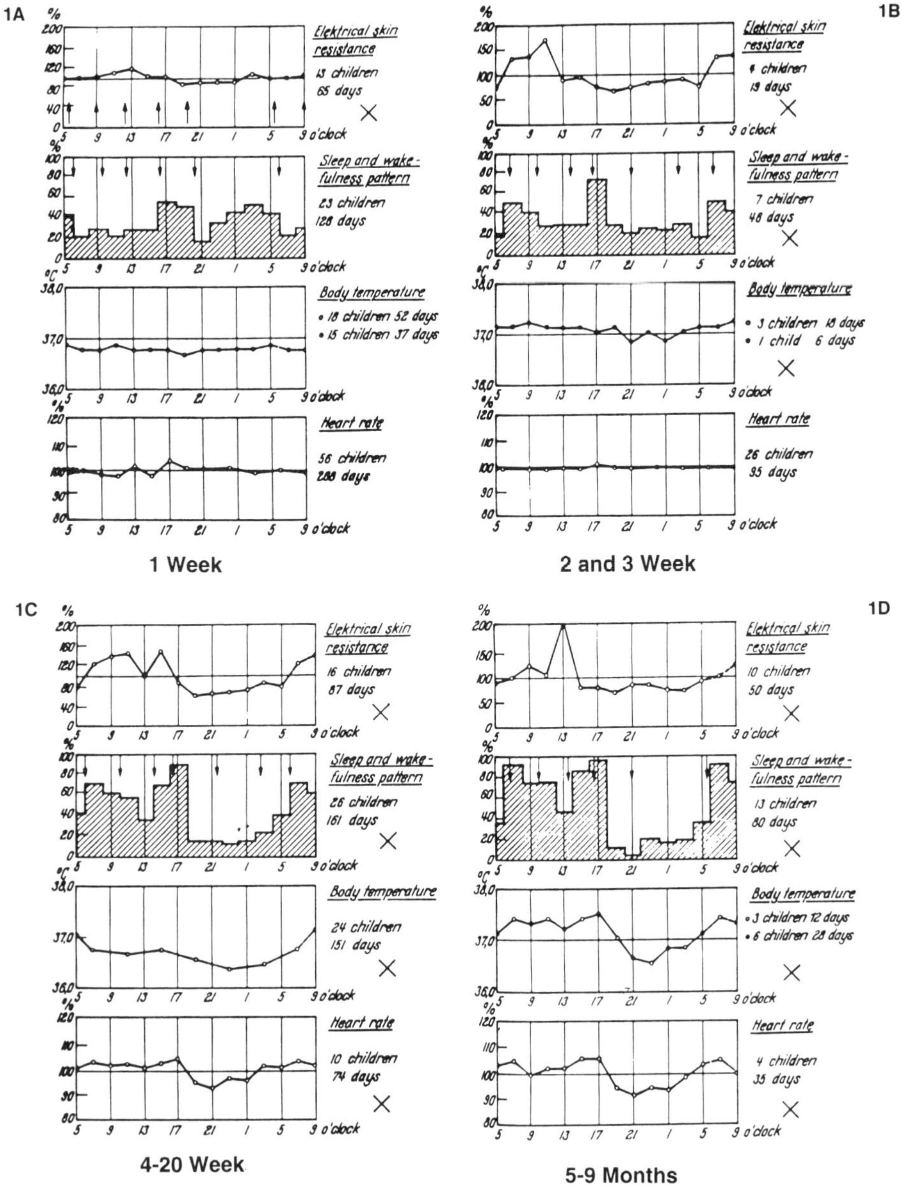

An adequate definition of hypertension for use with ABPM is also wanting. Current definitions[1] include the 24-hour mean exceeding the 95th percentile of age-adjusted norms based on Task Force, casual measurements and/or failure of mean sleep BPs to decrease by 10% from these daytime normals. Alternatively, elevations of BP load (percentage of BPs in a 24-hour period exceeding the 95th %ile

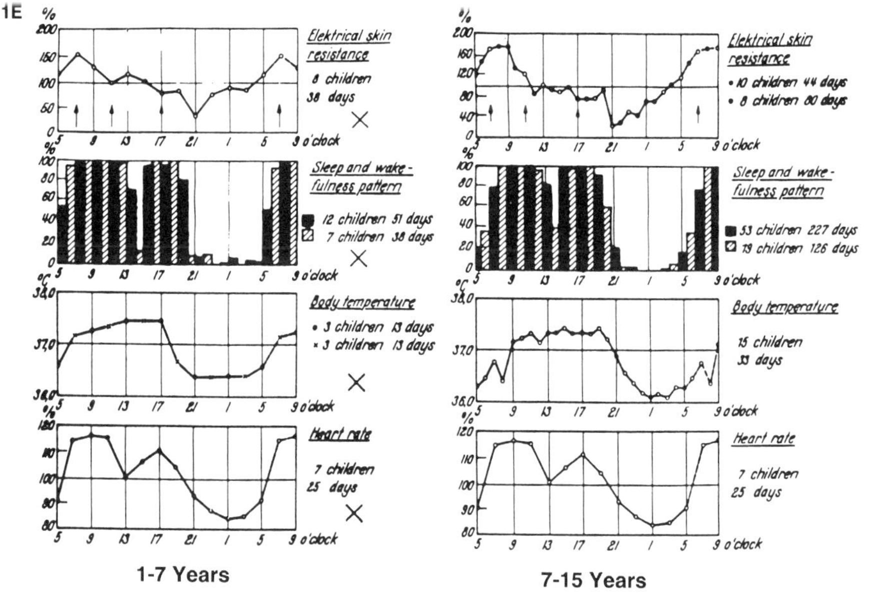

FIGURE 1. Development of circadian rhythms in children. Presented are six sets of graphs of patients at different ages, with hours of the day along the x-axis. *Arrows* along this axis represent feeding times. The X at the right of each graph represents statistical significance of circadian rhythm detection by cosinor analysis. During the first week of life, only electrical skin resistance demonstrates a significant circadian rhythm. Other variables such as sleep, body temperature, and heart rate (as well as urine output and sodium and potassium excretion, not shown) demonstrate no discernible circadian pattern. By weeks 2–3 of life, circadian patterns are demonstrated with skin resistance, sleep, and temperature. By 4–20 weeks of age and certainly by 5–9 months, children continue to have significant patterns for all variables with the same phase as earlier in life, but with a deeper amplitude, and MESOR (24–hour mean) increases with increasing age. The 1–7 and 7–15-year-old children continue to have significant patterns of these variables, now similar to those of an adult. Reproduced with permission from ref. 10.

for age) greater than 25% are often used.[9] For these definitions of hypertension, casual BP data from the Task Force,[1] and not normative data from the ABPM, are used. For none is the definition based on end-organ damage.

CIRCADIAN RHYTHMS IN INFANTS

The classic cross-sectional studies of circadian rhythms in infants of Hellbruegge *et al.*[10] were presented at a conference of the New York Academy of Sciences in 1964. Conclusions from this study of 297 children from the first week of life to age 15 years include six important points: (1) physiologic functions develop circadian rhythms independently of each other; (2) rhythms develop at different times after birth; (3) increased range of oscillation occurs in all physiologic functions with age; (4) an increase in oscillation can occur from increases in the upper width of oscillation during light (activity) or increases in lower oscillation during dark hours (sleep); (5) in man, monophasic rhythms or circadian rhythms originate out of polyphasic or ultradian ones, and; (6) maturation of the infant at birth is essential to rhythm development. Details of the study can be seen in FIGURE 1.

Recent studies have confirmed Hellbruegge et al.'s findings. Mirmiran and Kok[11] found in a group of 12 premature infants 29–35 weeks of gestational age a circadian pattern of heart rate in about half the subjects by 1–2 weeks of age. Gemelli *et al.*[12] studied BP and heart rate patterns in 21 term newborns at 4 days of age. Single cosinor analysis revealed few significant rhythms. None of the subjects had rhythms of heart rate, but males had greater BP variability than did females. Sitka *et al.*[13] (FIG. 2) examined circadian patterns in 17 infants on day 2 and at 4 weeks of age. Almost

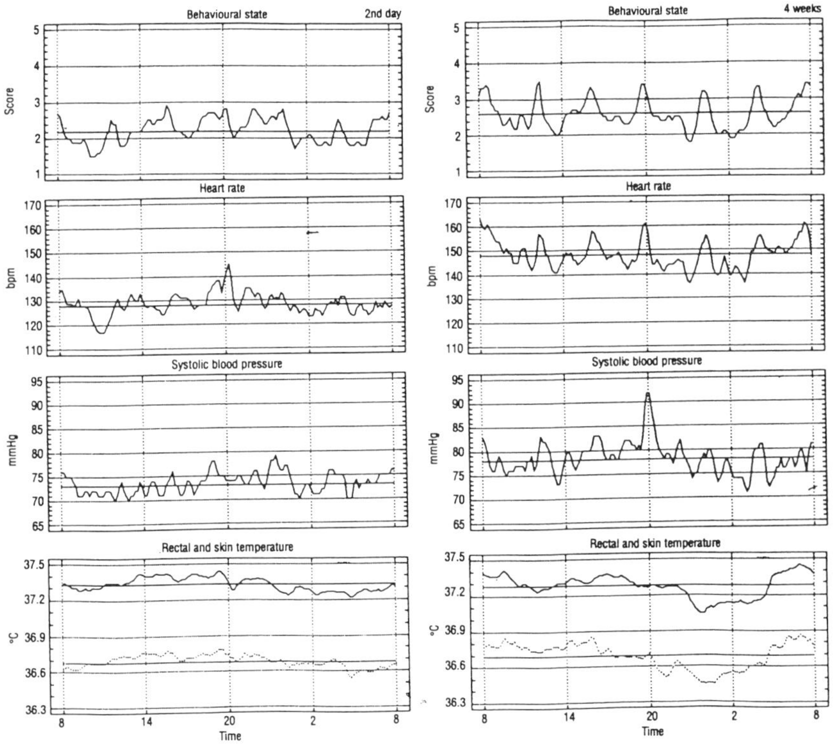

FIGURE 2. Rhythms of 24 hours in neonates. (**Left**) Rhythms in eleven 2-day-old neonates. Note that primarily low-amplitude ultradian rhythms are visualized. At 4 weeks of age (**right**), ultradian rhythms have larger amplitudes and there is the onset of a circadian rhythm for temperature. Reproduced with permission from ref. 13.

all subjects had rhythms in body temperature unrelated to activity, and greater than 50% had rhythms for SBP. None demonstrated a circadian rhythm in DBP. In all of these studies, activity, environment, and feeding times played a significant role in pattern development. However, the infant rhythm was distinct and could be separated from maternal rhythms.

In summary, 12- and 24-hour rhythms develop from higher frequency, ultradian rhythms at an early age. Statistically significant circadian rhythms are often absent at birth, especially with increasing prematurity, and rhythms observed are most likely a reflection of activity.[13,14] The circadian rhythms in children have the same pattern as those in adults, but with a lower 24-hour mean level and occasionally with circadian peaks (acrophases) shifted due to earlier sleep or awake times. Amplitudes of these various rhythms, especially BP, gradually increase in childhood, peaking in adolescence.

AMBULATORY BLOOD PRESSURE MONITORING IN CHILDREN

At least 30 articles[14–50] have documented that ABPM is more effective and accurate than intraarterial and mercury manometric BP measurements in the pediatric patient. Nineteen studies[16,17,22,23,26,29,31–33,35,37–39,41–43,46–48] have attempted to define normal circadian patterns of BP in children. The use of these studies as norms is limited because of inadequate numbers of subjects in the various age ranges and/or the lack of consideration of the subjects' heights. The necessary studies to define normal BP from ABPM have yet to be performed. The definition of day or night (or more appropriately awake and asleep) is problematic in many studies. The largest study attempting to provide normative data included children of various ages and maturations (and thus variable times of sleep and awakening) from 10 centers in Germany. Actigraphy (body movement-based) and diary definitions of sleep were not obtained. To ensure that only pure awake and asleep values were included, these investigators defined awake as 8 AM to 8 PM and sleep as midnight to 6 AM. The remaining BPs (25% of each day's data set) were eliminated. Elimination of these data is of great concern as these are the times of greatest transition in BP (FIG. 3).

Knowledge about the timing of the sleep/wake cycle is critical to the accurate determination of nocturnal hypertension and BP loads. We found significant differences in BP loads when the default time of sleep, as set by the proprietary software (usually 10 PM), was compared to actual sleep times, determined by wrist actigraphy.[51] In this study, a 9.9% difference in sleep SBP and 8.2% in sleep DBP loads was noted. During the awake period, the difference was 4.9% for SBP and 2.8% for BP.

CONSIDERATION OF ACTIVITY

In children, activity must be considered when interpreting ABPM values. Although a circadian BP pattern exists in children, it is frequently masked by activity.[13,14] In adults, mean 24-hour BP from ABPM is generally lower than casual measurements; in children, mean 24-hour BP tends to be higher than casual measurements.[16] This difference is thought to be due to activity.

Several studies in adults and children have demonstrated a statistically significant correlation of activity with BP. Neus *et al.*[17] found increased SBP with higher diary-reported activity levels (FIG. 4). Similarly, Van Egeren[52] found increased BP with increased activity as determined with actigraphy. Using neither diaries nor actigraphy, Yetman *et al.*[14] found increased rhythmicity of BP when heart rate (as a surrogate of activity) was used as a covariate in the analysis of BP data.

Blood pressure is not invariably correlated with activity. In 14 normal subjects from our center,[58] correlation coefficients for activity and SBP, DBP, mean arterial BP, and heart rate were 0.46, 0.53, 0.54, and 0.53, respectively. For a group of 14 renal transplant patients, these values were 0.06, 0.13, 0.2, and 0.55, respectively. Thus, heart rate correlated with activity, whereas none of the measures of BP did (FIG. 5). We are hopeful that with further study, such correlations may prove useful to identify various forms of hypertension.

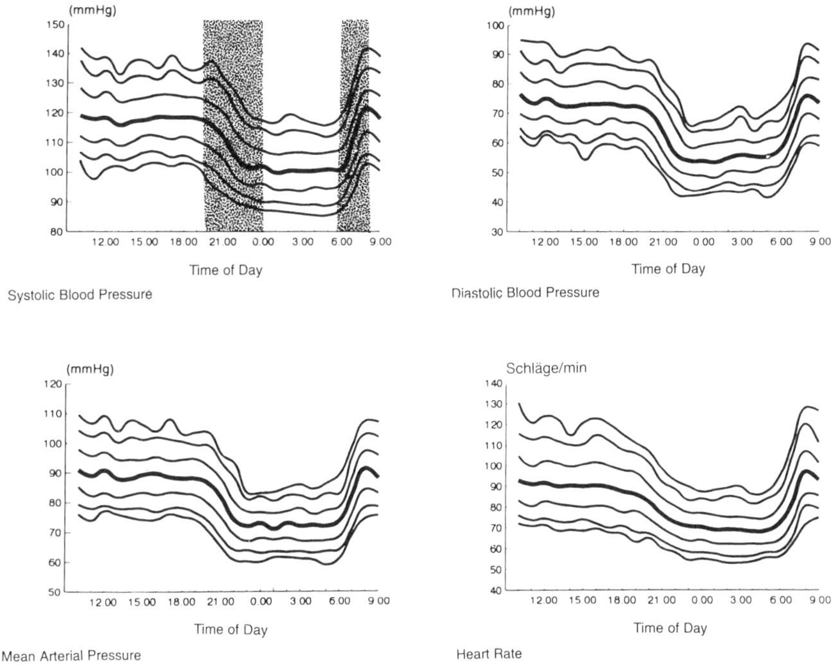

FIGURE 3. Normal values for 24-hour blood pressure in children. Two hundred ten children (111 girls and 99 boys aged 6–16 years) underwent ABPM for 24 hours at 20–30–minute intervals. Plotted for each graph are the 5, 10, 25, 50, 75, 90, and 95th percentile. In the *upper left hand graph, shaded areas* represent data that would be eliminated by an arbitrary definition of sleep and wake from 8 AM–8 PM wake and midnight to 6 AM sleep periods. Reproduced with permission from ref. 38.

AMBULATORY VERSUS CASUAL BLOOD PRESSURE MEASUREMENT IN CHILDREN

Whether ABPM will improve our ability over casual BP measurements to discriminate among various groups of children is unknown. Important clues to answer this question are beginning to emerge from the literature.

In most studies using casual BP measurements, no differences in racial groups prior to puberty have been identified.[54] Using ABPM, however, Harshfield *et al.*[21] demonstrated an increase in nocturnal BP in young black males as compared to white boys or girls of either ethnicity. This group[20] also found higher nocturnal BPs in black boys and a lower BP with improved fitness in both blacks and whites.

Several investigators have identified an impact of family history of hypertension on a child's BP. Wilson *et al.*[23] found higher DBP during school in patients with at least one parent with hypertension. These data are consistent with those from adults whose BP obtained during work correlate best to end-organ damage. Ravogli *et al.*[24] found that children with a positive family history of hypertension have higher ABPM values than do those without (FIG. 6). Berenson *et al.*[25] also found a positive family history of hypertension to be associated with higher BP in a cohort of children with high BPs and a cohort of children with low BPs. Children with a positive family history of hypertension have a higher amplitude of the 24-hour pattern of BP than do subjects without such a history.[25] Thus, ABPM may better discriminate BP changes related to race or a positive family history at a younger age than casual measurements.

ABNORMAL BLOOD PRESSURE PATTERNS IN CHILDREN

As has been documented for adults, abnormal BP patterns including stress, nocturnal, white coat, and amplitude hypertension have been seen in pediatric

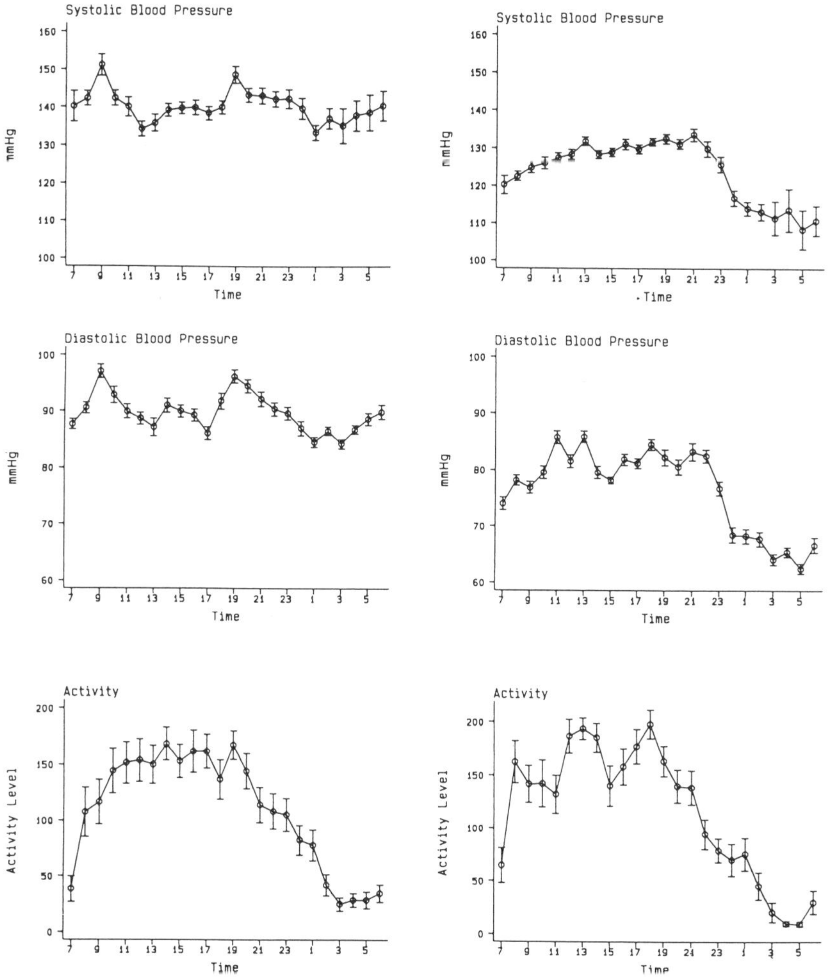

FIGURE 4. Relationship of activity to BP and heart rate in normal subjects and renal transplant patients. Twenty-four-hour variation in the SBP (*top*), DBP (*middle*), and wrist actigraphy level (*bottom*) in 14 diurnally active kidney transplant patients (*left*) and 14 active healthy controls (*right*). Patients exhibit little day-night variation in the hourly average of SBP and DBP in marked contrast to control subjects who exhibit the typical decline in BP at night when activity level is minimal. Reproduced with permission from ref. 53.

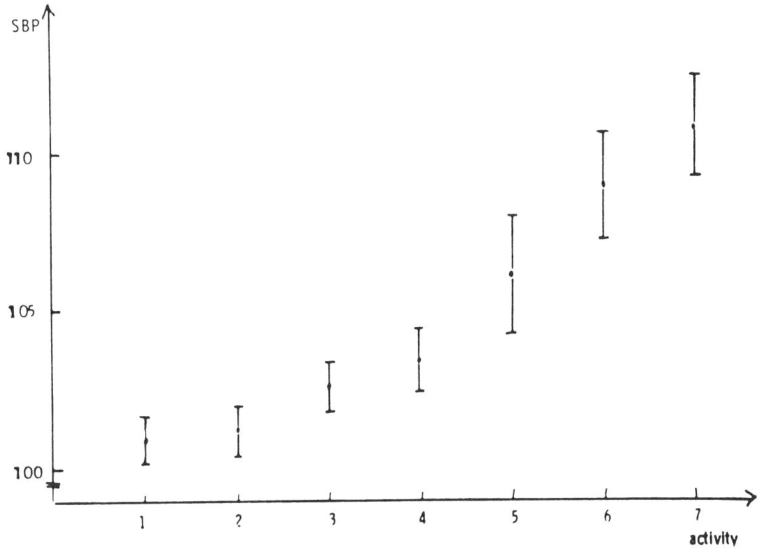

FIGURE 5. Effect of activity score on ambulatory BP values in children. One hundred children had ABPM for 24 hours with scoring of activity from 1 (very inactive) to 7 (physically active). Excellent correlation was noted between activity score and SBP. Reproduced with permission from ref. 17.

patients. At least three groups have shown excellent correlation between mental stress and elevated BP in adolescents.[26–28] Adolescents may be especially prone to stress hypertension. Modesti *et al.*[27] found higher BP reactivity to stress while playing video games in 13–16 year olds compared to children 5–8 and 9–12 years old.

The normal fall in SBP during sleep is 8–15 mm Hg SBP and 15–22 mm Hg for DBP.[16] Lack of this nocturnal fall in BP has been seen in children with high renin levels, renal diseases including chronic renal failure, end-stage renal disease, and renal transplantation, glycogen storage disease, and diabetes. We found in 52 patients with various types of renal disease an increase at night of 2 mm Hg in SBP and 1 mm Hg in DBP. Also observed in patients with renal disease were increasing SBP loads from the day to the night (load SBP-day 39% vs load SBP-night 44%; load DBP-day 30% vs load SBP-night 32%). By contrast, in our previous report in normal children,[16] we found that SBP loads decrease from 15% during the day to 10% at night and DBP loads from 10% to 7%. Harshfield *et al.*[29] examined the relationship of high, medium, and low renin states by casual BP measurements in a group of normotensive subjects. They found a trend of higher BP in the high renin group without statistically significant differences for day or night periods. Using ABPM, they found no difference between various renin levels during the awake period but a significant increase in sleeping BP for the high renin group. This difference is likely due to a marked increase in BP variance (i.e., high amplitude) in this group (FIG. 7).

White coat hypertension in children, that is, hypertension in the presence of health care professionals and normal at all other times, has been demonstrated by Hornsby *et al.*[30] who found that 70 (44%) of 159 pediatric patients with a positive family history of hypertension had normal mean 24-hour BP; 18% of 120 children with no family history of hypertension also had white coat hypertension. Thus, overall, 33% of these children had white coat hypertension. Interestingly, these children with white coat hypertension were heavier than children who did not demonstrate white coat hypertension.

Amplitude hypertension is predominately seen in adolescents[34] and may represent "borderline" hypertensives as defined by Julius *et al.*[55] These patients have marked BP variability with peaks and troughs of BP negating each other and yielding a normal mean BP. However, examination of the BP load shows that these patients are in the hypertensive range a large amount of the day. Some investigators feel that the total amount of time spent in the hypertensive range may be more important than the 24-hour mean or MESOR level[56] and may correlate better with end-organ damage than do mean BP values.[56]

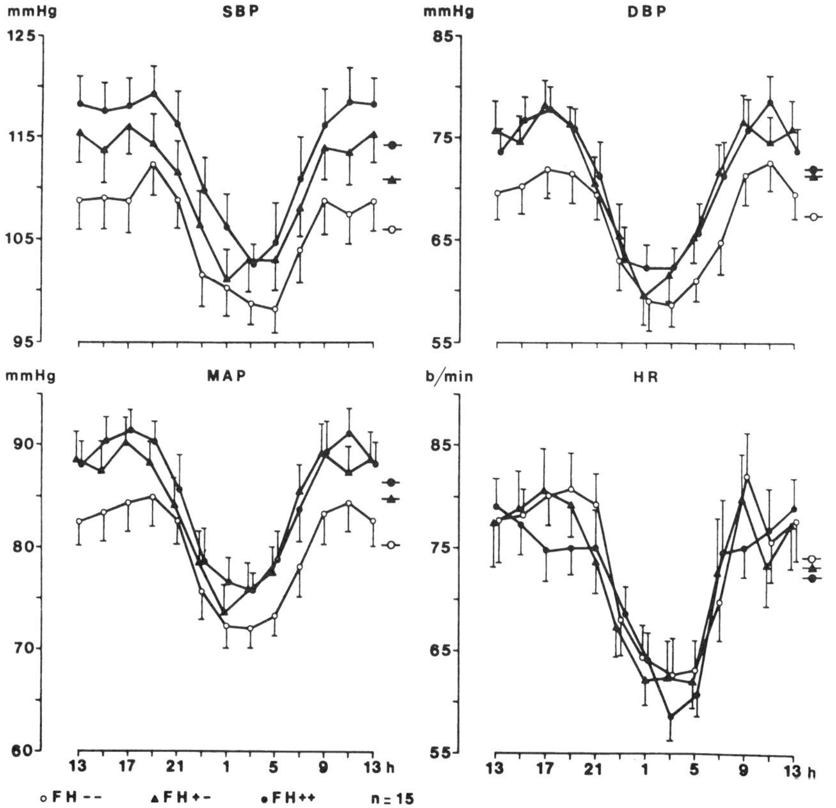

FIGURE 6. Effect of family history of hypertension on BP in 45 normotensive children. Line graphs showing SBP, DBP, MAP, and HR with data grouped in 2-hour subsegments against time. Subjects with two hypertensive parents (FH++) had higher BP values than did those with one hypertensive parent (FH+−) or normotensive parents (FH−). Reproduced with permission from ref. 24.

ANTIHYPERTENSIVE MEDICATIONS AND CHILDREN

Antihypertensive medications are rarely tested in children, as studies are expensive, time-consuming, aimed at a small market, and fraught with potential liabilities; moreover, companies are concerned that studies in children could confound the approval process. Ambulatory monitoring may allow more accurate studies in children with smaller numbers of patients to demonstrate efficacy. Furthermore, some have suggested that the ambulatory monitor may also eliminate the need for a placebo population.[57]

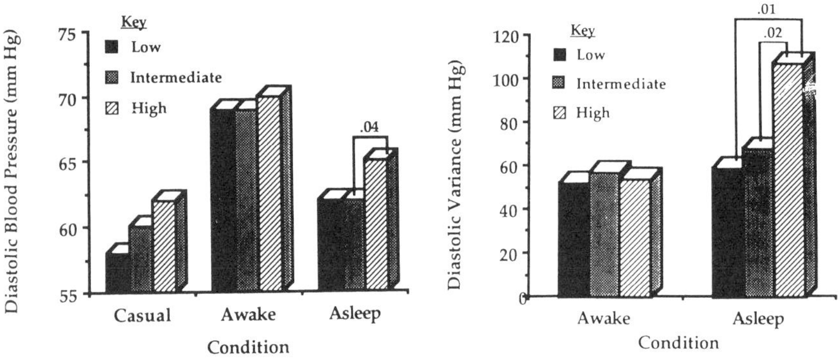

FIGURE 7. Comparison of BP and renin profile under varying conditions in 159 normotensive subjects. (**Left**) Diastolic blood pressure is compared to the three renin groups measured by casual BP (differences not significant). The highest renin-sodium group exhibits higher BP than do the lower groups using ABPM to conduct assessments. (**Right**) The highest renin-sodium profile group had greater sleeping BP variance than did the two groups with lower levels. High renin concentrations may be associated with a lack of normal nocturnal fall in BP. Similar findings (not shown) were seen for SBP. Reproduced with permission from ref. 29.

SUMMARY

Circadian rhythms of BP are established early in life. The ABPM coupled with an objective measure of activity is a powerful tool for the management of hypertension and is a sensitive tool for the dissection of the determinants of hypertension. The challenge for pediatricians in the future includes such important goals as establishing normal values for the circadian assessment of BP for a growing child, coupling of BP determinations with activity, using the ABPM to better understand BP tracking, studying the rhythms of families using the ABPM, educating the medical and lay communities of the importance of rhythmic phenomena, studying longitudinally patients with varying forms of hypertension, using the ABPM to test and optimize antihypertensive medications, and determining the proper diagnosis of hypertension in children based on end-organ damage rather than surrogate measures.

REFERENCES

1. REPORT OF THE SECOND TASK FORCE ON BLOOD PRESSURE CONTROL IN CHILDREN. 1987. Pediatrics **79:** 1–15.

2. DRIZD, T., A. L. DANNENBERG & A. ENGEL. 1986. Blood pressure in persons 18–74 years of age in 1976–1980 and trends in blood pressure from 1960 to 1980 in the United States. Vital Health Stat. **234:** 1–68.
3. KURTZ, T. W. 1995. Possible genetic lesions in experimental and clinical forms of essential hypertension. *In* Hypertension: Pathophysiology, Diagnosis and Management. 2nd Ed. J. H. Laragh & B. M. Brenner, Eds. Vol. **1:** 1281–1287. Raven Press. New York.
4. REPORT OF THE TASK FORCE ON BLOOD PRESSURE CONTROL IN CHILDREN. 1977. **59:** 797–820.
5. DE MAN, S. A., J. L. ANDRE, H. BACHMANN, D. E. GROBBEE, K. K. IBSEN & U. LAASEN. 1991. Blood pressure in childhood: Pooled findings of six European studies. J. Hypertens. **9:** 109–114.
6. SINAIKO, A. R. 1995. Childhood Hypertension. *In* Hypertension: Pathophysiology, Diagnosis and Management. 2nd Ed. J. H. Laragh & B. M. Brenner, Eds. Vol. **1:** 209–226. Raven Press. New York.
7. GILLMAN, M. W., N. R. COOK, B. ROSNER, D. A. EVANS, M. E. KEOUGH, J. O. TAYLOR & C. H. HENNEKENS. 1993. Identifying children at high risk for the development of essential hypertension. J. Pediatr. **122:** 837–846.
8. JOINT NATIONAL COMMITTEE ON DETECTION, EVALUATION, AND TREATMENT OF HIGH BLOOD PRESSURE. 1993. Fifth Report of the Joint National Committee on Detection, Evaluation & Treatment of High Blood Pressure. Arch. Intern. Med. **153:** 154–183.
9. ZACHARIAH, P. K., & W. E. SUMNER, 3RD. 1993. The clinical utility of blood pressure load in hypertension. Am. J. Hypertens. **6:** 194s–197s.
10. HELLBRUEGGE, T., J. E. LANG, J. RUTENFRANZJ, & K. STEHR. 1964. Circadian periodicity of physiological functions at different stages of infancy and childhood. Ann. N.Y. Acad. Sci. **117:** 361–373.
11. MIRMIRAN, M. & J. H. L. KOK. 1991. Circadian rhythms in early human development. Early Human Dev. **26:** 121–128.
12. GEMELLI, M., R. MANGANARO, C. MAMI, F. RANDO & F. DE LUCA. 1989. Circadian BP pattern in full-term newborn infants. Biol. Neonate **56:** 315–323.
13. SITKA, U., D. WEINERT, K. BERLE, B. RUMLER & A. SCHUH. 1994. Investigations of the rhythmic function of heart rate, blood pressure and temperature in neonates. Eur. J. Pediatr. **153:** 117–122.
14. YETMAN, R. J., M. S. WEST & R. J. PORTMAN. 1994. Chronobiologic evaluation of blood pressure in school children. Chronobiol. Int. **11:** 132–140.
15. SOERGEL, M., M. KIRSCHSTEIN, C. BUSCH, T. DANNE, J. GELLERMANNJ, R. HOLL, F. KRULL, H. REICHERT, G. S. REUSZ & W. RASCHER. 1995. Twenty-four hour ambulatory blood pressure values in healthy children and adolescents: A multicenter trial including 1,176 subjects. Pediatr. Nephrol. **9:** C53.
16. PORTMAN, R. J., R. J. YETMAN & M. S. WEST. 1991. Efficacy of 24-hour ambulatory blood pressure monitoring in children. J. Pediatr. **118:** 842–849.
17. NEUS, H., E. GOGLIN, W. LANGEWITZ & A. W. VON EIFF. 1984. Intermittent ambulatory blood pressure recording in children: Methodological aspects and influence of family history on hypertension. Klin. Wochenschr. **62:** 1039–1043.
18. PORTMAN, R. J. & R. J. YETMAN. 1994. Ambulatory blood pressure monitoring: Diagnosis and management of hypertension. Pediatr. Nephrol. **8:** 367–376.
19. RODRIGUEZ, O. J., R. J. YETMAN, M. S. WEST & R. J. PORTMAN. 1991. Unexpected diagnosis of nocturnal hypertension in pediatric renal patients by the ambulatory blood pressure monitor. Pediatr. Res. **29:** 350A.
20. HARSHFIELD, G. A., L. M. DUPAUL, B. S. ALPERT, J. V. CHRISTMAN, E. S. WILLEY, J. K. MURPHY & G. W. SOMES. 1990. Aerobic fitness and diurnal rhythm of blood pressure in adolescents. Hypertension **15:** 810–814.
21. HARSHFIELD, G. A., G. S. ALPERT, E. S. WILLEY, G. W. SOMES, J. K. MURPHY & I. M. DUPAUL. 1989. Race and gender influence on ambulatory blood pressure patterns of adolescents. Hypertension **14:** 598–603.
22. HARSHFIELD, G. A., G. ALPERT, D. PULLIAM, G. SOMES & D. WILSON. 1994. Ambulatory blood pressure in children and adolescents. Pediatrics **94:** 180–184.
23. WILSON, P. D., C. FERENCZ, C. DISCHINGER, J. I. BRENNER & S. L. ZEGER. 1988.

Twenty-four hour ambulatory blood pressure in normotensive and hypertensive adolescent children of hypertensive and normotensive parents. Am. J. Epidemiol. **127:** 946–954.

24. RAVOGLI, A., S. TRAZZI, A. VILLANI, E. MUTTI, C. CUSPIDI, L. SAMPIERI, L. DE AMBROGGI, G. PARATI, A. ZANCHETTI & G. MANCIA. 1990. Early 24-hour blood pressure elevation in normotensive subjects with parental hypertension. Hypertension **16:** 491–497.
25. BERENSEN, G. S., E. DALFERES, D. SAVAGE, L. S. WEBBER & W. BAO. 1993. Ambulatory blood pressure measurements in children and young adults selected by high and low casual blood pressure levels and parental history of hypertension. The Bogolusa Heart Study. Am. J. Med. Sci. **305:** 372–382.
26. LANGEWITZ, W., A. W. VON EIFF, E. GOGOLIN, H. NEUS, H. RUDDEL & R. SCHMIEDER. 1985. Reliability and validity of ambulatory blood pressure monitoring in children. Clin. Exp. Theory Pract. **A7:** 217–225.
27. MODESTI, P. A., I. PELA, I. CECIONI, G. F. GENSINI, G. G. NERI SERNERI & G. BARTOLOZZI. 1994. Changes in blood pressure reactivity and 24-hour blood pressure occurring at puberty. Angiology **45:** 443–450.
28. FIXLER, D. M., J. M. WALLACE, W. E. THORNTON & P. DIMMITT. 1990. Ambulatory BP monitoring in hypertensive adolescents. Am. J. Hypertens. **3:** 288–292.
29. HARSHFIELD, G. A., D. A. PULLIUM, B. S. ALPERT, F. B. STAPLETON, E. S. WILLEY & G. W. SOMES. 1991. Ambulatory blood pressure patterns in children and adolescents: Influence of renin-sodium profiles. Pediatrics **87:** 94–100.
30. HORNSBY, J. L., P. F. MONGAN, T. TAYLOR & F. A. TREIBER. 1991. "White Coat" hypertension in children. J. Fam. Pract. **33:** 617–623.
31. REUSS, G. S., M. HOBOR, T. TULASSAY, P. SALLAY & M. MILTENYI. 1994. 24-Hour blood pressure monitoring in healthy and hypertensive children. Arch. Dis. Child. **70:** 90–94.
32. BRENNER, J. I., P. C. DISCHINGER & C. FERENCZ. 1983. Twenty-four hour blood pressure monitoring in adolescents. Circulation **68:** 396.
33. TUCK, M. L., N. STERN & J. R. SOWERS. 1985. Enhanced 24-hour norepinephrine and renin secretion in young patients with essential hypertension: Relation with the circadian pattern of arterial blood pressure. Am. J. Cardiol. **55:** 112–115.
34. HALBERG, F., J. I. M. DRAYER, G. CONNELISSEN & M. A. WEBER. 1984. Cardiovascular reference data base for recognizing circadian Mesor and amplitude hypertension in apparently healthy men. Chronobiologia **11:** 275–298.
35. REICHERT, H. & W. HOFFMAN. 1990. Experiences with ambulatory blood pressure monitoring in childhood and adolescence. Klin. Paediatr. **202:** 408–412.
36. SOERGEL, M., A. MAISIN, A. AZANCOT-BENISTY & C. LOIRAT. 1992. Ambulante Blutdruckmessung bei nierentransplantieren Kindern Und Jugendlichen. Z. Kardiol. **81:** 67–70.
37. BALD, M. & W. RASCHER. 1994. 24-Hour blood pressure monitoring in children and adolescents. Monatsschr. Kinderheilkd. **142:** 57–62.
38. KRULL, F., T. BUCK, G. OFFNER & J. BRODEHLJ. 1994. Twenty-four hour blood pressure monitoring in healthy children. Eur. J. Pediatr. **152:** 555–558.
39. KRULL, F., J. H. H. OFFNER, K. FROEDE & T. BUCK. 1994. Twenty-four hour blood pressure monitoring in healthy children and adolescents. Monatschr. Kinderheilkd. **142:** 110–117.
40. HOLL, R. W., E. HEINZE, R. NEUMANN, R. TRAUB & W. M. TELLER. 1994. 24 Hour blood pressure monitoring in pediatrics: Results in healthy children as compared to patients with primary or secondary hypertension as well as diabetes mellitus type I. Monatsschr. Kinderheilkd. **142:** 272–278.
41. LURBE, E., J. REDON, Y. LIAO, J. TACONS, R. S. COOPER & V. ALVAREZ. 1994. Ambulatory blood pressure monitoring in normotensive children. J. Hypertens. **12:** 1417–1423.
42. JACOBY, A. C., D. E. FIXLER & E. J. TORRES. 1993. Limitations of an oscillometric ambulatory blood pressure monitor in physically active children. J. Pediatr. **122:** 231–236.
43. REICHERT, H., A. LINDINGER, O. FREY, J. MORTZEK, J. KEITER, C. BUSCH & W. HOFFMANN. 1995. Ambulatory blood pressure monitoring in healthy school children. Pediatr. Nephrol. **9:** in press.

44. DANIELS, S. R., M. H. LOGGIE, T. BURTON & S. KAPLAN. Difficulties with ambulatory blood pressure monitoring in children and adolescents. J. Pediatr. **111:** 397–400.
45. WAEBER, G., B. WAEBER, J. NUSSBERGER & H. R. BRUNNER. 1986. Ambulatory blood pressure monitoring in adolescent untreated hypertensive patients. Clin. Exp. Hypertens. **A8:** 611–614.
46. EGGER, M., M. G. BIANCHETTI, M. GNAEDINGER, R. KOBELT & O. OETLIKER. 1987. Twenty-four hour intermittent, ambulatory blood pressure monitoring. Arch. Dis. Child. **62:** 1130–1135.
47. CUGINI, P., I. GIRELLI, M. LATINI, A. A. COGLIATI, L. DI PALMA, P. BATTISTI, W. FELICI & L. TUCCIARONE. 1986. Chronobiology of blood pressure in childhood: Implications for tracking. Chronobiologia **15:** 291–298.
48. NISHIBATA, K., M. NAGASHIMA, A. TSUJI, S. HASEGAWA, N. NAGAI, M. GOTO & H. HAYASHI. 1995. Comparison of casual blood pressure and twenty-four hour ambulatory blood pressure in high school students. J. Pediatr. **127:** 34–39.
49. WAEBER, B., M. NIEDERBERGER, J. NUSSBERGER & H. R. BRUNNER. 1991. Ambulatory blood pressure monitoring in children, adolescents and elderly people. J. Hypertens. **9:** s72–s74.
50. LINGENS, N., M. SOERGEL, C. LOIRAT, C. BUSCH & K. SCHAERER. 1995. Ambulatory blood pressure monitoring in pediatric patients treated by regular haemodialysis and peritoneal dialysis. Pediatr. Nephrol. **9:** 167–172.
51. PORTMAN, R. J. & V. THOMAS. 1995. Comparison of ABPM data using actual sleep-wake time determined by actigraphy with default day-night times in patients with essential hypertension. Presented at NYAS Conference on Time Dependent Structure and Control of Arterial Blood Pressure. Ferrara, Italy, 1995.
52. VAN EGEREN, L. F. 1991. Monitoring activity and blood pressure. J. Hypertens. **9:** S25–S27.
53. THOMAS, V. W., R. J. YETMAN, B. D. KAHAN & R. J. PORTMAN. 1995. Lack of correlation of blood pressure with activity in renal transplant recipients. Chronobiol. Int. **12:** 419–426.
54. PORTMAN, R. J. & A. M. ROBSON. 1984. Controversies in pediatric hypertension in Mendoza, S. Pediatric Nephrology. *In* Contemporary Issues in Nephrology. B. M. Brenner, Ed. 265–296. Churchill-Livingstone, Inc. New York.
55. JULIUS, S., L. KRAUSE, N. J. SCHNORK, A. D. MEJIA, K. A. JONES, C. VAN DE VEN & E. H. JOHANSSEN. 1991. Hyperkinetic borderline hypertension in Tecumseh, Michigan. J. Hypertens. **9:** 77–84.
56. WHITE, W. B., H. M. DEY & P. SCHULMAN. 1989. Assessment of the daily blood pressure load as a determinant of cardiac function in patients with mild-moderate hypertension. Am. Heart J. **118:** 782–795.

Chronopharmacology of Hypertension

BJÖRN LEMMER

Institute of Pharmacology and Toxicology
Faculty of Clinical Medicine
Ruprecht-Karls-University Heidelberg
Maybachstr. 14–16
D-68169 Mannheim, Germany

Heart rate was among the earliest physiologic functions reported to be inconstant throughout 24 hours. At the beginning of the 17th century, daily variations as well as a rapid increase in pulse rate on awakening were already described.[1] In the 18th and 19th centuries detailed data on daily heart rate variation and pulse quality were reported,[2] and the pulse rate of a healthy subject as it is in the late afternoon was recommended by Joachim Quantz in 1752 to be an easily available "metronome" for a flute player.[3] Later, the invention of a "third hand" in the clock to measure the seconds allowed precise quantification of changes in pulse rate.[4] The symptoms of "office" or "white coat hypertension," nowadays of increasing scientific interest,[5,6] were also described more than 250 years ago by Christoph Hellwig and Théophile Bordeu when they reported pronounced changes in the patient's pulse on the appearance of the doctor.[7]

With the advent of plethysmographic devices it became apparent that blood pressure in healthy as well as in diseased persons may not be constant throughout 24 hours.[2] Zadek[8] was the first to present detailed data on daily variations ("Tagesschwankungen") in blood pressure, with an increase in the afternoon and a decrease at night. Even different types of hypertension were already verified in the 1920s by their different blood pressure profiles.[9,10]

TWENTY-FOUR–HOUR BLOOD PRESSURE PROFILE

Circadian fluctuations in blood pressure (BP) with higher daytime than nighttime values have now been well established in both normotensive and hypertensive patients. The typical circadian pattern of BP exhibits two daytime peaks (around 9 AM and 7 PM, respectively), a small afternoon nadir (around 3 PM), and a profound nocturnal drop (around 3 AM) as can be seen on FIGURE 1.[11,12] Though the BP level is higher in uncomplicated essential hypertensive patients, the circadian pattern is about the same as in normotensive subjects (FIG. 1). Recent intraarterial data in normal subjects confirm that the early morning rise in BP begins hours before awakening, so that its sole attribution to arousal seems highly unlikely.[13] Moreover, telemetric data in normotensive and hypertensive rats clearly gave evidence for a dominant endogenous component of BP rhythm under free-run conditions.[14]

In hypertensive patients mainly with secondary hypertension due, for example, to chronic renal failure, diabetes, Cushing's syndrome, exogenous glucocorticoid administration, renal and cardiac transplantation, and fatal familiar insomnia and in hypertensive pregnancies the nightly decrease in BP is reduced/abolished or even higher levels are found at night.[2]

The loss of a nocturnal BP fall is likely to determine a higher cardiovascular and cerebral risk correlating with an increased damage to target organs such as the heart, brain, and kidneys. This could indicate that restoration of a normal circadian BP pattern is likely to be of prognostic relevance and should be regarded as an additional therapeutic goal in antihypertensive drug treatment.

PATHOPHYSIOLOGY

Evidence indicates that pathophysiologic events within the cardiovascular system do not occur at random. Since several reviews have dealt with this topic,[2,15–18] only some details are presented here. The onset of myocardial infarction predominates between 6 AM and 12 noon. A very similar pattern in onset has been described for sudden cardiac death, stroke, ventricular arrhythmias, and arterial embolism. Also, the onset of myocardial ischemia, angina attacks, or silent ischemia is significantly more frequent during the daytime hours than at night, whereas the onset of angina attacks in variant angina peaks around 4 AM.

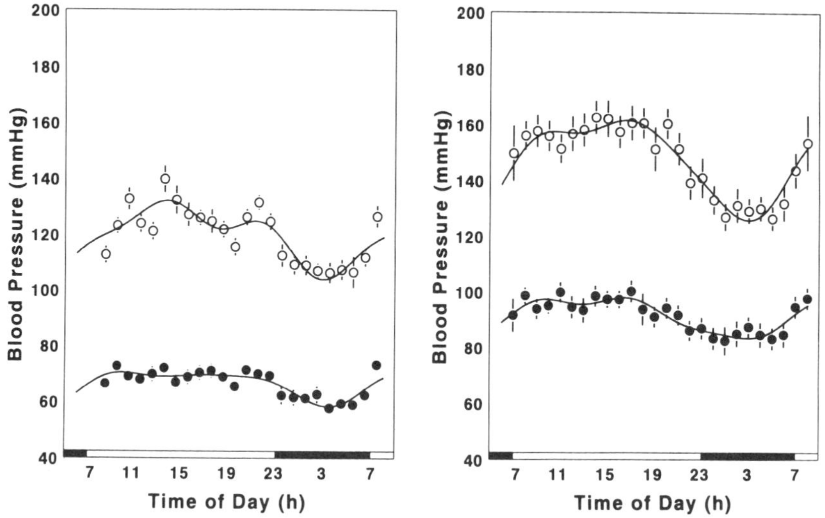

FIGURE 1. Twenty-four–hour blood pressure profiles of 10 normotensive (*left,* 20–30 years) and 10 untreated essentially hypertensive (*right,* 25–45 years) patients. Data from refs. 11, 12, and 29 were fitted by ABPM-FIT.[34]

ANTIHYPERTENSIVE DRUG TREATMENT

Drug treatment of hypertension includes different types of drugs such as diuretics, β- and α-adrenoceptor blocking drugs, calcium channel blockers, and converting enzyme inhibitors.

There are many studies on the effects of various antihypertensive drugs on the 24-hour BP profile.[2,19,20] In most studies, however, time of day of drug application was not a specific point of investigation. Since the drug's mechanisms of action and its target tissues, its half-life, galenic formulation as well as the circadian time of drug dosing may influence the degree and duration of the BP-lowering effect, it is difficult to draw final conclusions about the clinical importance of circadian time-dependent drug dosing. This issue can only be adequately addressed by cross-over studies comparing morning and evening drug dosing. The few data available on this aspect will mainly be considered in this review[21–30] (TABLE 1).

Ambulatory blood pressure monitoring (ABPM) is now generally regarded as the method of choice in evaluating BP profiles, and guidelines have been published.[31–33] However, the method of analyzing a set of 60–80 ABPM data is of additional importance. Profiles were generally compared by visual inspection, or only mean values for the whole 24-hour period or the daytime and nighttime periods were taken into account. Both procedures are unable to detect subtle disturbances in the circadian profile. Different types of rhythm analyses seem to produce better results. (For a review, see ref. 2). Using Fourier series,[34] fitting to the individual data seems to be advantageous.

Beta-Adrenoceptor Antagonists

Beta-adrenoceptor antagonists inhibit sympathetic nervous system function in organs supplied with β-adrenoceptors such as the heart, kidneys, and smooth muscles of the blood vessels and bronchi. Of great therapeutic importance is the decrease in heart rate, cardiac output, and cardiac oxygen consumption and the inhibition of renin release.

Concerning the circadian phase-dependency of the BP lowering effect of β-adrenoceptor antagonists in hypertensive patients, final conclusions are difficult to draw, because neither cross-over (morning versus evening) nor equieffective dose studies have been published. A resume of about 20 studies[20] shows that β-adrenoceptor antagonists, either β_1-selective or nonselective, with intrinsic agonist activity, do not either affect, reduce, or even abolish the rhythmic pattern in BP. However, there is a general tendency for β-adrenoceptor antagonists not to greatly affect nighttime values and to be less effective in reducing the early morning rise in BP.[35,36] Drugs with partial agonist activity, mainly pindolol, even increase heart rate at night.[37] Decreases in heart rate by propranolol and oxprenolol are also more pronounced during daytime hours.[38,39]

The circadian phase dependency in the dose-response relationship of antihypertensive subjects including β-adrenoceptor antagonists will be discussed. In conclusion, clinical data indicate that β-adrenoceptor mediated regulation of BP dominates during daytime hours and is of minor importance during the night and early morning hours. This correlates well with circadian rhythm in sympathetic tone as indicated by the rhythm in plasma noradrenaline and cAMP.[2]

Calcium Channel Blockers

Inhibition of calcium influx relaxes arterial smooth muscle, leading to vasodilatation mainly of the arteries. At the site of the heart this results in a negative inotropic effect. Coronary artery tone is reduced and vasospasms are attenuated, which is important in variant angina. With all calcium channel blockers vasodilatation occurs at lower concentrations than do the cardiodepressant effects. However, the difference between vasodilating and cardiodepressant effects is greater with the 1,4-dihydropyridines (e.g., nifedipine, nitrendipine, isradipine, and amlodipine) than with the verapamil- and diltiazem-like compounds. Thus, nifedipine-like compounds may lead to a reflexly induced increase in heart rate.

TABLE 1. Antihypertensive Drugs: Morning versus Evening Studies[a]

Drug	Dose mg/d	Dose Duration, Dosing Time	Patients (*n*), Diagnosis	Effect on 24-Hour Blood Pressure Day	Night	24-Hour Profile	Reference
Calcium Channel Blockers							
Amlodipine	5	4 wk	20, EH				21
		—AM		++	++	Preserved	
		—PM		++	++	Preserved	
Isradipine	5	4 wk	18, EH				22
		—7 AM		++	++	Preserved	
		—7 PM		++	++	Preserved	
Isradipine	5	4 wk	16, RH[b]				23
		—8 AM		++	+	Not normalized	
		—8 PM		+(+)	++(+)	Normalized	
Nifedipine	30 GITS	1 or 2 wk	10, EH				24
		—10 AM		++	++	Preserved	
		—10 PM		++	++	Preserved	
Nifedipine	10 i.r.	Single dose	12, NT				11
		—8 AM		+	+(+)	Preserved	
		—7 PM		+	+	Preserved	
Nitrendipine	20	4 wk	41, EH				25
		—7 AM		+	+	Preserved	
		—7 PM		+	+	Preserved	
Nitrendipine	10	3 days	6, EH				26
		—6 AM		++	++	Preserved	
		—6 PM		+	++	Changed	
ACE Inhibitors							
Benazepril	10	Single dose	10, EH				27
		—9 AM		+++	++	Preserved	
		—9 PM		+	++	Changed	
Captopril plus Hydrochlorothiazide	25 + 12.5	3 wk	13, EH				28
		—7–8 AM		++	++	Attenuated	
		—6–8 PM		+	++	Preserved	
Enalapril	10	Single dose	10, EH				29
		—7 AM		++	+	Preserved	
		—7 PM		++	+++	Changed	
		3 wk					
		—7 AM		++	(+)	Preserved	
		—7 PM		+	++	Changed	
Quinapril	20	4 wk	18, EH				30
		—8 AM		++	+	Preserved	
		—10 PM		++	++	Preserved	

Abbreviations: EH = essential (primary) hypertension; RH = renal (secondary) hypertension; NT = normotensive subjects.

[a]Effects of calcium channel blockers and ACE inhibitors on the 24-hour profile in blood pressure; only data obtained in cross-over studies (morning versus evening) are given. Drug dose, duration of treatment, circadian time of drug application, number of patients, the diagnosis as well as the effect on the BP profile are given.

[b]Abolished 24-hour blood pressure profile (non-dippers).

Studies with calcium channel blockers were also analyzed mainly by visual inspection of BP profiles or by linear analysis (as just described). In primary hypertensive subjects, three times daily dosing of nonretarded verapamil did not greatly change the BP profile, being, however, less effective at night.[40,41] A single morning dose of sustained-release verapamil showed good 24-hour BP control,[42] whereas a sustained-release formulation of diltiazem was less effective at night.[12] Dihydropyridine derivatives seem to reduce blood pressure to a varying degree during the day and at night; differences in half-life, drug formulation, and dosing interval may play a role. Seven studies using a cross-over morning versus evening design have been reported (TABLE 1). In essential hypertensive subjects amlodipine, isradipine, and nifedipine GITS (gastrointestinal therapeutic system) and in normotensive subjects immediate-release nifedipine did not affect the BP profile differently after once morning or once evening dosing, whereas in two studies with nitrendipine the profile was either preserved or disturbed after evening dosing (TABLE 1). In primary hypertensive patients twice daily nifedipine lowered BP throughout a 24-hour period.[11] Most interestingly, the disturbed BP profile (nondippers) in secondary hypertensive subjects due to renal failure was only normalized after evening but not after morning dosing of isradipine[23] (TABLE 1).

In conclusion, the effect of calcium channel blockers on the 24-hour BP profile of primary hypertensive patients greatly depends on the type of calcium channel blocker. Cross-over studies of morning versus evening dosing were presented only for dihydropyridines.

Converting Enzyme Inhibitors

These drugs are competitive inhibitors of the converting enzyme by which they reduce the conversion of angiotensin I into angiotensin II in plasma and tissues. Angiotensin II is a highly potent vasoconstrictor. Degradation of the vasodilating substance bradykinin is also reduced. Converting enzyme inhibitors reduce peripheral resistance, the formation of aldosterone, and aldosterone-mediated retention of water and sodium.

Converting enzyme inhibitors not only are effective antihypertensive drugs but also can increase the life expectancy in congestive heart failure. Several studies with converting enzyme inhibitors, given once in the morning or twice daily, showed that these drugs did not greatly modify the 24-hour BP pattern.[20] However, intraarterial studies with enalapril or ramipril showed that while causing sustained daytime reduction in BP, these drugs had only marginal effects on nighttime pressures.[36] Thus, the findings obtained with converting enzyme inhibitors in conventional clinical studies are not equivocal.

Four cross-over studies (single morning versus single evening dosing) with converting enzyme inhibitors in essential hypertensive patients were reported (TABLE 1). They demonstrated that evening dosing of benazepril and enalapril changed the BP profile, resulting in a more pronounced nightly decrease which was even observed after a 3-week single dose evening treatment with enalapril (TABLE 1). After chronic treatment with quinapril, evening dosing also resulted in a more pronounced effect than did morning dosing; the BP pattern, however, was not greatly modified (TABLE 1). A fixed combination of captopril and the diuretic hydrochlorothiazide only slightly reduced BP after both morning and evening dosing; the latter treatment, however, was less effective on the next day (TABLE 1).

Other Antihypertensive Drugs

Antihypertensives of other classes have rarely been studied in relation to possible circadian variation. Interestingly, once-daily morning dosing of the diuretics xipamide[43] and indapamide[44] reduced BP in essential hypertensive subjects without changing the 24-hour BP pattern. On twice-daily dosing the α-adrenoceptor antagonists indoramin[45] and prazosin[46] also did not change the BP profile. Recently, Pickering *et al.*[47] reported that a single nighttime dose of the α-adrenoceptor antagonist doxazosin reduced both systolic and diastolic BP throughout the day and night, but that the greatest reduction occurred in the morning hours. Because α-adrenoceptor blockade more effectively reduced peripheral resistance during the early morning hours than at other times of the day,[48] these findings point to the importance of α-adrenoceptor–mediated regulation of BP during this time of day.

CHRONOPHARMACOKINETICS

The pharmacokinetics of various calcium channel blockers in different galenic formulations were investigated in clinical studies in relation to circadian time of drug dosing (TABLE 2). In healthy volunteers the bioavailability of an immediate-release

TABLE 2. Chronopharmacokinetics of Antihypertensive Drugs[a]

Drug	Dose (mg/d, duration)	C_{max} (ng/ml) Morning	C_{max} (ng/ml) Evening	t_{max} (h) Morning	t_{max} (h) Evening	Ref.
Enalapril[b]	10, sd	33.8	41.9	4.4	4.5	29
	10, 3 w	46.7	53.5	3.5[c]	5.6	
Nifedipine ir[d]	10, sd	82.0[c]	45.7	0.4[c]	0.6	11, 17
Nifedipine sr	2 × 20, 1 w	48.5	50.1	2.3	2.8	11
Oxprenolol[e]	80, sd	507.0	375.0	1.0	1.1	39
Propranolol[f]	80, sd	38.6[c]	26.2	2.5	3.0	38
Propranolol	80, sd	68	60	2.3	2.7	51
Verapamil sr	360, 2 w	389.0	386.0	7.2[c]	10.6	50
Verapamil	80, sd	59.4[c]	25.6	1.3	2.0	49

Abbreviations: sd = single dose; w = week(s); ir = immediate-release preparation; sr = sustained-release preparation.

[a]Data were obtained in cross-over studies (morning versus evening: around 6–8 AM and 6–8 PM); sometimes up to six circadian times were studied. Only the parameters C_{max} = peak drug concentration and t_{max} = time to C_{max} are given.

[b]Pharmacokinetics of enalaprilat after enalapril application.

[c]Morning vs evening, p at least <0.05.

[d]Bioavailability significantly reduced after evening dosing.

[e]Significant difference in half-life.

[f]Pharmacokinetics of (−)-propranolol.

formulation of nifedipine was reduced by 40% after evening compared to morning dosing, with C_{max} being higher and t_{max} being shorter after morning dosing.[11,17] By contrast, pharmacokinetic parameters of a sustained-release formulation of nifedipine studied in primary hypertensive subjects did not depend on the circadian time of dosing.[11] Also, regular[49] as well as sustained-release verapamil[50] displayed higher C_{max} or shorter t_{max} values after morning dosing (TABLE 2).

Similar chronokinetics were reported after oral dosing of other antihypertensive drugs such as enalapril,[29] propranolol,[38,51] and oxprenolol[39] (TABLE 2). Interestingly, intravenously infused nifedipine (1.5 mg for 1 hour at 7 AM or 7 PM) did not show daily variations in its pharmacokinetics,[11] indicating that gastrointestinal mechanisms must be involved in the drug's chronokinetics after oral application. Because the drugs are mainly absorbed by passive diffusion, circadian effects on gastric emptying time and gastrointestinal perfusion may be the main determinants, resulting in circadian phase-dependent absorption. This assumption is supported by the observation of delayed gastric emptying time in the evening for solids[52] and a higher hepatic blood flow (representative of gastrointestinal tract perfusion) during the late night and early morning than around noon or in the afternoon.[53] Both mechanisms can explain higher C_{max} and shorter t_{max} values after oral ingestion of nonretarded drugs in the morning[54] (TABLE 2). Because some side effects of drugs used in the treatment of cardiovascular diseases, such as increased heart rate or bradycardia and orthostatic dysregulation, are linked to peak drug concentrations,[17,54,55] it seems worthwhile to investigate in more detail at different times of day circadian time-dependent pharmacokinetics simultaneously with the effects in hypertensive patients.

DOSE-RESPONSE RELATIONSHIP AND 24-HOUR BLOOD PRESSURE PROFILE

Studies of the effects of various antihypertensive drugs belonging to different drug classes indicate that different regulating mechanisms of the 24-hour BP pattern

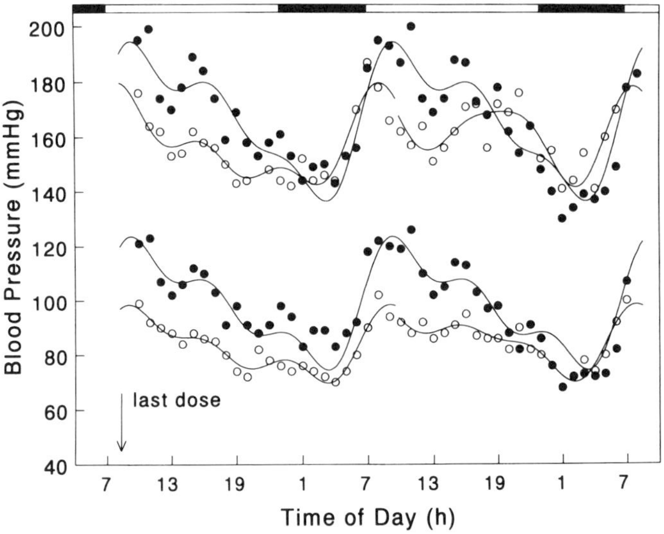

FIGURE 2. Blood pressure profiles in hypertensive patients after 6 weeks of treatment with atenolol 100 mg at 8 AM daily. Shown are the 48-hour profiles after the last dose. Closed symbols = pretreatment profiles; open symbols = profiles after atenolol. Shown are systolic (*upper part*) and diastolic (*lower part*) blood pressures. Data from ref. 36 were fitted by ABPM-FIT[34] and redrawn.

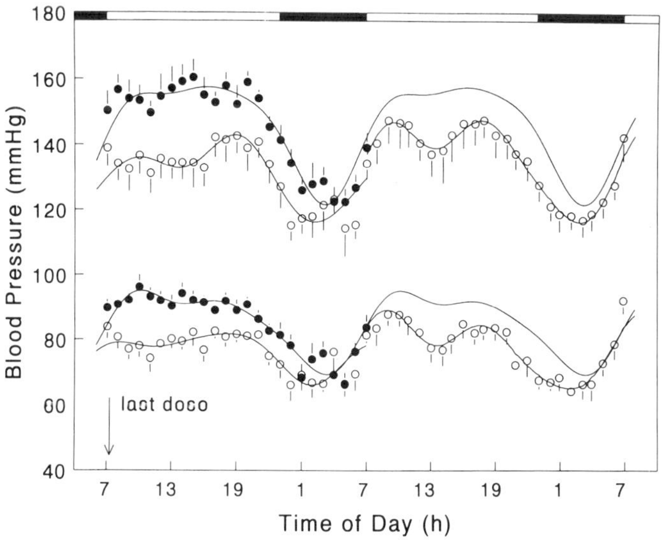

FIGURE 3. Blood pressure profiles in hypertensive patients after 3 weeks of treatment with 10 mg enalapril at 7 AM daily. Shown are the 48-hour profiles after the last dose. Closed symbols = pretreatment profiles; open symbols = profiles after enalapril. Shown are systolic (*upper part*) and diastolic (*lower part*) blood pressures. Data from ref. 29 were fitted by ABPM-FIT[34] and redrawn.

may predominate at certain times of the day. It would therefore be surprising if antihypertensive drugs did not affect the 24-hour BP profile differently. In line with this is the observation that a drug's pharmacokinetics may not predict its treatment profile, mainly when considering the correlation between time-to-peak drug concentration (t_{max}) with time-to-peak drug effect (T_{max}).[2,17,38,56]

Pickering *et al.*[47] reported that the peak treatment effect after nighttime dosing of doxazosin was later than that predicted from the drug's pharmacokinetics, an observation that supports similar findings on BP described for nifedipine[11] and enalapril[29] and on heart rate for propranolol.[38] On the other hand, after morning dosing of nifedipine[11] and propranolol,[38] t_{max} from kinetics coincided with time-to-peak effects (T_{max}), whereas with enalapril t_{max} was still slightly shorter than T_{max}.[29]

There is also support from experiments with five strains of normotensive and hypertensive rats in which the long-acting calcium channel blocker amlodipine,[57] the converting enzyme inhibitor enalapril,[58] the α-adrenoceptor antagonist doxazosin,[59] and the angiotensin II receptor antagonist losartan[60] differed greatly in efficacy and in dose- and circadian phase-dependency of their BP lowering effects as evaluated in unrestrained rats by telemetry.[61]

The degree and duration of the antihypertensive effect are generally studied on a 24-hour basis using ABPM. If at the end of this 24-hour period no significant BP lowering effect is observed, it is concluded that single-drug dosing does not cover an entire 24-hour period. However, some studies do not support this notion. In a study with the β-adrenoceptor blocking drug atenolol a reduction in daytime pressure was reported in line what was just summarized herein, but a greatly decreased effect was seen during the night and early morning[36] (FIG. 2). Thus, it was suggested that the agent produced inadequate 24-hour control of blood pressure.[36] However, as shown in FIGURE 2, the pattern of reduction persisted 1 day after the drug was withdrawn and ABPM was further monitored. A similar observation was made with once-morning dosing of the ACE inhibitor enalapril,[29] as presented in FIGURE 3. On the other hand, after once-daily morning dosing of a sustained-release formulation of diltiazem, the BP lowering effect was no longer observed on the first day the drug was withdrawn[12] (FIG. 4), that is, this agent obviously has a restricted duration of action. These data clearly further support the notion that different mechanisms of regulating BP may predominate at different times of day (see also Witte and Lemmer[62] and this volume). Therefore, it is wise to record ABPM for longer than 24 hours (36 to 48 hours) to avoid false conclusions.

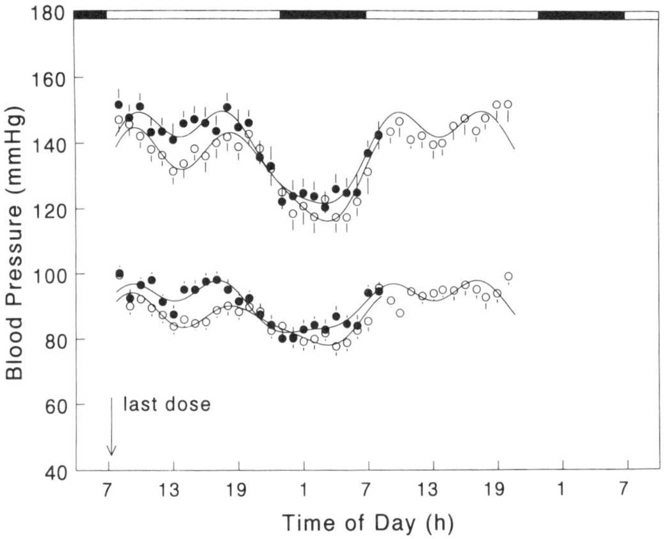

FIGURE 4. Blood pressure profiles in 16 hypertensive patients after 3 weeks of treatment with 240 mg sustained-release diltiazem at 8 AM daily. Shown are 36-hour profiles after the last dose. Closed symbols = pretreatment profiles; open symbols = profiles after diltiazem. Shown are systolic (*upper part*) and diastolic (*lower part*) blood pressures. Data from ref. 62 were fitted by ABPM-FIT.[34]

CONCLUSIONS

It can be concluded that in uncomplicated, essential hypertensive patients or those who show a nightly decrease in BP (dippers), antihypertensive drugs should generally be given in the morning. Whether cardiovascular risk during the morning hours is decreased by evening dosing of an α-adrenoceptor antagonist needs additional confirmation. In secondary hypertension (or in non-dippers) evening dosing may be advantageous in better normalizing the disturbed BP pattern. Possible chronokinetics seem less important for drug efficacy, with the exception that peak drug effects are often correlated with side effects. Finally, it is recommended that ABPM readings be performed for longer than 24 hours to determine the duration of action.

REFERENCES

1. STRUTHIUS, J. 1602. Ars sphygmica.: 127 pp. Königs. Basel, Switzerland.
2. LEMMER, B. & F. PORTALUPPI. 1996. Chronopharmacology of cardiovascular diseases. *In* Physiology and Pharmacology of Biological Rhythms. P. Redfern & B. Lemmer, Eds. Springer. Heidelberg, New York. In press.
3. QUANTZ, J. J. 1752. Versuch einer Anweisung die Flöte traversiere zu spielen.: 261. Voβ. Berlin, Germany.
4. ZIMMERMANN, J. G. 1793. Von der Erfahrung in der Arzneykunst. New edition.: 233. Edlen von Trattnern. Agram, Germany.
5. PICKERING, T. G., M. LEVENSTEIN, P. WALMSLEY FOR THE HYPERTENSION AND LIPID TRIAL STUDY GROUP. 1994. Differential effects of doxazosin on clinic and ambulatory pressure according to age, gender, and presence of white coat hypertension. Results of the HALT study. Am. J. Hypertens. **7:** 848–852.
6. MIDDEKE, M. & J. SCHRADER. 1994. Nocturnal blood pressure in normotensive subjects and those with white coat, primary, and secondary hypertension. Br. Med. J. **308:** 630–632.
7. LEMMER, B. 1995. White coat hypertension: Described more than 250 years ago. Am. J. Hypertens. **8:** 437–438.
8. ZADEK, J. 1881. Die Messung des Blutdrucks des Menschen mittels des Basch'schen Apparates. Z. Klin. Med. **2:** 509–551.
9. MÜLLER, C. 1921. Die Messung des Blutdrucks am Schlafenden als klinische Methode—Speciell bei der gutartigen (primären) Hypertonie und der Glomerulonephritis I. Acta Med. Scand. **55:** 381–442.
10. MÜLLER, C. 1921. Die Messung des Blutdrucks am Schlafenden als klinische Methode—Speciell bei der gutartigen (primären) Hypertonie und der Glomerulonephritis II. Acta Med. Scand. **55:** 443–485.
11. LEMMER, B., G. NOLD, S. BEHNE & R. KAISER. 1991. Chronopharmacokinetics and cardiovascular effects of nifedipine. Chronobiol. Int. **8:** 485–494.
12. LEMMER, B., U. SASSE, K. WITTE & R. HOPF. 1994. Pharmacokinetics and cardiovascular effects of a new sustained-release formulation of diltiazem. Naunyn-Schmiedeberg's Arch. Pharmacol. **349:** R141.
13. BROADHURST, P., G. BRIGDEN, P. DASATWGUPTA, A. LAHIRI & E. B. RAFTERY. 1990. Ambulatory intra-arterial blood pressure in normal subjects. Am. Heart J. **120:** 160–166.
14. WITTE, K. & B. LEMMER. 1995. Free-running rhythms in blood pressure and heart rate in normotensive and transgenic hypertensive rats. Chronobiol. Int. **12:** 237–247.
15. SMOLENSKY, M. H., S. E. TATAR, S. A. BERGMANN, J. G. LOSMAN, C. N. BARNARD, C. C. DACSO & I. A. KRAFT. 1976. Circadian rhythmic aspects of human cardiovascular function: A review by chronobiologic statistical methods. Chronobiologia **3:** 337–371.
16. LEMMER, B. 1989. Temporal aspects in the effects of cardiovascular active drugs in man. *In* Chronopharmacology—Cellular and Biochemical Interactions. B. Lemmer, Ed.: 525–542. Marcel Dekker. New York.
17. LEMMER, B., B. SCHEIDEL & S. BEHNE. 1991. Chronopharmacokinetics and chronopharmacodynamics of cardiovascular active drugs: Propranolol, organic nitrates, nifedipine. Ann. N.Y. Acad. Sci. **618:** 166–181.
18. WILLICH, S. N., M. MACLURE, M. MITTLEMAN, H. R. ARNTZ & J. E. MULLER. 1993. Sudden cardiac death: Support for a role of triggering in causation. Circulation **87:** 1442–1450.
19. LEMMER, B. 1991. The cardiovascular system and daily variation in response to antihypertensive and antianginal drugs. Recent advances. Pharmacol. Ther. **51:** 269–274.
20. STANTON, A. & E. O'BRIEN. 1994. Auswirkungen der Therapie auf das zirkadiane Blutdruckprofil. Kardio **3:** 1–8.
21. MENGDEN, T., B. BINSWANGER & S. GRUENE. 1992. Dynamics of drug compliance and 24-hour blood pressure control of once daily morning vs evening amlodipine. J. Hypertens. **10**(Suppl. 4): S136.
22. FOGARI, R., E. MALOCCO, F. TETTAMANTI, F. TETTAMANTI, A. E. GNEMMI & M. MILANI.

1993. Evening vs morning isradipine sustained release in essential hypertension: A double-blind study with 24 h ambulatory monitoring. Br. J. Clin. Pharmacol. **35:** 51–54.
23. PORTALUPPI, F., L. VERGNANI, R. MANFREDINI, E. C. DEGLI UBERTI & C. FERSINI. 1995. Time-dependent effect of isradipine on the nocturnal hypertension of chronic renal failure. Am. J. Hypertens. **8:** 719–726.
24. GREMINGER, P., P. M. SUTER, D. HOLM, R. KOBELT & W. VETTER. 1994. Morning versus evening administration of nifedipine gastrointestinal therapeutic system in the management of essential hypertension. Clin. Invest. **72:** 864–869.
25. MEILHAC, B., J. M. MALLION, A. CARRE, X. CHANUDET, L. POGGI, P. GOSSE & M. DALLOCHIO. 1992. Étude de l'influence de l'horaire de la prise sur l'effet antihypertenseur et la tolérance de la nitrendipine chez des patients hypertendus essentiels légers à modérés. Therapie **47:** 205–210.
26. UMEDA, T., S. NAOMI, T. IWAOKA, J. INOUE, M. SASAKI, Y. IDEGUCHI & T. SATO. 1994. Timing for administration of an antihypertensive drug in the treatment of essential hypertension. Hypertension **23**(Suppl. 1): 1211–1214.
27. PALATINI, P., L. MOS, M. MOTOLESE, P. MORMINO, M. DELTORRE, L. VAROTTO, E. PAVAN & A. E. PESSINA. 1993. Effect of evening versus morning benazepril on 24-hour blood pressure: A comparative study with continuous intraarterial monitoring. Int. J. Clin. Pharmacol. Ther. Toxicol. **31:** 295–300.
28. MIDDEKE, M., M. KLÜGLICH & H. HOLZGREVE. 1991. Chronopharmacology of captopril plus hydrochlorothiazide in hypertension: Morning versus evening dosing. Chronobiol. Int. **8:** 506–510.
29. WITTE, K., K. WEISSER, M. NEUBECK, E. MUTSCHLER, K. LEHMANN, R. HOPF & B. LEMMER. 1993. Cardiovascular effects, pharmacokinetics and converting enzyme inhibition of enalapril after morning versus evening administration. Clin. Pharmacol. Ther. **54:** 177–186.
30. PALATINI, P., A. RACIOPPA, G. RAULE, M. ZANINOTTI, M. PENZO & A. C. PESSINA. 1992. Effect of timing of administration on the plasma ACE inhibitory activity and the antihypertensive effect of quinapril. Clin. Pharmacol. Ther. **52:** 378–383.
31. ANLAUF, M., P. BAUMGART, I. FRANZ, B. KRÖNIG, W. MEYER-SABELLEK, M. MIDDEKE & J. SCHRADER. 1993. Ambulante Blutdruck-Langzeitmessung. Dtsch. med. Wchnschr. **118:** 1305–1306.
32. SHEPS, S. G., T. G. PICKERING, W. B. WHITE, M. A. WEBER, D. L. CLEMENT, L. R. KRAKOFF, F. H. MESSERLI & D. PERLOFF. 1994. Ambulatory blood pressure monitoring. J. Am. Coll. Cardiol. **23:** 1511–1513.
33. HEINTZ, B., B. LEMMER & H.-G. SIEBERTH. 1995. 24-Stunden-Blutdruckmessung. Dtsch. Ärztebl. **92:** 1521–1527.
34. LEMMER, B., P. ZUTHER, K. WITTE & A. MATTES. 1993. Programm zur automatischen Auswertung von 24-Stunden-Blutdruckprofilen ambulatorischer Blutdruckmeβgeräte: ABPM-FIT. Klin. Pharmakol. akt. **2:** 38/V5.
35. RAFTERY, E. B. 1983. The effects of beta-blocker therapy on diurnal variation of blood pressure. Eur. Heart J. **4:** 61–64.
36. GOULD, B. A. & E. B. RAFTERY. 1991. Twenty-four-hour blood pressure control: An intraarterial review. Chronobiol. Int. **8:** 495–505.
37. QUYYUMI, A. A., C. WRIGHT, L. MOCKUS & K. M. FOX. 1984. Effect of partial agonist activity in β blockers in severe angina pectoris: A double blind comparison of pindolol and atenolol. Br. Med. J. **289:** 951–953.
38. LANGNER, B. & B. LEMMER. 1988. Circadian changes in the pharmacokinetics and cardiovascular effects of oral propranolol in healthy subjects. Eur. J. Clin. Pharmacol. **33:** 619–624.
39. KOOPMANS, R., B. OOSTERHUIS, J. M. KAREMAKER, J. WEINER & G. J. VAN BOXTEL. 1993. The effect of oxprenolol dosage time on its pharmacokinetics and haemodynamic effects during exercise in man. Eur. J. Clin. Pharmacol. **44:** 171–176.
40. GOULD, B. A., S. MANN, H. KIESO, V. BALASUBRAMANIAN & E. B. RAFTERY. 1982. The 24-hour ambulatory blood pressure profile with verapamil. Circulation **65:** 22–27.
41. GOULD, B. A., R. S. HORNUNG, S. MANN, V. BALASUBRAMANIAN & E. B. RAFTERY. 1982.

Slow channel inhibitors verapamil and nifedipine in the management of hypertension. J. Cardiovasc. Pharmacol. **4:** 5369–5373.
42. CARUANA, M., M. HEBER, G. BRIDGEN & E. B. RAFTERY. 1987. Assessment of "once daily" verapamil for the treatment of hypertension using ambulatory, intra-arterial pressure recording. Eur. J. Clin. Pharmacol. **32:** 549–553.
43. RAFTERY, E. B., D. I. MELVILLE, B. A. GOULD, S. MANN & J. R. WHITTINGTON. 1981. A study of the antihypertensive action of xipamide using ambulatory intra-arterial monitoring. Br. J. Clin. Pharmacol. **12:** 381–385.
44. OCON, J. & J. MORA. 1990. Twenty-four-hour blood pressure monitoring and effects of indapamide. Am. J. Cardiol. **65:** 58H–61H.
45. GOULD, B. A., S. MANN, A. DAVIES, D. G. ALTMAN & E. B. RAFTERY. 1981. Indoramin: 24-hour profile of intra-arterial ambulatory blood pressure, a double-blind placebo controlled crossover study. Br. J. Clin. Pharmacol. **12**(Suppl.): 67s–73s.
46. WEBER, M. A., M. J. TONKON & R. C. KLEIN. 1987. Effect of antihypertensive therapy on the circadian blood pressure pattern. Am. J. Med. **82**(Suppl. 1A): 50–52.
47. PICKERING, T. G., M. LEVENSTEIN & P. WALMSLEY FOR THE HYPERTENSION AND LIPID TRIAL STUDY GROUP. 1994. Night-time dosing of doxazosin has peak effect on morning ambulatory blood pressure. Results of the HALT study. Am. J. Hypertens. **7:** 844–847.
48. PANZA, J. A., S. E. EPSTEIN & A. A. QUYYUMI. 1991. Circadian variation in vascular tone and its relation to alpha-sympathetic vasoconstrictor activity. N. Engl. J. Med. **325:** 986–990.
49. HLA, K. K., A. N. LATHAM & J. A. HENRY. 1992. Influence of time of administration on verapamil pharmacokinetics. Clin. Pharmacol. Ther. **51:** 366–370.
50. JESPERSEN, C. M., M. FREDERIKSEN, J. F. HANSEN, N. A. KLITGAARD & C. SORUM. 1989. Circadian variation in the pharmacokinetics of verapamil. Eur. J. Clin. Pharmacol. **37:** 613–615.
51. SEMENOWICZ-SIUDA, K., A. MARKIEWICZ & J. KORCZYNSKA-WARDECKA. 1984. Circadian bioavailability and some effects of propranolol in healthy subjects and liver cirrhosis. Int. J. Clin. Pharmacol. Ther. Toxicol. **22:** 653–658.
52. GOO, R. H., J. G. MOORE, E. GREENBERG & N. P. ALAZRAKI. 1987. Circadian variation in gastric emptying of meals in humans. Gastroenterology **93:** 515–518.
53. LEMMER, B. & G. NOLD. 1991. Circadian changes in estimated hepatic blood flow in healthy subjects. Br. J. Clin. Pharmacol. **32:** 627–629.
54. LEMMER, B. & B. BRUGUEROLLE. 1994. Chronopharmacokinetics—Are they clinically relevant? Clin. Pharmacokinet. **26:** 419–427.
55. BRUGUEROLLE, B. & B. LEMMER. 1993. Recent advances in chronopharmacokinetics: Methodological problems. Life Sci. **52:** 1809–1824.
56. LEMMER, B., B. SCHEIDEL, H. BLUME & H. J. BECKER. 1991. Clinical chronopharmacology of oral sustained-release isosorbide-5-mononitrate in healthy subjects. Eur. J. Clin. Pharmacol. **40:** 71–75.
57. MATTES, A. & B. LEMMER. 1991. Effects of amlodipine on circadian rhythms in blood pressure, heart rate and motility, a telemetric study in rats. Chronobiol. Int. **8:** 526–538.
58. LEMMER, B., K. WITTE, T. MAKABE, D. GANTEN & A. MATTES. 1994. Effects of enalaprilat on circadian profiles in blood pressure and heart rate of spontaneously and transgenic hypertensive rats. J. Cardiovasc. Pharmacol. **23:** 311–314.
59. MATTES, A. & S. BOESE. 1992. Circadian phase-dependent effects of doxazosin on blood pressure and heart rate of the rat. Naunyn-Schmiedeberg's Arch. Pharmacol. **345:** R82.
60. SCHNECKO, A., K. WITTE & B. LEMMER. 1995. Effects of the angiotensin II receptor antagonist losartan on 24-hour blood pressure profiles of primary and secondary hypertensive rats. J. Cardiovasc. Pharmacol. **26:** 214–221.
61. LEMMER, B., A. MATTES, M. BÖHM & D. GANTEN. 1993. Circadian blood pressure variation in transgenic hypertensive rats. Hypertension **22:** 97–101.
62. WITTE, K. & B. LEMMER. 1996. Rhythms in second messenger mechanisms. *In* Physiology and Pharmacology of Biological Rhythms. P. Redfern & B. Lemmer, Eds. Springer. Heidelberg, New York. In press.

Analysis of Blood Pressure Rhythms for Drug Efficacy Evaluation

GIUSEPPE GERMANÒ,[a] LUIGI FEDERICO, STEFANIA ANGOTTI, PAOLO CODISPOTI, MIRYAM MUSCOLO, SILVIA BRAVO, AND SILVIO DAMIANI[b]

I Clinica Medica and [b]Biometria
University "La Sapienza"
Rome, Italy

In evaluating the efficacy of antihypertensive drugs, the conventional use of a sphygmomanometer is supplemented with noninvasive 24-hour monitoring of blood pressure (BP). This 24-hour monitoring may be the only way to confirm whether the therapeutic effect occurs *completely,* for the entire 24 hours, and *correctly,* without disturbing the circadian BP rhythm. Higher levels of BP are known to occur in the late morning[1] and late afternoon[2] and lower levels in the early afternoon[3] and at night.[4] This study analyzes methods of evaluating the effects of antihypertensive drugs on BP circadian rhythm. Different methods have been proposed to analyze ambulatory BP monitoring data. The most commonly used methods compute average values from small time intervals, such as 1 hour,[5] and evaluate the net effects of the drug versus those of the placebo.

Another approach, based on the trend of hourly averages, determines the maximum reduction and the time at which the least effect is recorded,[6,7] with the principal aim of estimating the number of daily doses to be administered. Furthermore, it is useful to calculate the ratio between these two time intervals, termed the trough-to-peak ratio. For the correct determination of this ratio the placebo effect must be substracted.

Observations on a phenomenon, such as BP, that is moving through time generate an ordered set, known as a time series. The feature of time series analysis that distinguishes it from other statistical analyses is the explicit recognition of the importance of the sequential order in which the observations are made. Although in other clinical cases the observations are statistically independent, in time series successive observations are dependent, and the dependence is due to position in the sequence.[8] As experimental data obtained from BP ambulatory monitoring give origin to a time series, the use of a statistical method that fits a time series is necessary.

The cosinor method[9,10] considers a periodic model based on a trigonometric function. It is useful for investigating all phenomena with a cyclic pattern over time. The phenomena must be influenced only by one periodic-type factor such as the awake-sleep cycle. If, as with BP, such factors are multiple, a generalization of the cosinor method using a known trigonometric polynomial such as Fourier partial series can be applied.[11,12] The cosinor method and its generalization may be appropriate in comparing the different profiles of phenomena with a periodic trend.

[a]Address for correspondence: Giuseppe Germanò, MD, I Clinica Medica, Policlinico Umberto I, Viale del Policlinico, 00161 Rome, Italy (Fax: 39.6.8075393).

In the presence of discrepancies the method demonstrates which characteristic of the phenomenon is responsible for the discrepancy, that is, the mean level or the time at which the maximal value occurs.

We therefore reviewed these four different approaches in evaluating the time course of the effects of isradipine, a calcium channel blocker that exerts prolonged action when administered as a sustained release (SRO) 5-mg formulation once daily to control BP.[13–15]

PATIENTS AND METHODS

A total of 141 patients of both sexes (mean age 53 years, range 30–76) with mild to moderate essential hypertension were enrolled in this multicenter trial after a 2-week placebo run-in when their sphygmomanometric diastolic blood pressure, measured between 11 AM and 1 PM, was 96–110 mm Hg.

Treatment with isradipine SRO 5 mg/day administered between 8 and 9 AM was started. Each patient underwent ambulatory BP monitoring at the time of entry in the study and after 6 weeks of treatment. TM 2420 equipment (model 7, A&D Co. Ltd, Japan) was used. The cuff was placed on the left arm, and subjects were asked to keep their arm immobile during each cuff deflation. Recordings were performed every 15 minutes during the day (from 7 AM to 11 PM) and every 30 minutes during the night (from 11 PM to 7 AM). Thus, each recording consisted of 80 measurements unless artifacts occurred. Patients kept a diary in which they carefully noted daytime activities, meal times, bedtimes, quality of sleep, the time they woke up, and the time they took isradipine. Before starting each 24-hour recording, quality control was performed consisting of simultaneous measurement of BP with a mercury sphygmomanometer and an automatic pressurometer. If the differences were over 5 mm Hg, the cuff was repositioned or substituted or the device was substituted, as appropriate. After each recording, experimental data were checked thoroughly to identify outliers or artifacts.

Furthermore, we accepted as valid recordings that had less than 20% artifact. That is, each recording consisted of at least 64 measurements and not more than 80. Data were collected for at least 24 consecutive hours. However, any recording with gaps of 2 or more hours due to artifacts was discarded and the patient withdrawn from the study.

ANALYSES AND RESULTS

Of the 141 patients enrolled we considered only the 107 in whom 24-hour BP monitoring was performed successfully on two occasions. The 34 excluded patients did not undergo the second recording, or if they did have both recordings, the aforedescribed study criteria were not met.

Calculation of the hourly averages before and after drug administration is the most frequently used method of analysis. Visual inspection (FIG. 1) showed that active treatment changed BP without affecting BP circadian rhythm. The net effect was then estimated as the reduction of BP levels after therapy compared to those observed with placebo. However, in the study of a temporal series, this trend was jagged, and use of a "smoothing" technique with moving averages was therefore appropriate. For both systolic and diastolic blood pressure a weighted moving average with five terms was used. Decreases became evident after 4 AM, and from about 8–9 AM, when the drug was administered, decreases practically did not vary

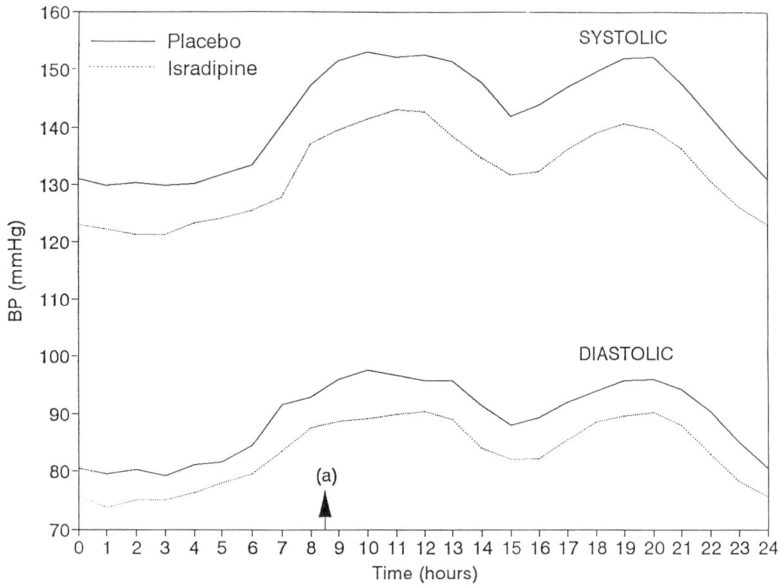

FIGURE 1. Hourly averages of systolic and diastolic blood pressure monitoring for 24 hours after placebo and after 6 weeks of isradipine SRO 5 mg therapy in 107 hypertensive patients. (a) = administration drug time.

until about 10 PM. Subsequently, the decreases became smaller and indicated reduced drug activity. This hypothesis no longer holds, however, after 4 AM.

The trough-to-peak ratio, the second technique used, was calculated and included every hourly average after the dose (corrected by the corresponding run-in value) divided by the lowest hourly average (corrected by the run-in value for the corresponding hour). Both systolic and diastolic blood pressure (FIG. 2) presented a constant reduction from 3 PM (time of peak) to 11 PM. However, after 11 PM higher trough-to-peak ratios paradoxically occurred and were due to a major reduction obtained with placebo. The negative percentages obtained just before the next dose are also difficult to interpret and cannot be attributed to the treatment.

The third approach evaluated, the cosinor method, produced unsatisfactory results (FIG. 3) as the maximal values recorded during placebo administration were greatly underestimated, nocturnal values were overestimated, and the absolute maximum occurred in proximity to the minimum relating to the postprandial dip.

We therefore applied the generalized cosinor model known as Fourier partial series. This model considers oscillations synchronized to periods of 24, 12, and 8 hours and so on. Preliminary analysis was devoted to evaluating individual curves and to choosing an adequate number of terms or harmonics to describe the pattern. We used a criterion that was based on analysis of variance and that allowed us to test the significance of every harmonic of Fourier partial series.[16] All terms up to the third harmonic were always significant. The fourth harmonic was significant at the limit of 5%, but the amplitude due to this harmonic was always smaller than 1 mm Hg, a value far smaller than the instrumental error, and its standard deviation was bigger than the corresponding amplitude; the relative contribution of this harmonic to BP dynamic was therefore ignored. For coherence and compatibility, the model was always curtailed to the third harmonic. The seven variables (one mean BP level, three amplitudes, and three phases) of the four models (for both systolic and diastolic blood pressure after placebo and after isradipine) were obtained as the arithmetic values estimated on individual patients.[17] Phases and amplitudes relating to both diurnal peaks (maxima) are reported in TABLE 1. The fit with the experimental data was excellent (FIG. 4), and thus the Fourier partial series was able to describe the daily trend of BP both before and after isradipine administration.

It was then necessary to evaluate with appropriate statistical tests whether the differences described by the preceding models were significant and thus attributable to the action of the medication or if they were negligible.

First, the null hypothesis on mean BP levels was tested using Student's paired t test.[18] The differences were always significant at the critical level of 0.1%, with 106 degrees of freedom for both systolic (5.390) and diastolic blood pressure (4.725). These results indicate that the placebo mean levels differed significantly from those observed after isradipine. The mean level after treatment was lower, supporting the hypothesis that isradipine induces an effective decrease in BP.

As regards the amplitudes, expressed in millimeter of mercury like the mean level, the null hypothesis was tested using Student's t test. The results obtained in the six tests (three for SBP and three for DBP) were always much lower than 1.982, the tabular value at the critical level of 5%; they were often even less than 1%. These results confirm that treatment with isradipine SRO 5 mg does not alter the BP profile.

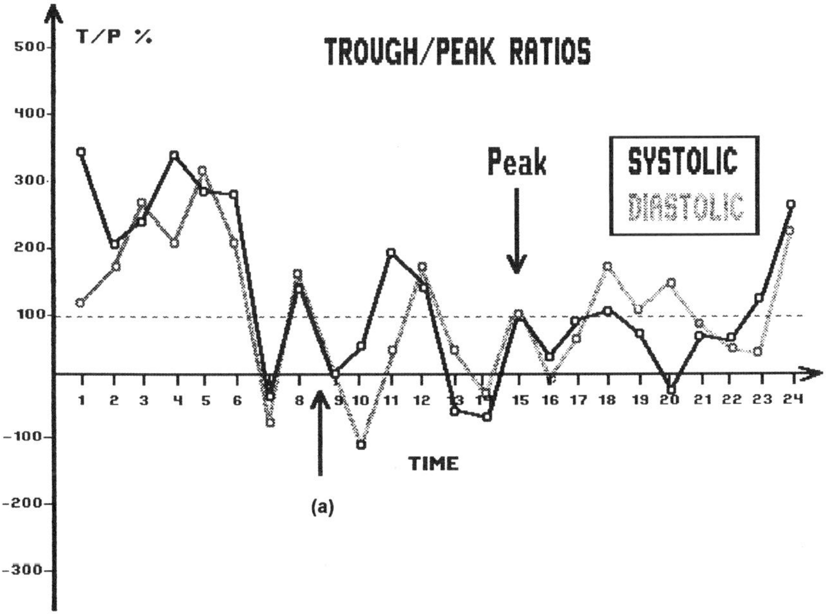

FIGURE 2. Trough-to-peak ratios calculated including every systolic and diastolic hourly average after the dose (corrected by corresponding run-in value) divided by the lowest hourly average = peak (corrected by run-in value for corresponding hour) in 107 hypertensive patients. (a) = administration drug time.

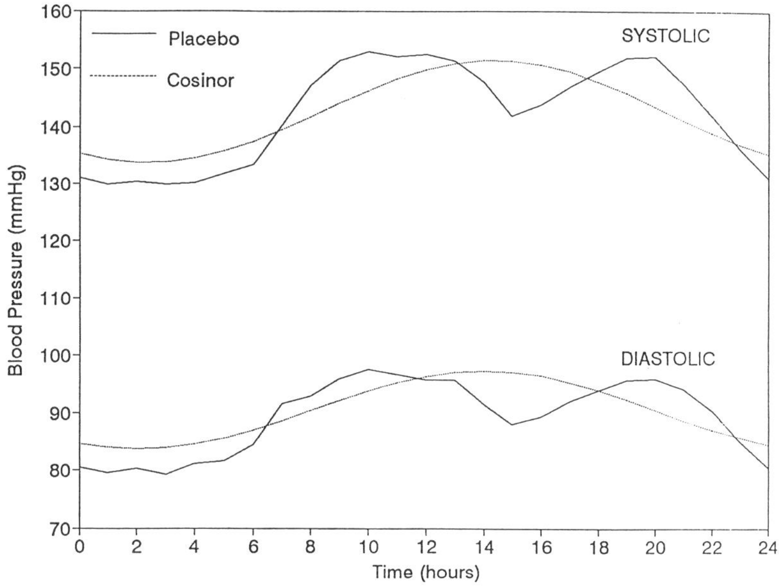

FIGURE 3. Systolic and diastolic blood pressure hourly averages and data evaluated by cosinor method after placebo in 107 hypertensive patients.

As the phases that appear in the Fourier partial series are circular variables, they cannot be analyzed with the usual statistical methods for linear variables. In these circumstances, Watson and William's test[19,20] is appropriate in evaluating the null hypothesis whether the phases relating to two or more BP profiles are similar. The test indicates that there were no appreciable phase-related variations and that any variations could be considered as entirely due to chance.

As shown in FIGURE 4, BP decreases and increases occurred with long, practically linear intervals. Such intervals can be singled out using a point of inflection in which the curve changes its concavity, determining the straight line passing through this point of inflection, that is, its trigonometric tangent. The two points limiting the length of the straight line are determined, imposing the condition that the differences between the BP value obtained from the curve and that obtained from the

TABLE 1. Phases and Amplitudes Relating to Both Maxima

	Mean Level (mm Hg)	Amplitudes (mm Hg) First Max	Amplitudes (mm Hg) Second Max	Phases (hours · minutes) First Max	Phases (hours · minutes) Second Max
		Placebo			
Systolic	142.6	9.9	12.3	10.10	19.16
Diastolic	90.6	8.8	9.1	10.32	19.42
		Sixth Week			
Systolic	129.0	10.4	7.7	10.24	19.12
Diastolic	87.4	7.1	6.9	11.06	19.46

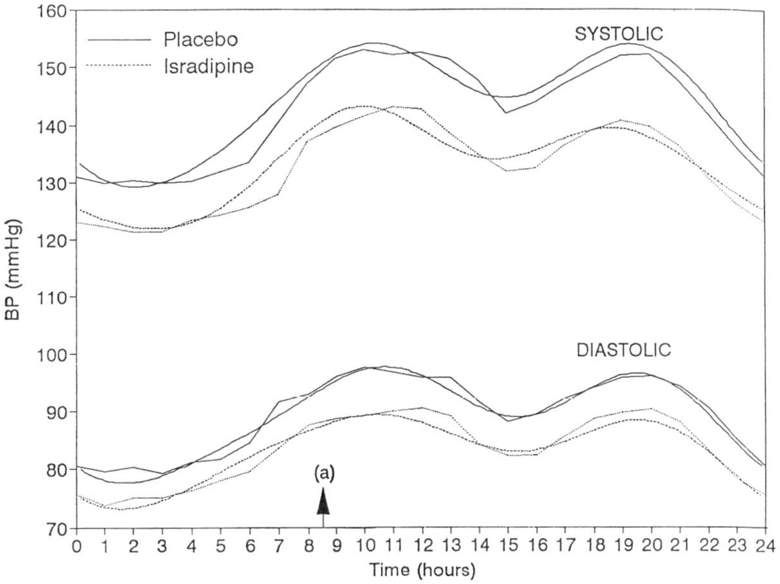

FIGURE 4. Systolic and diastolic blood pressure hourly averages and curves obtained by Fourier partial series curtailed at third term after placebo and after 6 weeks of isradipine SRO 5 mg therapy in 107 hypertensive patients. (a) = administration drug time.

straight line are equal to 1 mm Hg. The consequent approximation does not influence the value of the analysis results. The straight line slope, that is, the angular coefficient, represents the speed with which BP increases or decreases in the diurnal and nocturnal succession of the curves.[21] The nocturnal and periawakening intervals (TABLE 2), in which BP changed linearly, oscillated between 3 and 4 hours. Within these intervals BP rose by 13–16 mm Hg or fell by 14–19 mm Hg. The null hypotheses, 6 weeks of treatment versus placebo, for the intervals and ranges were tested using Student's *t* test, whereas for the speeds, as the coefficients are angular, a

TABLE 2. Speed, Interval, and Range of Blood Pressure Changes during Nocturnal Decline and Periawakening Increase

	Decrease		Increase	
	Base	Sixth Week	Base	Sixth Week
	Systolic Blood Pressure			
Speed (mm Hg/h)	−6.6	−4.4	4.9	4.7
Interval (h · m)	2.54	3.36	3.18	3.12
Range (mm Hg)	−19.0	−16.0	16.0	15.0
	Diastolic Blood Pressure			
Speed (mm Hg/h)	−5.9	−4.9	3.7	3.5
Interval (h · m)	3.06	2.54	4.00	3.48
Range (mm Hg)	−18.0	−14.0	15.0	13.0

different test comparable to Student's *t* test was applied.[22] The results were never significant.

CONCLUSIONS

Statistical tests of hourly or daytime and nighttime mean or median BP changes do not allow evaluation of the time course of the antihypertensive drug effect.[23,24] In addition, cross-sectional and longitudinal evidence suggests that the organ damage of hypertension depends not only on the arithmetic mean but also on the magnitude of BP oscillations.[25]

In an individual patient's profile and even more so when several subjects are considered together, BP variability may be too great to determine the exact time of maximal or minimal hemodynamic effect of a medication, and the calculation of the trough-to-peak ratio appears problematic.

In addition, the variable examined must be constant over time, whereas BP, on the contrary, has a notably nonlinear trend.[26] There is a well-founded risk that the maximal reduction expected is overestimated because of the addition of the BP cycle—at that time in a phase of reduction—to the true effect of the medication. On the contrary, the residual effect, at the end of 24 hours, is underestimated if from it are subtracted the most elevated values of the day observed in the morning.

Fourier analysis generally presents a smoothing effect due to the various phases and not to the various amplitudes, and it illustrates BP temporal patterns better than does the nonperiodic linear model.[27]

This method is generally adopted only for smoothing individual BP curves; when data from groups of patients are being analyzed, smoothing will occur automatically owing to the large number of subjects involved. This assumption is not correct because a particular kind of mean relative to the parameters of every equation is calculated. The smoothing effect obtained does not modify the true nyctohemeral cycle as is usually observed with common statistical tests.[28]

The medication tested was effective only at the mean level, the variables used to characterize the time course of its effect remaining unchanged. There was no significant phase shift of the curves, and BP rhythm was preserved.

Evaluation of the nocturnal period by the additional calculation of the speed at which BP first tended to fall until it reached the minimal daily value and then to rise just before and after awakening, demonstrated that isradipine did not modify the nocturnal fall, reducing only the highest excursion. Therefore, a favorable effect on the prevention of risk of cerebrovascular and cardiovascular acute events in the early morning[29] is hypothesized and could delay or prevent the development of left ventricular hypertrophy.[30]

SUMMARY

This study analyzes some methods of evaluating the effects of antihypertensive drugs on blood pressure circadian rhythm. We reviewed four different approaches: hourly averages, trough-to-peak ratio, cosinor method, and Fourier series applied to the same data to prove the time course of the effects of isradipine administered once daily. A total of 141 patients of both sexes (mean age 53 years, range 30–76) with mild to moderate essential hypertension were enrolled in this multicenter trial after a 2-week placebo run-in. Treatment with isradipine SRO 5 mg/day administered between 8 and 9 AM was started. Each patient underwent ambulatory BP monitoring at the time of entry and after 6 weeks of treatment.

Calculation of hourly averages showed decreases after 4 AM, and from about 8–9 AM, when the drug was administered, and the decreases practically did not vary until about 10 PM. Subsequently, the decreases became smaller and indicated reduced drug activity. However, this hypothesis no longer held after 4 AM. The trough-to-peak ratio was calculated including hourly averages after the dose divided by the lowest hourly average. Both systolic and diastolic blood pressure showed constant reduction from 3 PM (time of peak) to 11 PM. However, after 11 PM, higher trough-to-peak ratios paradoxically occurred due to a major reduction obtained with placebo, and the negative percentages just before the next dose cannot be attributed to treatment.

Applying the cosinor method, maximal values were greatly underestimated, nocturnal values were overestimated, and the absolute maximum occurred in proximity to the minimum relating to postprandial dip. The generalized cosinor model, known as Fourier partial series, was always curtailed to the third harmonic. Fourier analysis was able to describe the daily trend of BP both before and after isradipine administration.

We used statistical tests to determine if the differences described by the models were significant. The tests indicated significant mean level reductions after therapy and no appreciable amplitude and phase-related variations. The nocturnal and periawakening intervals, in which BP changed linearly, oscillated between 3 and 4 hours. Within these intervals BP rose by 13–16 mm Hg or fell by 14–19 mm Hg. The intervals, ranges, and speed differences were tested and were never significant. The medication tested was effective only at the mean level, the variables used to characterize the time course of its effect remaining unchanged. There was no significant phase shift of the curves, and BP rhythm was preserved.

REFERENCES

1. MILLAR-CRAIG, M., C. BISHOP & E. RAFTERY. 1978. Circadian variation of blood-pressure. Lancet **ii:** 795–797.
2. DEGAUTE, J. P., P. VAN DE BORNE, P. LINKOWSKI & E. VAN CAUTER. 1991. Quantitative analysis of the 24-hour blood pressure and heart rate patterns in young men. Hypertension **18:** 199–210.
3. GERMANO', G., S. DAMIANI, M. CIAVARELLA & V. CORSI. 1984. Detection of a diurnal rhythm in arterial blood pressure during antihypertensive therapy. Clin. Cardiol. **7:** 525–535.
4. VAN DEN MEIRACKER, A., A. MAN IN'T VELD, H. J. R. VAN ECK, G. WENTING & M. SCHALEKAMP. 1980. Determinants of short-term blood pressure variability: Effects of bed rest and sensory deprivation in essential hypertension. Am. J. Hypertens. **1:** 22–26.
5. WHITE, W. B., P. SCHULMAN, E. J. MCCABE & H. M. DEY. 1989. Average daily blood pressure, not office blood pressure, determines cardiac function in patients with hypertension. JAMA **261:** 873–877.
6. ROSE, M. & F. G. MCMAHON. 1990. Some problems with antihypertensive drug studies in the context of the new guidelines. Am. J. Hypertens. **3:** 151–155.
7. LIPICKY, R. 1991. The United States Food and Drug Administration guidelines on ambulatory blood pressure monitoring. J. Hypertens. **9**(Suppl. 8): 559–563.
8. ANDERSON, T. W. 1971. The Statistical Analysis of Time Series. J. Wiley. New York.
9. HALBERG, F., L. E. SCHEVING, E. LUCAS, G. CORNELISSEN, R. B. SOTHERN, J. HALBERG, J. CARTER, K. D. STRAUB & D. P. REDMOND. 1984. Chronobiology of human blood pressure in the light of static (room restricted) automatic monitoring. Chronobiologia **11:** 217–247.
10. HALBERG, F., E. HALBERG, R. C. HERMIDA-DOMINGUEZ, J. HALBERG, G. CORNELISSEN,

W. C. McCall, V. R. McCall, W. Maerz & F. Del Pozo-Guerrero. 1985. Chronobiologic blood pressure and heart rate self monitoring at home, workplace, school and elsewhere. *In* IEEE/7th Conf. Engineering in Medicine and Biology Society. :660–664 Chicago.

11. Damiani, S. 1987. Analisi biometrica delle registrazioni continue della pressione arteriosa. Atti Accad. Lancisiana Roma **31:** 162–168.
12. Germano', G. & S. Damiani. 1990. Peri-awakening rise of blood pressure revisited: A linear "speed" of increase in unit time. Progr. Clin. Biol. Res. **341:** 273–282 (part A). Wiley-Liss. New York.
13. Cihak, R., J. Widimsky, K. Lefflerova & P. Jansky. 1991. Effect of isradipine on 24-h blood pressure profile demonstrated by repeated monitoring. Am. J. Hypertens. **4:** 158s–160s.
14. Fitscha, P. & W. Meisner. 1993. Antihypertensive effect of isradipine on ambulatory and casual blood pressure. Am. J. Hypertens. **6:** 67S–69S.
15. Waeber, B., P. H. Soret, P. N. Poget & H. R. Brunner. 1992. Evaluation de l'efficacite' antihypertensive de l'isradipine SRO par enregistrament tensionnel en ambulatoire. Schweiz. Rund. Medizin/Praxis **81:** 1514–1517.
16. Bliss, C. I. 1970. Statistics in Biology. McGraw-Hill. New York.
17. Tong, Y. L. 1976. Parameter estimation in studying circadian rhythms. Biometrics **32:** 83–85.
18. Scheffé, H. 1970. The Analysis of Variance. J. Wiley. New York.
19. Mardia, K. V. 1972. Statistics of Directional Data. Academic Press. London.
20. Watson, G. S. & E. J. Williams. 1956. On the construction of significance tests on the circle and the sphere. Biometrika **43:** 344–352.
21. Germano', G., S. Damiani, S. Valentino, A. Caparra, C. Giarrizzo, A. Santucci, B. Pica & P. Antonini. 1994. Equal hypotensive effect but varied influence on 24-hour pressure with two different dosages of lisinopril. Adv. Ther. **11:** 248–261.
22. Armitage, P. 1971. Statistical Methods in Medical Research. Blackwell. London.
23. White, W. B. 1992. The role of ambulatory monitoring of the BP for assessment of antihypertensive agents. J. Clin. Pharmacol. **32:** 524–528.
24. Staessen, J., R. Fagard, P. Lijnen, L. Thijs, R. Van Hoof & A. Amery. 1991. Ambulatory BP monitoring in clinical trial. J. Hypertens. **9**(Suppl. 1): S13–S19.
25. Mancia, G., A. Frattola, A. Groppelli, S. Omboni, G. Parati, L. Ulian & A. Villani. 1994. Blood pressure reduction and end-organ damage in hypertension. J. Hypertens. **12**(Suppl. 8): S35–S42.
26. Damiani, S. & G. Germano'. 1985. Biometrical study of ambulatory blood pressure in normal subjects. *In* Blood pressure recording in the clinical management of hypertension. G. Germano', Ed. :22–33. L. Pozzi. Rome.
27. Imai, Y., M. Munakata & J. Hashimoto. 1993. Age-specific characteristics of nocturnal blood pressure in a general population in a community of Northern Japan. Am. J. Hypertens. **6:** 179S–183S.
28. Streitberg, B., W. Meyer-Sabellek & P. Baumgart. 1990. Smoothing twenty-four hour ambulatory blood pressure profiles: A comparison of alternative methods. J. Hypertens. **8:** 21–29.
29. Muller, J. E., G. H. Tofler & G. H. Stone. 1989. Circadian variation and triggers of onset of acute cardiovascular disease. Circulation **79:** 733–743.
30. Verdecchia, P., G. Schillaci, M. Guerrieri, C. Gatteschi, G. Benemio & F. Boldrini. 1990. Circadian blood pressure changes and left ventricular hypertrophy in essential hypertension. Circulation **81:** 528–536.

Drug Effects on Blood Pressure Rhythms in Essential Hypertension

TALMA ROSENTHAL[a]

Chorley Institute of Hypertension
Chaim Sheba Medical Center
Tel Hashomer
and
Sackler Faculty of Medicine
Tel Aviv University
Tel Aviv, Israel

Blood pressure and heart rate follow a similar circadian rhythm. They both climb rapidly during the morning, fall throughout the day to a low at around 3 AM, and rise quickly very early in the morning, with blood pressure peaking somewhere around 10 AM.[1] Indeed, cardiovascular events, including stroke,[2] myocardial infarction, and sudden cardiac death,[3] occur more frequently in the morning hours, making it imperative that these hours be well covered in any antihypertensive treatment regimen.

The late morning rise in blood pressure has been attributed to "household or chore-related" activities, the peaks at noon and 6 in the evening to mealtime, and the subsequent decline to leisure activities.[4] Variation between individuals in blood pressure response to the same activity is due to differing levels of baroreflex sensitivity. It has been well established that there is no important circadian rhythm of blood pressure that is independent of activity, be it working at a desk, walking, sleeping, or watching television. This phenomenon is confirmed by the observation of elevated blood pressure in shift workers throughout the night when they are active. In a recent study by Sternberg *et al.*,[5] bakery workers exhibited considerable flattening of normal blood pressure and heart rate curves, indicating that the sympathetic nervous system was active during the night. Antihypertensive treatment for such persons should give good coverage during these hours. In contrast to the active hours, even an afternoon nap decreases blood pressure significantly.[6]

The variability in blood pressure throughout the day can serve for diagnosing as well as for determining the optimal schedule for antihypertensive treatment. This variability might mean, for example, that the drug not only should reduce the mean 24-hour values, but should also curb the degree of fluctuation. There is probably no drug that can reduce blood pressure variability, and no drug that affects variability and not average blood pressure.[7–9] These two parameters, however, can be affected separately in animals[7]: variability is increased without affecting mean blood pressure when sino-aortic denervation is performed.

Even the newest antihypertensive drugs appear to reduce 24-hour mean blood pressure levels, but are not so effective in reducing 24-hour blood pressure variability.[9] Drugs developed to ensure a constant, uniform reduction in blood pressure over 24 hours could also be endowed with the capacity to prevent the increase in 24-hour

[a]Address for correspondence: Prof. Talma Rosenthal, Chorley Institute of Hypertension, Chaim Sheba Medical Center, Tel Hashomer 52621, Israel. tel: 972-3-5302624; fax: 972-3-5355428.

blood pressure fluctuations that sometimes follow the administration of short-acting antihypertensive agents.

The ideal antihypertensive treatment should effectively control blood pressure for 24 hours, give a smooth antihypertensive effect with decreased variability, interfere with the early morning surge in blood pressure while preserving the normal circadian pattern of blood pressure, take into account the less-than-desirable compliance associated with drug therapy, and avoid the reflex stimulation of the sympathetic nervous system.[10] Based on current information, antihypertensive drugs with a long duration of action are the most likely agents to achieve this.

Changing conditions such as additional drug intake and diet can also have a profound effect on the outcome of antihypertensive therapy. Altering drug combinations can induce variability; for example, the combination of chlorpromazine and small doses of captopril caused a reduction in supine blood pressure compared with chlorpromazine alone, and exaggerated the postural hypotension (FIG. 1).[11] Felodipine combined with an antiepileptic agent will give what appears to be a picture of poor blood pressure control, because the drug interaction results in a low felodipine level.[12] Grapefruit juice, on the other hand, enhanced the absorption of felodipine, giving better blood pressure control.[13]

A useful measure of the reduction in blood pressure over a 24-hour period is the trough-to-peak ratio (FIG. 2),[14] which has been proposed as an arithmetic indicator of the duration of the antihypertensive effect of a drug. It is used to evaluate both efficacy and duration of therapeutic effect.[14–16] The Food and Drug Administration found the trough-to-peak ratio particularly useful because it drives home the need for caution when reducing blood pressure: too drastic a reduction carries the risk of cerebral hypotension.

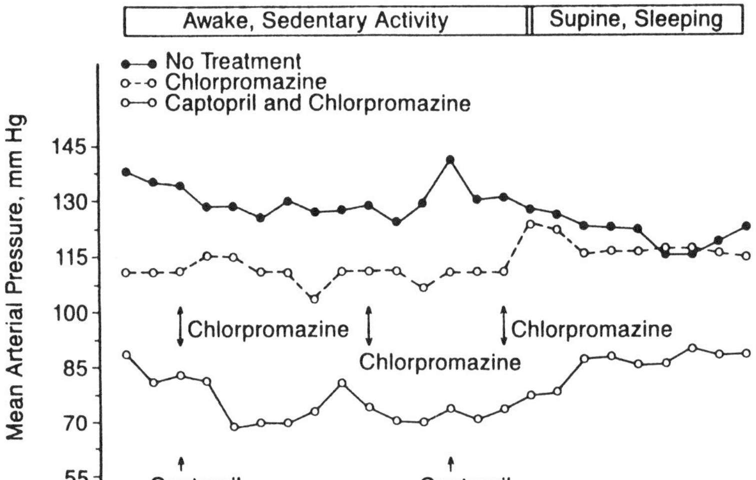

FIGURE 1. Twenty-four-hour profile of mean blood pressure on no therapy, chlorpromazine (200 mg three times daily), and chlorpromazine plus captopril (6.25 mg twice daily). (Reprinted, with permission, from White.[11])

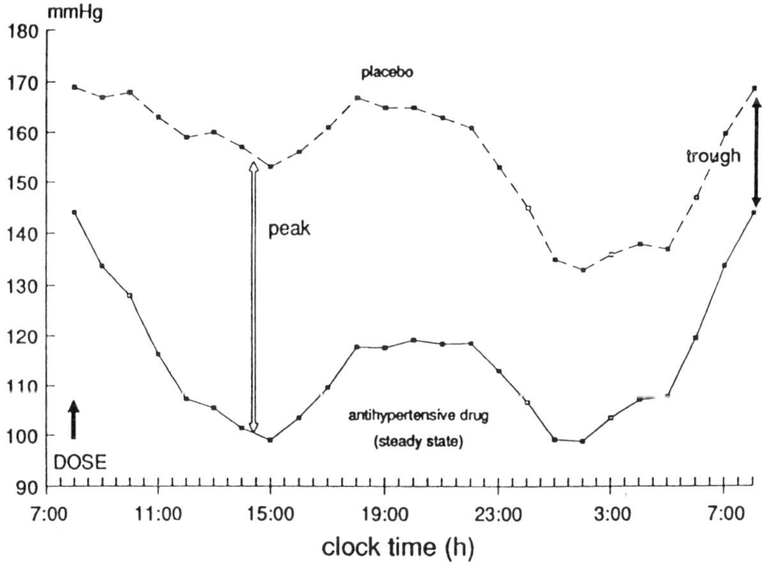

FIGURE 2. Trough-to-peak ratio presents the ratio between the minimal and maximal therapeutic effect (blood pressure reduction) of antihypertensive therapy. (Reprinted, with permission, from Elliott.[14])

A trough-to-peak ratio approaching 100% seems to indicate a drug's suitability for once daily administration, because it reflects the ability to maintain a consistent antihypertensive effect throughout the dosage interval and even beyond. An accurate trough-to-peak ratio for a particular drug cannot be determined without first evaluating the patient's reaction to a placebo period. Studies comparing two drugs, such as the Tel Hashomer group's evaluations of verapamil versus nifedipine[17] and felodipine versus nifedipine,[18] are informative about the response of a particular patient or group of patients to two different drug regimens. They do not, however, allow conclusions about trough-to-peak ratio, because they are conducted while patients have been on a particular drug therapy for several weeks with no intervening placebo period.

The form in which a drug is given, that is, capsule or slow or extended release tablet, as well as the dosage can have a major impact on the blood pressure curve. The curve of the plasma level of nifedipine ER is much smoother than that of a 10-mg capsule, making the former a better controller of blood pressure. Nifedipine GITS gives an even smoother curve.[19]

Difficulties of interpretation arise from the lack of a standardized methodology.[16] Different ranges of values are obtained within the same study or in different studies when different methods and drug doses are used. A case in point is the trough-to-peak ratio quoted for lisinopril, which ranges from 40–80%, namely, from unsatisfactory to highly satisfactory.[20]

Effective antihypertensive response that lasts even beyond the dose interval is obtained with drugs in which the relationship between concentration and effect is linear and not sigmoid.[21] The accumulating data on this topic underscore the importance of investigating the possibility of obtaining an additional reduction in cardiovascular risk for hypertensive patients by minimizing the net trough-to-peak effect of antihypertensive drugs.

Evidence points to the morning hours as the optimal time for cardioprotective medication.[22] Sundberg *et al.*[23] found similar blood pressure curves when atenolol was given at night for 1 week and in the morning for 1 week (FIG. 3); the evening dose did not reduce the nighttime blood pressure to any greater extent than did the morning dose. However, the morning rise in heart rate was clearly attenuated by the evening dose. Pickering *et al.*[24] found that nighttime dosing of dioxazosin had peak effect on morning ambulatory blood pressure. Palatini and co-workers,[25,26] on the other hand, found the antihypertensive effect of quinapril to be constant throughout the 24-hour period only when the drug was given at 10 PM (FIG. 4)[26]; when it was given in the morning, there was partial loss of effectiveness during nighttime and early morning hours. When Lemmer's group[27] gave 10 mg enalapril at 7 PM, there was a further significant decrease in nighttime blood pressure and a subsequent rise during the day, but no effect on elevated afternoon values. No difference in the 24-hour blood pressure monitoring records was found by Bernard-Fernier *et al.*[28] when diltiazem 30 mg was given in the morning or the evening for 4 weeks. Mengden and co-workers[29] studied the effect of amlodipine 5 mg given in the morning or in the evening and found no significant difference between the two schedules. Self-measurement, however, did document significantly greater blood pressure reduction in the morning administration group. Portaluppi and co-workers[30] found that giving isradipine 5 mg in a sustained release oral formulation either morning or evening to patients with chronic renal failure, who frequently exhibit nighttime hypertension, provided equally effective blood pressure control, but evening dosing had the added benefit of exerting an additional lowering effect at night. This is important in this population in which nocturnal hypotension does not occur.

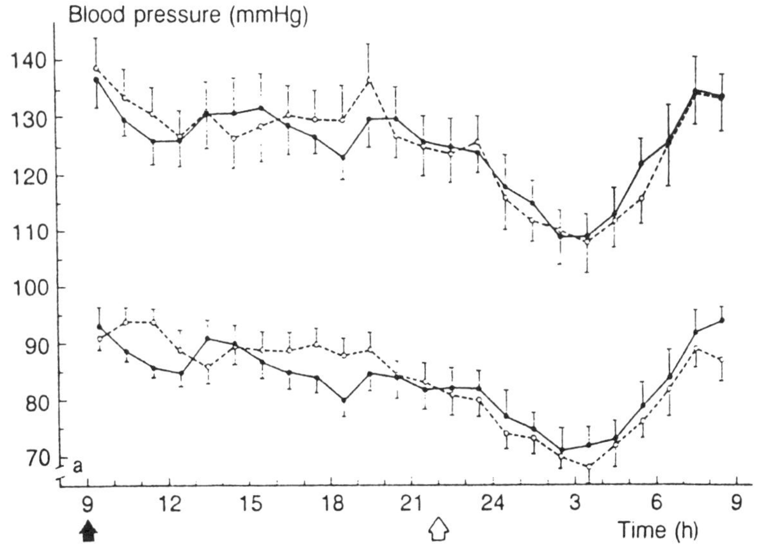

FIGURE 3. Twenty-four-hour blood pressure profiles in 10 hypertensive patients treated with once daily atenolol in the morning (●—●) and in the evening (○—○). *Arrows* denote the time when the drug was taken in the morning (▲) and when the drug was taken in the evening (△). (Reprinted, with permission, from Sundberg *et al.*[23])

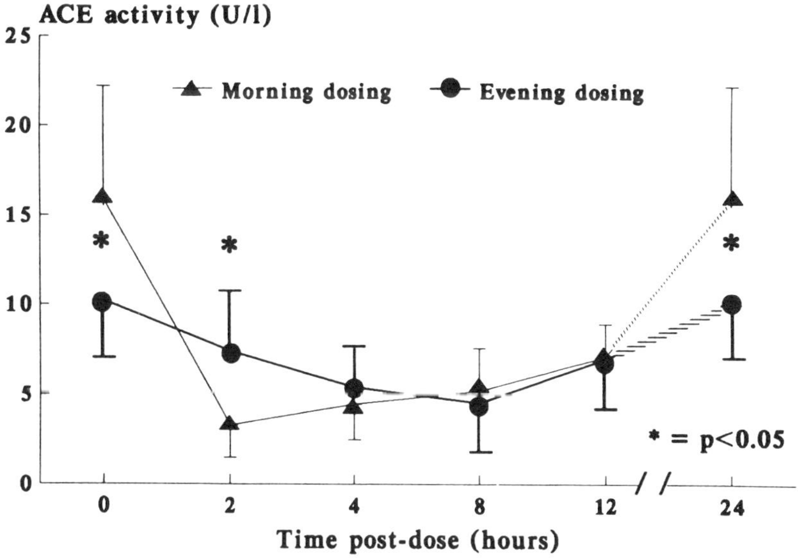

FIGURE 4. ACE activity. Pattern with the morning and evening administration of 20 mg of quinapril (n = 7). (Reprinted, with permission, from Palatini *et al.*[26])

Thus, optimal timing for cardioprotective medication may be in the evening, enabling the drug to begin acting in the early morning hours and work against the rise in blood pressure typical of those hours. On the other hand, a significant decrease in blood pressure which puts patients at risk of cardiac or cerebral ischemia[31,32] must be avoided. Giving antihypertensive medication in the evening to patients who receive monotherapy in the morning will provide valuable information about the blood pressure control achieved by the two approaches, and help evaluate any differences between them. Accumulating evidence on the optimal timing for antihypertensive treatment challenges the pharmaceutical industry to develop a therapy commensurate with the circadian rhythm of blood pressure and heart rate. Whether pharmacologic intervention can reduce cardiovascular risk attributable to diurnal increases in arterial pressure and heart rate also remains to be investigated. We still do not have a clear picture of the benefits and disadvantages of altering circadian patterns.

REFERENCES

1. Sirgo, M. A., R. J. Mills & V. DeQuattro. 1988. Effects of antihypertensive agents on circadian blood pressure and heart rate patterns. Arch. Intern. Med. **148:** 2547–2552.
2. Marler, J. R., T. R. Price, G. L. Clark *et al.* 1989. Morning increase in onset of ischemic stroke. Stroke **20:** 473–476.
3. Muller, J. E., G. H. Tofler & P. H. Stone. 1989. Circadian variation and triggers of onset of acute cardiovascular diseases. Circulation **79:** 733–743.
4. Clark, L. A., L. Denby, D. Pregibon, G. A. Harshfield, T. B. Pickering, S. Blank & H. J. Laragh. 1987. A quantitative analysis of the effects of activity and time of day on the diurnal variations of blood pressure. J. Chronic Dis. **40:** 671–681.
5. Sternberg, H., T. Rosenthal, A. Shamiss & M. Green. 1995. Altered circadian rhythm of blood pressure in shift workers. J. Hum. Hypertens. **9:** 349–353.

6. BURSZTYN, M., J. MEKLER, N. WACHTEL & D. BEN-ISHAY. 1994. Siesta and ambulatory blood pressure monitoring. Comparability of the afternoon nap and night sleep. Am. J. Hypertens. **7:** 217–221.
7. MANCIA, G., A. FRATTOLA, A. GROPPELLI, S. OMBONI, G. PARATI, L. ULIAN & A. VILLANI. 1994. Blood pressure reduction and end-organ damage in hypertension. J. Hypertens. **12**(Suppl. 8): S35–S42.
8. PESSINA, A. C. In discussion following MANCIA, G., A. FRATTOLA, S. GROPPELLI OMBONI, G. PARATI, A. ULIAN & A. VILLANI. 1994. Blood pressure reduction and end organ damage in hypertension. J. Hypertens. **12**(Suppl 8): S35–S42.
9. PARATI, G., A. RAVOGLI, A. FRATTOLA, A. GROPPELI, L. ULIAN, C. SANTUCCIU & G. MANCIA. 1994. Blood pressure variability: Clinical implications and effects of antihypertensive treatment. J. Hypertens. **12**(Suppl. 5): S35–S40.
10. MEREDITH, P. A., D. PERLOFF, G. MANCIA & T. PICKERING. 1995. Blood pressure variability and its implications for antihypertensive therapy. Blood Pressure **4:** 5–11.
11. WHITE, W. W. 1986. Hypotension with postural syncope secondary to the combination of chlorpromazine and captopril. Arch. Intern. Med. **146:** 1833–1834.
12. CAPEWELL, S., S. FREESTONE, J. A. CRITCHLEY, A. POTTAGE & L. F. PRESCOTT. 1987. Gross reduction in felodipine bioavailability in patients taking anticonvulsants. Proc. BPS April 8–10, 243–244.
13. BAILEY, D. G., J. D. SPENCE, C. MUNOZ & J. M. O. ARNOLD. 1991. Interaction of citrus juices with felodipine and nifedipine. Lancet **337:** 268–269.
14. ELLIOTT, H. L. 1994. Trough: peak ratio and twenty-four-hour blood pressure control. J. Hypertens. **12**(Suppl. 5): S29–S33.
15. ELLIOTT, H. L. & P. A. MEREDITH. 1995. Trough:peak ratio: Clinically useful or practically irrelevant? J. Hypertens. **13:** 279–283.
16. ELLIOTT, H. L. & P. A. MEREDITH. 1995. Analysis of trough:peak ratio and the assessment of anti-hypertensive drug action. J. Hum. Hypertens. **9:** 423–427.
17. SHAMISS, A., S. MEISEL, N. NUSSINOVITCH & T. ROSENTHAL. 1990. 24 Hour blood pressure monitoring comparing slow release verapamil and nifedipine in the treatment of mild to moderate hypertension. J. Hum. Hypertens. **4:** 465–468.
18. CAROLL, J., A. SHAMISS, D. ZEVIN, J. LEVI & T. ROSENTHAL. 1995. 24 Hour blood pressure monitoring during treatment with extended release felodipine vs. slow release nifedipine. J. Cardiovasc. Pharmacol. **26**(6): 974–977.
19. ZANCHETTI, A. for the Italian Nifedipine GITS Study Group. 1994. Trough and peak effects of a single dose of nifedipine gastrointestinal therapeutic system (GITS) as assessed by ambulatory blood pressure monitoring. J. Hypertens. **12**(Suppl. 5): S23–S27.
20. ZANNARD, F. 1993. Trandolapril: How does it differ from other angiotensin converting enzyme inhibitors? Drugs **46**(Suppl. 2): 172–182.
21. MEREDITH, P. A. & H. L. ELLIOTT. 1994. Antihypertensive treatment and trough:peak ratio: General considerations. J. Hypertens. **12**(Suppl. 8): S79–S83.
22. WILLICH, S. N., T. LINDERER, K. WEGSCHEIDER, A. LEIZOROVICZ, I. ALAMERCERY & R. SCHRODER. 1989. The ISAM Group. Increased morning incidence of myocardial infarction in the ISAM study. Absence with priiop β-adrenergic blockade. Circulation **80:** 853–858.
23. SUNDBERG, S., O. H. J. LUURILA, A. KOHVAKKA & A. GORDIN. 1991. The circadian heart rate but not blood pressure profile is influenced by the timing of β-blocker administration in hypertensives. Eur. J. Clin. Pharmacol. **40:** 435–436.
24. PICKERING, T. G., M. LEVENSTEIN, P. WALNSLEY for the Hypertension and Lipid Trial Study Group. 1994. Nighttime dosing of doxazosin has peak effects on morning ambulatory blood pressure. Results of the HALT study. Am. J. Hypertens. **7:** 844–847.
25. PALATINI, P. 1992. Can an angiotensin converting enzyme inhibitor with a short half-life effectively lower blood pressure for 24 hours? Am. Heart J. **123:** 1421–1425.
26. PALATINI, P., A. RACIOPPA, G. RAULE, M. Z. ZANINOTTO, M. PEUZO & A. C. PESSINA. 1992. Effects of timing of administration on the plasma ACE inhibitory activity and the antihypertensive effect of quinapril. Clin. Pharmacol. Ther. **52:** 378–383.
27. WITTE, K., K. WEISSER, M. NEUBECK, E. MUTSCHLER, K. LEHMANN, R. HOPF & B.

LEMMER. 1993. Cardiovascular effects, pharmacokinetics and converting enzyme inhibition of enalapril after morning versus evening administration. Clin. Pharmacol. Ther. **54:** 177–186.

28. BERNARD-FERNIER, M. F., J. F. RENUCCI, J. P. SICHE, B. VAISSE, L. POGGI & J. M. MALLION. 1994. Ambulatory measurement of blood pressure and efficacy of diltiazem. Influence of the timing of drug ingestion and initial blood pressure level. Ann. Cardiol. Angeiol. Paris **43:** 357–364.
29. MENGDEN, T., B. BINSWANGER, T. SPUHLER, B. WEISSER & W. VETTER. 1993. The use of self-measured blood pressure determinations in assessing dynamics of drug compliance in a study with amlodipine once a day. Morning vs. evening. J. Hypertens. **11:** 1403–1411.
30. PORTALUPPI, F., V. LUCIANA, R. MANFREDINI, E. C. DEGLI UBERTI & C. FERSINI. 1995. Time-dependent effect of isradipine on nocturnal hypertension in chronic renal failure. Am. J. Hypertens. **8:** 719–726.
31. JONSSON, P. V., L. A. PITSITZ, M. KELLEY & J. KOESTNER. 1990. Hypotensive responses to common daily activities in institutionalized elderly. Arch. Intern. Med. **150:** 1518–1524.
32. CRUICKSHANK, J. M. 1994. J curve in antihypertensive therapy—Does it exist? A personal point of view. Cardiovasc. Drugs Ther. **8:** 757–760.

Drug Effects on BP Rhythm in Secondary Hypertension[a]

MARTIN MIDDEKE

Rehabilitation Center Spreewald
D-03096 Burg, Germany

Secondary hypertension forms are believed to be curable by causal treatments such as angioplasty (PTA) or operation. Therefore, great effort is made to establish the diagnosis of secondary hypertension based on biochemical findings and imaging procedures. However, in renovascular hypertension, for example, causal therapy such as PTA is successful only in a minority of patients, and even in Conn's syndrome blood pressure normalized only in 68% of cases[1,2] after operation. For that reason antihypertensive drug therapy plays an important role in these hypertension forms. In renal hypertension it is even more important and frequently used. Chronopharmacologic aspects are of great importance in secondary hypertension because altered blood pressure rhythm is often found in these patients.[3–5]

DIAGNOSTIC ASPECTS

Although blood pressure rhythms are not yet considered when hypertension is to be defined as primary or secondary, we often find a blunted 24-hour blood pressure profile or even inverse circadian rhythm in secondary hypertension[3–5] in contrast to primary hypertension.

This phenomenon was observed in 1922 by Katsch and Pansdorf[6] using an extended stethoscope tube to measure blood pressure through a wall without disturbing their patients during nighttime rest. They reported an abnormal nocturnal blood pressure "movement" in five patients with end-stage renal insufficiency with high nighttime blood pressure values.

We can now monitor our patients' blood pressure around the clock using ambulatory blood pressure monitoring (ABPM). An international database from 24 clinical research units provides data on normal values of 24-hour blood pressure monitoring including nighttime blood pressure in normotensive subjects and hypertensive patients.[7] Today, ABPM is necessary especially in secondary hypertension to evaluate the circadian blood pressure profile. It is a prerequisite for individualization of antihypertensive therapy based on the patient's circadian blood pressure fluctuations.

The same patterns of circadian blood pressure rhythm as those in primary and secondary hypertension can be observed in models of spontaneously hypertensive rats with primary hypertension showing a normal circadian blood pressure rhythm and in transgenic rats (harboring a mouse Renin gene) with secondary hypertension showing inverse circadian blood pressure rhythm.[8]

[a] Dedicated to Professor H. Holzgreve in honor of his 60th birthday.

PROGNOSTIC ASPECTS

Besides its diagnostic aspects, disturbed circadian blood pressure rhythm has important prognostic implications. The loss of the nocturnal blood pressure fall is likely to indicate a higher risk because of the prolonged time over which the increased blood pressure impact is exerted on target organs. This view is supported by recent data showing that the nighttime blood pressure level and the entire nocturnal blood pressure fall significantly correlate with the decline in creatinine clearance, glomerular filtration rate,[9,10] left ventricular hypertrophy, and cardiovascular events[11–14] independent of the mean 24-hour blood pressure level. Our own data show a significant correlation between the nocturnal blood pressure fall and creatinine clearance over a wide range, from 15–90 ml/min (FIG. 1A and B).

CHRONOPHARMACOLOGIC STUDIES

Different dosing times to study the influence on blood pressure rhythm were first used in 1991 in normotensive subjects[15] and in those with primary hypertension.[16] Recently the first chronopharmacologic study was performed in patients with secondary hypertension and high nighttime blood pressure[17]; the calcium channel blocker isradipine (sustained-release formulation) was applied in the morning and evening in a double-blind cross-over study in 16 patients with chronic renal failure for 4 weeks. Both isradipine treatments were equally effective in reducing mean 24-hour blood pressure levels. However, the evening regimen showed a more pronounced effect during the night. Only the evening administration reset the normal synchronization of the 24-hour blood pressure and heart rate profiles. Similar results were obtained with doxazosin (4 mg) in six patients with chronic renal failure; evening dosing was more successful in restoring normal circadian blood pressure rhythm.[18] Our own results in hypertensive diabetic patients[19] treated with the long-acting ACE inhibitor trandolapril[20–25] are similar. Twenty patients were treated according to their 24-hour blood pressure rhythm; 10 patients with normal rhythms received 2 mg trandolapril as a single morning dose (7 AM) and 10 patients with abnormal rhythm (blunted or inverse circadian rhythm) received the same dose in the evening (7 PM). Despite a significant difference in blood pressure rhythm in both groups, mean values for 24-hour daytime and nighttime blood pressure were significantly reduced with both treatments (TABLE 1). Circadian blood pressure rhythm was preserved in dippers with morning dosing and was normalized in non-dipping patients with evening dosing.

All patients included in the three studies (TABLE 1) had mild to moderate hypertension. In these patients monotherapy was sufficient to normalize mean blood pressure values and circadian blood pressure rhythm. It seems reasonable to prefer long-acting drugs or formulations in these patients. In more severe forms of secondary hypertension chronopharmacologic studies are not yet performed. However, from clinical experience we determined that these patients need combination therapy consisting of three, four, and even more antihypertensive drugs to control their blood pressure. In persistent nocturnal hypertension the addition of Clonidine or an alpha-blocker in a single evening dose can induce nighttime blood pressure reduction.

Drug effects on blood pressure rhythm in secondary hypertension depend on the time of dosing and probably on the duration of action and the kind and severity of the underlying disease.

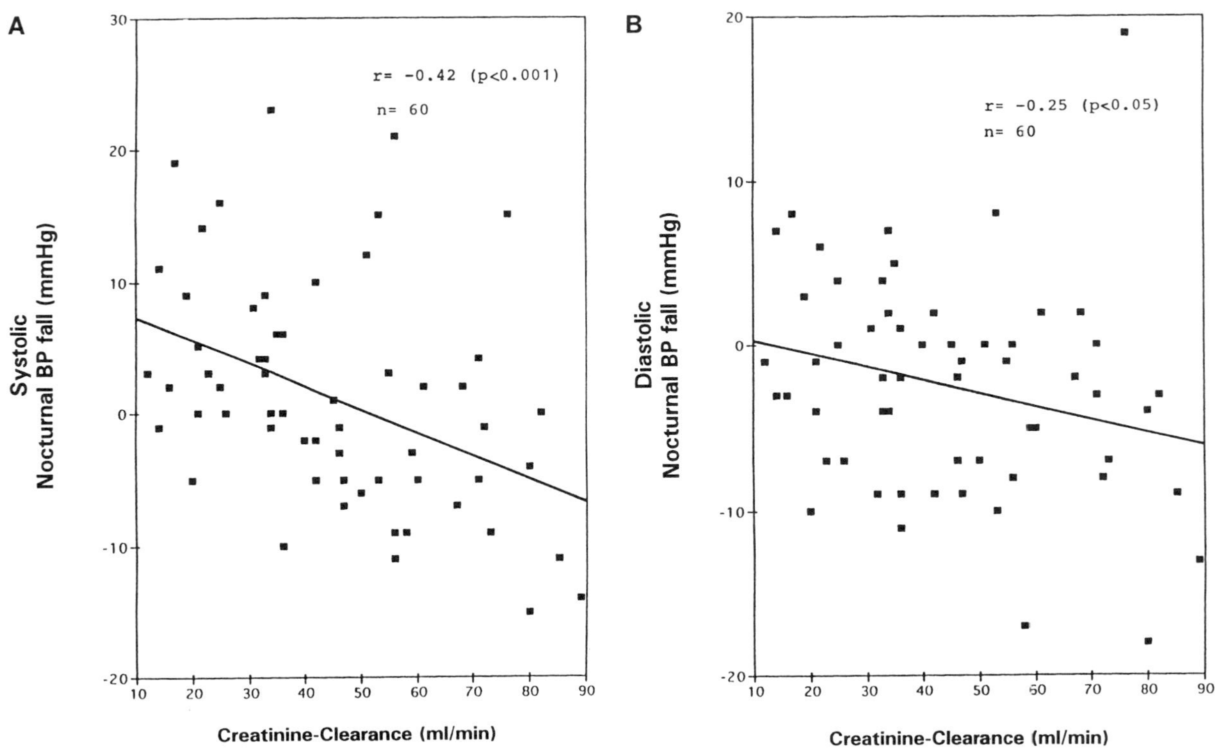

FIGURE 1. Correlation between renal function (creatinine clearance) and nocturnal decrease in systolic (**A**) and diastolic (**B**) blood pressure in 60 patients with diabetic and hypertensive nephropathy.

TABLE 1. Chronopharmacologic Studies: Morning versus Evening Dosing in Patients with Secondary/Nocturnal Hypertension[a]

Hypertension		Patients			Dosing	Effect on 24-Hour Blood Pressure			
Type	Severity	(*n*)	Drug Dose Duration	Design	Time	Day	Night	24-Hour Profile	Reference
Chronic renal failure	Moderate	16	Isradipine, sustained release, 5 mg, 4 wk	Cross-over	8 AM	++	+	Not normalized	Portaluppi *et al.*[17]
					8 PM	++	++	Normalized	
Chronic renal failure	Moderate	6	Doxazosin, 4 mg, 4 wk	Cross-over	8 AM	++	+	Not normalized	Portaluppi[18]
					8 PM	++	++	Normalized	
Hypertensive diabetics	Mild	21	Trandolapril, 2 mg, 4 wk	Dippers vs	7 AM	++	+	Preserved	Middeke[19]
				Non-dippers	7 PM	++	+	Normalized	

[a] +, ++, +++ = rating of drug effects.

DISCUSSION

Antihypertensive drug therapy in secondary hypertension is often necessary and frequently used, because only a minority of patients with different forms of secondary hypertension can be cured by operation or angioplasty. Most patients, especially those with renal hypertension, depend on drug therapy. In renal hypertension there is a very small margin between primary and secondary hypertension in many patients: How should we define a hypertensive diabetic patient, for example, who has normal serum creatinine values, no proteinuria, but an inverse circadian blood pressure profile? Without regard to blood pressure rhythm this patient would be classified as having primary hypertension. However, serum creatinine is not a good parameter of kidney function, and we can speculate that this patient may already have morphologic abnormalities on renal biopsy, leading to altered circadian blood pressure regulation. In this situation there is no indication for renal biopsy. However, this is a worthy example for reconsidering the definition of secondary hypertension. Circadian blood pressure rhythm should be taken into account when hypertension is classified and defined, because of important prognostic and therapeutic implications.[26–28] Hypertensive diabetic persons are particularly prone to renal insufficiency. The progression of nephropathy can be delayed by effective antihypertensive therapy.[29] The conventional goal of antihypertensive treatment in these patients is not only normalization of blood pressure. A pressure of 130–135/80–85 mm Hg has been recommended for many years.[29] In addition, patients with an inverse/abnormal blood pressure rhythm should be treated so as to achieve a normal blood pressure rhythm. This is possible with different dosing times depending on the kind of circadian blood pressure profile (TABLES 1 and 2). The first chronopharmacologic studies show that evening dosing could have advantages in patients with chronic renal failure and in hypertensive diabetic patients with abnormal circadian blood pressure profiles.

TABLE 2. Stepped Care Program of Antihypertensive Drug Therapy in Patients with Inverse Circadian Blood Pressure Profile/Secondary Hypertension

Normal daytime bp/increased nighttime bp:
Evening dosing/monotherapy
Mild daytime hypertension/inverse bp profile:
Evening dosing/long-acting drug/monotherapy
Mild to moderate hypertension/blunted circadian bp profile:
Morning and evening dosing/monotherapy or combination
Severe hypertension/blunted or inverse bp profile:
Morning and evening combination therapy
Additional evening dosing of clonidine or alpha-blocker in refractory nighttime hypertension

At this time it is possible to conclude that antihypertensive drug therapy in secondary hypertension (1) is often necessary and frequently used (especially in chronic renal failure); (2) is important to prevent target organ damage; (3) has to reduce daytime and nighttime blood pressure; and (4) has to restore a normal circadian blood pressure rhythm.

In patients with mild to moderate secondary hypertension, antihypertensive monotherapy is sufficient as either morning or evening dosing depending on the circadian blood pressure profile. In more severe forms of hypertension combination therapy is necessary. A stepped up care program to treat patients with secondary hypertension with respect to the severity of hypertension and the circadian blood pressure rhythm is proposed in TABLE 2. Ambulatory blood pressure monitoring is a prerequisite for individualization of antihypertensive therapy and for achievement of optimal drug effects in secondary hypertension.

SUMMARY AND CONCLUSIONS

Antihypertensive drug treatment is necessary in most patients with secondary hypertension. Only a small percentage of cases can be cured by operation or angioplasty. Because blunted or reversed nocturnal blood pressure fall is frequently found in patients with secondary hypertension, time of day of drug application is of special interest in these patients. The time structure of blood pressure over 24 hours or even longer can be obtained using ambulatory blood pressure monitoring. This is a prerequisite for individualization of antihypertensive therapy and for achievement of optimal drug effects. Chronopharmacologic aspects are playing an increasingly important role in the treatment of secondary hypertension, especially in renal forms. Recently, the first studies in this field were completed, isradipine and doxazosin in chronic renal failure and trandolapril in hypertensive diabetic subjects. Results show that restoration of normal nocturnal blood pressure fall and constitution of normal circadian rhythm are possible in patients with mild to moderate renal hypertension after evening dosing as compared to morning dosing with monotherapy. In more severe hypertension combination therapy including multiple dosing with special attention to evening application is necessary.

In conclusion, optimal drug effects and individualization of antihypertensive treatment in patients with secondary forms of hypertension can be achieved using a stepped care program including chronopharmacologic regimens to restore normal circadian rhythm on a normotensive level.

REFERENCES

1. DÜSING, R., G. HENNECKE & H. VETTER. 1991. Blutdruckverhalten nach Korrektur der zugrundeliegenden Funktionsstörung: Nierenarterienstenose vs. primärer Aldosteronismus. Hochdruck **11:** 54–55.
2. RAMSEY, L. E. & P. C. WALLER. 1990. Blood pressure response to percutaneous transluminal angioplasty for renovascular hypertension: An overview of published series. Br. Med. J. **300:** 569–572.
3. MIDDEKE, M., E. MIKA, M. A. SCHREIBER, B. BECK, B. WÄCHTER & H. HOLZGREVE. 1989. Ambulante indirekte Blutdrucklangzeitmessung bei primärer und sekundärer Hypertonie. Klin. Wochenschr. **67:** 713–716.
4. MIDDEKE, M. & J. SCHRADER. 1994. Nocturnal blood pressure in normotensive subjects and those with white coat, primary, and secondary hypertension. Br. Med. J. **308:** 630–632.
5. PORTALUPPI, F., L. MONTANARI, M. MASSARI, V. DI CHIARA & M. CAPANNA. 1991. Loss of nocturnal decline of blood pressure in hypertension due to chronic renal failure. Am. J. Hypertens. **4:** 20–26.
6. KATSCH, G. & H. PANSDORF. 1992. Die Schlafbewegung des Blutdrucks. Münch. med. Wochenschr. **69:** 1715–1718.
7. THIJS, L., J. STAESSEN, E. O'BRIEN, A. AMERY, N. ATKINS, P. BAUMGART, P. DE CORT, J.-P. DEGAUTE, P. DOLENC, R. DE GAUDEMARIS, I. ENSTRÖM, R. FAGARD, P. GOSSE, S. GOURLAY, H. HAYASHI, Y. IMAI, G. JAMES, T. KAWASAKI, E. KUSCHNIR, I. KUWAJIMA, L. LINDHOLM, L. LIU, F. MACOR, G. MANCIA, B. MCGRATH, M. MIDDEKE, J. MING, S.

OMBONI, K. OTSUKA, P. PALATINI, G. PARATI, C. PIEPER, P. VERDECCHIA, P. ZACHARIAH & W. ZHANG. 1995. The ambulatory blood pressure in normotensive and hypertensive subjects: Results from an international database. The Netherlands J. Med. **46:** 106–114.
8. LEMMER, B., A. MATTES, M. BÖHM & D. GANTEN. 1993. Circadian blood pressure variation in transgenic hypertensive rats. Hypertension **22:** 97–101.
9. DEL ROSSO, G., L. AMOROSE, A. SANTOFERRARA, B. FIEDERLING, L. DI LIBERATO & A. ALBERTAZZI. 1994. Impaired blood pressure nocturnal decline and target organ damage in chronic renal failure. J. Hypertens. **12**(Suppl. 3): S15.
10. TIMIO, M., S. LOLLI, C. VERDURA, C. MONARCA, F. MERANTE & E. GUERRINE. 1993. Circadian blood pressure changes in patients with chronic renal insufficiency: A prospective study. Renal Failure **15:** 231–237.
11. KUWAJIMA, I., Y. SUZUKI, T. SHIMOSAWA, A. KANEMARU, S. HOSHINO & K. KURAMOTO. 1992. Diminished nocturnal decline in blood pressure in elderly hypertensive patients with left ventricular hypertrophy. Am. Heart J. **123:** 1307–1311.
12. RIZZONI, D., M. L. MUIESAN, G. MONTANI, R. ZULLI, S. CALEBICH & E. AGABITI-ROSEI. 1992. Relationship between initial cardiovascular structural changes and daytime and nighttime blood pressure monitoring. Am. J. Hypertens. **5:** 180–186.
13. VERDECCHIA, P., G. SCHILLACI, M. GUERRIERI, C. GATTESCHI, G. BENEMIO, F. BOLDRINI & C. PORCELLATI. 1990. Circadian blood pressure changes and left ventricular hypertrophy in essential hypertension [see comments]. Circulation **81:** 528–536.
14. VERDECCHIA, P., C. PORCELLATI, G. SCHILLACI, C. BORGIONI, A. CIUCCI, M. BATTISTELLI, M. GUERRIERI, C. GATTESCHI, I. ZAMPI, A. SANTUCCI, C. SANTUCCI & G. REBOLDI. 1994. Ambulatory blood pressure: An independent predictor of prognosis in essential hypertension. Hypertension **24:** 793–801.
15. LEMMER, B., G. NOLD, S. BEHNE & R. KAISER. 1991. Chronopharmacokinetics and cardiovascular effects of nifedipine. Chronobiol. Int. **8:** 485–494.
16. MIDDEKE, M., M. KLÜGLICH & H. HOLZGREVE. 1991. Chronopharmacology of captopril plus hydrochlorothiazide in hypertension: Morning versus evening dosing. Chronobiol. Int. **8:** 506–510.
17. PORTALUPPI, F., L. VERGNANI, R. MANFREDINI, E. C. DEGLI UBERTI & C. FERSINI. 1995. Time-dependent effect of isradipine on the nocturnal hypertension of chronic renal failure. Am. J. Hypertens. **8:** 719–726.
18. PORTALUPPI, F. 1995. Time-dependent effects of doxazosin on 24-hour blood pressure profiles of essential and secondary hypertension. World Conference on Chronobiology and Chronotherapeutics. Ferrara, September 1995.
19. MIDDEKE, M. & C. VON BIELINSKI. 1995. Trandolapril in hypertensive diabetics: Effects of different dosing times. Submitted.
20. ARNER, P., A. WADE, P. ENGFELDT, M. MOUREN, J. P. STEPNIEWSKI, E. STULTAN, T. BRYCE & B. LENFANT. 1994. Pharmacokinetics and pharmacodynamics of trandolapril after repeated administration of 2 mg to young and elderly patients with mild-to-moderate hypertension. J. Cardiovasc. Pharmacol. **23:** 44–49.
21. BACKHOUSE, C. I., B. OROFIAMMA & N. C. PAULY. 1994. Long-term therapy with trandolapril, a new nonsulfhydryl ACE inhibitor, in hypertension: a multicenter international trial. J. Cardiovasc. Pharmacol. **23:** 86–90.
22. CESARONE, M. R., M. T. DE SANCTIS, G. LAURORA, L. AMBROSOLI, C. MARELLI & G. BELCARO. 1994. Effects of trandolapril on 24-h ambulatory blood pressure in patients with mild-to-moderate essential hypertension. J. Cardiovasc. Pharmacol. **23:** 65–72.
23. CORBIER, A., D. LECAQUE, J. SECCHI, B. DEPOUEZ & G. HAMON. 1994. Effects of 4 weeks of treatment with trandolapril on renal hypertension and cardiac and vascular hypertrophy in the rat. J. Cardiovasc. Pharmacol. **23:** 26–29.
24. DANIELSON, B., S. QUERIN, P. LAROCHELLE, E. SULTAN, M. MOUREN, T. BRYCE, J. P. STEPNIEWSKI & B. LENFANT. 1994. Pharmacokinetics and pharmacodynamics of trandolapril after repeated administration of 2 mg to patients with chronic renal failure and healthy control subjects. J. Cardiovasc. Pharmacol. **23:** 50–59.
25. MEYER, B. H. & N. C. PAULY. 1994. Double-blind comparison of the efficacy and safety of

trandolapril 2 mg and hydrochlorothiazide 25 mg in patients with mild-to-moderate essential hypertension. J. Cardiovasc. Pharmacol. **23:** 77–80.
26. FUJISE, Y. & M. MIYAHARA. 1975. Comparison of the prognosis of hypertension associated with chronic glomerulonephritis with that of essential hypertension. Jpn Circ. J. **39:** 793–798.
27. O'BRIEN, E., J. SHERIDAN & K. O'MALLEY. 1988. Dippers and non-dippers. Lancet **i:** 397.
28. PALANTINI, P., M. PENZO, A. RACIOPPA, E. ZUGNO, G. GUZZARDI, M. ANACLERIO & A. C. PESSINA. 1992. Clinical relevance of nighttime blood pressure and of daytime blood pressure variability. Arch. Intern. Med. **152:** 1855–1860.
29. MOGENSEN, C. E. 1989. The effect of blood pressure intervention on renal function in insulin-dependent diabetes. Diabetes & Metab. **15:** 343–351.

Ambulatory Blood Pressure Monitoring

Application to Clinical Medicine and Antihypertension Medication Trials

MICHAEL H. SMOLENSKY[a,b]
AND FRANCESCO PORTALUPPI[c]

[a] *University of Texas-Houston School of Public Health*
and
[b] *Hermann Chronobiology Center*
Houston, Texas

[c] *Hypertension Unit*
First Institute of Internal Medicine
University of Ferrara
Ferrara, Italy

The biology and function of the cardiovascular system are highly organized in time over 24 hours and other time periods.[1,2] The day-night pattern in blood pressure (BP), in particular, results from endogenous rhythms in neurohumoral events[3] plus superimposed effects of physical and mental loads which are cyclic in occurrence and intensity during the 24 hours.[4–12] In diurnally active individuals, BP is higher during daytime activity and lowest during nocturnal sleep; this is true for normal and most essential hypertensive persons.[13–20] This pattern is evident also in normal[16] and hypertensive[21] bedridden patients as well as those with a fixed heart rate during the 24-hour period.[22] The typical BP pattern is characterized by two peaks, a very prominent one in the morning coincident with the commencement of daytime activity and a second in the early evening. Generally, there is slight nadir in BP in the afternoon and a profound dip nocturnally. The amplitude of the variations, which is slightly greater for diastolic than systolic BP, ranges between 10 and 15% of the daytime mean.[18,19,23–27]

Under normal conditions, the autonomic nervous system plays a major role in mediating the BP circadian variation; sympathetic drive is dominant during the day, and parasympathetic drive prevails at night.[28–36] Hemodynamic mechanisms are also operative. So-called rapid autoregulatory mechanisms continuously control tissue blood flow, venous return, and cardiac output, while slow volume regulatory mechanisms moderate the effective circulating blood volume and absolute plasma volume. All such physiologic adjustments influence BP from one moment to another.[37] Circadian rhythms in catecholamine, cortisol, aldosterone, and renin concentration and activity, among other hormones and physiologic variables, all contribute to the 24-hour BP profile.[12]

The well-described day-night pattern of BP may be altered in disease; for example, the level around which variation in time is expressed may be modified to values higher or lower than normal reference limits, and/or the phasing of the

[b] Address for correspondence: Michael H. Smolensky, University of Texas-Houston, School of Public Health, Rm. 442, 1200 Herman Pressler, P.O. Box 20186, Houston, TX 77225. Tel: 713-500-9237; fax: 713-500-9221.

24-hour pattern may be changed. In certain circumstances the rhythm may even be obliterated. Nonetheless, in most hypertensive patients the circadian BP rhythm is preserved with daytime peaks and nocturnal nadir, thus reflecting the persistence of the normal temporal structure of the cardiovascular system, even though the 24-hour mean is shifted higher. Only in a minority of cases, usually in secondary forms or in complicated essential hypertension, is the temporal pattern of BP disrupted and the 24-hour profile altered.

TECHNOLOGICAL/METHODOLOGICAL ISSUES OF BP ASSESSMENT BY AMBULATORY BLOOD PRESSURE MONITORING

Noninvasive ambulatory blood pressure monitoring (ABPM) was introduced more than 30 years ago. Over the years, the technology has greatly improved. Today, a variety of ABPM devices are commercially marketed. Although they differ by the method in which they assess BP—microphonics, oscillometrics, or ECG gating, most are reliable to use, convenient to wear, and accurate in measurement.[38] Although standard protocols for their validation have been established,[39,40] standardization of measurement frequency during the 24 hours and duration of assessment, 24 versus 48 hours, across user groups have yet to be achieved. In clinical practice, for example, European and Asian physicians commonly measure BP every 10–15 minutes during both day and night. North American physicians tend to measure BP less often, every 20–30 minutes during the day and less frequently, every 30–60 minutes, during the night. The length of the monitoring span is not thought to significantly influence the 24-hour BP pattern when assessments are done at least every 30 minutes.[41] Some recommend sampling at 15-minutes intervals over 2 consecutive days to take into account day-to-day differences in environmental influences on BP.[42] In clinical settings, patient compliance to monitorings lasting longer than 24 hours is frequently poor; however, in research and drug trials, compliance is better. Universal standardization of the exact endpoints and methods of data analysis and interpretation is also lacking among user groups. Some rely on day, night, and/or 24-hour BP levels, whereas others prefer BP load (percentage of BP values that exceed established normative limits during designated time periods).[43,44] In addition, some are interested in the magnitude of the morning BP rise, because it is hypothesized that it constitutes an important risk factor for acute coronary events in the morning at the commencement of daily activity.[45]

Relatively few studies have addressed the reproducibility of BP findings by ABPM. Mancia *et al.*,[46] by direct comparisons of data sets from the same subjects, reported the standard deviations of the differences in diastolic BP between repeated measurements to be in the order of 6–7 mm Hg. This is disappointingly close to that of conventional sphygmomanometric measurements obtained during three different clinic visits as recommended in the international guidelines for diagnosis.[47–52] Other studies show the mean between-visit correlations to amount to 0.76 for ABPM versus 0.70 for casual office BP determinations.[47–49,53–55]

The between-day variability in ABPM findings is due in large part to day-to-day differences in physical activity, emotional load, and the environmental/occupational setting in which monitorings take place. Some have argued that the advantage derived from ABPM by the acquisition of multiple measurements during a 24-hour period is offset by the disadvantage that measurements are obtained in uncontrolled environments from unrestrained subjects. However, concerns have been raised about the degree to which casual clinical BP measurements are representative. The accuracy of casual BP readings made in clinical settings is often compromised. They are seldom performed under optimal conditions, being compromised by reliance on inaccurate techniques, unblinded measurements, nonrandom-zero sphygmomanometers, and "white-coat" effects (BP increase only in the presence of health care professionals), among others. In actuality, ABPM values exhibit particularly small between-day variation when assessment is done under comparable conditions in individuals studied in controlled research settings or patients monitored in comparable environments, for example, during work or school days.[56] The use of wrist activity monitoring devices,[57,58] in addition, makes possible better and more accurate interpretation of ABPM data in both diagnostic studies and drug trials, because it allows the variation in activity level and pattern on BP profiles within and between 24-hour monitorings to be taken into account.

Intermittent BP measurement provides only a brief "snapshot" picture of the BP level during the day, during the time when the clinic is open. In contrast, ABPM gathers numerous readings throughout the day and night. How can the 24-hour BP times series data best be summarized and interpreted? Interpretation of ABPM data too often is restricted to parameters derived only from linear analysis, such as day, night, and 24-hour means. Common indices of BP variability typically include range, standard deviation, standard error of the mean, root mean square, and fractiles. This approach is inadequate, because it disregards a great deal of useful clinical information related to BP variability or stability, circadian pattern, and excess over selected BP limits (load).

Analysis of the temporal pattern and its waveform may be performed objectively by the cosinor method,[59] partial Fourier series usually limited to the first four or five harmonics of the fundamental 24-hour component,[6,60–62] spline functions,[63] cumulative sums,[64] square-wave fit,[65] and spectral analysis.[66] The advantages and limitations of these different approaches are currently a matter of active debate and study.[66–68] Further research is required to fully assess the prognostic and therapeutic implications of the information and indices obtained from the different analyses before specific procedures can be recommended for application to the ABPM time series.

The proper application of statistical procedures to ascertain indices of BP variability and profile throughout the 24 hours is dependent on the density, frequency, and duration of data collection. For example, the use of Fourier analysis to determine periodic components in discontinuous data series as a mathematical method of describing the BP profile is often inadequate because of aliasing errors. The low sampling frequency of conventional ABPM devices results in excessive smoothing of daytime BP peaks and nighttime troughs.[66] The error intrinsic to discontinuous time series data is even of greater importance when the goal is to determine the frequency domain of BP variability. Instruments that sample BP continuously are necessary to accomplish the accurate description of BP variability over time. In this regard, new noninvasive (for example, the Fingerpres[69] and Portapres[70]) technology makes possible the continuous measurement of finger BP over prolonged periods of time. Finally, in some patients, such as those with carotid sinus syncope and pacemaker syndromes, it is advantageous to simultaneously record both ECG and BP using instrumentation now commercially available.

BLOOD PRESSURE INFORMATION PROVIDED BY AMBULATORY BLOOD PRESSURE MONITORING

Reliance on casual intermittent clinical BP measurement alone may be inadequate to base diagnostic decisions and to properly evaluate the therapeutic benefit of antihypertensive interventions. ABPM provides different information about BP than does conventional single, once-a-day casual BP determinations. ABPM provides several types of information. Most commonly, ABPM is used to determine:

- The average BP level during the day, night, or 24 hours before and/or during treatment with antihypertensive medications;
- The 24-hour BP profile (diurnal vs nocturnal hypertension vs non-dipping nighttime pattern);
- The presence or absence of white coat hypertension or effect;
- The short-term variability in BP.

The first three types of information are of importance to clinical medicine, that is, in the diagnosis of hypertension and the outcome assessment of antihypertensive treatments. However, the absence of studies demonstrating the cost efficiency of ABPM currently limits greater use of this technology in health care settings. The fourth application of ABPM is presently restricted to patient research.

Traditionally, interest has focused on the actual BP level during the day, night, or 24-hour period. Yet, there is no universal agreement on how to use ABPM data to make the diagnosis of hypertension. Can the 24-hour, as opposed to the daytime, mean be relied on for the differential diagnosis of normotension versus hypertension? If the answer is yes, what reference should be used? Should it be 140/90 mm Hg, 135/85, or others? Some investigators suggest reliance on BP load—the percentage of assessments exceeding established normative values—to establish the diagnosis of hypertension.[43,44] BP load can be determined only by 24-hour ABPM study; single or infrequently conducted self-assessments of BP during the activity span do not permit calculation of BP load. In general, patients are judged to be hypertensive if the BP load exceeds 25%.[44] BP loads in excess of 40% are predictive of end-organ injury.[43]

Software packages make possible the summary of BP data for specific time spans during the day and night. However, there is disagreement as to the definition of day versus night in terms of clock time. Software programs of some companies define, a priori, daytime to be between 6 AM and 10 PM and nighttime between 10 PM and 6 AM. Those of other companies divide the 24-hour period into 12-hour day and nighttime spans. In all cases, the assumption is made that patients adhere to a social routine of diurnal activity and nocturnal rest, which is not always the case. We, as do other investigators, rely on wrist actigraphy[57,58] and patient diaries to determine the clock times of activity and sleep of each patient. This, in turn, enables us to precisely determine the activity and sleep span BP levels individually.

DIURNAL BLOOD PRESSURE PATTERN

The BP profile of uncomplicated primary hypertension is characterized by elevated values during daytime activity and reduced ones during nocturnal sleep. ABPM studies reveal that most patients experience an accelerated morning rise in BP coincident with the onset of daily activity. Circadian rhythms in neuroendocrine function and balance favor vasoconstrictive and pressure-increasing effects at this time, as reviewed elsewhere in these proceedings.[12] The change in posture and activity level occurring with termination of nighttime sleep and commencement of daytime activity results in a rapid rise in serum catecholamines. This significantly contributes to this functional state.

It is unclear if the rise in BP commences before the termination of nocturnal sleep or if it is triggered entirely by a change in posture and commencement of diurnal activity. Millar-Craig *et al.*,[18] using invasive monitoring, noted that BP begins to rise in the early morning hours during sleep. Others[19–23] reported the rise to be dependent only on a change in posture and activity pattern in the morning. Through the years, the prewaking rise in BP has been alternately denied[71,72] and confirmed.[24–26] Difficulty in determining the precise timing of morning arousal in individual patients complicates the interpretation of morning BP profiles and most likely explains the differences in findings among the various investigations. A recent study compared the morning BP change in patients arising from bed and immediately engaging in activity with that of persons remaining in bed and inactive after awakening from nocturnal sleep.[73] The BP rise was significant in those who were immediately active, implicating its relationship to physical activity.

It has been postulated that the morning BP surge exhibits two distinct patterns, one in which the rise is gradual before awakening, a pattern common to younger persons, and another in which the rise is initiated by physical activity, a pattern common to elderly persons.[74,75] Age-related differences in sleep architecture may play a role in the morning pattern of BP. BP varies during sleep; after declining during deep sleep (stages 3–4), it rises to higher levels.[17] REM sleep, which preferentially occurs during the final one-third of the night in younger persons, prevails during the pre-waking period and could contribute to the morning BP surge. In elderly persons, sleep architecture is different; it tends to be disrupted with REM stages more evenly distributed throughout the night. Thus, the cause of the morning rise in BP in older patients is not likely to be dependent on REM-related phenomena.

ABPM clearly reveals and quantifies the morning surge in BP. Muller *et al.*[45] hypothesize that the shearing force associated with the morning BP increase may be capable of inducing rupture of a vulnerable coronary plaque. If the rupture is minor, the consequence may be a nonocclusive thrombus, resulting in asymptomatic or unstable angina or non-Q-wave myocardial infarction. If the rupture is major with an occlusive thrombus, acute myocardial infarction and sudden cardiac death are possible morning-time events.

NOCTURNAL DIPPING AND NON-DIPPING BLOOD PRESSURE PATTERNS

The nocturnal (sleep-time) decline equal to at least 10% of the daytime mean level constitutes a prominent characteristic of the BP profile of primary hypertension.[76] Moderation or reversal of the fall of BP overnight, which occurs secondary to specific medical conditions, is abnormal. These medical conditions include orthostatic autonomic failure,[77] Shy-Drager syndrome,[78] Alzheimer-type dementia,[80] vascular dementia,[79] cerebral atrophy,[81] cerebrovascular disease,[76–80] ischemic arterial disease after endarterectomy,[87] neurogenic hypertension,[82–86] fatal familial insomnia,[89] diabetes,[88] catecholamine-producing tumors,[96–100] Cushing's syndrome,[101] glucocorticoid therapy,[102–103] mineralocorticoid excess syndromes,[104–106] Addison's disease,[107] pseudohypoparathyroidism,[108] sleep apnea syndrome,[109] normotensive and hypertensive asthma,[110] chronic renal failure,[104,111–122] severe hypertension,[123] salt-sensitive essential hypertension,[124] pregnancy-induced hypertension,[125] toxemia of pregnancy,[126–127] essential hypertension with left ventricular hypertrophy,[76,128] organ transplantation secondary to congestive heart failure,[137–140] immunosuppressive drugs,[103,129–136] and recombinant human erythropoietin therapy.[141] Nocturnal hypertension is estimated to be present in about 10–20% of those diagnosed with essential hypertension and as many as 60–80% of those with secondary hypertension.[27,76] The reproducibility of nocturnal hypertension as a temporal pattern, in terms of between-visit correlation, is about 0.75 in essential hypertension. In our experience, it is even higher in secondary forms of hypertension, such as chronic renal failure.[115,142]

Nocturnal hypertension most likely results from alterations in the circadian time structure of the autonomic nervous system and neurohumoral functions which are normally involved in central and/or peripheral BP regulation. Imbalance of sympathetic versus parasympathetic activity during the 24-hour period is a major determinant of nocturnal hypertension. This is the cause of the nocturnal hypertension observed in neurogenic dysautonomias as well as diabetes and chronic renal failure in which alterations in the circadian pattern of BP are minimal[95] or absent[143] before the onset of autonomic neuropathy. Dysfunction of the hypothalamic-pituitary-adrenocortical axis may be responsible for some forms of nocturnal hypertension as demonstrated by the close correlation between a change in the circadian pattern of BP and that in cortisol and corticotropin in fatal familial insomnia,[89] Cushing's syndrome,[101] and high-dose glucocorticosteroid-treated patients.[102] Catecholamine excess due to a tumor of the adrenal medulla does not obliterate the nocturnal decline of BP unless the plasma concentration is extremely high.[100] Alterations in plasma volume may be involved in the genesis of nocturnal hypertension in certain conditions, such as renal insufficiency. Although left ventricular hypertrophy has been associated with nocturnal hypertension, cardiac changes are more likely the consequence rather than the cause of the loss of nocturnal decline in BP.

The evidence is quite convincing that morbid and mortal cardiovascular events—myocardial angina and infarction, sudden cardiac death, arrhythmias, fatal pulmonary thromboembolism, and ischemic cerebrovascular accidents—are of greatest risk in the morning during the initial hours of daytime activity. Circadian rhythms in critical biological functions plus 24-hour cycles in the occurrence and intensity of environmental triggers explain the elevated frequency of these clinical events, as reviewed elsewhere in this volume.[144]

The impact of nocturnal hypertension is beginning to be appreciated clinically. Non-dipping and elevated nocturnal BP profiles are associated with increased risk of hypertension-induced cardiac,[76,128,145–149] cerebral,[83,84,150] or renal[121,122,151] organ injury. Indeed, the correlation between night (sleep-time) BP load and left ventricular hypertrophy is high. Prospective investigations now in progress, such as the European Multicenter Trial of Office versus Ambulatory BP Monitoring,[152] Syst-Eur study side project,[153] Ambulatory BP and Treatment of Hypertension,[154] and Italian SAMPLE[155] studies further address the relationship between cardiovascular morbidity and mortality and indices of the 24-hour BP profile. It is worthy of mention that the results of initial prospective studies by Perloff *et al.*,[156,157] conducted over 10 years ago, demonstrate improved prognostic prediction of ABPM plus clinical BP values in comparison with the latter only. A recently completed investigation by Verdecchia *et al.*[158] shows that cardiovascular morbidity is greater in female non-dippers than dippers with ambulatory hypertension. Left ventricular structure seems to be more load-dependent in women than men with hypertension as demonstrated by higher echocardiographic indices of left ventricular hypertrophy in female, but not male, non-dippers than dippers.[149]

THERAPEUTIC IMPLICATIONS OF THE BLOOD PRESSURE CIRCADIAN RHYTHM

Hypertension is of great clinical significance because of the role it plays in inducing organ damage. Traditionally, treatment of hypertension has relied on medications and dosing schedules that achieve rather constant (homeostatic) drug levels throughout the 24-hour period. However, BP is not constant over the 24 hours. Thus, at certain times of the day, such as at the commencement of daily activity when BP surges, the drug level derived from conventional dosage forms is likely to be lower than needed relative to biological requirements; conversely, at other times, such as overnight when BP cycles to a minimum, the drug level is likely to be greater than required relative to biological need. One theoretical means of improving the management of primary hypertension is to vary medication level in synchrony with the day-night pattern in BP. Theory has been transformed into practice with the introduction of a new chronotherapy of verapamil by an American pharmaceutical company. Covera-HS (Searle Pharmaceuticals, USA) makes use of modern drug delivery technology to time-modulate verapamil drug levels during the 24-hour period. Following bedtime dosing, drug release is inhibited for 4–5 hours, the amount of time required to dissolve the tablet's built-in internal "delay coat." Thereafter, fluids within the gastrointestinal tract that are absorbed into the core of the tablet expand the osmotically active drug push layer, thereby forcing the release of medication through two laser-drilled micropores. This controlled-onset, extended-release drug delivery system results in a circadian rhythm-adapted plasma concentration of verapamil that varies in synchrony with the 24-hour profile of BP.

The day-night pattern in BP, as discussed earlier, is dependent on circadian rhythms in various neuroendocrine, neurohumoral, and hemodynamic variables and factors. Such rhythm dependencies constitute the theoretical basis for optimizing the efficiency and/or safety of present and future antihypertensive medications by means of specific circadian chronotherapeutic delivery profiles.

Circadian rhythms of the gastrointestinal tract and vital organs are capable of affecting the absorption, biotransformation, and elimination of antihypertensive medications when ingested at different times during the 24-hour period as illustrated, for example, by propranolol and nifedipine.[159] Moreover, the effect of cardiovascular-active medications may differ according to the circadian time of their dosing due to rhythms in drug-targeted tissues.[159–164] In this regard, plasma concentrations of medications are not necessarily predictive of biological effect, as illustrated by the day-night difference in the propranolol dose–heart rate and BP response relationship.[165] Recent studies clearly demonstrate that the effects of conventional once-a-day antihypertensive formulations, such as isradipine and quinapril, are quite different depending on their time of ingestion; the nighttime BP level is reduced more with evening than morning drug administration.[142,166] Even though administration–time-dependent effects have been the subject of an increasing number of publications (see ref. 164 for review), few investigations have been specifically designed to properly address circadian time-dependencies of BP-lowering medications. Thus, our knowledge is far from complete.

BLOOD PRESSURE CIRCADIAN RHYTHM AND CLINICAL DRUG TRIALS

BP is a convenient and cost-efficient surrogate measure of the effectiveness of antihypertensive medications. The reduction and normalization of BP level and pattern are assumed to translate into diminished risk of hypertension-induced injury of targeted tissues and organs. ABPM technology offers several advantages over the traditional methods used to conduct drug trials.[38] In the past, clinical trials of antihypertensive agents relied entirely upon intermittent BP determinations to qualify patients for investigative protocols and to assess the therapeutic efficiency of medications. In-clinic BP determinations, which run the risk of white coat effects, might result in the qualification of white coat hypertensive subjects, that is, persons who exhibit elevated BP only in the presence of health professionals. Based on recently published studies, as many as 20–30% of patients involved in clinical trials of antihypertensive medications could actually be normotensive.[38] Twenty-four–hour ABPM identifies those individuals, so the evaluation of drug effects is done on true hypertensive subjects only. ABPM also enables us to accurately assess the effect of antihypertensive medications with minimal risk of white coat effect as a function of the circadian time of dosing. Conventional intermittent clinical measurements are capable of neither identifying the different (diurnal versus nocturnal) types of hypertension nor assessing drug effects on each. Only ABPM can accomplish this.

The use of ABPM in drug trials, however, poses a new set of questions for the pharmaceutical industry and drug regulatory agencies. Intermittent BP measurements determine the effect of antihypertensive medications in terms of mean values obtained during the daytime and/or specific times of the day or night typically corresponding to the peak and trough of blood drug concentration. ABPM precisely defines the duration and extent of the BP-lowering effect of medications according to the circadian dosing time. It provides measurement of the effect of medications on 24-hour, daytime and nighttime BP and BP load. Moreover, it enables the quantification of the BP surge at the commencement of diurnal activity and BP decline during sleep. Finally, ABPM reveals the effect of treatment on the 24-hour BP pattern. It is hypothesized that the control of the morning surge in BP reduces one's risk of cardiac and cerebral ischemic events.[45] However, at this time, clinical and epidemiological evidence in support of this hypothesis is lacking. Thus, although several of the aforementioned endpoints are currently of extreme academic interest, current data do not yet warrant their inclusion by regulatory agencies as meaningful indices of drug assessment.

CONCLUSIONS

Ambulatory blood pressure monitoring with calibrated and noninvasive devices is now widely accepted as a means of detecting hypertension and describing BP profile and variability.[166–170] Problems arise, however, in the analysis of the time series data sampled. BP profiles are typically examined visually for subjective, rather than objective, impression. Linear and rhythm analyses, which provide meaningful objective indices of clinical relevance, are required to describe the attributes of BP profile and shape. The information derived from ABPM studies of BP pattern and variability in different pathophysiologic states has advanced our understanding of rhythm dependencies of the mechanisms and treatment of hypertension. Future applications of this method are expected to further enhance the research of hypertension and to better assess the efficiency of antihypertensive medications.

SUMMARY

The wide use of ABPM has resulted in greater appreciation of the circadian time structure of BP variability and its clinical relevance. It is now recognized that the day-night change in BP results from an interplay of circadian rhythms in neurohumoral mechanisms coupled with temporal patterns in physical activity and mental load. The composite effect and balance of these endogenous and exogenous cyclic phenomena give rise to elevated BP during diurnal activity and reduced BP during nighttime repose in both normotension and uncomplicated essential hypertension. The balance is frequently disturbed in complicated and secondary forms of hypertension causing gross alteration of the 24-hour BP profile. ABPM also reveals the efficiency of antihypertensive treatment throughout the 24 hours and as a function of drug administration time. The pharmacokinetics and/or pharmacodynamics of antihypertensive medications have been demonstrated to vary with ingestion time. Such time-dependencies arise from circadian rhythms in BP and underlying mechanisms. The effect of antihypertensive medications is not simply superimposed upon endogenous bioperiodicities. Rhythms in neurohumoral mechanisms of BP control may modulate treatment effect. Certain aspects of the shape of the 24-hour BP profile, such as the magnitude of the morning surge and nocturnal decrease, have been implicated as determinants of morbid and mortal cardiovascular events. One large clinical multicenter investigation, known as the CONVINCE (Controlled Onset Verapamil Investigation of Clinical Endpoints) trial, is aimed at assessing the impact (cardiovascular morbidity and mortality) of verapamil chronotherapy over standard diuretic or beta anatagonist treatment in hypertensive patients with at least one risk factor of coronary heart disease. ABPM will help ascertain to what extent depression of the morning surge in BP relates to reduction in cardiac morbidity and mortality in this as well as other such trials. In any event, the importance of ABPM and the indices derived from its application are just beginning to be appreciated and explored.

REFERENCES

1. SMOLENSKY, M. H., S. E. TATAR, S. A. BERGMAN, J. G. LOSMAN, C. N. BARNARD, C. C. DASCO & I. A. KRAFT. 1977. Circadian rhythmic aspects of cardiovascular function. A review by chronobiologic statistical methods. Chronobiologia **3:** 337–371.
2. LEMMER, B. 1992. Cardiovascular chronobiology and chronopharmacology. *In* Biologic Rhythms in Clinical and Laboratory Medicine. Y. Touitou & H. Haus, eds.: 419–427. Springer-Verlag. Berlin.
3. CLARK, L. A., L. DENBY, D. PREGIBON *et al.* 1987. A quantitative analysis of the effects of activity and time of day on the diurnal variations of blood pressure. J. Chronic Dis. **40:** 671–681.
4. SUNDBERG, S., A. KOHVAKKA & A. GORDIN. 1988. Rapid reversal of circadian blood pressure rhythm in shift workers. J. Hypertens. **6:** 393–396.
5. BAUMGART, P., P. WALGER, G. FUCHS, K. G. DORST, H. VETTER & K. H. RAHN. 1989. Twenty-four-hour blood pressure is not dependent on endogenous circadian rhythm. J. Hypertens. **7:** 331–334.
6. CHAU, N. P., J. M. MALLION, R. DE GAUDEMARIS *et al.* 1989. Twenty-four-hour ambulatory blood pressure in shift workers. Circulation **80:** 341–347.
7. JAMES, G. D. & T. G. PICKERING. 1993. The influence of behavioral factors on the daily variation of blood pressure. Am. J. Hypertens. **6:** 170S–173S.
8. PIEPER, C., K. WARREN & T. G. PICKERING. 1993. A comparison of ambulatory blood pressure and heart rate at home and work on work and non-work days. J. Hypertens. **11:** 177–183.
9. DEGAUTE, J. P., E. VAN CAUTER, P. VAN DE BORNE & P. LINKOWSKI. 1994. Twenty-four-hour blood pressure and heart rate profiles in humans. A twin study. Hypertension **23:** 244–253.
10. GOTO, T., K. YOKOYAMA, T. ARAKI *et al.* 1994. Identical blood pressure levels and slower heart rates among nurses during night work and day work. J. Hum. Hypertens. **8:** 11–14.
11. STERNBERG, H., T. ROSENTHAL, A. SHAMISS & M. GREEN. 1995. Altered circadian rhythm of blood pressure in shift workers. J. Hum. Hypertens. **9:** 349–353.

12. PORTALUPPI, F., L. VERGNANI, R. MANFREDINI & C. FERSINI. 1996. Endocrine mechanisms of blood pressure rhythms. Ann. N.Y. Acad. Sci. **783:** 113–131.
13. HILL, L. 1898. On rest, sleep and work and the concomitant changes in the circulation of the blood. Lancet **1:** 282–285.
14. SNYDER, F., A. HOBSON, D. F. MORRISON & F. GOLDFRANK. 1964. Changes in respiration, heart rate and systolic blood pressure in human sleep. J. Appl. Physiol. **19:** 417–422.
15. BRISTOW, J. D., A. J. HONOUR, T. G. PICKERING & P. SLEIGHT. 1969. Cardiovascular and respiratory changes during sleep in normal and hypertensive subjects. Cardiovasc. Res. **3:** 476–485.
16. REINBERG, A., J. GHATA, F. HALBERG *et al.* 1970. Rythmes circadiens du pouls, de la pression arterielle, des excretions urinaires en 17-hydroxycorticosteroides catecholamines et potassium chez l'homme adulte sain, actif et au repos. Ann. Endocrinol. (Paris) **31:** 277–287.
17. COCCAGNA, G., M. MANTOVANI, F. BRIGNANI, A. MANZINI & E. LUGARESI. 1971. Laboratory note. Arterial pressure changes during spontaneous sleep in man. Electroencephalogr. Clin. Neurophysiol. **31:** 277–281.
18. MILLAR-CRAIG, M. W., C. N. BISHOP & E. B. RAFTERY. 1978. Circadian variation of blood-pressure. Lancet **1:** 795–797.
19. LITTLER, W. A. 1979. Sleep and blood pressure: Further observations. Am. Heart J. **97:** 35–37.
20. HALBERG, F., E. HALBERG, J. HALBERG & F. HALBERG. 1984. Chronobiologic assessment of human blood pressure variation in health and disease. Steinkopff. Darmstadt.
21. TUCK, M. L., N. STERN & J. R. SOWERS. 1985. Enhanced 24-hour norepinephrine and renin secretion in young patients with essential hypertension: Relation with the circadian pattern of arterial blood pressure. Am. J. Cardiol. **55:** 112–115.
22. DAVIES, A. B., B. A. GOULD, P. M. CASHMAN & E. B. RAFTERY. 1984. Circadian rhythm of blood pressure in patients dependent on ventricular demand pacemakers. Br. Heart J. **52:** 93–98.
23. FLORAS, J. S., J. V. JONES, J. A. JOHNSTON, D. E. BROOKS, M. O. HASSAN & P. SLEIGHT. 1978. Arousal and the circadian rhythm of blood pressure. Clin. Sci. Mol. Med **4 (Suppl. 1):** S395–S397.
24. BROADHURST, P., G. BRIGDEN, P. DASGUPTA, A. LAHIRI & E. B. RAFTERY. 1990. Ambulatory intra-arterial blood pressure in normal subjects. Am. Heart J. **120:** 160–166.
25. GOULD, B. A. & E. B. RAFTERY. 1991. Twenty-four-hour blood pressure control: An intraarterial review. Chronobiol. Int. **8:** 495–505.
26. DEGAUTE, J. P., P. VAN DE BORNE, P. LINKOWSKI & E. VAN CAUTER. 1991. Quantitative analysis of the 24-hour blood pressure and heart rate patterns in young men. Hypertension **18:** 199–210.
27. MIDDEKE, M. & J. SCHRADER. 1994. Nocturnal blood pressure in normotensive subjects and those with white coat, primary, and secondary hypertension. Br. Med. J. **308:** 630–632.
28. COWLEY, A. W., JR., J. F. LIARD & A. C. GUYTON. 1973. Role of baroreceptor reflex in daily control of arterial blood pressure and other variables in dogs. Circ. Res. **32:** 564–576.
29. LINSELL, C. R., S. L. LIGHTMAN, P. E. MULLEN, M. J. BROWN & R. C. CAUSON. 1985. Circadian rhythms of epinephrine and norepinephrine in man. J. Clin. Endocrinol. Metab. **60:** 1210–1215.
30. KASTING, G. A., D. L. ECKBERG, J. M. FRITSCH & C. L. BIRKETT. 1987. Continuous resetting of the human carotid baroreceptor-cardiac reflex. Am. J. Physiol. **252:** R732–736.
31. PARATI, G., M. DI RIENZO, G. BERTINIERI *et al.* 1988. Evaluation of the baroreceptor-heart rate reflex by 24-hour intra-arterial blood pressure monitoring in humans. Hypertension **12:** 214–222.
32. FURLAN, R., S. GUZZETTI, W. CRIVELLARO *et al.* 1990. Continuous 24-hour assessment of

the neural regulation of systemic arterial pressure and RR variabilities in ambulant subjects. Circulation **81:** 537–547.
33. HAYANO, J., Y. SAKAKIBARA, M. YAMADA *et al.* 1990. Diurnal variations in vagal and sympathetic cardiac control. Am. J. Physiol. **258:** H642–646.
34. GOLDSMITH, R. L., J. T. BIGGER, R. C. STEINMAN & J. L. FLEISS. 1992. Comparison of 24-hour parasympathetic activity in endurance-trained and untrained young men. J. Am. Coll. Cardiol. **20:** 552–558.
35. HARTIKAINEN, J., I. TARKIAINEN, K. TAHVANAINEN, M. MANTYSAARI, E. LANSIMIES & K. PYORALA. 1993. Circadian variation of cardiac autonomic regulation during 24-h bed rest. Clin. Physiol. **13:** 185–196.
36. VAN DE BORNE, P., H. NGUYEN, P. BISTON, P. LINKOWSKI & J. P. DEGAUTE. 1994. Effects of wake and sleep stages on the 24-h autonomic control of blood pressure and heart rate in recumbent men. Am. J. Physiol. **266:** H548–554.
37. VEERMAN, D. P., B. P. M. IMHOLZ, W. WIELING, K. H. WESSELING & G. A. VAN MONTFRANS. 1995. Circadian profile of systemic hemodynamics. Hypertension **26:** 55–59.
38. PICKERING, T. G. 1991. Ambulatory Monitoring and Blood Pressure Variability. London. Science Press.
39. WHITE, W. B., A. S. BERSON, C. ROBBINS *et al.* 1993. National standard for measurement of resting and ambulatory blood pressures with automated sphygmomanometers. Hypertension **21:** 504–509.
40. O'BRIEN, E., J. PETRIE, W. LITTLER *et al.* 1993. An outline of the revised British Hypertension Society protocol for the evaluation of blood pressure measuring devices. J. Hypertens. **11:** 677–679.
41. DI RIENZO, M., G. GRASSI, A. PEDOTTI & G. MANCIA. 1983. Continuous vs intermittent blood pressure measurements in estimating 24-hour average blood pressure. Hypertension **5:** 264–269.
42. PORTALUPPI, F., L. MONTANARI, M. CAPANNA & M. FERLINI. 1989. Chronobiologic vs. sphygmomanometric assessment of hypertension in a hospital setting. Clin. Cardiol. **12:** 380–383.
43. WHITE, W. B., H. M. DEY, & P. SCHULMAN. 1989. Assessment of the daily blood pressure load as a determinant of cardiac function in patients with mild-to-moderate hypertension. Am. Heart J. **118:** 782–795.
44. ZACHARIAH, P. K., S. G. SHEPS, D. M. ILSTRUP *et al.* 1988. Blood pressure load: A better determinant of blood pressure. Mayo Clin. Proc. **63:** 1085–1091.
45. MULLER, J. E., G. H. TOFLER & P. H. STONE. 1989. Circadian variation and triggers of onset of acute cardiovascular disease. Circulation **79:** 733–743.
46. MANCIA, G., S. OMBONI, G. PARATI, S. TRAZZI & E. MUTTI. 1992. Limited reproducibility of hourly blood pressure values obtained by ambulatory blood pressure monitoring: Implications for studies on antihypertensive drugs. J. Hypertens. **10:** 1531–1535.
47. TRAZZI, S., E. MUTTI, A. FRATTOLA, B. IMHOLZ, G. PARATI & G. MANCIA. 1991. Reproducibility of non-invasive and intra-arterial blood pressure monitoring: Implications for studies on antihypertensive treatment. J. Hypertens. **9:** 115–119.
48. VERDECCHIA, P., G. SCHILLACI, F. BOLDRINI, M. GUERRIERI, I. ZAMPI & C. PORCELLATI. 1991. Quantitative assessment of day-to-day spontaneous variability in non-invasive ambulatory blood pressure measurements in essential hypertension. J. Hypertens. **9 (Suppl. 6):** S322–323.
49. REEVES, R. A., F. H. LEENEN & C. D. JOYNER. 1992. Reproducibility of nurse-measured, exercise and ambulatory blood pressure and echocardiographic left ventricular mass in borderline hypertension. J. Hypertens. **10:** 1249–1256.
50. SILAGY, C. A., J. J. MCNEIL, S. FARISH, P. I. MCCLOUD & B. P. MCGRATH. 1992. Components of blood pressure variability in the elderly and effects on sample size calculations for clinical trials. Am. J. Hypertens. **5:** 449–458.
51. STAESSEN, J., C. J. BULPITT, E. O'BRIEN *et al.* 1992. The diurnal blood pressure profile. A population study. Am. J. Hypertens. **5:** 386–392.
52. THIJS, L., A. AMERY, D. CLEMENT *et al.* 1992. Ambulatory blood pressure monitoring in elderly patients with isolated systolic hypertension. J. Hypertens. **10:** 693–699.

53. DES COMBES, B. J., M. PORCHET, B. WAEBER & H. R. BRUNNER. 1984. Ambulatory blood pressure recordings. Reproducibility and unpredictability. Hypertension **6:** C110–115.
54. JAMES, G. D., T. G. PICKERING, L. S. YEE, G. A. HARSHFIELD, S. RIVA & J. H. LARAGH. 1988. The reproducibility of average ambulatory, home, and clinic pressures. Hypertension **11:** 545–549.
55. GIACONI, S., C. PALOMBO, A. GENOVESI-EBERT, C. MARABOTTI, D. VOLTERRANI & S. GHIONE. 1988. Long-term reproducibility and evaluation of seasonal influences on blood pressure monitoring. J. Hypertens. 6 (Suppl. 4): S64–66.
56. COATS, A. J., A. RADAELLI, S. J. CLARK, J. CONWAY & P. SLEIGHT. 1992. The influence of ambulatory blood pressure monitoring on the design and interpretation of trials in hypertension. J. Hypertens. **10:** 385–391.
57. STEWART, M. J., H. BROWN & P. L. PADFIELD. 1993. Can simultaneous ambulatory blood pressure and activity monitoring improve the definition of blood pressure? Am. J. Hypertens. **6:** 174S–178S.
58. MULLANEY, D. J., D. F. KRIPKE & S. MESSIN. 1980. Wrist-actigraphic estimation of sleep time. Sleep **3:** 83–92.
59. HALBERG, F. 1969. Chronobiology. Ann. Rev. Physiol. **31:** 675–725.
60. GERMANÒ, G., S. DAMIANI, M. CIAVARELLA, A. APPOLLONI, A. FERRUCCI & V. CORSI. 1984. Detection of a diurnal rhythm in arterial blood pressure in the evaluation of 24-hour antihypertensive therapy. Clin. Cardiol. **7:** 525–535.
61. MATTES, A., K. WITTE, W. HOHMANN & B. LEMMER. 1991. PHARMFIT—A nonlinear fitting program for pharmacology. Chronobiol. Int. **8:** 460–476.
62. THIJS, L., J. STAESSEN & R. FAGARD. 1992. Analysis of the diurnal blood pressure curve. High Blood Press. **1:** 17–28.
63. STREITBERG, B., W. MEYER-SABELLEK & P. BAUMGART. 1989. Statistical analysis of circadian blood pressure recordings in controlled clinical trials. J. Hypertens. **7** (Suppl. 3): S11–17.
64. STANTON, A., J. COX, N. ATKINS, K. O'MALLEY & E. O'BRIEN. 1992. Cumulative sums in quantifying circadian blood pressure patterns. Hypertension **19:** 93–101.
65. IDEMA, R. N., E. S. GELSEMA, G. J. WENTING *et al.* 1992. A new model for diurnal blood pressure profiling. Square wave fit compared with conventional methods. Hypertension **19:** 595–605.
66. PARATI, G., M. DI RIENZO, S. OMBONI, P. CASTIGLIONI, A. FRATTOLA & G. MANCIA. 1993. Spectral analysis of 24 h blood pressure recordings. Am. J. Hypertens. **6:** 188S–193S.
67. STAESSEN, J. A., R. FAGARD, L. THIJS & A. AMERY. 1993. Fourier analysis of blood pressure profiles. Am. J. Hypertens. **6:** 184S–187S.
68. WU, M. C. 1993. Analysis of blood pressure variability. Am. J. Hypertens. **6:** 198S–200S.
69. PARATI, G., R. CASADEI, A. GROPPELLI, M. DI RIENZO & G. MANCIA. 1989. Comparison of finger and intra-arterial blood pressure monitoring at rest and during laboratory testing. Hypertension. **13:** 647–655.
70. IMHOLZ, B. P., G. J. LANGEWOUTERS, G. A. VAN MONTFRANS *et al.* 1993. Feasibility of ambulatory, continuous 24-hour finger arterial pressure recording. Hypertension **21:** 65–73.
71. ATHANASSIADIS, D., G. J. DRAPER, A. J. HONOUR & W. I. CRANSTON. 1969. Variability of automatic blood pressure measurements over 24 hour periods. Clin. Sci. **36:** 147–156.
72. MANCIA, G., A. FERRARI, L. GREGORINI *et al.* 1983. Blood pressure and heart rate variabilities in normotensive and hypertensive human beings. Circ. Res. **53:** 96–104.
73. KHOURY, A. F., P. SUNDERAJAN & N. M. KAPLAN. 1992. The early morning rise in blood pressure is related mainly to ambulation. Am. J. Hypertens. **5:** 339–344.
74. SUZUKI, Y., I. KUWAJIMA, K. MITANI & K. KURAMOTO. 1993. The relation of morning rise in blood pressure to pressure response to exercise and mental stress. Jpn. J. Geriatr. **30:** 841–848.
75. KUWAJIMA, I., K. MITANI, M. MIYAO, Y. SUZUKI, K. KURAMOTO & T. OZAWA. 1995. Cardiac implications of the morning surge in blood pressure in elderly hypertensive patients: Relation to arising time. Am. J. Hypertens. **8:** 29–33.
76. VERDECCHIA, P., G. SCHILLACI, M. GUERRIERI *et al.* 1990. Circadian blood pressure

changes and left ventricular hypertrophy in essential hypertension. Circulation **81:**528–536.
77. MANN, S., D. G. ALTMAN, E. B. RAFTERY & R. BANNISTER. 1983. Circadian variation of blood pressure in autonomic failure. Circulation **68:** 477–483.
78. MARTINELLI, P., G. COCCAGNA, N. RIZZUTO & E. LUGARESI. 1981. Changes in systemic arterial pressure during sleep in Shy-Drager syndrome. Sleep **4:** 139–146.
79. TOHGI, H., K. CHIBA & M. KIMURA. 1991. Twenty-four-hour variation of blood pressure in vascular dementia of the Binswanger type. Stroke **22:** 603–608.
80. OTSUKA, A., H. MIKAMI, K. KATAHIRA *et al.* 1990. Absence of nocturnal fall in blood pressure in elderly persons with Alzheimer-type dementia. J. Am. Geriatr. Soc. **38:** 973–978.
81. TOMINAGA, M., T. TSUCHIHASHI, H. KINOSHITA, I. ABE & M. FUJISHIMA. 1995. Disparate circadian variations of blood pressure and body temperature in bedridden elderly patients with cerebral atrophy. Am. J. Hypertens. **8:** 773–781.
82. STOICA, E. & O. ENULESCU. 1983. Inability to deactivate sympathetic nervous system in brainstem infarct patients. J. Neurol. Sci. **58:** 223–234.
83. SHIMADA, K., A. KAWAMOTO, K. MATSUBAYASHI & T. OZAWA. 1990. Silent cerebrovascular disease in the elderly. Correlation with ambulatory pressure. Hypertension **16:** 692–699.
84. SHIMADA, K., A. KAWAMOTO, K. MATSUBAYASHI, M. NISHINAGA, S. KIMURA & T. OZAWA. 1992. Diurnal blood pressure variations and silent cerebrovascular damage in elderly patients with hypertension. J. Hypertens. **10:** 875–878.
85. MATSUMURA, K., I. ABE, M. FUKUHARA *et al.* 1993. Attenuation of nocturnal BP fall in essential hypertensives with cerebral infarction [letter]. J. Hum. Hypertens. **7:** 309–310.
86. SANDER, D. & J. KLINGELHÖFER. 1994. Changes of circadian blood pressure patterns after hemodynamic and thromboembolic brain infarction. Stroke **25:** 1730–1737.
87. ASMAR, R. G., P. L. JULIA, V. L. MASCAREL, J. N. FABIANI, A. BENETOS & M. E. SAFAR. 1994. Ambulatory blood pressure profile after carotid endarterectomy in patients with ischaemic arterial disease. J. Hypertens. **12:** 697–702.
88. FRANKLIN, S. S., J. R. SOWERS & U. BATZDORF. 1986. Relationship between arterial blood pressure and plasma norepinephrine levels in a patient with neurogenic hypertension. Am. J. Med. **81:** 1105–1107.
89. PORTALUPPI, F., P. CORTELLI, P. AVONI *et al.* 1994. Diurnal blood pressure variation and hormonal correlates in fatal familial insomnia. Hypertension **23:** 569–576.
90. HORNUNG, R. S., R. F. MAHLER & E. B. RAFTERY. 1989. Ambulatory blood pressure and heart rate in diabetic patients: An assessment of autonomic function. Diabet. Med. **6:** 579–585.
91. WIEGMANN, T. B., K. G. HERRON, A. M. CHONKO, M. L. MAC DOUGALL & W. V. MOORE. 1990. Recognition of hypertension and abnormal blood pressure burden with ambulatory blood pressure recordings in type I diabetes mellitus. Diabetes **39:** 1556–1560.
92. LURBE, A., J. REDON, J. M. PASCUAL, J. TACONS, V. ALVAREZ & D. C. BATLLE. 1993. Altered blood pressure during sleep in normotensive subjects with type I diabetes. Hypertension **21:** 227–235.
93. FOGARI, R., A. ZOPPI, G. D. MALAMANI, P. LAZZARI, M. DESTRO & L. CORRADI. 1993. Ambulatory blood pressure monitoring in normotensive and hypertensive type 2 diabetes. Prevalence of impaired diurnal blood pressure patterns. Am. J. Hypertens. **6:** 1–7.
94. IKEDA, T., T. MATSUBARA, Y. SATO & N. SAKAMOTO. 1993. Circadian blood pressure variation in diabetic patients with autonomic neuropathy. J. Hypertens. **11:** 581–587.
95. CHAU, N. P., B. BAUDUCEAU, X. CHANUDET, P. LARROQUE & D. GAUTIER. 1994. Ambulatory blood pressure in diabetic subjects. Am. J. Hypertens. **7:** 487–491.
96. ISSHIKI, T., N. AKATSUKA, H. TSUNEYOSHI & H. OKA. 1986. Periodic fluctuation of blood pressure in a case of norepinephrine secreting extra-adrenal pheochromocytoma. Jpn. Heart J. **27:** 437–442.
97. OISHI, S., M. SASAKI, M. OHNO, T. UMEDA & T. SATO. 1988. Periodic fluctuation of blood

pressure and its management in a patient with pheochromocytoma. Case report and review of the literature. Jpn. Heart J. **29:** 389–399.

98. IMAI, Y., K. ABE, Y. MIURA *et al.* 1988. Hypertensive episodes and circadian fluctuations of blood pressure in patients with phaeochromocytoma: Studies by long-term blood pressure monitoring based on a volume-oscillometric method. J. Hypertens. **6:** 9–15.
99. DABROWSKA, B., T. FELTYNOWSKI, B. WOCIAL, W. SZPAK & W. JANUSZEWICZ. 1990. Effect of removal of phaeochromocytoma on diurnal variability of blood pressure, heart rhythm and excretion of catecholamines. J. Hum. Hypertens. **4:** 397–399.
100. STATIUS VAN EPS, R. G., A. H. VAN DEN MEIRACKER, F. BOOMSMA, A. J. MAN IN'T VELD & M. A. D. H. SCHALEKAMP. 1994. Diurnal variation of blood pressure in patients with catecholamine-producing tumors. Am. J. Hypertens. **7:** 492–497.
101. IMAI, Y., K. ABE, S. SASAKI *et al.* 1988. Altered circadian blood pressure rhythm in patients with Cushing's syndrome. Hypertension **12:** 11–19.
102. IMAI, Y., K. ABE, S. SASAKI *et al.* 1989. Exogenous glucocorticoid eliminates or reverses circadian blood pressure variations. J. Hypertens. **7:** 113–120.
103. VAN DE BORNE, P., M. GELIN, J. VAN DE STADT & J. P. DEGAUTE. 1993. Circadian rhythms of blood pressure after liver transplantation. Hypertension **21:** 398–405.
104. IMAI, Y., K. ABE, S. SASAKI *et al.* 1992. Circadian blood pressure variation in patients with renovascular hypertension or primary aldosteronism. Clin. Exp. Hypertens [A] **14:** 1141–1167.
105. WHITE, W. B. & C. MALCHOFF. 1992. Diurnal blood pressure variability in mineralocorticoid excess syndrome. Am. J. Hypertens. **5:** 414–418.
106. VEGLIO, F., G. PINNA, R. MELCHIO *et al.* 1993. Twenty-four-hour power spectral analysis by maximum entropy method of blood pressure in primary hyperaldosteronism. Blood Press. **2:** 189–196.
107. FALLO, F., G. FANELLI, A. CIPOLLA, C. BETTERLE, M. BOSCARO & N. SONINO. 1994. 24-hour blood pressure profile in Addison's disease. Am. J. Hypertens. **7:** 1105–1109.
108. BRICKMAN, A. S., N. STERN & J. R. SOWERS. 1990. Circadian variations of catecholamines and blood pressure in patients with pseudohypoparathyroidism and hypertension. Chronobiologia **17:** 37–44.
109. TILKIAN, A. G., C. GUILLEMINAULT, J. S. SCHROEDER, K. L. LEHRMAN, F. B. SIMMONS & W. C. DEMENT. 1976. Hemodynamics in sleep-induced apnea. Studies during wakefulness and sleep. Ann. Intern. Med. **85:** 714–719.
110. FRANZ, I. W., D. ERB & U. TONNESMANN. 1992. Gestorte 24-Stunden-Blutdruckrhythmik bei normotensiven und hypertensiven Asthmatikern. Z. Kardiol. **81** (Suppl. 2): 13–16.
111. HEBER, M. E., A. LAHIRI, D. THOMPSON & E. B. RAFTERY. 1989. Baroreceptor, not left ventricular, dysfunction is the cause of hemodialysis hypotension. Clin. Nephrol. **32:** 79–86.
112. MIDDEKE, M., E. MIKA, M. A. SCHREIBER, B. BECK, B. WACHTER & H. HOLZGREVE. 1989. Ambulante indirekte Blutdrucklangzeitmessung bei primärer und sekundärer Hypertonie. Klin. Wochenschr. **67:** 713–716.
113. BAUMGART, P., P. WALGER, M. GERKE, K. G. DORST, H. VETTER & K. H. RAHN. 1989. Nocturnal hypertension in renal failure, haemodialysis and after renal transplantation. J. Hypertens. **7**(Suppl. 6): S70–71.
114. PORTALUPPI, F., L. MONTANARI, M. FERLINI & P. GILLI. 1990. Altered circadian rhythms of blood pressure and heart rate in non-hemodialysis chronic renal failure. Chronobiol. Int. **7:** 321–327.
115. PORTALUPPI, F., L. MONTANARI, M. MASSARI, V. DI CHIARA & M. CAPANNA. 1991. Loss of nocturnal decline of blood pressure in hypertension due to chronic renal failure. Am. J. Hypertens. **4:** 20–26.
116. ROSANSKY, S. J. 1991. Nocturnal hypertension in patients receiving chronic hemodialysis [letter]. Ann. Intern. Med. **114:** 96.
117. MIDDEKE, M., M. KLUGLICH & H. HOLZGREVE. 1991. Circadian blood pressure rhythm in primary and secondary hypertension. Chronobiol. Int. **8:** 451–459.
118. BAUMGART, P., P. WALGER, S. GEMEN, M. VON EIFF, H. RAIDT & K. H. RAHN. 1991. Blood pressure elevation during the night in chronic renal failure, hemodialysis and after renal transplantation. Nephron **57:** 293–298.

119. HAYASHI, T., T. SHOJI, E. KITAMURA, N. OKADA, I. NAKANISHI & Y. TSUBAKIHARA. 1993. Circadian blood pressure pattern in patients with chronic glomerulonephritis. Nippon Jinzo Gakkai Shi **35:** 233–237.
120. TORFFVIT, O. & C. D. AGARDH. 1993. Day and night variation in ambulatory blood pressure in type 1 diabetes mellitus with nephropathy and autonomic neuropathy. J. Intern. Med. **233:** 131–137.
121. TIMIO, M., S. LOLLI, C. VERDURA, C. MONARCA, F. MERANTE & E. GUERRINI. 1993. Circadian blood pressure changes in patients with chronic renal insufficiency: A prospective study. Ren. Fail. **15:** 231–237.
122. DEL ROSSO, G., L. AMOROSO, A. SANTOFERRARA, B. FIEDERLING, L. DI LIBERATO & A. ALBERTAZZI. 1994. Impaired blood pressure nocturnal decline and target organ damage in chronic renal failure [abstr.]. J. Hypertens. **12**(Suppl. 3): S15.
123. SHAW, D. B., M. S. KNAPP & D. H. DAVIES. 1963. Variations in blood pressure in hypertensives during sleep. Lancet **1:** 797–798.
124. DE LA SIERRA, A., M. DEL MAR LLUCH, A. COCA *et al.* 1995. Assessment of salt sensitivity in essential hypertension by 24-h ambulatory blood pressure monitoring. Am. J. Hypertens. **8:** 970–977.
125. OLOFSSON, P. 1995. Characteristics of a reversed circadian blood pressure rhythm in pregnant women with hypertension. J. Hum. Hypertens. **9:** 565–570.
126. REDMAN, C. W., L. J. BEILIN & J. BONNAR. 1976. Reversed diurnal blood pressure rhythm in hypertensive pregnancies. Clin. Sci. Mol. Med. **3:** 687s–689s.
127. BEILIN, L. J., J. DEACON, C. A. MICHAEL *et al.* 1983. Diurnal rhythms of blood pressure, plasma renin activity, angiotensin II and catecholamines in normotensive and hypertensive pregnancies. Clin. Exp. Hypertens. [B] **2:** 271–293.
128. KUWAJIMA, I., Y., SUZUKI, T. SHIMOSAWA, A. KANEMARU, S. HOSHINO & K. KURAMOTO. 1992. Diminished nocturnal decline in blood pressure in elderly hypertensive patients with left ventricular hypertrophy. Am. Heart J. **123:** 1307–1311.
129. SOERGEL, M., A. MAISIN, A. AZANCOT-BENISTY & C. LOIRAT. 1992. Ambulante Blutdruckmessung bei nierentransplantierten Kindern und Jugendlichen. Z. Kardiol. **81**(Suppl. 2): 67–70.
130. TALER, S. J., S. C. TEXTOR, V. J. CANZANELLO, D. J. WILSON, R. H. WIESNER & R. A. KROM. 1995. Loss of nocturnal blood pressure fall after liver transplantation during immunosuppressive therapy. Am. J. Hypertens. **8:** 598–605.
131. REEVES, R. A., A. P. SHAPIRO, M. E. THOMPSON & A. M. JOHNSEN. 1986. Loss of nocturnal decline in blood pressure after cardiac transplantation. Circulation **73:** 401–408.
132. DART, A. M., J. K. YEOH, G. L. JENNINGS, J. D. CAMERON & D. S. ESMORE. 1992. Circadian rhythms of heart rate and blood pressure after heart transplantation. J. Heart Lung Transplant. **11:** 784–792.
133. SEHESTED, J., F. THOMAS, M. THORN *et al.* 1992. Level and diurnal variations of hormones of interest to the cardiovascular system in patients with heart transplants. Am. J. Cardiol. **69:** 397–402.
134. VAN DE BORNE, P., M. LEEMAN, G. PRIMO & J. P. DEGAUTE. 1992. Reappearance of a normal circadian rhythm of blood pressure after cardiac transplantation. Am. J. Cardiol. **69:** 794–801.
135. IDEMA, R. N., A. H. VAN DEN MEIRACKER, A. H. BALK, E. BOS, M. A. SCHALEKAMP & A. J. MAN IN'T VELD. 1994. Decreased circadian blood pressure variation up to three years after heart transplantation. Am. J. Cardiol. **73:** 1006–1009.
136. IDEMA, R. N., A. H. VAN DEN MEIRACKER, A. H. BALK, E. BOS, M. A. SCHALEKAMP & A. J. MAN IN'T VELD. 1994. Abnormal diurnal variation of blood pressure, cardiac output, and vascular resistance in cardiac transplant recipients. Circulation **90:** 2797–2803.
137. CARUANA, M. P., A. LAHIRI, P. M. CASHMAN, D. G. ALTMAN & E. B. RAFTERY. 1988. Effects of chronic congestive heart failure secondary to coronary artery disease on the circadian rhythm of blood pressure and heart rate. Am. J. Cardiol. **62:** 755–759.
138. PORTALUPPI, F., L. MONTANARI, M. FERLINI, L. VERGNANI, B. BAGNI & E. C. DEGLI UBERTI. 1992. Differences in blood pressure regulation of congestive heart failure,

before and after treatment, correlate with changes in the circulating pattern of atrial natriuretic peptide. Eur. Heart J. **13:** 990–996.
139. VAN DE BORNE, P., M. ABRAMOWICZ, S. DEGRE & J. P. DEGAUTE. 1992. Effects of chronic congestive heart failure on 24-hour blood pressure and heart rate patterns: A hemodynamic approach. Am. Heart J. **123:** 998–1004.
140. SUZUKI, Y., I. KUWAJIMA, A. KANEMARU *et al.* 1992. The cardiac functional reserve in elderly hypertensive patients with abnormal diurnal change in blood pressure. J. Hypertens. **10:** 173–179.
141. VAN DE BORNE, P., C. TIELEMANS, J. L. VANHERWEGHEM & J. P. DEGAUTE. 1992. Effect of recombinant human erythropoietin therapy on ambulatory blood pressure and heart rate in chronic haemodialysis patients. Nephrol. Dial. Transplant. **7:** 45–49.
142. PORTALUPPI, F., L. VERGNANI, R. MANFREDINI, E. C. DEGLI UBERTI & C. FERSINI. 1995. Time-dependent effect of isradipine on the nocturnal hypertension of chronic renal failure. Am. J. Hypertens. **8:** 719–726.
143. VAN DE BORNE, P., C. TIELEMANS, F. COLLART, J. L. VANHERWEGHEM & J. P. DEGAUTE. 1993. Twenty-four-hour blood pressure and heart rate patterns in chronic hemodialysis patients. Am. J. Kidney Dis. **22:** 419–425.
144. MANFREDINI, R., M. GALLERANI, F. PORTALUPPI & C. FERSINI. 1996. Relationships of the circadian rhythms of thrombotic, ischemic, hemorrhagic, and arrhythmic events to blood pressure rhythms. Ann. NY Acad. Sci. **783:** 141–158.
145. KOBRIN, I., W. OIGMAN, A. KUMAR *et al.* 1984. Diurnal variation of blood pressure in elderly patients with essential hypertension. J. Am. Geriatr. Soc. **32:** 896–899.
146. RIZZONI, D., M. L. MUIESAN, G. MONTANI, R. ZULLI, S. CALEBICH & E. AGABITI-ROSEI. 1992. Relationship between initial cardiovascular structural changes and daytime and nighttime blood pressure monitoring. Am. J. Hypertens. **5:** 180–186.
147. PALATINI, P., M. PENZO, A. RACIOPPA *et al.* 1992. Clinical relevance of nighttime blood pressure and of daytime blood pressure variability. Arch. Intern. Med. **152:** 1855–1860.
148. VERDECCHIA, P., G. SCHILLACI, C. GATTESCHI *et al.* 1993. Blunted nocturnal fall in blood pressure in hypertensive women with future cardiovascular morbid events. Circulation **88:** 986–992.
149. SCHMIEDER, R. E., J. K. ROCKSTROH, F. AEPFELBACHER, B. SCHULZE & F. H. MESSERLI. 1995. Gender-specific cardiovascular adaptation due to circadian blood pressure variations in essential hypertension. Am. J. Hypertens. **8:** 1160–1166.
150. O'BRIEN, E., J. SHERIDAN & K. O'MALLEY. 1988. Dippers and nondippers. Lancet **2:** 397.
151. BIANCHI, S., R. BIGAZZI, G. BALDARI, G. SGHERRI & V. M. CAMPESE. 1994. Diurnal variations of blood pressure and microalbuminuria in essential hypertension. Am. J. Hypertens. **7:** 23–29.
152. CLEMENT, D. L., M. DE BUYZERE & D. D. DUPREZ. 1991. Ambulatory blood pressure and prognosis: Summary of ongoing studies. J. Hypertens. **9**(Suppl. 8): S51–53.
153. STAESSEN, J., A. AMERY, D. CLEMENT *et al.* 1992. Twenty-four hour blood pressure monitoring in the Syst-Eur trial. Aging (Milano) **4:** 85–91.
154. STAESSEN, J. & A. AMERY. 1993. APTH—a trial on ambulatory blood pressure monitoring and treatment of hypertension: Objectives and protocol. Acta Cardiol. **48:** 25–42.
155. MANCIA, G., A. ZANCHETTI, E. AGABITI-ROSEI *et al.* 1992. Prognostic value of ambulatory blood pressure. The SAMPLE study. High Blood Press. **1:** 297–301.
156. PERLOFF, D., M. SOKOLOW & R. COWAN. 1983. The prognostic value of ambulatory blood pressures. JAMA **249:** 2792–2792.
157. PERLOFF, D., M. SOKOLOW, R. M. COWAN & R. P. JUSTER. 1989. Prognostic value of ambulatory blood pressure measurements: Further analyses. J. Hypertens. **7**(Suppl. 3): S3–10.
158. VERDECCHIA, P., C. PORCELLATI, G. SCHILLACI *et al.* 1994. Ambulatory blood pressure. An independent predictor of prognosis in essential hypertension. Hypertension **24:** 793–801.
159. LEMMER, B., B. SCHEIDEL & S. BEHNE. 1991. Chronopharmacokinetics and chronopharmacodynamics of cardiovascular active drugs. Propranolol, organic nitrates, nifedi-

pine. *In* Temporal Control of Drug Delivery. W. J. M. Hrushesky, R. Langer & F. Theeuwes, Eds. Ann. N.Y. Acad. Sci. **618:** 166–181.

160. DECOUSUS, H. A., M. CROZE, F. A. LEVI *et al.* 1985. Circadian changes in anticoagulant effect of heparin infused at a constant rate. Br. Med. J. **290:** 341–344.
161. MATTES, A. & B. LEMMER. 1991. Effects of amlodipine on circadian rhythms in blood pressure, heart rate, and motility: A telemetric study in rats. Chronobiol. Int. **8:** 526–538.
162. LEMMER, B., A. MATTES, M. BOHM & D. GANTEN. 1993. Circadian blood pressure variation in transgenic hypertensive rats. Hypertension **22:** 97–101.
163. LEMMER, B., K. WITTE, T. MAKABE, D. GANTEN & A. MATTES. 1994. Effects of enalaprilat on circadian profiles in blood pressure and heart rate of spontaneously and transgenic hypertensive rats. J. Cardiovasc. Pharmacol. **23:** 311–314.
164. LEMMER, B. & F. PORTALUPPI. 1996. Chronopharmacology of cardiovascular diseases. *In* Physiology and Pharmacology of Biological Rhythms. P. H. Redfern & B. Lemmer, Eds. Springer-Verlag. Heidelberg. In press.
165. LANGNER, B. & B. LEMMER. 1988. Circadian changes in the pharmacokinetics and cardiovascular effects of oral propranolol in healthy subjects. J. Clin. Pharmacol. **33:** 619–624.
166. PALATINI, P. 1992. Can an angiotensin-converting enzyme inhibitor with a short half-life effectively lower blood pressure for 24 hours? Am. Heart J. **123:** 1421–1425.
167. NATIONAL HIGH BLOOD PRESSURE EDUCATION PROGRAM COORDINATING COMMITTEE. 1990. National High Blood Pressure Education Program Working Group report on ambulatory blood pressure monitoring. Arch. Intern. Med. **150:** 2270–2280.
168. 1993 GUIDELINES FOR THE MANAGEMENT OF MILD HYPERTENSION: Memorandum from a World Health Organization/International Society of Hypertension Meeting. Guidelines Sub-Committee. J. Hypertens. **11:** 905–918.
169. FIFTH REPORT OF THE JOINT NATIONAL COMMITTEE ON DETECTION, EVALUATION, AND TREATMENT OF HIGH BLOOD PRESSURE (JNC V). 1993. Arch. Intern. Med. **153:** 154–183.
170. AMERICAN COLLEGE OF PHYSICIANS. 1993. Automated ambulatory blood pressure devices and self-measured blood pressure monitoring devices: Their role in the diagnosis and management of hypertension. Ann. Intern. Med. **118:** 889–892.
171. SHEPS, S. G., D. L. CLEMENT, T. G. PICKERING *et al.* 1994. Ambulatory blood pressure monitoring. Hypertensive Diseases Committee, American College of Cardiology. J. Am. Coll. Cardiol. **23:** 1511–1513.

Ambulatory Blood Pressure Monitoring in Clinical Trials

LESZEK BIENIASZEWSKI,[a] JAN A. STAESSEN,[b]
LUTGARDE THIJS, AND ROBERT FAGARD

Hypertension and Cardiovascular Rehabilitation Unit
Department of Molecular and Cardiovascular Research
University of Leuven
Leuven, Belgium

Blood pressure is characterized by high variability, such that single or even multiple readings taken at a certain point in time may only partially reflect a subject's blood pressure over the day.[1] In addition, the presence of an observer often leads to overestimation of pressure, so-called white coat hypertension.[2,3] These problems have stimulated the use of noninvasive recorders which enable repeated blood pressure measurements in ambulatory subjects at programmable intervals during a period of 24 hours or longer.[4–8] Some aspects related to the use of ambulatory blood pressure monitoring in clinical therapeutic trials are discussed.

DIAGNOSIS OF HYPERTENSION

Randomized clinical trials on the treatment of hypertension are frequently preceded by a run-in period during which the diagnosis of hypertension needs to be confirmed. In addition, interventions or medications are frequently titrated during randomized treatment in order to achieve normotension in one or more treatment groups. Despite the possible advantages of ambulatory over clinic pressure measurements, experts do not agree on reference values for ambulatory pressure,[9] which may be used as operational cut-off points for both diagnosing hypertension and optimizing treatment. Several proposals have been reported.[10–15] However, before these threshold values can be widely recommended, the relation between ambulatory blood pressure and the incidence of cardiovascular morbidity and mortality needs to be further clarified in prospective studies beyond the currently available evidence.[16,17]

ASSESSMENT OF THE BLOOD PRESSURE LOWERING EFFECT

Until recently most pharmacologic studies in the field of hypertension relied exclusively on clinic measurements of arterial pressure to prove efficacy. Often a blood pressure reduction observed at one particular moment during the day without reference to the time of drug administration was considered sufficient to prove

[a] Address for correspondence: Leszek Bieniaszewski, MD, PhD, Department of Hypertension and Diabetology, Medical Academy of Gdansk, Debinki 7, 80-211 Gdansk, Poland (tel./fax: +48-58-471481).

[b] Address for reprint requests: Jan A. Staessen, MD, PhD, Klinisch Laboratorium Hypertensie, Inwendige Geneeskunde-Cardiologie, U.Z. Gasthuisberg, Herestraat 49, B-3000 Leuven, Belgium (tel. +32-16-34.36.31; fax: +32-16-34.37.66).

therapeutic efficacy throughout the day. In view of the technical possibilities offered nowadays by noninvasive ambulatory blood pressure monitoring, and in view of the demand for longer acting antihypertensive drugs with once or twice daily dosing schemes, clinical trials will increasingly use ambulatory pressure to assess blood pressure lowering activity.

Effect of Placebo on Ambulatory Blood Pressure

The decrease in clinic pressure during treatment with a placebo is usually explained by the variability of blood pressure, by regression to the mean, and by habituation of the patient to both the clinic environment and the observer who is measuring the blood pressure. Because of the placebo effect, most clinical trials on antihypertensive drugs in which blood pressure was measured by conventional sphygmomanometry have included a placebo control.

Whether ambulatory blood pressure measurements decrease on placebo remains debated. Intraarterially measured ambulatory blood pressure was demonstrated to remain at the same level, when hypertensive patients were placed on placebo for 6 weeks.[18] Along similar lines, in a 6-week study with a noninvasive recording technique,[19] ambulatory pressure fell only slightly during the initial recording hours, such that the average blood pressure over 24 hours stayed unaffected. On balance, most publications currently favor the view that ambulatory pressure is not subject to a placebo effect.[18–36] However, few long-term studies on this subject have been published. A recent Syst-Eur report based on 112 patients with isolated systolic hypertension and limited to the placebo arm of the trial[37] showed that after a median follow-up of 1 year the clinic systolic pressure fell ($p < 0.05$) by 6.6 mm Hg and the 24-hour and daytime systolic pressures by 2.4 and 2.6 mm Hg, respectively. The corresponding decreases in diastolic pressure were smaller, averaging only 1.4 ($p = 0.06$), 1.1, and 0.7 mm Hg.

The issue of whether ambulatory measurements need to be corrected for placebo is not without importance, as it inherently touches the design of clinical trials in the field of hypertension. Some experts[38] feel that ambulatory monitoring eliminates only observer bias and expectation[39] and does not remove regression to the mean and patient-related factors that contribute to the placebo effect. Other researchers defend the point of view that ambulatory blood pressure monitoring would make control observations on placebo superfluous.[9] If this were true, trials with the new technique could investigate antihypertensive interventions just by comparing blood pressure levels before and after therapy. The Syst-Eur findings[37] suggest that the latter approach may lead to overestimation of the true antihypertensive effect and should therefore be abandoned. This point of view was recently confirmed in a reanalysis of the Syst-Eur data, which included patients randomized to placebo as well as patients allocated to active treatment.[40]

Effect of Treatment on Ambulatory Blood Pressure

The decision to treat hypertensive patients with pharmacologic agents is mainly based on conventional blood pressure measurements, usually performed during clinic visits or in the physician's office. Although guidelines for the management of hypertension[41] are based not only on blood pressure, but also on age, duration of follow-up, outcome of nonpharmacologic treatment, and the presence of cardiovascular complications and risk factors, most patients are advised to start antihypertensive treatment when their clinic diastolic blood pressure remains 95 mm Hg or above. On the other hand, an estimated 20% of subjects with elevated office blood pressure also have so-called white coat hypertension based on the observed difference between clinic and ambulatory blood pressure.[42] Recent studies have assessed the effect of drug treatment on ambulatory blood pressure in such patients. Weber *et al.*[43] and Lacourcière *et al.*[44] selected patients with essential hypertension whose diastolic blood pressure was at least 95 mm Hg after a 4-week placebo run-in period. They were then treated for 8–12 weeks with the calcium antagonists diltiazem[43] or isradipine.[44] In patients whose baseline 24-hour diastolic blood pressure average was below 90 mm Hg, this ambulatory pressure did not change significantly with treatment; 24-hour systolic pressure remained unchanged[43] or fell only slightly.[44] The statistical method to define the cut-off value, used to subdivide the patients into groups with high or low ambulatory blood pressure, however, was not described. In a more recent study[45] baseline ambulatory blood pressure below which pressure, on average, does not decrease during active treatment was defined as pressure at which the regression line between on-treatment pressure and blood pressure in the run-in period intersects the line of identity. In this study[45] ambulatory diastolic blood pressure did not decrease in response to antihypertensive therapy, which was guided by conventional blood pressure measurements, when its initial values were $\leq$88 mm Hg (upper limit: 93 mm Hg) for the day, $\leq$73 mm Hg (78 mm Hg) for the night, and $\leq$81 mm Hg (86 mm Hg) for the whole 24-hour period.

Thus, in patients with white coat hypertension antihypertensive treatment reduces the clinic but not the ambulatory blood pressure.

Trough-to-Peak Ratio

The quest for long-acting antihypertensive agents is vigorously persued by the pharmaceutical industry,[46] because once-daily administration is thought to increase patient compliance[47,48] and because of the prevailing hurt still unproven[49] hypothesis that to reduce the cardiovascular complications of hypertension, blood pressure must be evenly lowered throughout the whole day. Against this background, the trough-to-peak ratio was originally conceptualized because of the growing concern that relatively short-acting antihypertensive drugs might be administered in inappropriately large doses to prolong the dosage interval.[48,50–53] If such drugs were prescribed, peak depressor action would be substantially larger than the trough effect, increasing the likelihood of hypotension and of wide swings in the blood pressure level at some time during the day. For instance, a regimen of isradipine 5 mg twice daily has been shown to produce a pronounced fall in blood pressure within 2 hours of drug intake.[54] If the two daily doses were halved in an attempt to prevent the early peak hypotensive response, blood pressure control was lost during most of the dosage interval.[54]

Conventional sphygmomanometry allows the determination of the trough-to-peak ratio.[52,55] However, the number of blood pressure readings that an auscultating observer can practicably obtain in a group of patients constitutes a major limiting factor. As a consequence, using conventional blood pressure measurements, peak depressor action is usually determined at an arbitrarily defined time point, which is often chosen to coincide with peak concentrations of the antihypertensive agent.[56] This approach does not fully account for the interindividual variation in the timing of the peak depressor effect. Its occurrence is dependent not only on between-subject and circadian[57] variation in the pharmacokinetics of the antihypertensive agent, but also on between-subject differences in the mechanisms, which sustain the elevated blood pressure through the day, such as α-adrenergic tone[58,59] and the activity of the renin axis.[60] In view of these considerations, ambulatory monitoring is increasingly employed to determine the trough-to-peak ratio in clinical trials.

In calculating the trough-to-peak ratio, it is critical to subtract the placebo effect.[50,52,53,61] Some investigators failed to do this and miscalculated the ratios.[50] The apparent trough-to-peak ratio can also be manipulated by changing the resolution of the diurnal profile.[62] Moreover, not only the point estimate of the ratio is relevant. Its error term and confidence interval are rarely reported and yet reflect the precision by which advisory boards may need to be guided.

Recently, the surface ratio[63] was considered as an alternative to the trough-to-peak ratio in assessing the duration of action of antihypertensive agents, because it tended to be characterized by higher within-subject reproducibility. Moreover, the consistency of the novel index was significantly better than that of the trough-to-peak ratio. However, neither the trough-to-peak ratio nor the surface ratio should ever be considered in isolation, but viewed in relation to the absolute treatment-induced decrease in blood pressure at the peak and at the end of the dosage interval.

VALIDATION OF RECORDERS

The procedures required for the validation of ambulatory blood pressure monitors have been thoroughly standardized.[64,65] Obviously, monitors used in clinical research should have successfully passed these validation tests. Moreover, when ambulatory blood pressure monitoring is to be used in special patient populations, such as older subjects or pregnant women,[66–68] or in special conditions, such as exercise,[69–71] a specific demonstration of accuracy in these defined subgroups and conditions is necessary.[72,73]

SAMPLE SIZE

Ambulatory monitoring of blood pressure facilitates clinical trials not only by identifying and excluding patients whose blood pressure is elevated only in the clinic environment, but also by improving the repeatability of the blood pressure estimates. Indeed, in most pharmacologic studies of antihypertensive agents the anticipated decrease in blood pressure is only modest, representing a change of about 5–10% from the initial level.[22] This modest fall must be viewed against the magnitude of the variability in blood pressure. Fluctuations of 40% in blood pressure can readily occur during the day.[22]

A common strategy to enhance the statistical power of clinical trials is to enroll a large number of subjects. An alternative approach, which was made possible by ambulatory blood pressure monitoring, is to improve the reproducibility of blood pressure readings so that the variability of the measurements becomes smaller than the anticipated therapeutic effect to be tested. The benefit of ambulatory blood pressure monitoring lies in the avoidance of the so-called white coat effect[2,3] and in the greater number of readings obtained over a longer period as compared with measurements by an observer in the hospital or at home. In a Belgian population study the blood pressure of 24 healthy subjects was measured five times consecutively on the occasion of three separate home visits. Intervals between these visits averaged 9 days (range 1–25) and 352 days (range 254–430), respectively. On the second and third home visits 24-hour ambulatory blood pressure was also recorded. This study[74] showed that increasing the number of observer-made pressure readings reduced the standard deviation of the short- and long-term differences in the systolic, but not in the diastolic blood pressure (TABLE 1). Measurements of ambulatory pressure over a complete 24-hour period, as compared with five observer-made measurements, produced a decrease in the standard deviation of the within-subject differences between duplicate measurements by 48% for systolic and by 41% for diastolic pressure. These findings are in agreement with those of the Oxford group,[22] who demonstrated that the standard deviation of the difference between two estimates of the diastolic pressure decreased from 12.3 mm Hg for a single reading of the clinic pressure to 6.3 mm Hg for the average of at least 20 readings of the daytime ambulatory pressure.

Technical innovation, leading to a decrease in the variability of an outcome measure in a trial, may produce a substantial reduction in the sample size required to prove the hypothesized experimental effect with a given significance and power. For instance, in a cross-over trial assessing the efficacy of an antihypertensive agent, a 50% reduction in the standard deviation of the blood pressure difference between two measurements will result in a decrease in the required sample size by 75%, on the assumption that the other relevant statistical parameters remain unchanged. However, by contrast with what is often perceived,[38,75] the advantage of the higher

TABLE 1. Differences between Repeated Measurements of Blood Pressure in 29 Subjects

	Short-Term		Long-Term	
	9 (1–25)		352 (254–430)	
Interval (days)	SBP	DBP	SBP	DPB
Home Visit				
1 Reading (mm Hg)	+0.5 ± 10.8	−2.5 ± 6.9	+0.9 ± 14.6	−3.7 ± 7.9
5 Readings (mm Hg)	−1.4 ± 7.4	−2.1 ± 5.5	+0.1 ± 10.8	−1.6 ± 8.4
Ambulatory Pressure				
24-Hour (mm Hg)	—	—	+1.8 ± 6.7	+0.6 ± 5.0
10 AM–8 PM (mm Hg)	—	—	−0.5 ± 8.1	−0.9 ± 7.0
0 AM–6 AM (mm Hg)	—	—	+2.6 ± 10.8	+0.5 ± 8.1

Abbreviations: SBP = systolic blood pressure; DBP = diastolic blood pressure.

reproducibility of ambulatory monitoring is lost in trials with a parallel-group design because then between-subject, rather than within-subject, variability is driving the test statistics. A smaller sample size is also not possible when effects on the full course of blood pressure through the day are being examined, because of the large variability of hourly blood pressure means and hence low reproducibility. Moreover, patients with essential hypertension constitute a heterogeneous population in which the high blood pressure is sustained by diverse pathophysiologic mechanisms. In view of this heterogeneity, sample size cannot be unlimited reduced if a clinic needs to stay representative of the hypertension population in general. Nonresponders should therefore also not be excluded.

SUMMARY

Ambulatory blood pressure monitoring is increasingly used in clinical trials. The preliminary threshold values proposed for diagnosing hypertension and adjusting treatment based on ambulatory monitoring cannot yet be widely recommended because they have not been validated in prospective studies.

The trough-to-peak or surface ratios may be useful instruments for assessing the duration of action of antihypertensive drugs. Trials with ambulatory monitoring just as clinical experiments based on conventional sphygmomanometry need to be properly controlled, because ambulatory blood pressure measurement is not completely devoid of a placebo effect. Ambulatory compared with conventional blood pressure measurements are characterized by higher reproducibility which makes it possible to reduce sample size in cross-over but not in parallel group trials.

Finally, ambulatory monitors used in clinical research should have successfully passed one of the standardized validation protocols.

REFERENCES

1. Armitage, P. & G. A. Rose. 1966. The variability of measurements of casual blood pressure. I. A laboratory study. Clin. Sci. **30:** 325–335.
2. Padfield, P. L., G. Jyothinagaram, I. M. McGinley & D. M. Watson. 1991. Reversal of the relationship between heart rate and blood pressure in phaeochromocytoma: A non-invasive diagnostic approach? J. Hum. Hypertens. **5:** 501–504.
3. Mancia, G., G. Bertinieri, G. Grassi, G. Parati, G. Pomidossi, A. Ferrari, L. Gregorini & A. Zanchetti. 1983. Effects of blood pressure measurement by the doctor on patient's blood pressure and heart rate. Lancet **ii:** 695–698.
4. Kennedy, H. L., N. E. Padgett & M. J. Horan. 1979. Performance reliability of the Del Mar Avionics non-invasive ambulatory blood pressure instrument in clinical use. Ambul. Electrocardiol. **1:** 13–17.
5. Casadei, R., G. Parati, G. Pomidossi, A. Groppelli, S. Trazzi, M. Di Rienzo & G. Mancia. 1988. 24-Hour blood pressure monitoring: Evaluation of Spacelabs 5300 monitor by comparison with intra-arterial blood pressure recording in ambulant subjects. J. Hypertens. **6:** 797–803.
6. Hope, S. L., E. Alun-Jones & P. Sleight. 1988. Validation of the accuracy of the Medilog ABP non-invasive blood-pressure monitor. J. Ambul. Monit. **1:** 39–51.
7. Meyer-Sabellek, W., K. L. Schulte & R. Gotzen. 1989. Technical possibilities and limits of indirect automatic twenty-four-hour blood pressure devices. J. Hypertens. **7**(Suppl. 3): S1–S24.
8. O'Brien E., F. Mee, N. Atkins & K. O'Malley. 1991. Evaluation of the SpaceLabs 90202 non-invasive ambulatory recorder according to the AAMI Standard and BHS criteria. J. Hum. Hypertens. **5:** 223–226.
9. Mancia, G., M. Di Rienzo & G. Parati. 1993. Ambulatory blood pressure monitoring use in hypertension research and clinical practice. Hypertension **21:** 510–524.
10. O'Brien, E. T., J. Murphy, A. Tyndall, N. Atkins, F. Mee, G. McCarthy, J. Staessen, J. Cox & K. O'Malley. 1991. Twenty-four-hour ambulatory blood pressure in men and women aged 17 to 80 years: The Allied Irish Bank Study. J. Hypertens. **9:** 355–360.
11. Staessen, J. A., R. H. Fagard, P. J. Lijnen, L. Thijs, R. Van Hoof & A. K. Amery. 1991. Mean and range of the ambulatory pressure in normotensive subjects from a meta-analysis of 23 studies. Am. J. Cardiol. **67:** 723–727.
12. Staessen, J., E. O'Brien, A. Amery, N. Atkins, P. Baumgart, P. De Cort, J. P. Degaute, P. Dolenc, R. De Gaudemaris, I. Enström, R. Fagard, P. Gosse, S. Gourlay, H. Hayashi, Y. Imai, G. James, T. Kawasaki, E. Kuschnir, I. Kuwajima, L. Lindholm *et al.* 1994. The ambulatory blood pressure in normotensive compared with hypertensive subjects: Results from a large international database. CVD Epidemiol. Newsletter **49:** 49.
13. Mancia, G., S. Omboni, A. Ravogli, G. Parati & A. Zanchetti. 1995. Ambulatory blood pressure monitoring in the evaluation of the antihypertensive treatment: Additional information from a large data base. Blood Press. **4:** 159.
14. Staessen, J., C. J. Bulpitt, R. Fagard, P. Lijnen, G. Mancia, E. O'Brien, G. Vyncke & A. Amery. 1991. Reference values for the ambulatory pressure and pressure measured at home: A population study. Acta Clin. Belg. **46:** 269.
15. Sega, G., C. Bravi, G. Cesana, F. Valagussa, G. Mancia & A. Zanchetti. 1994. Ambulatory and home blood pressure normality: The Pamela Study. J. Cardiovasc. Pharmacol. **23**(suppl 5): S12–15.
16. Perloff, D., M. Sokolow, R. M. Cowan & R. P. Juster. 1989. Prognostic value of ambulatory blood pressure measurements: Further analyses. J. Hypertens. **7**(Suppl. 3): S3–S10.
17. Mann, S., M. W. Millar Craig & E. B. Raftery. 1985. Superiority of 24-hour

measurement of blood pressure over clinic values in determining prognosis in hypertension. Clin. Exp. Hypertens. **A7:** 279–281.
18. GOULD, B. A., S. MANN, A. B. DAVIES, D. G. ALTMAN & E. B. RAFTERY. 1981. Does placebo lower blood pressure? Lancet **ii:** 1377–1384.
19. MUTTI, E., S. TRAZZI, S. OMBONI, G. PARATI & G. MANCIA. 1991. Effect of placebo on 24-h non-invasive ambulatory blood pressure. J. Hypertens. **9:** 361–364.
20. THIJS, L., A. AMERY, D. CLEMENT, J. COX, P. DE CORT, R. FAGARD, G. FOWLER, C. GUO, G. MANCIA, R. MARIN, E. O'BRIEN, K. O'MALLEY, P. PALATINI, G. PARATI, J. PETRIE, A. RAVOGLI, J. ROSENFELD, J. STAESSEN & J. WEBSTER. 1992. Ambulatory blood pressure monitoring in elderly patients with isolated systolic hypertension. J. Hypertens. **10:** 693–699.
21. ANTIVALLE, M., S. LATTUADA, A. SALVAGGIO,M. PARAVICINI, M. RINDI & A. LIBRETTI. 1990. Placebo effect and adaptation to noninvasive monitoring of BP. J. Hum. Hypertens. **4:** 633–637.
22. CONWAY, J. & A. COATS. 1989. Value of ambulatory blood pressure monitoring in clinical pharmacology. J. Hypertens. **7**(Suppl. 3): S29–32.
23. CONWAY, J., J. JOHNSTON, A. COATS, V. SOMERS & P. SLEIGHT. 1988. The use of ambulatory blood pressure monitoring to improve the accuracy and to reduce the number of subjects in clinical trials of antihypertensive agents. J. Hypertens. **6:** 111–116.
24. COATS, A. J. S., A. RADAELLI, S. J. CLARK, J. CONWAY & P. SLEIGHT. 1992. The influence of ambulatory blood pressure monitoring on the design and interpretation of trials in hypertension. J. Hypertens. **10:** 385–391.
25. COATS, A. J. S., J. CONWAY, V. K. SOMERS, J. E. ISEA & P. SLEIGHT. 1989. Ambulatory pressure monitoring in the assessment of antihypertensive therapy. Cardiovasc. Drugs Ther. **3:** 303–311.
26. DUPONT, A. G., P. VAN DER NIEPEN & R. O. SIX. 1987. Placebo does not lower ambulatory blood pressure. Br. J. Clin. Pharmacol. **24:** 106–109.
27. STAESSEN, J., R. FAGARD, P. LIJNEN, L. THIJS, R. VAN HOOF & A. AMERY. 1991. Ambulatory blood pressure monitoring in clinical trials. J. Hypertens. **9**(Suppl. 1): S13–S19.
28. PALATINI, P., A. C. PESSINA & C. DAL PALÙ. 1992. The Hypertension and Ambulatory Recording Venetia Study (HARVEST). A trial on the predictive value of ambulatory monitoring for the development of fixed hypertension in patients with borderline hypertension. High Blood Pressure Cardiovasc. Prev. **3:** 11–18.
29. STAESSEN, J., C. J. BULPITT, E. O'BRIEN, J. COX, R. FAGARD, A. STANTON, L. THIJS, S. VAN HULLE, G. VYNCKE & A. AMERY. 1992. The diurnal blood pressure profile: A population study. Am. J. Hypertens. **5:** 386–392.
30. PALATINI, P., M. PENZO, G. GUZZARDI, M. ANACLERIO & A. C. PESSINA. 1991. Ambulatory blood pressure monitoring in the assessment of antihypertensive treatment: 24-h blood pressure control with lacidipine once a day. J. Hypertens. **9**(Suppl. 3): S61–S66.
31. PARAN, E., M. LANDAU-SALZBERG, Y. KOBRIN & R. VISKOPER. 1993. Effect of placebo on office and 24 hour noninvasive ambulatory blood pressure measurements. J. Hum. Hypertens. **7:** 567–570.
32. WEBER, M. A., J. I. M. DRAYER, F. A. WYLE & J. L. DE YOUNG. 1982. Reproducibility of the whole-day blood pressure pattern in essential hypertension. Clin. Exp. Hypertens. **A4:** 1377–1390.
33. DRAYER, J. I. M., M. A. WEBER, J. L. DEYOUNG & D. D. BREWER. 1983. Long-term BP monitoring in the evaluation of antihypertensive therapy. Arch. Intern. Med. **143:** 898–901.
34. BELLET, M., J. Y. PAGNY, G. CHATELLIER, P. CORVOL & J. MÉNARD. 1987. Evaluation of slow release nicardipine in essential hypertension by casual and ambulatory blood pressure measurements. Effects of acute versus chronic administration. J. Hypertens. **5:** 599–604.
35. POGGI, L., B. VAISSE, F. BERNARD & P. AGABRIEL. 1987. Etude de l'effet placebo par mesure ambulatoire non sanglante de la pression artérielle. Arch. Mal. Coeur **80:** 1031–1036.
36. PARATI, G., G. POMIDOSSI, R. CASADEI, S. TRAZZI, A. RAVOGLI, A. ZANCHETTI & G.

MANCIA. 1988. Evaluation of the antihypertensive effect of celiprolol by ambulatory blood pressure monitoring. Am. J. Cardiol. **61:** 27C–33C.
37. STAESSEN, J. A., L. THIJS, D. CLEMENT, C. DAVIDSON, R. FAGARD, A. LEHTONEN, G. MANCIA, P. PALATINI, E. T. O'BRIEN, G. PARATI, J. WEBSTER, A. AMERY & ON BEHALF OF THE SYST-EUR INVESTIGATORS. 1994. Ambulatory pressure decreases on long-term placebo treatment in older patients with isolated systolic hypertension. J. Hypertens. **12:** 1035–1039.
38. MANSOOR, G. A. & W. B. WHITE. 1994. Contribution of ambulatory blood pressure monitoring to the design and analysis of antihypertensive therapy trials. J. Cardiovasc. Risk **1:** 136–142.
39. SASSANO, P., G. CHATELLIER, P. CORVOL & J. MENARD. 1987. Influence of observer's expectation on the placebo effect in blood pressure trials. Curr. Ther. Res. **41:** 305–312.
40. STAESSEN, J. A., L. THIJS, L. BIENIASZEWSKI, E. T. O'BRIEN, P. PALATINI, C. DAVIDSON, J. DOBOVISEK, M. JÄÄSKIVI, T. LAKS, A. LEHTONEN, H. VANHANEN, J. WEBSTER, R. FAGARD & ON BEHALF OF THE SYST-EUR INVESTIGATORS. 1995. Ambulatory monitoring uncorrected for placebo overestimates antihypertensive action. Hypertension. In press.
41. THE GUIDELINES SUBCOMMITTEE. 1989. Guidelines for the management of mild hypertension: Memorandum from a WHO/ISH meeting. Bull. WHO **67:** 498.
42. THE SCIENTIFIC COMMITTEE. 1990. Consensus document on non-invasive ambulatory blood pressure monitoring. J. Hypertens. **8**(Suppl. 6): S135–140.
43. WEBER, M. A., D. G. CHEUNG, W. F. GRAETTINGER & J. L. LIPSON. 1988. Characterization of antihypertensive therapy by whole-day blood pressure monitoring. JAMA **259:** 3281–3285.
44. LACOURCIÈRE, Y., L. POIRIER, D. DION & P. PROVENCHER. 1990. Antihypertensive effect of isradipine administered once or twice daily on ambulatory blood pressure. Am. J. Cardiol. **65:** 467–472.
45. FAGARD, R., E. BIELEN, J. STAESSEN, L. THIJS & A. AMERY. 1993. Response of ambulatory blood pressure to antihypertensive therapy guided by clinic pressure. Am. J. Hypertens. **6:** 648–653.
46. STAESSEN, J. A., H. CELIS, L. THIJS, R. FAGARD & A. K. AMERY. 1994. Efficacy of antihypertensive drugs given once a day: The calcium antagonist revisited. J. Hypertens. **12**(Suppl. 8): S107–S115.
47. GREENBERG, R. N. 1984. Overview of patient compliance with medication dosing: A literature review. Clin. Ther. **6:** 592–599.
48. EISEN, S. A., D. K. MILLER, R. S. WOODWARD, E. SPITZNAGEL & T. R. PRZYBECK. 1990. The effect of prescribed daily dose frequency on patient medication compliance. Arch. Intern. Med. **150:** 1881–1884.
49. CLEMENT, D. L., M. DE BUYZERE & D. DUPREZ. 1994. Prognostic value of ambulatory blood pressure monitoring. J. Hypertens. **12:** 857–864.
50. ROSE, M. & F. G. MCMAHON. 1990. Some problems with antihypertensive drug studies in the context of the new guidelines. Am. J. Hypertens. **3:** 151–155.
51. MEREDITH, P. A. & H. L. ELLIOTT. 1994. FDA Guidelines on trough:peak ratios in the evaluation of antihypertensive agents. J. Cardiovasc. Pharmacol. **23**(Suppl. 5): S26–S30.
52. MEREDITH, P. A. 1994. Trough/peak ratios for antihypertensive agents. The issue in perspective. Drugs **48:** 661–666.
53. ELLIOTT, H. L. 1994. Trough:peak ratio and twenty-four-hour blood pressure control. J. Hypertens. **12**(Suppl. 5): S29–S33.
54. STOKES, G. S., G. M. SHENFIELD, H. J. JOHNSTON, J. F. MARWOOD, F. DUNAGAN & J. BOUTAGY. 1990. Timing of blood pressure measurements in determining anomalies in duration of effect of an antihypertensive drug: Assessment of isradipine. J. Cardiovasc. Pharmacol. **15**(Suppl. 1): S65–S69.
55. MÉNARD, J., G. CHATTELIER, M. DAY & L. VAUR. 1994. Self-measurement of blood pressure at home to evaluate drug effects by the trough:peak ratio. J. Hypertens. **12**(Suppl. 8): S21–S25.
56. CAPPUCCIO, F. P., N. D. MARKANDU, D. R. J. SINGER & G. A. MACGREGOR. 1993. Amlodipine and lisinopril in combination for the treatment of essential hypertension: Efficacy and predictors of response. J. Hypertens. **11:** 839–847.

57. JESPERSEN, C. M., M. FREDERIKSEN, J. FISCHER HANSEN, N. A. KLITGAARD & C. SORUM. 1989. Circadian variation in the pharmacokinetics of verapamil. Eur. J. Clin. Pharmacol. **37:** 613–615.
58. PICKERING, T. G., M. LEVENSTEIN, P. WALMSLEY & FOR THE HYPERTENSION AND LIPID TRIAL STUDY GROUP. 1994. Nighttime dosing of doxazosin has peak effect on morning ambulatory blood pressure. Results of the HALT Study. Am. J. Hypertens. **7:** 844–847.
59. PANZA, J. A., S. E. EPSTEIN & A. A. QUYYUMI. 1991. Circadian variation in vascular tone and its relation to α-sympathetic vasoconstrictor activity. N. Engl. J. Med. **325:** 986–990.
60. PORTALUPPI, F., B. BAGNI, E. DEGLI UBERTI, L. MONTANARI, R. CAVALLINI, G. TRASFORINI, A. MARGUTTI, M. FERLINI, M. ZANELLA & M. PARTI. 1990. Circadian rhythms of atrial natriuretic peptide, renin, aldosterone, cortisol, blood pressure and heart rate in normal and hypertensive subjects. J. Hypertens. **8:** 85–95.
61. ELLIOTT, H. L. & P. A. MEREDITH. 1994. Methodological considerations in calculation of the trough:peak ratio. J. Hypertens. **12**(Suppl. 8): S3–S7.
62. STAESSEN, J. A., L. BIENIASZEWSKI, F. BUNTINX, H. CELIS, E. T. O'BRIEN, R. VAN HOOF & R. FAGARD. 1995. The trough-to-peak ratio as an instrument to evaluate antihypertensive drugs. Hypertension. **26:** 942–949.
63. BIENIASZEWSKI, L., J. A. STAESSEN, G. BYTTEBIER, P. W. DE LEEUW, T. VAN HEDENT, R. FAGARD & ON BEHALF OF THE APTH INVESTIGATORS. 1995. Trough-to-peak versus surface ratio in the assessment of antihypertensive agents. Blood Press. **4:** 350–357.
64. O'BRIEN, E., J. PETRIE, W. LITTLER, M. DE SWIET, P. L. PADFIELD, K. O'MALLEY, M. JAMIESON, D. ALTMAN, M. BLAND & N. ATKINS. 1991. The British Hypertension Society protocol for the evaluation of automated and semi-automated blood pressure measuring devices with special reference to ambulatory systems. J. Ambul. Monit. **4:** 207–228.
65. WHITE, W. B., A. S. BERSON, C. ROBBINS, M. J. JAMIESON, M. L. PRISANT, E. ROCCELLA & S. G. SHEPS. 1993. National standard for measurement of resting and ambulatory blood pressures with automated sphygmomanometers. Hypertension **21:** 504–509.
66. HALLIGAN, A., E. O'BRIEN, K. O'MALLEY, F. MEE, N. ATKINS, R. CONROY, J. J. WALSHE & M. DARLING. 1993. Twenty-four-hour ambulatory blood pressure measurement in a primigravid population. J. Hypertens. **11:** 869–873.
67. CLARK, S., G. J. HOFMEYR, A. J. S. COATS & C. W. G. REDMAN. 1991. Ambulatory blood pressure monitoring during pregnancy: Validation of the TM-2420 monitor. Obstet. Gynecol. **77:** 152–155.
68. O'BRIEN, E., F. MEE, N. ATKINS, A. HALLIGAN & K. O'MALLEY. 1993. Accuracy of the SpaceLabs 90207 blood pressure measuring system in normotensive pregnant women determined by the British Hypertension Society Protocol. J. Hypertens. **11**(Suppl. 5): 282–283.
69. WHITE, W. B., P. LUND-JOHANSEN & P. OMVIK. 1990. Assessment of four ambulatory blood pressure monitors and measurements by clinicians versus intraarterial blood pressure at rest and during exercise. Am. J. Cardiol. **65:** 60–66.
70. GOULD, B. A., R. S. HORNUNG, D. G. ALTMAN, P. M. M. CASHMAN & E. B. RAFTERY. 1985. Indirect measurement of blood pressure during exercise testing can be misleading. Br. Heart J. **53:** 611–615.
71. HENSCHEL, A., F. DE LA VEGA & H. L. TAYLOR. 1954. Simultaneous direct and indirect blood pressure measurements in man at rest and work. J. Appl. Physiol. **5:** 506–508.
72. O'BRIEN, E., J. PETRIE, W. LITTLER, M. DE SWIET, P. L. PADFIELD, D. G. ALTMAN, M. BLAND, A. COATS & N. ATKINS. 1993. Short report: An outline of the revised British Hypertension Society protocol for the evaluation of blood pressure measuring devices. J. Hypertens. **11:** 677–679.
73. COATS, A. J. S. & S. J. CLARK. 1992. Validation of ambulatory monitors in special populations. Am. J. Hypertens. **5:** 664–669.
74. STAESSEN, J. A., C. J. BULPITT, R. FAGARD, P. J. LIJNEN, G. MANCIA, E. T. O'BRIEN, G. VYNCKE & A. AMERY. 1990. Ambulatory blood pressure and pressure measured at home: A population study. CVD Epidemiol. Newsletter **46:** 32.
75. PICKERING, T. G. & E. T. O'BRIEN. 1991. Second International Consensus Meeting on twenty-four-hour blood pressure measurement: Concensus and conclusions. J. Hypertens. **9**(Suppl. 8): S2–S6.

Defining the Regulatory Role of Ambulatory Blood Pressure Monitoring

RAYMOND JOHN LIPICKY AND NORMAN STOCKBRIDGE[a]
FOR THE ABPM STEERING COMMITTEE

Food and Drug Administration
Center for Drug Evaluation and Research
Office of Drug Evaluation I
Division of Cardio-Renal Drug Products
5600 Fishers Lane
Rockville, Maryland 20857

The Division of Cardio-Renal Drug Products is the review division of the Food and Drug Administration (Food & Drug Administration, Center for Drug Evaluation and Research, Office of Drug Evaluation I) that reviews applications for marketing approval for antihypertensive drugs.

Studies that figure most prominently in the regulatory approval of antihypertensive drugs have been randomized, parallel group, dose-ranging, placebo-controlled, outpatient trials in which the end point has been sparse sampling of cuff blood pressure, usually done at the end of a dosing interval and measured in the unnatural setting of the clinic or physician's office.

It is routinely observed that over a period of weeks of study, subjects receiving placebo in such trials demonstrate a progressive decline in blood pressure which is difficult to distinguish from that seen in active treatment groups with respect to either its magnitude or its change with time. This happens even when there is a single-blind run-in period in which all subjects receive placebo for several weeks and must additionally demonstrate a stable (nondecreasing) blood pressure for up to 1 or 2 weeks. This phenomenon has been attributed to various mechanisms, including regression to the mean.

Regardless of the explanation and ignoring for the moment the need to distinguish drug-related side effects, this phenomenon alone requires a parallel placebo arm in trials intended to quantitate the magnitude of the drug effect. Currently, antihypertensive drug approval depends solely on the quantification of an antihypertensive effect; showing a reduction of morbidity or mortality is not required. Consequently, our division is concerned with the quantitative aspects of the measurement itself and with the algorithms used to decide that the central tendencies for blood pressure among treatment groups differ in some nonrandom way. The antihypertensive effect of a given drug regimen at trough (e.g., just before the first dose of a dosing interval) has historically been the variable of principal interest.

We would like an antihypertensive drug development program to characterize the relationship between magnitude of antihypertensive effect and dose and the relationship between dose and duration of effect. In practice, firms often characterize the relationship between dose and the magnitude of effect at trough. Of course, this is some complicated function that depends, at least in part, on the relationship between dose and duration of effect.

[a] Address for correspondence: Norman Stockbridge, M.D., Ph.D., Food and Drug Administration, 5600 Fishers Lane, HFD-110, Rockville, MD 20857 (tel: 301 594–5329; fax: 301 594–5494).

THE PROBLEM

It is well known that not only is the magnitude of drug effect not constant over an entire dosing interval, but also the blood pressure itself is not constant over time. Aside from beat-to-beat variability, blood pressure is subject to diurnal variation. So, to simply describe the drug effect over the entire inter-dosing interval (which is a minimal descriptor that many would think appropriate) is complicated. At the least it requires measurement of blood pressure at times other than just before the next dose. In principle, the more measurements, the better would be such a description. In addition, it requires assumptions about what the blood pressure over a dosing interval would have been in the active treatment group had no active drug been administered. Although not a straightforward description to obtain, ambulatory blood pressure monitoring (ABPM) would appear to be a logical means of providing some relevant data.

With occasional exceptions, the ABPM trials that we have reviewed have not fulfilled such expectations. In recent years, a number of applications have included reports of clinical trials in which blood pressure data were obtained both by office cuff measurement and with ABPM. Two recent applications each relied exclusively on ABPM. However, such studies have not figured prominently in most regulatory decision processes. By and large, ABPM studies have not shown drug effect varying as a function of dose, have not lent themselves to defining a dosing interval, and have relied, for purposes not only of data presentation but also of inference, on nothing more than graphic representation.

From our perspective the methods used for representing a group's central tendency are more of a determinant of how the data are interpreted than is the effect of the drug. This appears true whether one is simply representing the group's behavior as a function of time of day or representing the group by some quantitative single descriptor (e.g., mean ABPM). To our knowledge, no established, standard technique exists for analyzing such data, at least none that we recognize or accept. We recognize that there are groups or individuals that are working towards establishing such standardization.

THE PROJECT

The Division of Cardio-Renal Drug Products and the owners of ambulatory blood pressure clinical trial data have begun a research project intended to better define the role of ABPM in a clinical development program for an antihypertensive drug. As part of this project, the Division is encouraging corporate owners of ABPM clinical data to contribute those data to a database that exists within the Division's physical space and control.

The project is being undertaken with the assistance of Dr. Sandra Garrett of Medifacts, Ltd., a provider of ABPM services, under an arrangement known as a Cooperative Research and Development Agreement. Such cooperative agreements between Federal government entities and private industry are encouraged under the Technology Transfer Act. In addition to Dr. Garrett and the Cardio-Renal Division, corporate sponsors who have contributed ABPM data comprise an ABPM Steering Committee that meets periodically with the Division to discuss the development of the database, to propose research questions, and to view, discuss, and critique interim results.

To date (October, 1995), 26 companies have submitted at least placebo-group casual, office cuff blood pressure data from 5,133 subjects from 32 randomized, parallel, placebo-controlled, clinical trials, and a total of about 2,500 subjects had at least one 24-hour ambulatory recording. In addition, academic sponsors of SYST-EUR and HARVEST have also agreed to make their data available (not in the preceding totals). Accumulation of data is continuing.

The key question being addressed with these data concerns the magnitude of the placebo effect, and its variance, in ABPM. We think the variance of this quantity determines how large a clinical trial is needed to detect a given magnitude of drug effect in a clinical trial whose end point is measured with ABPM.

Often, comparisons between active and control groups are based on a single quantity, that is, some measure of central tendency for diastolic pressure over the inter-dosing interval. Some sponsors have used a 24-hour mean computed with all measurements equally weighted.

Many other algorithms can be and have been devised to assess central tendency, many of which have the likely desirable characteristic of weighting samples according to the period of time they represent. It is also reasonable, although less commonly considered in drug development programs, to consider also the time-of-day (usually daytime/nighttime).

One goal of this project is to explore the effects of such algorithms on the variance in estimates of central tendency in ABPM data and on the resulting power calculations for sizing future trials. One should also be able to characterize how demographic factors will affect such power calculations.

Although there are no data to support the need for round-the-clock blood pressure control to provide benefit in reducing morbidity and mortality, antihypertensive drugs approved on the basis of blood pressure reduction are expected to show that effect throughout the inter-dosing interval. Thus, a drug with a large effect that wanes to nothing within a few hours might distinguish itself from placebo in 24-hour mean effects, but it would, nevertheless, be considered inappropriate as once-a-day therapy. Currently, decisions about appropriate dosing interval are made based on effects at peak (or at some arbitrary time after dosing) and at trough. Sponsors occasionally have done analyses that compared hourly means in ABPM to determine the last time point statistically distinguishable between treatment groups. In our judgment, such measurements and/or estimates leave much to be desired. ABPM has the potential of making such quantitation more evaluable. Modelless comparisons (e.g., unweighted means, point-by-point comparisons) have little power, because there are few measurements in any given time window in all subjects. It would be far better to use all of a subject's data to characterize the whole diurnal response curve. It appears, and has not been shown to be incorrect, that the coefficients from each subject can then be combined to characterize the population-based response characteristics for the group.

Therefore, a secondary goal of our project is to compare model-based descriptions of the time course of blood pressure, as measured with ambulatory monitoring. Factors to be taken into account in such comparisons include the number of free parameters and the fidelity of models with respect to those aspects of blood pressure that are perhaps predictive of clinical outcome.

A problem with such comparisons is that no one knows the relative importance of nighttime falls, morning rises, or the lability of blood pressure in response to normal daily activities and stresses. We look forward to a time when a database similar to ours, but including outcome data, can be used to address this issue.

SUMMARY

Our division is in a privileged position to study the usefulness of ABPM in the clinical development of antihypertensive agents. Most trials involving ABPM are commercially sponsored, and in due course, much of that data is presented to the Division. As such studies become more prominent in drug development, the Agency must make decisions on how such data are analyzed and interpreted. These should be seen as this project's short-term goals. In the longer term, the accumulation of ABPM data from clinical trials will be a valuable resource that researchers can use to address questions about what aspects of hypertension are most associated with prognosis and which treatments affect those aspects.

Closing Remarks

FRANCESCO PORTALUPPI

Hypertension Unit
First Institute of Internal Medicine
University of Ferrara
Ferrara, Italy

The topic covered in this conference has important scientific, medical, and social implications and is an example of the multidisciplinary and yet unitary nature of modern chronobiology. Twenty-five leading scientists from 10 nations gathered in Ferrara to present firsthand information on the new findings related to the temporal structure of arterial blood pressure and its clinical and therapeutic consequences. A poster session was devoted to papers submitted and selected for presentation. Scientifically, the caliber of the meeting was high. Blood pressure variability and its correlates were examined by a variety of approaches, and information from a multitude of disciplines was integrated into the program.

The first area of discussion was the definitions and types of temporal patterns and their relative contribution to overall variability of blood pressure. Both animal and human models were presented. The neural, respiratory, vascular, renal, endocrine, and behavioral mechanisms of blood pressure rhythms were then analyzed. From presentations given during the first day of the conference, it became clear that recent studies have added considerably to our knowledge, but many relevant aspects remain unexplored, especially those regarding nocturnal variations in cardiovascular physiology and the temporal relationships between the different neurohormonal and hemodynamic mechanisms. The loss of the physiologic temporal balance between different regulatory mechanisms was recognized as a possible pathogenetic mechanism in hypertension—established already in many secondary forms—but further investigation of these relationships is also needed.

During the second day, epidemiologic evidence was presented suggesting that rhythms of blood pressure (in addition to the levels) may be of prognostic relevance. This is probable, even though still unproven, given the relationships between the rhythms of blood pressure, on the one hand, and acute cardiovascular and cerebral events, on the other. In addition, review of the various pathologic conditions in which blood pressure patterns are altered raised the question of whether and when the goal of treatment should be to reset the normal blood pressure pattern in essential and secondary forms of hypertension.

The key moment of the entire conference was the discussion of the now compelling evidence that the time course of drug absorption and distribution need not coincide with the time course of drug effects on blood pressure. This was made apparent by reviews of chronopharmacologic data available in experimental animals as well as by changes in the blood pressure patterns of essential and secondary forms of hypertension induced by current antihypertensive medications. Participants agreed on the conclusion that drug effect is not simply a result of drug concentration per se; it can be influenced by the previously discussed time-dependent mechanisms of blood pressure regulation. Many unsolved questions, all of potential practical impact and warranting further research, were also raised: How can blood pressure rhythms be modified in a time-specified manner? Is individualized chronotherapy of hypertension possible? Is it worthwhile, and when?

In dealing with the rationale and the methods of chronotherapy, a final crucial issue remained to be addressed: the endpoint parameters to be derived by blood pressure monitoring and the statistical methods that should be used in the interpretation of data collected with this technique. The last session of the conference covered these aspects, and it closed with the presentation of the regulatory perspective of the U.S. Food and Drug Administration.

The conference was noteworthy for the wide variety of participants from different scientific fields and disciplines, industries, and regulatory agencies (internists, cardiologists, nephrologists, neurologists, pneumologists, endocrinologists, diabetologists, pediatricians, epidemiologists, pharmacologists, physiologists, veterinarians, statisticians, biotechnologists, engineers, and representatives of the pharmaceutical, drug and biomedical equipment industries). This permitted coverage of the topics with an uncommonly broad range of considerations and perspectives that enhanced the thoroughness and interest of the discussions.

Most of the attending scientists would hardly define themselves as chronobiologists, even though all of them have been involved in research pertaining to the temporal aspects of life. By virtue of their special expertise, they offered new ideas and insights to the relatively young discipline of chronobiology. At the same time, I believe that the meeting was an occasion for them to better realize the contribution that a chronobiologic approach can offer to their work.

Allow me to conclude by saying that Ferrara also did a great job. It offered its visitors the architectural design of Biagio Rossetti, buildings erected by Leon Battista Alberti, paintings of the fourteenth century and then Guercino, Boldini, Carrà, De Chirico, and De Pisis. In this town you can still walk through the intersecting alleys of the medieval village and see the old castle and the almost intact city walls, while the immortal poems of Boiardo, Tasso, and Ariosto come to mind together with the gentle notes of Frescobaldi's madrigals. This town has grown together with its university, first sponsored by Duke Leonello of the Este family and then officially instituted by Pope Bonifacio IX with a charter dated 1391. When we committed ourselves to realizing a conference that would live up to everyone's expectations as well as helping our guests enjoy a pleasant visit, we were aware that we were continuing the traditions of scientific and professional excellence of our university as well as the culture and customs of our city and country. I wish to thank all those who made this possible, despite all imaginable difficulties, by adopting this conference, understanding its significance, and contributing to its success in all ways.

Corresponding Pulse Pressure and Arterial Compliance Variations during Ambulatory Monitoring

TODD J. BRINTON,[a] JOEL M. NEUTEL, SHIU-SHIN CHIO, E. DANIEL WALLS, LI-CHENG TAI, STANLEY S. FRANKLIN, DAVID H.G. SMITH, AND MICHAEL A. WEBER

University of California, Irvine
Orange, California 92668

Pulse Metric, Incorporated
San Diego, California 92121

Measurements of arterial compliance may provide clinically valuable information about the state of the cardiovascular system. These measurements may be useful in evaluating changes in the structural and functional components of the arterial wall. However, evaluation of arterial compliance measurements may be complicated by the dependent relationship with arterial blood pressure. Therefore, this study evaluates the relation between arterial compliance and blood pressure to better understand the time-dependent control of blood pressure and the relation between arterial structure and function.

SUBJECTS AND METHODS

The study population consisted of 14 subjects undergoing diagnostic evaluation for hypertension. The 14 patients consisted of 10 males and 4 females, all caucasian, ranging in age from 30–80 years (mean ± SEM = 53 ± 4). Blood pressure and arterial compliance measurements were determined using the Dynapulse 5000A ambulatory monitor. This new technology provides not only arterial pressures but also a graphic display of the oscillometric waveform. Average 24-hour ambulatory pressures for systolic, diastolic, pulse, and mean arterial pressure in the 14 patient sample were 136 ± 4, 70 ± 3, 66 ± 3, and 90 ± 4, respectively. Arterial compliance measurements were derived by analysis of the oscillometric waveform based on a new physical model. The model assumes a straight tube brachial artery and T-tube central aorta system and that the systolic phase of the suprasystolic oscillometric waveform and the diastolic phase of the subdiastolic oscillometric waveform reflect systolic and diastolic aortic pressure, respectively. Furthermore, the method assumes that arterial compliance is derived by radial compression and expansion of the artery due to the oscillating blood pressure. By theoretic design, arterial compliance was

[a] Address for correspondence: Todd J. Brinton, Pulse Metric, Inc., 6190 Cornerstone Court East, Ste. 103, San Diego, CA 92121 (tel: 619/546-9461; fax: 619/546-9470).

calculated at mean arterial pressure as:

$$C = \frac{\pi^{2}*D^{2}*[D + L_c]}{[(dP/dt)_{pp}*T_{pp}]}$$

in which $(dP/dt)_{pp}$ = amplitude from peak positive to peak negative dP/dt and T_{pp} = time interval between peak positive and peak negative dP/dt.[1] An estimate of brachial artery diameter (D) was generated by a mathematical model in which the average size of the vessel in an adult at mean arterial pressure (MAP) was scaled for body surface area.

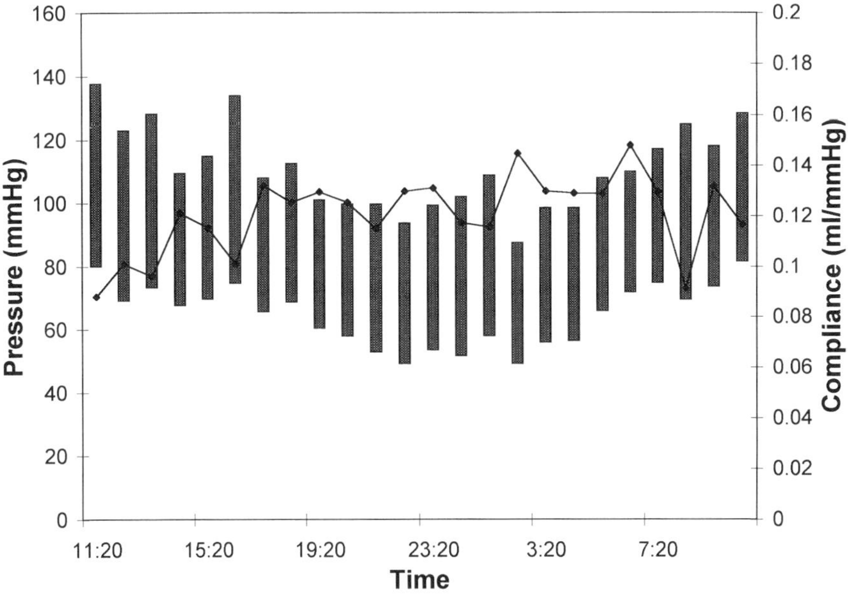

FIGURE 1. Twenty-four–hour ambulatory blood pressure and arterial compliance profile. Graph represents blood pressure and arterial compliance variations during circadian cycle. Time is plotted on the abscissa, and pressure (pulse pressure represented by a *solid bar*) and arterial compliance are plotted on the ordinate axes.

RESULTS

Linear regression analysis was performed between arterial compliance and simultaneous pressures for each individual patient. Average correlations for the 14 patient sample were pulse pressure ($r = -0.79$), systolic blood pressure ($r = -0.58$), mean arterial blood pressure ($r = -0.36$), and diastolic blood pressure ($r = 0.22$). Pulse pressure consistently exhibited the strongest linear regression correlation with arterial compliance, and the relationship was observed throughout each patient's 24-hour cycle (FIG. 1). Multivariate stepwise regression analysis was also performed to evaluate independent factors on arterial compliance in total individual measurements ($n = 798$) (TABLE 1). Pulse pressure was the strongest predictor of arterial compliance in the model.

TABLE 1. Multivariate Stepwise Regression Analysis of Independent Factors in Arterial Compliance.[a]

Statistics	Pulse Pressure	Mean Arterial Pressure	Diastolic Pressure	Body Surface Area	Age (yr.)
STD co-eff	−0.71	−0.36	0.29	0.50	0.15
p value	$p < 0.001$	$p < 0.001$	$p < 0.001$	$p < 0.001$	$p < 0.001$

[a]Analysis performed on 798 independent measurements.

CONCLUSION

The results of this study suggest that besides mean arterial pressure (MAP), pulse pressure is a strong inverse determinant of arterial compliance. Due to the established relation between arterial compliance and pulse pressure, differences in measured compliance in patients with matched pulse pressures should be a strong indication of underlying cardiovascular disease.

REFERENCES

1. BRINTON, T. J., B. COTTER, D. L. BROWN, P. BADDOUR, A. VOUNG, S. S. CHIO, C. CALISI & A. N. DEMARIA. 1994. A new non-invasive method for obtaining arterial pressure waveforms: Assessment of vascular compliance and validation with catheter data (abstr.). Circulation **90:** 1445.

Quantal Estimation of Hypertension and Hypertensive Damage in the Chronodiagnostic Evaluation of Ambulatory Blood Pressure Monitoring

P. CUGINI

Medical Semeiology and Methodology
University of Rome "La Sapienza"
Rome, Italy

Ambulatory blood pressure monitoring (ABPM) is currently still interpreted by the WHO criteria for determining the diagnosis of hypertension. Accordingly, the severity of hypertension is estimated by the highest BP value detectable over 140/90 mm Hg during the day-night span. Such a method of interpretation is inappropriate because the highest supranormal BP value may be repeatedly detectable, causing the hypertensive status to be more or less severe depending on its persistence. Assuming that the severity of hypertension depends on both the magnitude and the duration of supranormal BP values, one can argue that the reliable interpretation of the ABPM has to be based on a "quantal measure," which allows us to counivocally grade the hypertensive excess in quantity and temporality, that is, as a "quantum of hypertension." As a matter of fact, a quantum may be regarded as the measurable magnitude of a given event in the unitary time of its occurrence.

QUANTAL ESTIMATOR OF HYPERTENSION

As hypertension occurs over the 24 hours, its quantal estimator will be the measure of supranormal BP values over the time of their persistence, which can be operatively performed by estimating the integral of the area subtended by the time-qualified BP values in excess over the time of their excess. Such an integral estimate, which for simplicity may be named "baric hyperimpact," represents the hypertensive status as a quantum of hypertension, and returns the severity of hypertension in mm Hg∗hour(s):minute.

METHODS FOR ESTIMATING BARIC HYPERIMPACT

Methodologically, baric hyperimpact can be calculated (FIG. 1) by discrete (top panels) or analytic (bottom panels) methods of integration, respectively, using raw time-qualified BP values (BP chronogram) or the waveform profile which optimally fits raw BP data (BP cosinorgram). Importantly, both the discrete and the analytic methods show two hybrid procedures with respect to reference limits which are used for comparison. The two discrete hybrid procedures (left and right top panels) consist of a comparison of the individual systolic (S) and diastolic (D) BP chronograms, respectively, with the corresponding WHO time-unvarying upper reference

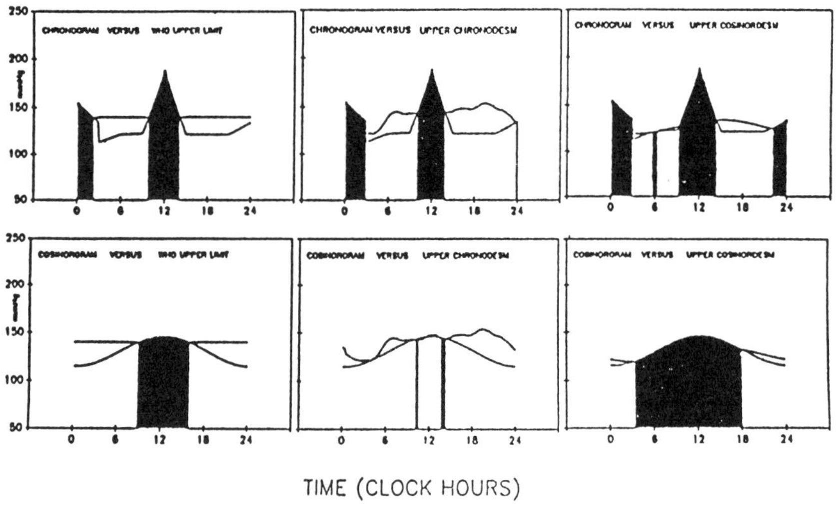

FIGURE 1. The six methods for measuring the baric hyperimpact as the area underlying the time-qualified blood pressure values which are higher than normal with respect to one of the upper reference limits adopted for their comparison in order to make a diagnosis of hypertension. The methods can be defined as "discrete" (*top panels*) or "analytic" (*bottom panels*) depending on the use of systolic and diastolic blood pressure values as they are (1) measured by the noninvasive ambulatory monitor (chronogram) and (2) calculated by their best fitting cosine curve (cosinorgram). Reference limits can be distinguished into "monodesms," "chronodesms," and "cosinordesms" depending on the fact that they represent the upper cutoff for systolic and diastolic blood pressure, respectively, as (1) time-unvarying fixed standards given by WHO; (2) time-varying standards for the raw time-qualified values; and (3) time-varying standards for the oscillatory circadian curve. Depending on the concordance of the comparison, the methods can be distinguished into "hybrid" and "nonhybrid."

limits (upper monodesms), or the corresponding upper reference limits for the optimal oscillatory curve (upper cosinordesms). Conversely, the discrete nonhybrid technique (middle top panel) consists of a comparison of the individual SBP and DBP chronograms with the corresponding upper reference limits for the time-qualified discrete values (upper chronodesms). The two analytic hybrid procedures (left and middle top panels) consist of a comparison of the individual SBP and DBP cosinorgrams, respectively, with the corresponding WHO upper monodesms, or the corresponding upper chronodesms. The analytic nonhybrid technique (right bottom panel) refers to a comparison of the individual SBP and DBP cosinorgrams with the corresponding upper cosinordesms.

DISCRETE VERSUS ANALYTIC NONHYBRID METHODS

From a statistical viewpoint, the hybrid techniques are considered unsuitable because of the intrinsic discordance between the individual series and their standard limits of reference. Therefore, the question is, Which is the best technique between the discrete and analytic nonhybrid methods?. It is important to keep in mind that the discrete nonhybrid procedure returns a measure that varies greatly as a function of the density and equidistance of the raw time-qualified BP values (FIG. 2, top panel), whereas the analytic nonhybrid procedure is less influenced by the sampling protocol (FIG. 2, bottom panel), justifying its adoption for quantifying the baric hyperimpact in an individual ABPM.

CONCLUSION

As a final comment, it is important to note that quantal estimation of hypertension is formally concordant with the physical mechanism that causes the hypertensive damage to target organs. As a matter of fact, any mechanical damage to structures that are resistant to a given baric load invariably depends on the quantity and duration of an excessive load. In other words, any stress damage is by definition a quantal damage that is related to the integration over time of the baric hyperimpact. Therefore, mechanical damage to target organs created by hypertension depends in itself on the parietal effect of the quantum of hypertension which is estimated by the baric hyperimpact.

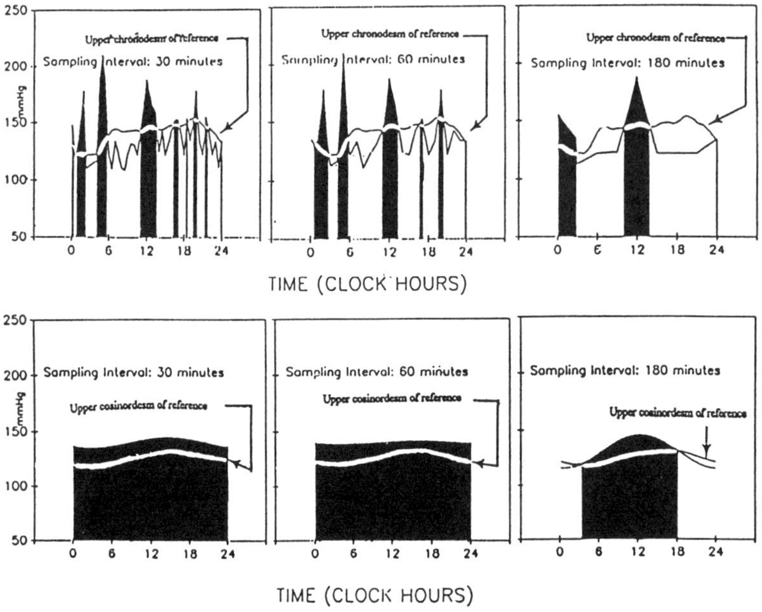

FIGURE 2. The effect of sampling protocol on the estimation of the baric hyperimpact via discrete (*top panels*) and analytic (*bottom panels*) nonhybrid methods of integration. By changing the density and temporal distance of blood pressure values over the 24-hour span, the discrete integration method returns a very variable estimate of the area subtended by supranormal values. Such an area is less variable if its estimate is performed by the analytic method of integration.

REFERENCES

1. HALBERG, F., J. K. LEE & W. NELSON. 1978. Time-qualified reference intervals-chronodesms. Experientia **34:** 713–716.
2. NELSON, W., G. CORNELISSEN, D. HINKLEY, C. BINGHAM & F. HALBERG. 1983. Construction of rhythm-specified reference intervals and regions, with emphasis on "hybrid" data, illustrated for plasma cortisol. Chronobiologia **10:** 179–193.

A Portable Data Logger for Recording Behavioral and Environmental Determinants of Blood Pressure during Ambulatory Monitoring

SERGIO GHIONE,[a] STEFANO DONATI,
LAURETTA SALERNO, GRAZIANO PALAGI,
ANDREA RIPOLI, DANIELE FRANCHI,
AND REMO BEDINI

CNR Institute of Clinical Physiology
Pisa, Italy

The main source of variability in arterial blood pressure (BP) is a host of environmental stimuli and behavioral and emotional activities or responses that occur in everyday life. Up to 80% of overall blood pressure variability, as quantified by the variance of all 24-hour blood pressure values, may be due to these factors, residual variability being due to rhythmic fluctuations intrinsic to control of the cardiovascular system.[1]

Despite these considerations, in clinical practice behavioral and environmental data are rarely related or integrated with BP measurement. The importance of situational factors in ambulatory blood pressure monitoring (ABPM), however, was indicated in studies by Pickering and coworkers[2,3] and other investigators including ourselves.[4–7]

The main difficulty in adding situational information to BP data is the implied use of diary and/or dedicated monitor. Diaries, especially if standardized, may provide useful information,[2,4,7] but have several limitations: they require a high degree of cooperation and understanding by the patient and need further work for data checking and input. However, dedicated instruments are designed for the acquisition of only a single type of information (typically, body motion by means of a portable piezoelectric accelerometer,[5,6] whereas multidimensional information would be needed.

We[8,9] recently reported on the design and preliminary experience with a portable computer (prototype) that functions as: (1) a simplified diary for direct data input by the patient; and (2) a data logger for automatic recording of various kinds of information by means of appropriate sensors (FIG. 1). The prototype contains a miniaturized microprocessor (ESP8680, Dovatron) running on MSDOS, an LCD CGA graphic display with a resolution of 192 × 128 dots, a six-key keyboard, and a signaling loudspeaker. An analogic front-end to the external sensors is interfaced to the processor by a 12-bit, 8-multiplexed channel, analog-to-digital converter. Two auxiliary connectors link it to a host computer via RS232 and to a standard keyboard. We here report on the application of this prototype in recording body posture, talking, and external temperature in conjunction with continuous blood pressure

[a] Address for correspondence: Dr. Sergio Ghione, CNR Institute of Clinical Physiology, Via Savi 8, 56100 Pisa, Italy (tel: -39 50 583111; fax: -39 50 553461).

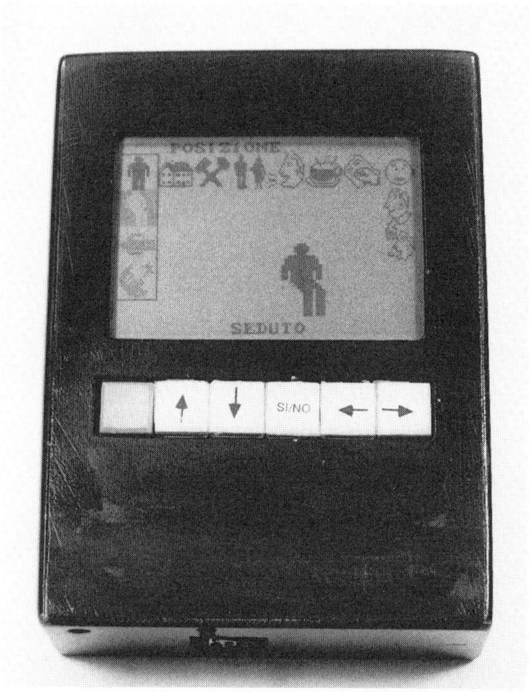

FIGURE 1. Prototype developed as a data logger for direct input of the diary by the patient (by means of user-friendly icons shown on the display) and for automatic acquisition of external variables (in this case, body posture, external temperature, and voice).

recording in normal healthy volunteers. Measurements were taken with a Finapress positioned on a movable dolly, which permitted fair mobility of the subject, in rooms with and without air conditioning. Body posture was recorded by application of a piezoelectric sensor to the ankle, measuring the height of the tip of a water column inside a capillary tube fixed along the leg of the subject. Talking and external temperature, respectively, were recorded by applying a shielded microphone to the neck and by a thermistor.

RESULTS AND DISCUSSION

A typical 1-hour acquisition is shown in FIGURE 2. As demonstrated, marked variability in blood pressure and to a lesser extent in heart rate, was observed which could in part be related to postural changes or to talking. In a preliminary group of five normal adult males, statistically significant effects were observed by two-way ANOVA for talking and posture on systolic and diastolic blood pressure and heart rate. The amount of variability in blood pressure and heart rate that could be "explained" by body posture, external temperature, and talking was estimated as the coefficient of determination R^2 by multiple regression analysis, applying a linear regression model employing "dummy" variables for category data (posture and talking).[10] On average, 38.2 ± 14.2%, 51.5 ± 15.9%, and 30.5 ± 18.0% of the variability, respectively, of systolic and diastolic blood pressure and heart rate could be accounted for by the three predictors, body posture, external temperature, and talking.

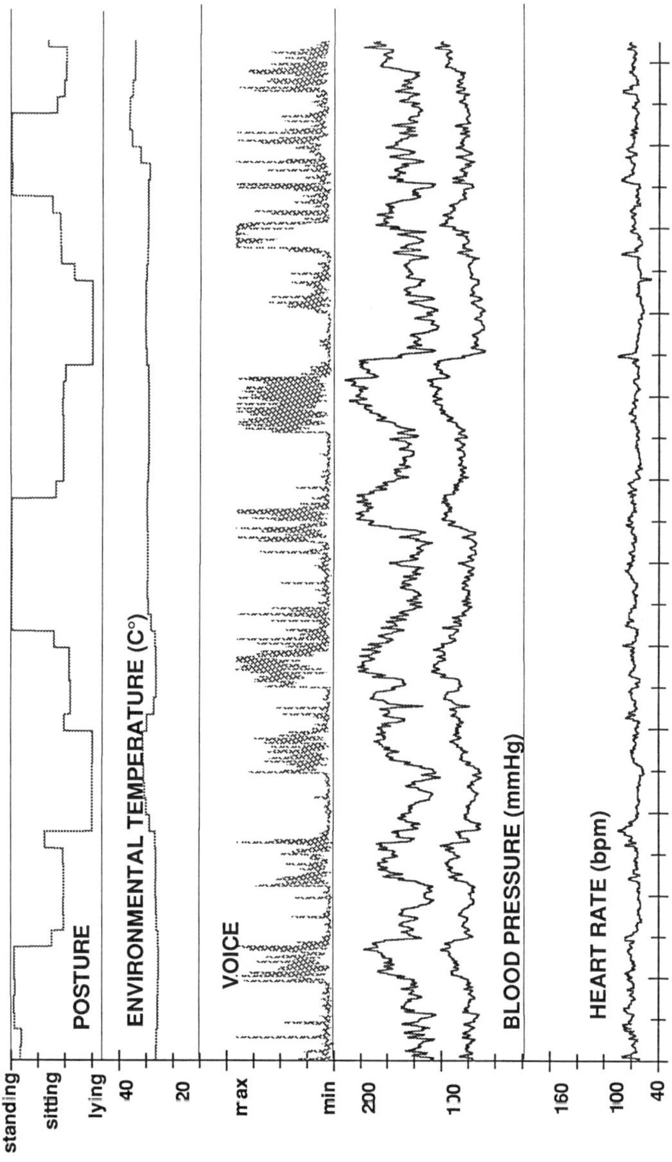

FIGURE 2. Plot of simultaneous acquisition (*bottom to top*) of heart rate, blood pressure, voice, environmental temperature, and posture. Heart rate and blood pressure were recorded by Finapress; voice, environmental temperature, and posture by the data logger.

REFERENCES

1. Mancia, G., M. DiRienzo & G. Parati. 1993. Ambulatory blood pressure monitoring: Use in hypertension research and clinical practice. Hypertension **21:** 510–524.
2. Pickering, T. G. 1991. Ambulatory Monitoring and Blood Pressure Variability. Science Press. London.
3. Pickering, T.G., W. Gerin, G. D. James, C. Pieper, Y. L. Schlussel & P. L. Schnall. 1992. The effects of occupational and domestic stress on the diurnal rhythm of blood pressure. *In* Temporal Variations of the Cardiovascular System. T. F. H. Schmidt, B. T. Engel & G. Blümchen, Eds.: 305–317. Springer-Verlag. Berlin
4. Ghione, S., C. Rosa, L. Salerno, G. Ceccanti & D. Volterrani. 1993. Assessment of physical and mental activity using a standardized computer-assisted diary during ambulatory blood pressure monitoring. J. Hypertens. **11** (suppl. 5): S284–S285.
5. Gretler, D., G. Carlson, A. Montano & M. Murphy. 1993. Diurnal blood pressure variability and physical activity measured electronically and by diary. Am. J. Hypertens. **6:** 127–133.
6. Van Egeren, L. F. 1991. Monitoring activity and blood pressure. J. Hypertens. **9** (suppl. 8): S25–S27.
7. Van-Egeren, L. F. 1992. Effects of behavioral rhythms on blood pressure rhythms. *In* Temporal Variations of the Cardiovascular System. T. F. H. Schmidt, B T. Engel & G. Blümchen, Eds.: 283–296. Springer-Verlag. Berlin.
8. Ghione, S., L. Salerno, R. Bedini, D. Franchi & G. Palagi. 1994. Physical and mental activities recorded by a special-purpose portable computer diary (PCD) for ambulatory blood pressure monitoring (ABPM). 15th Scientific Meeting of the International Society of Hypertension, Melbourne. J. Hypertens. **12:** S21.
9. Franchi, D., S. Ghione, G. Palagi, A. Belardinelli, S. Donati & R. Bedini. 1995. Multi-sensored palm-top data logger for ambulatory blood pressure long-term monitoring. XIII International Symposium on Biotelemetry. Williamsburg, Virginia, March 27–31.
10. Slinker, B. & S. Glantz. 1988. Multiple linear regression is a useful alternative to traditional analysis of variance. Am. J. Physiol. **255:** R353–R367.

The Dispersion Quotients: A New Approach for Evaluating the Blood Pressure Profile in Essential versus Cushing's Syndrome Hypertensive Patients

JAVIER GIMENEZ,[a] PABLO STIEFEL[b,c]
MARÍA LUISA MIRANDA,[b] ALFONSO LEAL-CERRO,[d]
OVIDIO MUÑIZ,[b] ENCARNACIÓN PAMIES,[b]
MARÍA DEL VAL MARTIN-SANZ,[b] JOSÉ VILLAR,[b]
AND JOAQUÍN CARNEADO[b]

[a]*S. de Medicina Intensiva*
Hospital Militar "Vigil de Quiñones"
Seville, Spain

[b]*Unidad de HTA y lípidos del S. de Medicina Interna and*
[d]*S. de Endocrinología*
Hospital Universitario Virgen del Rocío
Seville, Spain

Since the presence or absence of a nocturnal decrease in blood pressure (BP) may suggest an increased risk of visceral complications or the existence of secondary hypertension, several methods for evaluating the BP profile have been described. Nevertheless, a universally accepted system has not yet been established.[1–3]

This study tests different dispersion quotients (DQ) to estimate differences between the hourly mean and the mean of all readings in a 24-hour period during ambulatory BP recording. These quotients may be employed for systolic (SBP-DQ) or diastolic (DBP-DQ) blood pressure readings and may refer to the entire 24-hour period or to subperiods "morning" (m), "afternoon" (a), or "night" (n). The DQ is defined as:

$$\sqrt{\Sigma\,(\mathrm{HM} - \text{24-h M})^2/t}$$

where Σ means summatory (successive sums), HM is the mean of readings obtained each hour, 24-h M is the mean of all readings in a 24-hour period, and t is the time expressed in hours (24 for the entire record and 8 if referring to the morning, afternoon, or night periods). The morning period was between 8 AM and 3:59 PM, afternoon between 4 and 11:59 PM, and night between midnight and 7:59 AM.

For testing the mentioned DQs, we studied 20 essential and 17 secondary (Cushing's syndrome) hypertensive patients. The diagnosis of Cushing's syndrome was made by measuring the cortisol level in 24-hour urine samples (higher than 150 μg/24 hours in two different measurements). The etiologic diagnosis was based on

[c]Address for correspondence: Pablo Stiefel, MD, Unidad de HTA y Lípidos, C.S. Virgen del Rocío, Hospital General, Pta Baja, Avda Manuel Siurot s/n, 41013 Seville, Spain.

plasma corticotropin levels and magnetic nuclear resonance or computed tomography of the adrenal or pituitary glands, according to corticotropin levels. Five patients were diagnosed as having hypertension of adrenal origin (four adenomas and one macronodular hyperplasia) and 12 patients were diagnosed as having pituitary microadenoma. The essential hypertensive patients had mild to moderate hypertension and no visceral damage. All patients in both groups had a recent diagnosis and were free from drugs able to alter blood pressure measurements or its circadian rhythm. The observed shape of the blood pressure curve was not taken as an exclusion or an inclusion criterion in any group. Data were evaluated by comparing means for independent samples (Student's t test). The normality of the populations was estimated according to the Kolomogorow test.

Mean age and body mass index were not statistically different in both groups (in essential and secondary hypertensives, respectively, values were 40.6 ± 12.6 vs 39.5 ± 7.3 years, and 26.9 ± 4.4 vs 26.0 ± 2.5 kg/m^2). Also, mean ambulatory systolic or diastolic blood pressure during the 24-hour period were similar in both groups (values in essential vs Cushing's syndrome hypertensives were 142.4 ± 8.1 vs 140.1 ± 13.3 mm Hg for systolic and 90.7 ± 5.2 vs 89.1 ± 12.0 mm Hg for diastolic blood pressure, p = NS). The percentage of abnormal readings (higher than 140/90 mm Hg) of blood pressure measurements was not statistically different (values in essential vs Cushing's syndrome hypertension were 53.8 ± 18.5 vs 53.1 ± 30.3% for systolic and 52.8 ± 14.5 vs 46.5 ± 29.5% for diastolic blood pressure).

The standard deviation of the mean of all readings in the 24-hour period (measure of BP variability) was lower in the Cushing's syndrome group (in mm Hg: 12.3 ± 3.2 vs 16.2 ± 3.9, $p < 0.001$, for systolic and 11.7 ± 2.4 vs 14.1 ± 3.5, $p < 0.05$, for diastolic blood pressure).

All studied DQs were lower in the Cushing's syndrome group (TABLE 1). These differences were especially significant ($p < 0.0001$) for systolic BP during the entire period (DQ-SBP) and during the night (nDQ-SBP) and for diastolic BP ($p < 0.0005$) also during the night (nDQ-DBP), thus indicating the absence of a normal nocturnal dip of BP during the night in patients with Cushing's syndrome.

Patients whose DQ-SBP was lower than 12.5 mm Hg or whose nDQ-DBP were lower than 14 mm Hg showed high positive and negative predictive value (PPV, NPV), sensitivity (S_e), and specificity (S_p) pertaining to the Cushing's syndrome group (0.75, 0.88, 0.88, 0.75; 0.86, 0.82, 0.77, 0.90; and 0.78, 0.84, 0.82, 0.80, respectively). The mentioned quotients also showed higher PPV, NPV, S_e, and S_p values than did other more conventional methods for evaluating blood pressure profiles (TABLE 2).

TABLE 1. Results Obtained with the Studied Quotients in Both Populations[a]

Studied Quotients (mm Hg)	Hypertensive Patients (n = 20)	Patients (n = 17) with Cushing's Syndrome
DQ-SBP	15.6 ± 4.2	10.2 ± 2.9[d]
DQ-DBP	13.7 ± 3.7	10.3 ± 2.7[b]
mDQ-SBP	12.3 ± 6.1	8.8 ± 3.5[a]
mDQ-DBP	10.8 ± 5.6	9.2 ± 2.9
aDQ-SBP	11.2 ± 3.8	8.6 ± 2.9[a]
aDQ-DBP	10.1 ± 2.4	9.6 ± 4.0
nDQ-SBP	20.5 ± 6.3	11.8 ± 5.0[d]
nDQ-DBP	17.8 ± 5.6	10.9 ± 4.5[c]

Abbreviations: DQ = dispersion quotient; m = morning; a = afternoon; n = night.
[a]Values are means ± SD; [a]$p < 0.05$; [b]$p < 0.005$; [c]$p < 0.0005$; [d]$p < 0.0001$.

TABLE 2. Positive and Negative Predictive Value, Sensitivity, and Specificity Pertaining to the Cushing's Syndrome Group from Different Methods Testing Day/Night Variations in Blood Pressure

	PPV	NPV	S_e	S_p
DQ-SBP < 12.5 mm Hg	0.75	0.88	0.88	0.75
nDQ-SBP < 14 mm Hg	0.86	0.82	0.77	0.90
nDQ-DBP < 14 mm Hg	0.78	0.84	0.82	0.80
AR-SBP < 9 mm Hg	0.66	0.81	0.82	0.65
AR-DBP < 9 mm Hg	0.63	0.80	0.82	0.60
AR-ARPSBP < 15%	0.59	0.73	0.76	0.55
AR-ARPDBP < 12%	0.52	0.64	0.71	0.45
DN ratio SBP < 1.05 mm Hg	0.73	0.72	0.80	0.64
DN ratio DBP < 1.08 mm Hg	0.66	0.70	0.66	0.70

Abbreviations: DQ = dispersion quotient; nDQ = night's dispersion quotient; AR = absolute reduction from daytime to nighttime; ARPSBP and ARPDBP = abnormal reading percentage of systolic and diastolic blood pressure; DN ratio = daytime:nighttime ratio.

In conclusion, DQs are alternative methods for evaluating BP patterns in hypertensive subjects. In our patients, they were an easy and quick method for differentiating circadian BP patterns in essential and secondary (Cushing's syndrome) hypertensive subjects. Their use in evaluating the possibility of secondary hypertension or the future risk of target organ damage in essential hypertension should be investigated in the future.

REFERENCES

1. STREITBERG, B. & W. MEYER-SABELLEK. 1990. J. Hypertens. **8**(Suppl. 6): S21–S27.
2. MACHNING, T., G. ENGELS, S. SCHÖBEL, J. GELLERT & K. BACHMAN. 1993. Am. J. Hypertens. **6:** 248–252.
3. STATON A., J. COX, N. ATKINS, K. O'MALLEY & E. O'BRIEN. 1992. Hypertension **19:** 93–101.

Effects of Timing of Salt Intake to 24-Hour Blood Pressure and Its Circadian Rhythm[a]

KAZUE ITOH, TERUKAZU KAWASAKI,[b]
AND PIETRO CUGINI[c]

Nakamura Gakuen University
Fukuoka 814, Japan

[b]*Institute of Health Science*
Kyushu University
Kasuga 816, Japan

[c]*Institute of II Medical Clinic*
University of Rome "La Sapienza"
Rome 00161, Italy

This study investigates whether salt intake at lunch time or dinner time can determine variations in the 24-hour blood pressure (BP) pattern. The idea was generated by the observation that a given caloric intake at breakfast or dinner can only determine a consistent shift in many biologic rhythms along with a different adipogenic effect.[1–3]

MATERIALS AND METHODS

The study was conducted on seven clinically healthy normotensive females (20–21 years of age) who volunteered with informed consent to have noninvasive monitoring of their BP for two consecutive days, changing weekly their normal daily salt intake (12 g/day) for three experimental sessions: (1) regular distribution of dietary salt at the three meals (Stage R); (2) two thirds of dietary salt at lunch time (Stage LH); and (3) two thirds of dietary salt at dinner time (Stage DH). Daily amounts of nutrients were similar at each stage. The BP values were recorded every 15 or 30 minutes over 48 hours by an ambulatory automated device (model ABPM-630, Colin Electronics, Japan). Systolic (S) and diastolic (D) BP time series were analyzed by chronobiologic procedures.[4]

RESULTS

The circadian rhythm of SBP and DBP was invariably preserved, its rhythmometric parameters, mesor, amplitude, and acrophase, being comparable each day of monitoring in each session. The circadian mesor, amplitude, and acrophase of SBP and DBP in Stage R averaged, respectively, 105 ± 3/61 ± 2 mm Hg, 4 ± 3/3 ± 2 mm

[a]This study was published in J. Nutri. Sci. Vitaminol. **40:** 459–466, 1994 as a full original paper.

Hg, and 17:04 ± 2:20/16:32 ± 1:52 h:min. These estimates showed the following changes and significant differences, respectively, in Stages LH and DH, that is, 107 ± 3/63 ± 2 vs 103 ± 3*/60 ± 1† mm Hg; 4 ± 3/5 ± 4 vs 6 ± 3/6 ±2 mm Hg; 15.40 ± 2:16/15:24 ± 2:04 vs 16:36 ± 1:32†/16.28 ± 0:36 h:min*, where * = $p < 0.01$ and † = $p < 0.001$ on Bingham's test.[4]

DISCUSSION

This study reveals that SBP and DBP circadian rhythm undergoes two distinct changes as a function of the reapportionment of dietary salt intake: (1) an increase in mesor and anticipation in phase in Stage LH; and (2) a decrease in mesor and a resetting in phase in Stage DH. These findings are in keeping with two conclusions: (1) dietary salt intake has entraining effects on human BP circadian rhythm, and (2) BP in human beings has a circadian susceptibility to dietary salt intake. Both conclusions corroborate that arterial hypertension can better be prevented and treated by reapportioning salt intake so that two thirds is consumed at dinner.

REFERENCES

1. HIRSCH, E., E. HALBERG, F. HALBERG, F. C. GOETZ, D. CRESSEY, H. WENDT, R. SOTHERN, E. HAUS, P. STONEY, D. MINORS, G. D. ROSEN, B. HILL, M. HILLEREN & K. GARRET. 1975. Body weight change during one week on a single daily 2000-calorie meal consumed as breakfast (B) or dinner (D). Chronobiologia **2** (Suppl.1): 31–32.
2. NELSON, W., L. SCHEVING & F. HALBERG. 1975. Circadian rhythms in mice allowed single daily "meal" at different stages of LD (12-12) lighting regimen. J. Nutr. **105:** 71–184.
3. GRAEBER, R. C., R. GATTY, F. HALBERG & H. LEVINE. 1978. Human eating behaviour: Preferences, consumption patterns, and biorhythms. Food Science Laboratory, US Army Natik Research and Development Command, Natik, Massachusetts.
4. BINGHAM, C., B. ARBOGAST, G. CORNELISSEN, J. K. LEE & F. HALBERG. 1982. Inferential statistical methods for estimating and computing cosinor parameters. Chronobiologia **9:** 397–439.

Circadian Blood Pressure Pattern in Clinically Healthy Young Nepalese

TERUKAZU KAWASAKI, KAZUE ITOH,[a]
PIETRO CUGINI,[b] KEIKO UEZONO,
AND GERMAINE CORNELISSEN[c]

Institute of Health Science
Kyushu University
Kasuga 816, Japan

[a]*Nakamura Gakuen University*
Fukuoka 814, Japan

[b]*Institute of II Medical Clinic*
University of Rome "La Sapienza"
Rome 00161, Italy

[c]*Chronobiology Laboratories*
University of Minnesota
Minneapolis, Minnesota 55455

This study demonstrates how human lifestyle influences the 24-hour blood pressure (BP) pattern by investigating clinically healthy subjects recruited from Nepal and Japan. These two countries were chosen for two reasons: (1) the consistent difference in lifestyle (rural and industrialized, respectively), and (2) their common high sodium dietary intake.[1]

MATERIALS AND METHODS

The study consisted of 42 Nepalese and Japanese clinically healthy subjects, matched for gender (19/2 vs 19/2 males and females), age (25 ± 5 vs 25 ± 3 years), height (165 ± 5 vs 167 ± 5 cm), weight (60 ± 7 vs 60 ± 6 kg),and casual systolic (S) and diastolic (D) BP (116 ± 9/72 ± 5 vs 118 ± 11/69 ± 8 mm Hg), who voluntarily gave informed consent to undergo noninvasive ambulatory BP monitoring during an ordinary work day in the spring. The Nepalese worked in a rural setting in manual occupations; the Japanese were occupied in an industrial context as clerks. Each subject was asked to record activity/resting time and meal timing schedule and to answer a nutritional questionnaire in order to identify the differences in lifestyle. The individual BP series were analyzed according to chronobiologic methods.[2]

RESULTS

The arising and retiring times of the Nepalese were significantly earlier (5:50 ± 0:36 vs 7:03 ±0:42 h:min, $p < 0.001$; 21:45 ± 0:26 vs 00:43 ± 0:53 h:min, $p < 0.001$). The SBP and DBP were not significantly different in their daily mean (112 ± 6 vs 115 ± 8 and 65 ± 5 vs 66 ± 6 mm Hg), trough value (102 ± 6 vs 104 ± 8 and 56 ± 5 vs 58 ± 6 mm Hg), peak value (124 ± 6 vs 129 ± 8 and 82 ± 5 vs 78 ± 6 mm Hg), range (66 ± 13 vs 62 ± 14 and 49 ± 9 vs 43 ± 9 mm Hg), and coefficient of variation (12.8 ± 3.1 vs 11.4 ± 2.1 and 16.8 ± 3.8 vs 15.4 ± 2.6%). Importantly, the SBP and DBP circadian rhythm was coherent with the arising resting scheme, showing significant anticipation of its oscillatory phase in the Nepalese (14:00 ± 0:28 vs 16:12 ± 0:40 and 14:00 ± 0:24 vs 16:08 ± 0:21 h:min, $p < 0.001$ and $p < 0.001$).

DISCUSSION

This investigation clinically demonstrates that the activity/resting scheme in human beings acts as a synchronizer of the BP circadian rhythm. The arising/resting alternation in this study is the expression of a lifestyle that depends on the rural or industrialized context in which the investigated subjects were living. Therefore, through its social routines and occupational activities, human modernization apparently acts on human BP by modulating its circadian phase of oscillation. This implies that forced industrialization may be responsible for a "social lag syndrome," a term proposed by one of us[3] to indicate a dyschronism of phase that occurs in biologic circadian rhythms when human beings are faced with an abrupt change in their daily lifestyle because of newly imposed social cues, including occupational routines. However, adaptation to an industrialized and metropolitan lifestyle may have even more consistent consequences on the BP circadian pattern.[4]

REFERENCES

1. Kawasaki, T., K. Itoh, K. Uezono, T. Ogaki, Y. Yoshimizu, S. Kobayashi, T. Osaka, M. Ogata, S. Dhungel, S. Sharma & G. P. Acharya. 1993. Investigation of high salt intake in a Nepalese population with low blood pressure. J. Hum. Hypertens. **7:** 131–140.
2. Bingham, C., B. Arbogast, G. Cornelissen, J. K. Lee & F. Halberg. 1982. Inferential statistical methods for estimating and comparing cosinor parameters. Chronobiologia **9:** 397–439.
3. Cugini, P. 1988. Institutions of medical chronobiology. *In* Recentia Medica.: 1730–1806. USES. Florence (in Italian).
4. Otsuka, K., H. Watanabe, G. Cornelissen, M. Shinoda, K. Uezono, T. Kawasaki & F. Halberg. 1990. Gender, age and circadian blood pressure variation of apparently healthy rural vs. metropolitan Japanese. Chronobiologia **17:** 253–265.

Ambulatory Blood Pressure Monitoring in Normotensive Smokers and Nonsmokers

TERUKAZU KAWASAKI, KEIKO UEZONO, PIETRO CUGINI,[a] KAZUE ITOH,[b] AND GERMAINE CORNELISSEN[c]

Institute of Health Science
Kyushu University
Kasuga 816, Japan

[a] *Institute of II Medical Clinic*
University of Rome "La Sapienza"
Rome 00161, Italy

[b] *Nakamura Gakuen University*
Fukuoka 814, Japan

[c] *Chronobiology Laboratories*
University of Minnesota
Minneapolis, Minnesota 55455

This study investigates 24-hour blood pressure (BP) and heart rate (HR) patterns in *normotensive* smokers to detect how chronic smoking affects the cardiovascular apparatus in conditions of apparent normotension, the reported investigations dealing with the binomium "smoking-hypertension."[1,2]

MATERIALS AND METHODS

Fifty clinically healthy male subjects volunteered with informed consent to undergo noninvasive monitoring of their BP and HR for 24 hours. Twenty-five participants (34 ± 8 years; 168.1 ± 4.1 cm, 59.7 ± 8.3 kg) were chronic smokers who consumed an average of 25.4 cigarettes per day. The other 25 subjects were age-, height-, and weight-matched nonsmokers (34 ± 7 years; 169.0 ± 4.9 cm, 61.9 ± 7.0 kg), who had an identical lifestyle. Ambulatory BP monitoring was conducted on a work day in winter. Time-qualified systolic (S) and diastolic (D) BP and HR values were analyzed via numeric and analytic methods of chronobiometry.[3]

RESULTS

Smokers had statistically significantly higher SBP and DBP (118 ± 4.4 vs 115 ± 4.2 mm Hg and 70 ± 3.0 vs 67 ± 3.1 mm Hg; $p < 0.05$ and $p < 0.01$, respectively) during the day but not at night (109 ± 5.6 vs 107 ± 6.1 mm Hg and 62 ± 4.9 vs 61 ± 4.8 mm Hg). Both smokers and nonsmokers had statistically significant SBP ($p < 0.001$ and $p < 0.001$), DBP ($p < 0.001$ and $p < 0.001$), and HR ($p < 0.001$ and $p < 0.001$) circadian rhythm. Importantly, smokers had a statistically significantly higher amplitude of the second (12-hour period) harmonic wave fitting the 24-hour SBP and DBP values (3.1 ± 1.26 vs 1.96 ± 1.28 mm Hg, and 2.87 ± 1.00 vs 1.89 ± 0.97 mm Hg; $p < 0.01$ and $p < 0.001$, respectively).

DISCUSSION

The subjects who smoked more than 20 cigarettes per day had a significant increase in BP values concomitant with the time in which they smoked. This increase leaves BP values apparently normal and is operative for at least 15–16 hours per day, being due to structured interference with BP regulation, that is, amplification of the 12-hour period harmonic component which sustains BP periodicity around the clock. Whether the actual increase in diurnal BP is the prelude to future hypertension cannot be predicted; however, its occurrence per se is seen as a negative condition, provoking a tensiogenic response of the cardiovascular apparatus and exerting tensiogenic repercussion on target organs. Importantly, this latter effect is occult as the smokers appear to be normotensive.

REFERENCES

1. Mann, S. J., G. D. James, R. S. Wang & T. G. Pickering. 1991. Elevation of ambulatory systolic blood pressure in hypertensive smokers: A case-control study. JAMA **265:** 2226–2228.
2. Asmar, R. G., X. J. Gired, M. Brahimi, A. Safavian & M. E. Safar. 1992. Ambulatory blood pressure measurement, smoking and abnormalities of glucose and lipid metabolism in essential hypertension. J. Hypertens. **10:** 181–187.
3. Halberg, F., E. A. Johnson, W. Nelson, W. Runge & R. Sothern. 1972. Autorhythmometry procedures for physiologic self-measurements and their analysis. Physiol. Teacher. **1:** 1–11.

Reproducibility of Nocturnal Blood Pressure Reduction Rate and the Prevalence of "Non-Dippers" Using 48-Hour Ambulatory Blood Pressure Monitoring in Patients with Essential Hypertension

YASUAKI MOCHIZUKI, MITSUAKI OKUTANI, HIROSHI IWASAKI, ISAO KOHNO, SUNAO MOCHIZUKI, KEN UMETANI, HIROYUKI ISHII, HIROSHI IJIRI, SADAYOSHI KOMORI, AND KOHJI TAMURA

The Second Department of Medicine
Yamanashi Medical University
Yamanashi, Japan

We evaluated the circadian reproducibility of the nocturnal reduction rate in blood pressure (BP) and the prevalence of non-dippers among hypertensive patients.

METHODS

The study population consisted of 253 untreated patients (aged 22–80 years, mean 53) with mild to moderate essential hypertension. The patients' BP and heart rate (HR) during daily activities were measured for 48 hours every 30 minutes by a

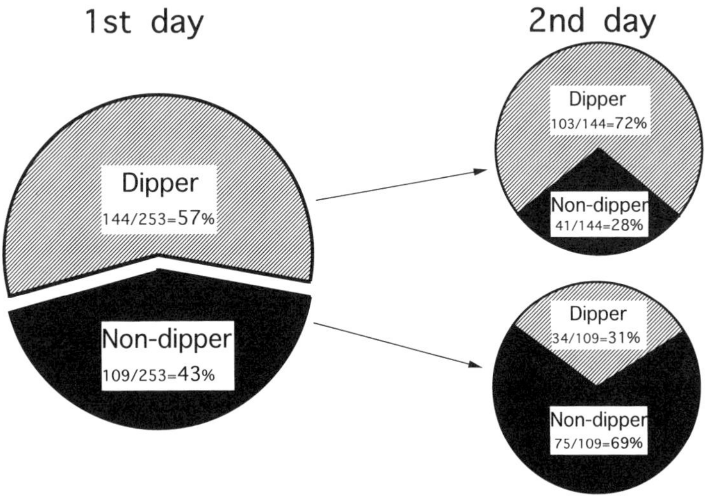

FIGURE 1. Prevalence of dippers and non-dippers.

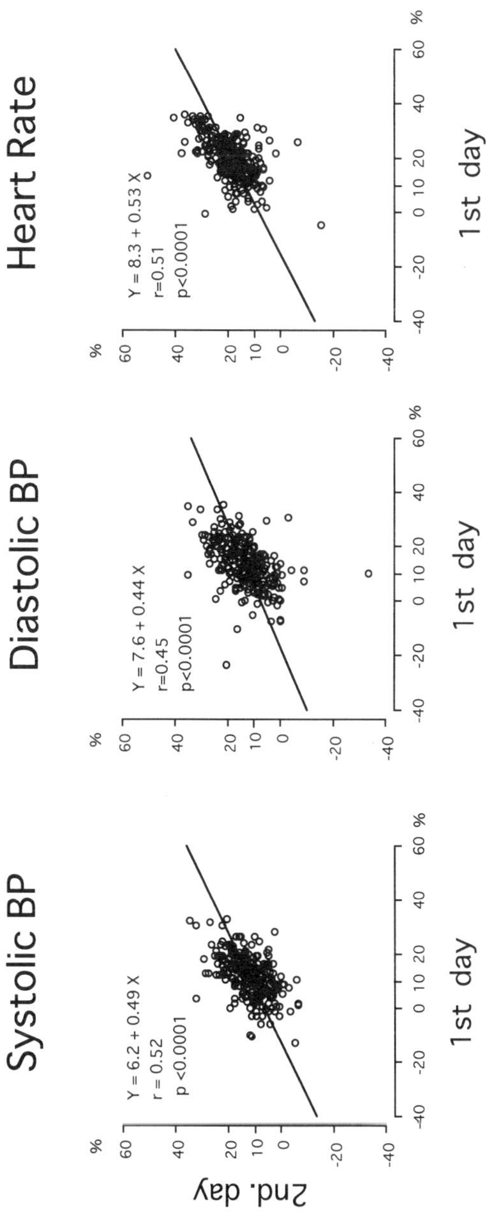

FIGURE 2. Concordance of nocturnal BP reduction rate between the first and second days.

noninvasive BP monitoring system (ABPM-630, Colin, Japan). Mean 48-hour daytime (awake) and nighttime (sleeping) systolic BP, diastolic BP, and HR values were analyzed by reviewing the patients' diaries. The nocturnal BP reduction rate was calculated according to Verdecchia's formula.[1] Subjects then were divided into two groups according to the presence (dipper) or absence (non-dipper) of a significant reduction in nocturnal BP (> 10%).

RESULTS

The first and second days did not differ in mean 24-hour systolic and diastolic BP during the day and at night.

The nocturnal reduction rates in BP were similar on the first and second days. The nocturnal reduction rate in HR on the first day was significantly higher than that on the second. The prevalence of dippers and non-dippers was 57% and 43% on the first day. On the second day, the prevalence of non-dippers among first-day dippers and that of dippers among first-day non-dippers was 28% and 31% (FIG. 1). Therefore, the prevalence of dippers and non-dippers on both days was 41% and 30%, respectively. FIGURE 2 shows a significant concordance in the nocturnal BP reduction rate between the 2 days ($r = 0.52$ and .45; $p < 0.0001$).

DISCUSSION

No significant differences in 24-hour daytime and nighttime mean BP were noted between the 2 days. Therefore, the reproducibility of the nocturnal BP reduction rate was similar on the first and second days. The prevalence of dippers and non-dippers on both days was 41% and 30%, respectively and that of "dipper to non-dipper" or "non-dipper to dipper" was 29%. Many subjects had nocturnal BP reduction rates of about 10%. Thus, the BP variation pattern differed between the two days.

CONCLUSION

There were three groups: "dippers and dippers," "dipper or non-dippers," and "non-dippers and non-dippers." The 48-hour recording, which included two circadian periods, was better than the 24-hour recording in detecting "non-dipper and non-dippers."

REFERENCE

1. VERDECCHIA, P., G. SCHILLACI, M. GUERRERI *et al.* 1990. Circadian blood changes and left ventricular hypertrophy in essential hypertension. Circulation **81:** 528–536.

Reproducibility of Ambulatory Blood Pressure Measurement in Renal Hypertension

Dippers and Non-Dippers

B. LIEBISCH, J. KLETZMAYR, F. WEBBER, AND B. SCHNEIDER

Department of Nephrology and Dialysis
University of Vienna Medical School
Vienna, Austria

As nocturnal dipping of blood pressure (BP) is associated with less BP target organ damage and less cardiovascular risk, this study assesses the reproducibility of dipping at the often recommended level of minus 10% of the mean daytime BP values.

PATIENTS AND METHODS

Forty patients (15 female and 25 male, aged 50 ± 12 years, 38 of them with antihypertensive therapy) with renal hypertension treated in the nephrologic ambulance underwent ambulatory BP measurement performed with the Spacelabs 90207 device (oscillometric method). The measurement was carried out over a 48-hour period with 20-minute intervals during daytime (6 AM–10 PM) and 30-minute intervals during nighttime (10 PM–6 AM).

RESULTS

A total of 4,771 measurements were performed; 59.6 measurements on average were obtained in each 24-hour period per patient, 74% of them during daytime. No significant difference between the number of measurements on days 1 and 2 was noted.

TABLE 1 lists BP values. The differences in daytime values measured on days 1 and 2 are statistically significant during the day ($p > 0.005$ for systolic and $p > 0.002$ for diastolic BP) but not at night.

TABLE 2 shows the number of patients dipping (defined as a fall in BP of at least 10% of the daytime mean).

TABLE 1. Blood Pressure Values (mean ± standard deviation; mm Hg)

	Systolic BP		Diastolic BP	
	Day 1	Day 2	Day 1	Day 2
Daytime	152 ± 14.2	149 ± 14.1	94 ± 10.2	92 ± 11.8
Nighttime	144 ± 19.6	144 ± 18.4	84 ± 11.8	84 ± 11.5

TABLE 2. Number of Dippers

Days	Systolic BP	Mean BP	Diastolic BP
1	20	18	18
2	15	15	17

CONCLUSION

The smaller number of "dippers" on the second day can be explained by the significant higher daytime (6 AM–10 PM) values on day 1. The simple definition of dipping using percentage of mean differences in BP needs further clarification.

REFERENCES

1. VERDECCHIA, P., G. SCHILLACI, M. GUERRIERI, C. GATTESCHI, G. BENEMIO & F. BOLDRINI, *et al.* 1990. Circadian blood pressure changes and left ventricular hypertrophy in essential hypertension. Circulation **81:** 528–536.
2. HANSEN, K. W., P. L. POULSEN & C. E. MOGENSEN. 1994. Ambulatory blood pressure and abnormal albuminuria in type I diabetic patients. Kidney Int. (Suppl.) **45:** 134–140.
3. VERDECCHIA, P. & C. PORCALLATI. 1995. Prognostic value of ambulatory blood pressure monitoring. J. Hypertens. **13:** 373–374.

Blood Pressure and Atmospheric Pressure

M. WEINBACHER, B. MARTINA, T. BART, J. DREWE,
P. GASSER, AND K. GYR

Medical Outpatient Department
University Clinic
Basel, Switzerland

Roche Pharma AG
Reinach, Switzerland

Cardiac output and peripheral resistance are the main determinants of blood pressure (BP). They are influenced by various factors, such as genetic predisposition, endogenous influences of the renin-angiotensin-system and sympathetic nervous system, dietary influences, as well as different environmental conditions. Few reports exist on the relation of atmospheric pressure and blood pressure.

Kunes *et al.*[1] reported a significant negative near-linear correlation between environmental temperature and blood pressure in 2,000 hypertensive patients. This result was confirmed by Bruce *et al.*[2] in a retrospective analysis of data from the British Regional Heart Study, with an 0.5 mm Hg decrease in blood pressure for each 1°C rise in room temperature. Filikowski[3] found a rise in blood pressure in hypertensive subjects on days with a considerable decrease in atmospheric pressure during an expedition to the subantarctic zone. Some studies assessed the influence of atmospheric pressure and weather variables on the occurrence of clinical events (subarachnoid hemorrhage[4] and pulmonary embolism[5]).

We performed ambulatory blood pressure measurements (ABPM) before and 6 weeks after treatment with 5 mg cilazapril and 12.5 mg hydrochlorothiazide in 25 outpatients with mild to moderate arterial hypertension (mean diastolic 24-hour BP > 90 and ≤ 115 mm Hg; mean age 49 years, range 33–68). The mean of three blood pressure measurements in each hour of ABPM was related to the atmospheric pressure reading of the corresponding hour (these data were provided by the local meteorologic institute of Basel). We found a weak negative correlation between systolic blood pressure and atmospheric pressure ($r = -0.46$; $p = 0.0001$) in nonresponders (i.e., mean diastolic daytime BP > 90 mm Hg or reduction < 5 mm Hg after 6 weeks of treatment). The same weak negative correlation was present in the morning hours as well as in the nighttime in nonresponders. Diastolic BP, mean arterial BP, and systolic BP in responders did not correlate with atmospheric pressure (FIG. 1).

The pathogenesis of this relation is still unclear. A loss in capability of endogenous regulation of blood pressure might be the cause for an increasing dependence of blood pressure on environmental conditions in nonresponders. As the interference of temperature and humidity was not considered in this study, extended analysis is needed to investigate this weak negative correlation between blood pressure and atmospheric pressure in a larger population.

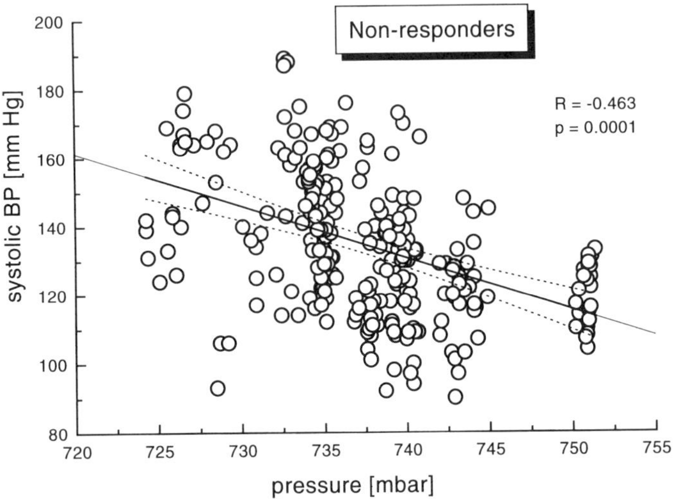

FIGURE 1. Weak negative correlation between systolic blood pressure and atmospheric pressure in hypertensive subjects (nonresponders to medical treatment).

REFERENCES

1. KUNES J., J. TREMBLAY, F. BELLAVANCE & P. HAMET. 1991. Influence of environmental temperature on the blood pressure of hypertensive patients in Montreal. Am. J. Hypertens. **4:** 422–426.
2. BRUCE, N., J. ELFORD, G. WANNAMETHEE & A.G. SHAPER. 1991. The contribution of environmental temperature to geographic variations in blood pressure. J Hypertens. **9:** 851–858.
3. FILIKOWSKI, J. 1983. Influence of meteorological conditions of various climatic zones on arterial pressure of participants in a sea voyage. Bull. Inst. Marit. Trop. Med. Gdynia 31–40.
4. JEHLE, D., R. MOSCATI, J. FRYE & N. REICH. 1994. The incidence of spontaneous subarachnoid hemorrhage with change in barometric pressure. Am. J. Emerg. Med. **12:** 90–91.
5. SCOTT, J.A., E.L. PALMER, A.J. FISCHMAN & H.W. STRAUSS. 1992. Meteorologic influences on the frequency of pulmonary embolism. Invest. Radiol. **27:** 583–586.

Circadian Blood Pressure Patterns in Pregnant Women with Intrauterine Growth Retardation

ANDREA L. TRANQUILLI,[a] ELISABETH GARBATI, HERBERT VALENSISE,[b] GIUSEPPE G. GARZETTI, AND CARLO ROMANINI[b]

Department of Obstetrics and Gynecology
University of Ancona
Ancona, Italy

[b]*Department of Obstetrics and Gynecology*
University of Tor Vergata
Rome, Italy

Intrauterine growth retardation (IUGR) frequently complicates hypertensive pregnancy and preeclampsia.[1] Reduced placental perfusion, which is associated with preeclampsia, results in reduced oxygenation of the placental vascular bed; the subsequent hypoxemia and lack of substrates produces IUGR. Apart from a preeclamptic background, congenital and chromosomal anomalies and intrauterine infections can cause IUGR. Excluding those main causes, the pathogenesis and clinical features of "idiopathic" IUGR still remain obscure.

This paper assesses maternal blood pressure in pregnancy complicated by unexplained, isolated, "idiopathic" IUGR.

PATIENTS AND METHODS

We reviewed 24-hour ambulatory blood pressure patterns in 36 normotensive women who demonstrated intrauterine growth retardation on serial ultrasound scans and neonatal birthweight lower than the fifth centile for gestational age at delivery (NIUGR group).

Seventy-eight normotensive nonproteinuric women whose fetuses showed normal intrauterine growth (NAGA group) and neonatal weight appropriate for gestational age at delivery were used as controls. All women studied were definitely normotensive at the time of the recording, none showed blood pressure values higher than or equal to 140/90 mm Hg. Twenty-four–hour blood pressure was recorded at 30-minute intervals by oscillometry (SpaceLabs 90207, SpaceLabs, Inc, Redmond, Washington) between 24 and 28 weeks before knowing if fetal growth was normal or retarded. Clinicians were blinded as to the results and results were interpreted off-line only after delivery to identify prediction of adverse outcome.

The 24-hour mean was chosen as the measure of centrality; it expresses the average level of BP and describes quantitative differences.[2] Analysis of day-night and

[a]Address for correspondence: Andrea L. Tranquilli, MD, Department of OB/GYN, University of Ancona at Salesi Hospital, Via F. Corridoni 11, 60123 Ancona, Italy (fax +39 +71 36 276).

activity-rest rhythm was performed both by calculating the mean difference between daytime/activity (6 AM–10 PM) and nighttime/rest (10 PM–6AM) measures and by using the cosinor method. This is the most frequently used strategy for smoothing or modeling the 24-hour BP curve and is equivalent to a Fourier periodic series with only one harmonic within a 24-hour period.[2,3]

All differences were tested statistically with analysis of variance. Predictivity was also calculated for centrality measures, using Cohen's k index.

RESULTS

Circadian blood pressure pattern of pregnant women with idiopathic IUGR significantly differs from that of pregnant women with normal fetal growth. All the centrality measures of blood pressure are significantly elevated in women with IUGR than in the controls, as outlined in TABLE 1. A circadian rhythm is present in both groups, but pregnant women with IUGR show a delayed acrophase.

After reviewing and comparing all the recordings in which IUGR was present, we also calculated the best balanced cutoff for each significant measure, to predict the emergence of IUGR in normotensive pregnant women at 24–28 weeks. Results, presented in TABLE 2, indicate the levels of automated ambulatory recorded blood pressure below which pregnant women are at very low risk of developing IUGR.

COMMENTS

In our study, women with idiopathic fetal growth retardation had higher mean diastolic and systolic blood pressure than did the controls, although values were in the normotensive range.

The delayed acrophase peak in chronobiologic analysis outlines an altered circadian regulation, with increased blood pressure at a time of the day when blood pressure begins its fall towards nocturnal levels. Although within normotensive

TABLE 1. Measures of 24-Hour Ambulatory Blood Pressure Monitoring Performed at 24–28 Weeks of Gestation

	Controls	IUGR[a]	p
Diastolic Arterial Blood Pressure			
Mesor (mm Hg)	66.7 ± 6.2	71.9 ± 6.4	<0.001
Daytime mean (mm Hg)	69.3 ± 6.4	73.3 ± 6.0	<0.01
Nighttime mean (mm Hg)	59.3 ± 6.6	67.3 ± 8.6	<.001
Acrophase (clock hours)	14:12	15:42	<0.001
Amplitude (mm Hg)	5.8 ± 0.7	6.0 ± 0.9	NS
Significance ($p <$)	0.00001	0.00003	
Systolic Arterial Blood Pressure			
Mesor (mm Hg)	112.5 ± 8.3	119.0 ± 7.2	<0.001
Daytime mean (mm Hg)	114.6 ± 8.5	120.6 ± 7.1	<0.001
Nighttime mean (mm Hg)	105.3 ± 9.2	115.9 ± 8.6	<0.001
Acrophase (clock hours)	14:28	15:52	<0.001
Amplitude (mm Hg)	5.5 ± 0.6	5.7 ± 0.7	NS
Significance ($p <$)	0.00001	0.0001	

[a]Intrauterine growth retardation.

TABLE 2. Predictivity towards Intrauterine Growth Retardation of 24-Hour Ambulatory Blood Pressure Monitoring Performed at 24–28 Weeks of Gestation

Measure	Cutoff	Sensitivity	Specificity	Positive Predictive Value	Negative Predictive Value	K	Accuracy
Diastolic Blood Pressure							
24-Hour mean	75	33	90	60	75	0.42	72
	67	78	55	44	85	0.18	62
Daytime mean	73	33	75	38	71	0.09	62
	67	100	40	43	100	0.17	59
Nighttime mean	71	33	100	100	77	1.00	79
	60	78	70	54	88	0.33	72
Systolic Blood Pressure							
24-Hour mean	119	56	85	63	81	0.46	76
Daytime mean	121	56	85	63	81	0.46	76
Nighttime mean	118	56	100	100	83	1.00	86
	108	78	70	54	88	0.33	72

limits, the altered pressure pattern may act on placental perfusion, resulting in fetal growth retardation.

REFERENCES

1. SPINILLO, A., M. STRONATI, A. OMETTO, E. FAZZI, G. LANZI & S. GUASCHINO. 1993. Infant neurodevelopmental outcome in pregnancies complicated by gestational hypertension and intra-uterine growth retardation. J. Perinat. Med. **21:** 195–203.
2. PICKERING, T. G. 1991. Ambulatory Monitoring and Blood Pressure Variability. Science Press. London.
3. HALBERG, F., J. I. DRAYER, G. CORNELISSEN & M. A. WEBER. 1984. Cardiovascular reference data-base for recognizing circadian MESOR- and amplitude hypertension in apparently healthy men. Chronobiologia **11:** 275–298.

Index of Contributors